A

OF

MACHINE DRAWING AND DESIGN.

A POCKET-BOOK FOR MECHANICAL ENGINEERS.

By DAVID ALLAN LOW. Fcap. 8vo, 740 pages, gilt edges, rounded corners. With over 1000 specially prepared Illustrations. Price 7s. 6d.

Press Opinions

"If we turn from such general matters to something more specific, we find the same brief, almost curt, but still effective, treatment. Thus the formulæ and figures which the locomotive engineer needs to have at his finger tips are all given clearly, but without any waste of words.

". . . One of the good characteristics of Mr. Low's work is the wide use he has made of valuable authorities. . . . In the section on heat there is some clever work in definitions and their elucidation. We cannot devote more space to the consideration of this little volume, and probably we have already said more than enough to show that we rate it very highly."—*The Engineer.*

"Opinions may differ as to what shape and size may be carried in a pocket with comfort, but we venture to think that opinion will not differ much as to the merits of its contents, for it is without doubt one of the very best and 'up-to-date' pocket-books which have been published."—*The Railway Engineer.*

"Everything which in the literary way that Mr. Low puts his hand to has finality and reliability. . . . Certainly this pocket-book of tables and rules is no exception to the rule.

"We have nothing but praise for the volume, and mechanical engineers will find in it a boon."—*Industries and Iron.*

"Although the last few years have seen several additions to an already fairly large number of engineers' pocket-books, it is safe to say that Mr. Low's recently-issued work merits the first place among modern works of this character."—*The Mechanical World.*

"Any work by Mr. David Allan Low would find ready acceptance among engineers and engineering students, and we have no hesitation in saying that the pocket-book for mechanical engineers is altogether admirable and excellent. . . ."—*The Science and Art of Mining.*

"Will certainly take a prominent place amongst works of a similar character. . . . This pocket-book is very freely illustrated, and will be widely appreciated when it becomes known."—*English Mechanic.*

"This is an altogether admirable work of the most complete kind.

"We have been through the book with great care . . . and conclude by confidently recommending it to mechanical engineers of every grade in the profession."—*Invention.*

"The author is to be congratulated on having produced a pocket-book for mechanical engineers which will be found indispensable, and will, we feel sure, be adopted in every drawing office and workshop as a standard book of reference."—*The Steamship.*

"It is a mine of valuable information presented in a terse form, easily understood by engineers.

"The book is beautifully printed and the type is astonishingly clear."—*Scientific American.*

"A lack of space alone prevents us from giving this book the extended notice which it deserves, for it is a complete and reliable work, worthy of a prominent place among the works intended for mechanical engineers."—*American Engineer.*

"A pocket-book for mechanical engineers, which is the most complete work of that nature yet produced in compact form.

"In illustrations, letterpress, paper, and general make-up the work is a credit to the author and to the publishers."—*Toronto Globe.*

"This is not a 'scissors and paste' production, as many engineering manuals are, but a genuinely valuable pocket-book for mechanical engineers.

"We can cordially recommend this useful work to all mechanical engineers."—*British Journal of Commerce.*

". . . Is really one of the best of its kind."—*Leeds Mercury.*

". . . Is welcomed as an important addition to the class of literature dealing with engineering work. . . . Much care and thought have been exercised in the compilation of the work, which may be confidently regarded as a reliable aid to the theoretical and practical engineer."—*Nottingham Daily Guardian.*

"Professor Low has probably produced the best work of its kind, the most comprehensive, most reliable, and by far the most illustrated."—*Newcastle Daily Chronicle.*

"To railway and hydraulic engineers, boiler-makers, or machinery constructors the work will be invaluable."—*The Dundee Advertiser.*

LONGMANS, GREEN, AND CO.

A MANUAL

OF

MACHINE DRAWING AND DESIGN

BY

DAVID ALLAN LOW

(WHITWORTH SCHOLAR), M. I. MECH. E.
PROFESSOR OF ENGINEERING, EAST LONDON TECHNICAL COLLEGE, LONDON

AND

ALFRED WILLIAM BEVIS

(WHITWORTH SCHOLAR), M. I. MECH. E.
HEAD ORGANISER OF MANUAL AND PRACTICAL INSTRUCTION AND DRAWING TO THE
COMMISSIONERS OF NATIONAL EDUCATION, IRELAND

NEW EDITION, REVISED AND ENLARGED

(*NINTH IMPRESSION*)

LONGMANS, GREEN, AND CO.
39 PATERNOSTER ROW, LONDON
NEW YORK AND BOMBAY
1903

PREFACE.

In this work the authors have attempted to provide:—(1.) A large number of dimensioned illustrations which may serve as good drawing examples for students, examples ranging in difficulty from the simplest machine detail to a set of triple-expansion marine engines. (2.) Illustrations and descriptions of a great variety of machine details, which may assist the designer in selecting the form of detail best suited to his purpose. (3.) Many rules and tables of proportions, based on scientific principles or on numerous examples from actual practice, which may be useful to the experienced designer for the sake of comparison with the results of his own practice, and which may, to some extent at least, take the place of the well-filled notebook and collection of designs usually possessed by the experienced designer, but which the young engineer or draughtsman can scarcely be expected to have. (4.) Numerous examples showing the application of the principles of mechanics to the calculation of the proportions of parts of machines.

The illustrations given are very numerous, and they have all been specially prepared for this work from working drawings, and the authors have been at great trouble to obtain examples representing the best modern practice in machine design. The authors would here acknowledge their great indebtedness to the many engineers and engineering firms throughout the country who have generously given them drawings and much valuable information, which they feel sure will prove useful to students, draughtsmen, and engineers. They would also record their indebtedness to the leading engineering papers, and to the published Proceedings of the various engineering societies, English and American, for particulars of examples of modern practice, which they have either incorporated directly, or have made use of in drawing up the numerous rules and tables which occur throughout the work.

In the introductory chapter, besides several brief articles on drawing appliances and the making of working drawings, there is a collection of problems in practical geometry which are very often required in machine drawing; but the student must not imagine that the amount of geometry there given is all that he will require; in fact, as machine drawing is simply the application of practical geometry to the representation of machines, it is evident that a thorough knowledge of the latter subject will be of immense advantage in the study and practice of the former

PREFACE.

The brief statements of various principles and rules of mechanics given in Chapter II. are inserted more for the sake of reference for the student than for the purpose of teaching those principles, and the student who wishes to succeed with the subject of machine design will do well to make a special study of mechanics.

The numerous dimensioned illustrations, which may be used as drawing examples, are not, as a rule, accompanied by any definite instructions as to what views should be drawn or what scales should be used. The student should, however, not be content with simply reproducing the views shown to the dimensions marked on them, but should project from them other views and sections. This is of the greatest importance, and is always insisted on by good teachers. The student must, of course, select his scale from a consideration of the dimensions of the object he is to represent, the number of views he is to show, and the size of his drawing-paper.

The great importance of the power of making intelligible freehand sketches of machine details can scarcely be overrated, and the student should practise this art, not only from illustrations in books, but from the actual machine details themselves. Fully dimensioned freehand sketches, made by the student himself, from actual machines or machine details, form excellent examples for drawing practice. Such sketches should be made in a notebook kept for the purpose, and no opportunity should be lost of inserting a sketch of any design which may be new to the student, always putting on the dimensions if possible. The student should also record in his notebook any new rules for proportioning machines which he may meet with in the course of his reading, or intercourse with others, whether subordinates or superiors in his profession, taking care, however, in each case to state the source of such information for his future guidance and for reference.

In conclusion, the authors would say that they have in this work, upon which they have expended a large amount of thought and labour, made use not only of the great experience of many engineers as recorded in the engineering press and the Proceedings of the engineering societies, but also of their own not inconsiderable experience in the workshops, the drawing-office, and the class-room, and they hope that the result of their efforts may be that many young engineers who wish to advance themselves in mechanical engineering—one of the most interesting and important of professions—may not only have their knowledge and power extended, but may be assisted in learning to combine science with practice.

D. A. L.
A. W. B.

January 1893.

CONTENTS.

CHAP.		PAGE
I.	INTRODUCTION	1
II.	VARIOUS PRINCIPLES OF MECHANICS	18
III.	STRENGTH AND NATURE OF MATERIALS USED IN MACHINE CONSTRUCTION	26
IV.	SCREWS, BOLTS, AND NUTS	46
V.	KEYS	65
VI.	COTTERS	73
VII.	PIPES AND PIPE JOINTS	79
VIII.	SHAFTING AND SHAFT COUPLINGS	94
IX.	SUPPORTS FOR SHAFTS	115
X.	BELT GEARING	134
XI.	ROPE GEARING	154
XII.	WIRE-ROPE GEARING	162
XIII.	FRICTION GEARING	169
XIV.	TOOTHED GEARING	173
XV.	CRANKS, CRANKED SHAFTS, AND ECCENTRICS	193
XVI.	CONNECTING-RODS	205
XVII.	CROSS-HEADS AND GUIDES	218
XVIII.	PISTONS AND PISTON-RODS	229
XIX.	STUFFING-BOXES	239
XX.	VALVES	247
XXI.	RIVETED JOINTS	268
XXII.	STEAM BOILERS	285
XXIII.	STEAM ENGINES—GENERAL DIMENSIONS	319
XXIV.	EXAMPLES OF TRIPLE-EXPANSION MARINE ENGINES	338
XXV.	EXAMPLE OF LOCOMOTIVE ENGINE	368
	INDEX	401

A MANUAL

OF

MACHINE DRAWING AND DESIGN.

CHAPTER I.

INTRODUCTION.

1. Drawing-Paper.—Smooth hand-made paper is the best for machine drawing, but it is the most expensive. Cartridge-paper is much cheaper, and may be obtained of various qualities. Valuable drawings which may be subjected to rough usage should be made on paper mounted on cotton or linen. The names and sizes of sheets of drawing-paper are given in the following table:—

Name of Paper.	Size.
Demy	20 × 15 inches
Medium	22 × 17 ,,
Royal	24 × 19 ,,
Imperial	30 × 22 ,,
Atlas	34 × 26 ,,
Double Elephant	40 × 27 ,,
Antiquarian	52 × 31 ,,

The sizes given in the above table must be taken as approximate only. In practice they will be found to vary as much as one inch in some cases. Cartridge-paper is made in sheets of various sizes, and also in rolls. Tracing-paper and tracing-cloth are generally made in rolls.

2. Drawing-Boards.—The best engineer's drawing-boards are made of well-seasoned pine, and are constructed as shown in Figs. 1 and 2. The cross-bars A B on the back of the board are fastened to the latter by screws, all of which, except the central one in each bar, pass through slotted holes in the bars. These slotted holes are provided with brass washers as shown in detail at (*a*), Fig. 2. By this arrangement the cross-bars prevent the board from warping, but they do not interfere with the expansion or contraction of the board due to changes in the amount of moisture in the atmosphere. The central screw in each bar being fixed to the bar and to the board, prevents the bar becoming displaced as a whole in the direction of its length. In order to lessen the tendency of the board to warp, longitudinal saw-cuts or grooves are made on the back about two inches apart, and of a depth equal to half the thickness of the board. A detail drawing of one of these grooves is shown at (*b*), Fig. 2. In the best boards the left-hand edge is provided with a slip of ebony let into the board. By this means the edge of the board is more likely to remain true, and the stock of the tee-

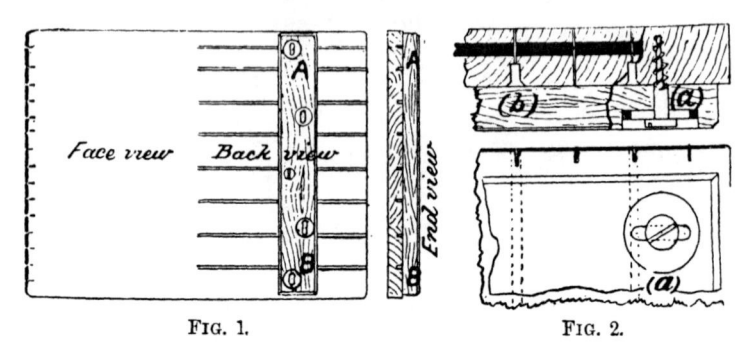

FIG. 1.   FIG. 2.

square works more smoothly on it. Sometimes the bottom edge, or edge nearest the draughtsman, is also fitted with an ebony slip. To prevent the ebony slip on the left-hand edge of the board from interfering with the expansion or contraction of the latter, the slip is cut across at intervals of about two inches after it is glued into the board.

3. Tee-Squares.—The most reliable form of tee-square is that in which the blade is fixed to the stock, but not sunk into it. The best squares are made of mahogany, and have the working edges of the stock and blade faced with ebony. Fig. 3 shows the most approved form of tee-square. The blade is fastened to the stock by screws, and is further secured by two "steady pins," which fit tightly into holes through the blade and stock.

4. Set-Squares.—These should be made of pear-tree or vulcanite. Framed set-squares are not so liable to get out of truth as the plain wooden ones, but they are objectionable on account of their greater thickness. The shortest edges of the set-squares should not be less than 5 or 6 inches long.

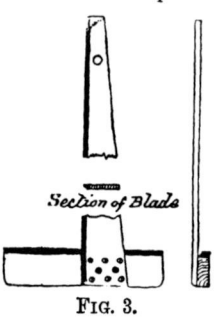

FIG. 3.

5. Various Drawing Instruments.—All compasses should have round steel points, and if possible, needle points. The dividers should have a screw adjustment. Small steel spring bows are very useful for drawing small circles. The drawing-pens for inking-in should have one nib hinged, so that it may be lifted up for cleaning.

For mechanical drawing, pencils marked H or HH are the best, although some draughtsmen prefer even harder pencils. For freehand and rough work, or for lining in a drawing which is not to be inked-in, pencils marked HB or F are most suitable. Pencils for mechanical drawing should be sharpened with a chisel-point; but for freehand

INTRODUCTION.

drawing the point should be round. Care should be taken not to wet the point of the pencil, as the lines are then very difficult to rub out.

The drawing-paper for working drawings is generally secured to the board by drawing-pins, having steel points and brass heads with bevelled edges. The steel point should not come quite through the head. The paper for finished or pictorial drawings, or for drawings upon which there is a large amount of work, are often "stretched" on the board— that is, it is damped and then glued or gummed at its edges to the board.

6. Scales.—The scales most frequently used in machine drawing range from one inch to a foot to full size. The smaller scales are used for the general views of large machines. The working drawings of details should be to as large a scale as convenient, and if possible should be full size. The best scales are made of ivory, the next quality are made of boxwood, and the cheapest are made of thin cardboard. Cardboard scales, if properly made and varnished, do very well for ordinary drawings, and if used carefully last a long time. The divisions of the scale should be marked down to the edge of the ivory, boxwood, or cardboard, so that measurements may be made by applying the scale direct to the drawing. On one side of the ivory, boxwood, or cardboard there should not be more than two scales, one to each edge. A large number of scales crowded together causes confusion, and the draughtsman is very liable to make mistakes by taking dimensions with the wrong scale.

7. Inking-in.—The best Indian-ink should be used for inking-in a drawing. Common ink corrodes the drawing-pens, and it does not dry so quickly as properly prepared Indian-ink. The most economical form in which to buy Indian-ink is the solid cake. Liquid Indian-ink is sold in bottles, but it is much more expensive than the solid cake, and is generally more difficult to procure of good quality. To make the ink ready for use from the solid cake, place a little clean water in a clean palette and rub the end of the cake in this until the liquid is sufficiently black, which must be tested by drawing a line with a pen containing a little of it. The pen may be filled by means of a brush or a narrow strip of stiff paper, but not by dipping the pen into the ink.

In cases where there are arcs of circles and straight lines touching one another, ink-in the arcs first, then the straight lines; in this way it is easier to hide the joints.

After being used the drawing-pens should be carefully cleaned.

8. Colouring.—The best water-colours should be used in colouring mechanical drawings. The colour should be rubbed down in a clean palette containing water, and the tint should be light. To colour a given surface, first damp it by going over it with a brush and clean water, then dry with clean blotting-paper to take off any superfluous water. Now go over the surface with a brush containing the colour required, and work it so as to have the unfinished edge of each portion, as it is coloured, as short as possible. If the brush does not hold enough of colour to go over the whole of the surface to be coloured, refill the brush before it is quite empty, and be careful to stir the colour in the palette frequently with the brush, to prevent it settling down. Sable brushes are the best.

but they are expensive. Good camel's-hair brushes do very well, and they are very much cheaper than sable.

The following table shows the colours which have been adopted by engineers to represent different materials :—

Table showing colours used to represent different materials.

Material.	Colour.
Cast-iron	Payne's grey or neutral tint.
Wrought-iron	Prussian blue.
Steel	Purple (mixture of Prussian blue and crimson lake).
Brass	Gamboge, with a little sienna, or a very little red added.
Copper	A mixture of crimson lake and gamboge, the former colour predominating.
Lead	Light Indian-ink with a very little indigo added.
Brickwork	Crimson lake and burnt sienna,
Firebrick	Yellow and Vandyke brown.
Grey stones	Light sepia or pale Indian-ink, with a little Prussian blue added.
Brown freestone	Mixture of pale Indian-ink, burnt sienna, and carmine.
Soft woods	For ground-work, pale tint of sienna.
Hard woods	For ground-work, pale tint of sienna with a little red added. For graining woods, use darker tint, with a greater proportion of red.

9. Dimensions on Drawings.—A good working drawing must not only be drawn to scale, but it should also be fully dimensioned. If a drawing is not dimensioned, the workman must get his sizes from the drawing by applying his rule or a suitable scale, an operation which takes time, and is very liable to result in error.

Some draughtsmen place the figures for the dimensions so that they

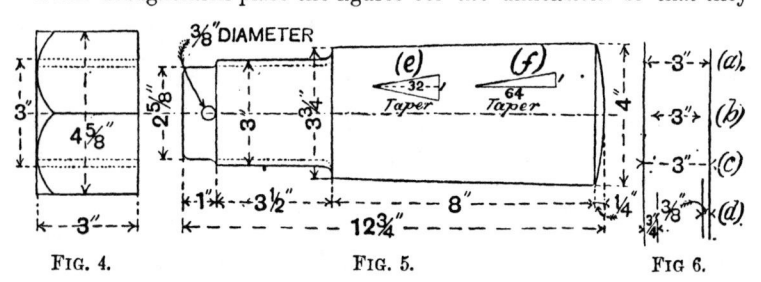

FIG. 4.   FIG. 5.   FIG 6.

are all upright or all read one way, as in Fig. 4. Others again prefer the figures at right angles to the distance or dimension lines, as shown in Fig. 5. In marking the distance between two lines, be careful to put the arrow-points up to the lines as shown at (*a*) Fig. 6, and not as shown at (*b*). The arrow-points should also be put inside as shown at (*a*), and not outside as shown at (*c*), unless the lines are very close together, as shown at (*d*).

All dimensions which a workman may require should be put on the drawing, so that no measurement of the drawing or calculation is required. For example, it is not enough to give the lengths 1″, 3½″, 8″ and ¼″ of the different parts of the bolt in Fig. 5, but the length over all, 12¾″, which is the sum of these lengths, should also be marked.

The amount of taper on a piece may be indicated by drawing a triangle as shown at (*e*) or (*f*). The triangle at (*e*) shows a taper of 1 in 32 *on the diameter*—that is, the diameter varies at the rate of 1 inch in 32 inches, or ¼ inch in 8 inches. The triangle (*f*) shows a taper of 1 in 64 *on the radius*.

10. Hints on the Finishing of Working Drawings.—After the drawing has been carefully made in pencil, it should be inked-in with Indian-ink. The parts in section should then be coloured, the colours used indicating the materials of which the parts are made, according to the table given on page 4. Parts which are round may also be lightly shaded with the brush and colours to suit the materials. The centre lines which are already drawn in pencil must now be inked-in with coloured ink. Red ink prepared by rubbing down crimson lake, or blue ink prepared by rubbing down Prussian blue, are the best for centre lines. The centre lines when inked-in with coloured ink should be thin continuous lines, and not dotted lines. The distance or dimension lines should also be thin continuous lines inked-in with coloured ink, but the colour for the dimension lines should not be the same as the colour for the centre lines. The arrow-heads at the ends of the dimension lines, and also the figures for the dimensions, look best in black ink, and they should be made with a common writing-pen, and not with a drawing-pen. The dimensions should be put on neatly. Many a good drawing has its appearance spoiled through being slovenly dimensioned.

In getting up the drawings for a complicated machine, the best mode of procedure is as follows:—First determine the chief dimensions, and then make a general drawing of the whole, leaving out the smaller details. Now get out the working drawings of all the details, taking the larger and more important ones first. Lastly, different views of the complete machine should be made from the detail drawings, so as to ensure that all the details fit in properly with one another.

11. Shade Lines.—From a pictorial point of view the appearance of a machine drawing is improved by putting in what are called shade lines. These are thick lines which are placed on the right-hand and bottom edges of a projecting piece, Fig. 7, or on the left-hand and top edges of a recess, Fig. 8. In the case of a circle, the shade line is terminated by a diameter inclined at 45°, as shown in Fig. 9. Straight shade lines should be of uniform thickness, but curved ones like those in Fig. 9 should finish off tapered at the ends.

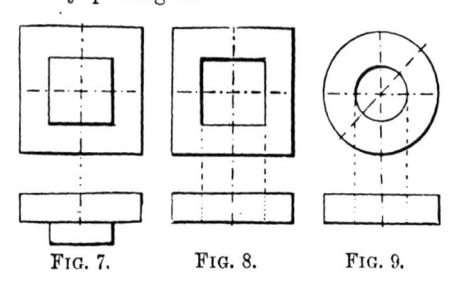

FIG. 7.   FIG. 8.   FIG. 9.

The shade lines on a drawing which is inked-in should be put on after the colouring, because the thick lines would run when touched with the damp brush.

12. Lettering.—The following are examples of letters and figures which may be used on drawings for titles, headings, and dimensions :—

A B C D E F G H I J K L M N O P Q R S T U V W X Y Z
1 2 3 4 5 6 7 8 9 0

A B C D E F G H I J K L M N O P Q R S T U V W X Y Z
1 2 3 4 5 6 7 8 9 0

A B C D E F G H I J K L M N O P Q R S T U V W X Y Z
1 2 3 4 5 6 7 8 9 0

Italic letters such as the following are most suitable for remarks on drawings :—

a b c d e f g h i j k l m n o p q r s t u v w x y z

13. Use of Set-Squares.—The chief use of set-squares is in the drawing of lines parallel or perpendicular to one another. Figs. 10 and 11 show two methods of drawing lines parallel to a given line AB. The

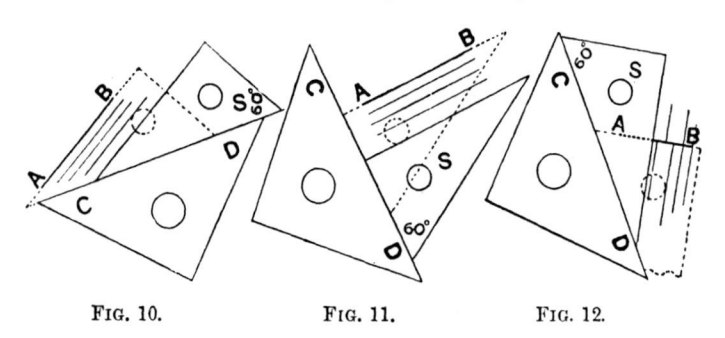

FIG. 10.   FIG. 11.   FIG. 12.

set-square S is first placed in the position shown by the dotted lines, and a straight-edge CD, or another set-square, is placed as shown. By sliding S on the edge of CD the required parallel lines may be drawn. To draw lines perpendicular to a given line AB, Fig. 12, first place the set-square in the position shown by the dotted lines, then slide S on CD as before. Fig. 13 shows another arrangement for the same purpose. The set-square and straight-edge are first placed in the positions shown by the dotted lines. The set-square is then moved into the position S, and the straight-edge is next placed in the position CD. By sliding S on CD the required lines perpendicular to AB may be drawn. The methods shown in Figs. 11 and 13 are the best when the first and last of the lines to be drawn are far apart.

14. **To Divide a Line into any Number of Equal Parts.**—The best practical method of dividing a straight line or an arc of a circle into a number of equal parts is the method of trial with the dividers. In applying this method care should be taken not to push the steel points of the dividers through the drawing-paper. To mark the divisions on the line, use a sharp pencil, placing the point of the pencil against one point of the dividers, which point should be raised slightly above the paper, the other point being on the line. Now place the first point on the line (it ought to land on the mark just made with the pencil), and swing the other round, and mark the next division, and so on, care being taken that there is always one point of the dividers on the line.

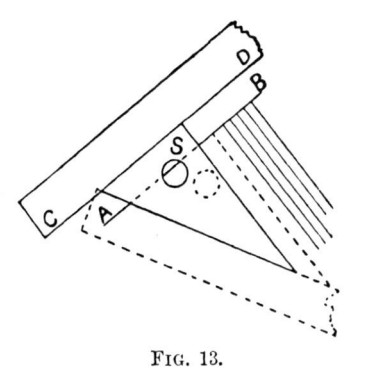

FIG. 13.

15. **To Bisect the Angle between two Straight Lines.**—CASE I., Fig. 14.—The given lines AB, AC, intersect within the paper. With centre A and any convenient radius describe the arc DE. With centres D and E and any convenient radius describe arcs to cut one another at F. The line AF will bisect the angle between AB and AC.

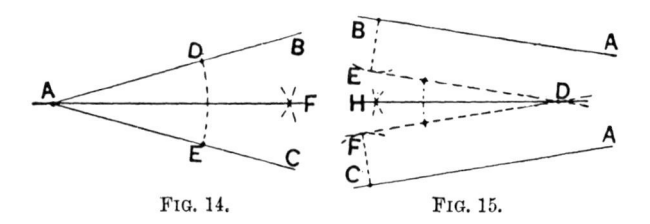

FIG. 14.   FIG. 15.

CASE II., Fig. 15.—The point of intersection of the given lines AB, AC, is inaccessible. Draw DE and DF parallel to AB and AC respectively, and at equal distances from them, so that the point D is accessible. The line DH, which bisects the angle EDF, is the line required.

16. **To Describe a Circle or an Arc of a Circle through Three Given Points.**—CASE I., Fig. 16.—Bisect two of the lines joining the given points by lines at right angles to them, and let these bisecting lines meet at O. O is the centre, and OA, OB, or OC the radius of the required circle.

CASE II., Fig. 17.—In this case the centre of the required circle is inaccessible. With centres A and B describe arcs BH and AK. Join AC and BC, and produce these lines to meet the arcs at H and K. Mark off short arcs H1 and K1 equal to one another. The intersection P of the lines A1 and B1 is another point on the arc required. In like

8 MACHINE DRAWING AND DESIGN.

manner other points may be found as shown. A fair curve drawn through these points will be the arc required.

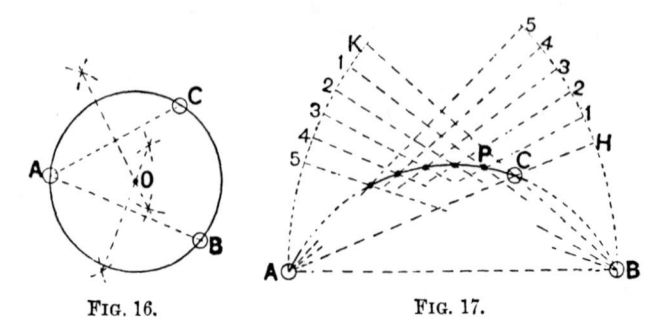

FIG. 16.

FIG. 17.

17. To Draw a Tangent to a Given Circle.—The fact to be remembered in problems on tangents to circles is that the tangent is at right angles to the radius or diameter passing through the point of contact. If the point of contact P, Fig. 18, is given, join P with O, the centre of the circle, and draw PR perpendicular to OP. If the tangent is

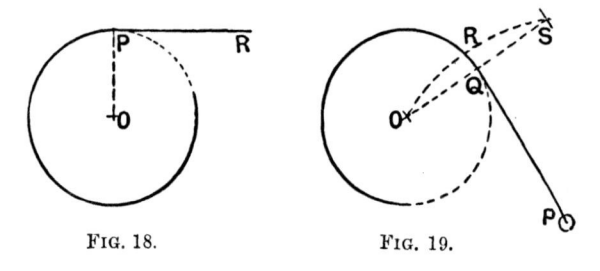

FIG. 18.

FIG. 19.

required to pass through a given point P, Fig. 19, outside the circle, with centre P and radius PO, describe the arc ORS, with centre O and radius equal to the diameter of the circle, cut this arc at S, join OS, meeting the circle at Q. PQ is the required tangent, and Q is the point of contact. It is very important to determine exactly the point of contact.

18. To Draw a Tangent to Two Given Circles.—Figs. 20 and 21.—Make DE = BC. Draw a circle with centre A and radius AE. Draw BF a tangent to this circle, F being the point of contact. Draw the line AF meeting the given circle, whose centre is A at H. Draw BK parallel to AH. HK is a tangent to both the given circles.

19. To Draw a Circle to Touch Two Given Circles.—The constructions for problems on circles touching one another are mostly based on two facts—first, the line joining the centres of two circles which touch one another, or that line produced, passes through the point of contact; second, the distance between the centres of the two circles is equal either to the sum or to the difference of their radii.

In Figs. 22, 23, and 24, A and B are the centres of the given circles. Make CD and FE each equal to the radius of the circle which is

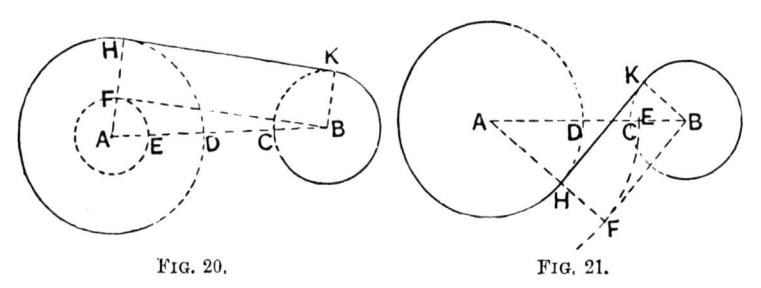

FIG. 20. FIG. 21.

required to touch the given circles. With centre A and radius AD describe the arc DO. With centre B and radius BE describe the arc

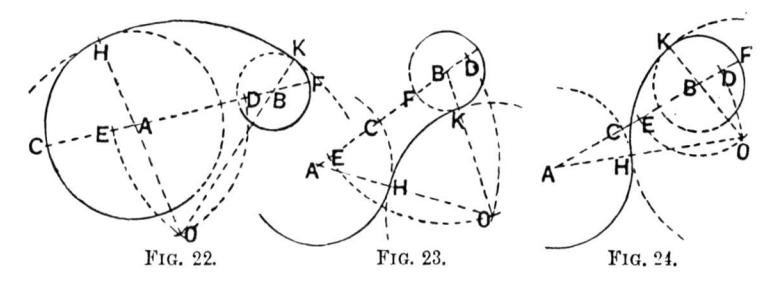

FIG. 22. FIG. 23. FIG. 24.

EO, cutting the former arc at O. Join O with A and B, and produce these lines if necessary. O is the centre of the required circle, and H and K are the points of contact.

20. **To Draw the Inscribed and Escribed Circles of a Triangle.**—The inscribed circle of a triangle is the one which touches each of the sides. An escribed circle touches one side of the triangle and the other two produced. When a circle touches each of two straight lines, its centre lies on the line bisecting the angle between them. By applying this, the centres of the inscribed and escribed circles may be found.

21. **To Draw a Regular Polygon on a Given Line.**—Let C5, Fig. 25, be the given line. With centre C and radius C5 describe the semicircle O 1, 2, 3, 4, 5, and divide this into as many equal parts as there are sides in the polygon to be drawn. Join C with 2, 3, 4, etc. With centre 2 and radius 2C describe an arc cutting C3 produced at D. With centre D and the same radius describe an arc cutting C4 produced at E, and so on.

FIG. 25.

22. **The Ellipse.**—In most cases where an ellipse has to be constructed in drawings of machines the two axes are known, and the best practical

method for drawing the ellipse is that shown in Fig. 26. AB is the major axis and CD the minor axis. EF is a strip of paper marked at P, M, and N, such that NP is equal to OA and MP equal to OC. If the slip is placed so that M is on the major axis and N on the minor axis, P will be a point on the ellipse. This point may be marked on the drawing-paper with a pencil. By moving the slip round, as many points on the ellipse as are desired may be found. A fair curve drawn through these points will complete the construction.

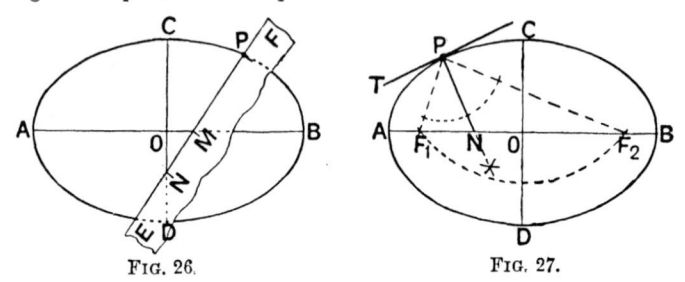

FIG. 26.   FIG. 27.

The foci of an ellipse are found by describing (Fig. 27) an arc with C as centre and radius equal to OA, cutting AB at F_1 and F_2.

To draw the normal to an ellipse at a point P, join P with the foci F_1, F_2, and bisect the angle F_1PF_2 by a line PN. PN is the normal to the ellipse at P, and a line PT perpendicular to PN is the tangent.

23. The Parabola.—The method illustrated in Fig. 28 for drawing a parabola is often useful. AB is the axis, A the vertex, and CBD a double ordinate. Complete the rectangle CDEF. Divide AE into any number of equal parts, and divide ED into the same number of equal parts. Through the points 1, 2, 3, etc., on AE draw parallels to AB. Join A with the points 1', 2', 3', etc., on ED. A curve drawn through the intersections of these lines, as shown, will be one half of the curve. The other half of the curve is found in the same way.

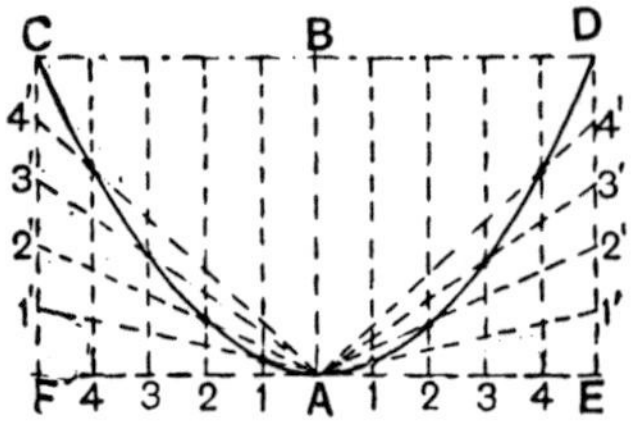

FIG. 28.

24. The Hyperbola.—The form of the hyperbola of most frequent occurrence in mechanical engineering is that known as the rectangular hyperbola. The construction for drawing this curve is shown in Fig. 29. OX and OY are two axes at right angles to one another. Let A be a known point on the curve. Draw AQ parallel to OX, and AB and QN perpendicular to OX. Join OQ, meeting AB at R. Draw RP parallel to

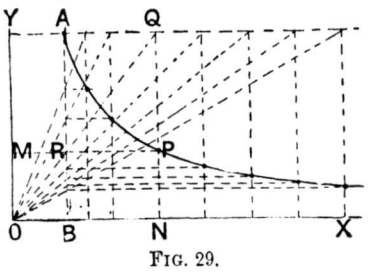

FIG. 29.

INTRODUCTION.

OX, meeting QN at P. P is another point on the hyperbola, and by repeating this construction other points may be found as shown.

This curve has the property that if from any point P on the curve, perpendiculars PN and PM be drawn to the axes, the product PN × PM is always the same, wherever the point P may be taken.

25. Cycloidal Curves.—If a circle be made to roll along a line and

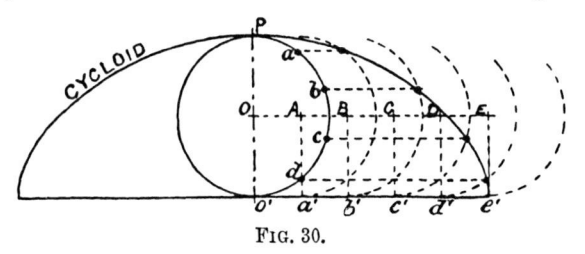

FIG. 30.

remain in the same plane, a point in the circumference of the rolling circle will describe a *cycloidal curve*. The rolling circle is called a *generating circle*, and the line along which it rolls is called a *director*. If the director is a straight line, the curve is called a *cycloid*. If the director is a circle, the curve described is called an *epicycloid* or a *hypo-*

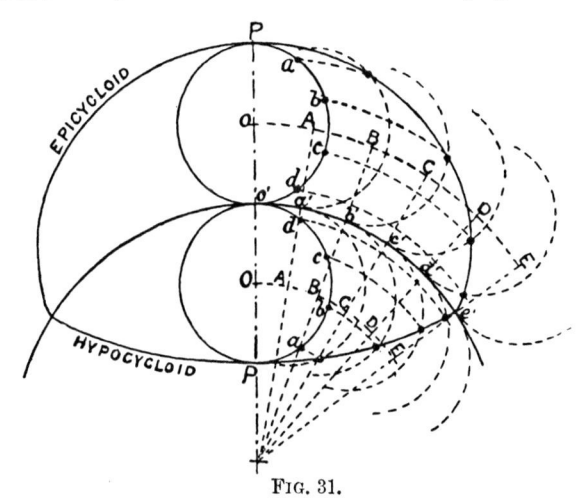

FIG. 31.

cycloid, according as the generating circle rolls on the outside or inside of the director.

The constructions for drawing the cycloid, epicycloid, and hypocycloid are shown in Figs. 30 and 31. Make the straight line or arc $o'e'$ equal to half of the circumference of the generating circle. Divide $o'e'$ into any number of equal parts, and divide the semicircle Pco' into the same number of equal parts. Draw the normals a'A, b'B, etc., to meet a line or arc OE parallel to or concentric with $o'e'$ at the points

A, B, etc. With centres A, B, etc., describe arcs of circles to touch $o'e'$. Draw through the points a, b, c, etc., lines parallel to $o'e'$, or arcs concentric with the arc $o'e'$, to meet the arcs whose centres are at A, B, C, etc.

The intersections of these determine points on the curve required.

The hypocycloid becomes a straight line when the diameter of the generating circle is equal to the radius of the director.

The same hypocycloid may be described by either of two generating circles whose diameters are together equal to the diameter of the director.

26. Involute of a Circle.—If a flexible line be wound round a circle, and the part which is off the circle be kept straight, any point in it will describe a curve called the *involute* of the circle. The involute is a particular case of the epicycloid. If the generating circle of an epicycloid be increased in size until its diameter becomes infinite, the circle becomes a straight line, and a point on this line will describe the involute of the director circle if the line is made to roll on that circle.

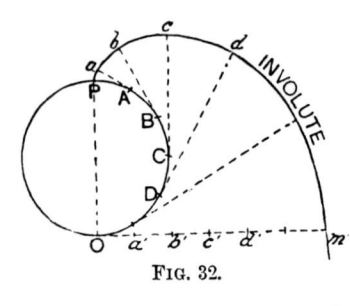

To draw the involute to a given circle OCP (Fig. 32), draw the tangent Om', and make Om' equal in length to half the circumference of the circle. Divide the semicircle OCP into any number of equal parts, and divide Om' into the same number of equal parts. At the points A, B, C, etc., draw tangents to the circle, and make Aa, Bb, Cc, etc., equal to Oa', Ob', Oc', etc., respectively. Pabc . . . m' is a portion of the curve required, which may be extended to any length.

Fig. 32.

27. True Length of a Line from its Plan and Elevation.—It frequently happens that the true length of some line or edge of a solid is not shown by either the plan or elevation, as in Fig. 33, which shows two views of a wedge. In both plan and elevation the sloping edges are shown shorter than their real lengths. To find the true length of a sloping edge, erect a perpendicular at one end of the plan, and make it equal in length to the height of one end of the elevation above the other. The line joining the top of this perpendicular with the other end of the plan is the true length required.

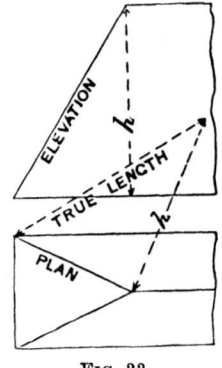

28. Plane Sections of Solids.—Several examples are here given showing how plane sections of solids may be drawn. Fig. 34 shows how to determine a section of an angle iron. In Fig. 35 is shown a bolt-head which is partly square and partly conical. The curves a' and b' are plane sections of a cone. Fig. 36 shows a hexagonal bolt-head, the end of which is turned to a spherical shape. The curves in this example are

Fig. 33.

INTRODUCTION. 13

plane sections of a sphere. An example of frequent occurrence in machine drawing is that shown in Fig. 37, where we have plane sections

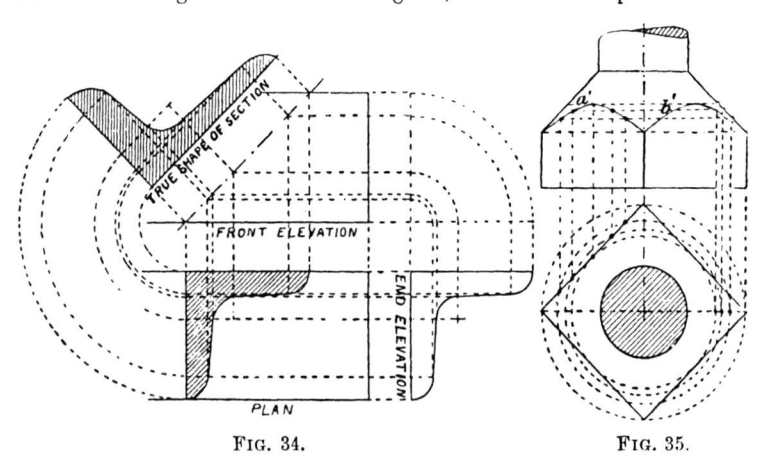

FIG. 34.

FIG. 35.

of a surface of revolution—that is, a surface turned in an ordinary lathe. In all these examples the construction lines are fully shown, and further explanation is unnecessary.

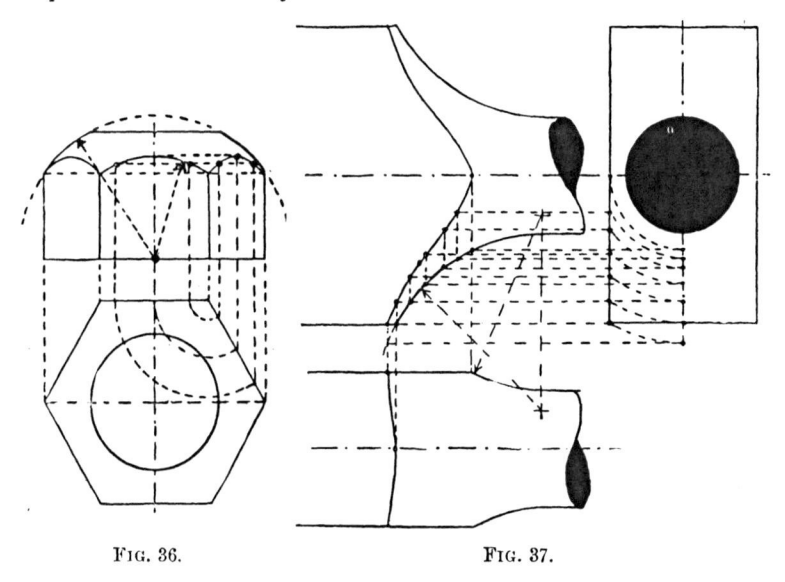

FIG. 36.

FIG. 37.

29. Development of the Surface of a Cylinder.—A surface is said to be developed when it is "folded back on one plane without tearing or creasing at any point." The development of a right circular cylinder is a rectangle whose base is equal in length to the circumference of the base of the cylinder, and whose height is equal to the length of the cylinder. In Fig. 38 the rectangle ABCD is the development of the surface of the cylinder, whose plan and elevation are shown. In order to determine the development of any line on the surface of the cylinder, a number of lines on the surface of the cylinder and parallel to its axis are represented. The points 1, 2, 3, etc., are the plans of these lines, and their positions on the development are shown at 1', 2', 3', etc. Suppose the cylinder to be cut by a plane represented by the line Ae', then the curve AEB would be the top boundary-line of the development of the surface of the portion of the cylinder AD$f'e'$. The construction for determining the curve AEB is simple, and will be readily understood from the figure.

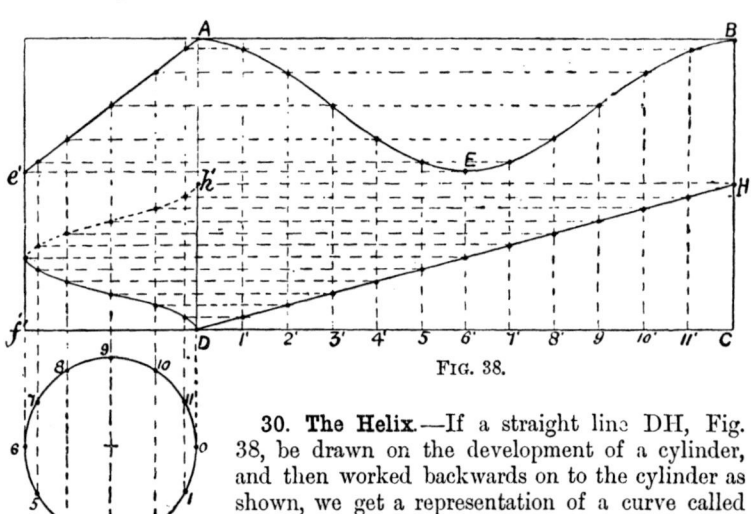

FIG. 38.

30. The Helix.—If a straight line DH, Fig. 38, be drawn on the development of a cylinder, and then worked backwards on to the cylinder as shown, we get a representation of a curve called the *helix*. This curve is important in connection with screws. The angle HDC is the *inclination* of the helix. HC or Dh', the height to which the curve rises in going once round the cylinder, is called the *pitch*. To draw the elevation of the helix it is not necessary to draw the development. Divide the pitch Dh' into the same number of equal parts that the plan is divided into, and through the divisions draw lines parallel to Df' as shown, to meet the perpendiculars from the plan.

31. Screw Threads and Spiral Springs.—The edges of screw threads are helices, and these are drawn as explained in the preceding article. Examples of these are shown in Figs. 39 and 40. In Fig. 39 the screw is single threaded, while in Fig. 40 it is double threaded. Fig. 41 shows a spiral spring made of square steel. This is drawn in the same way as the square-threaded screw in Fig. 40. To draw the spiral spring shown in Fig. 42, first draw the helix which is the centre line of the spiral. Next draw circles with their centres on the helix, and having a diameter

FIG. 39.—Triangular Thread (single).

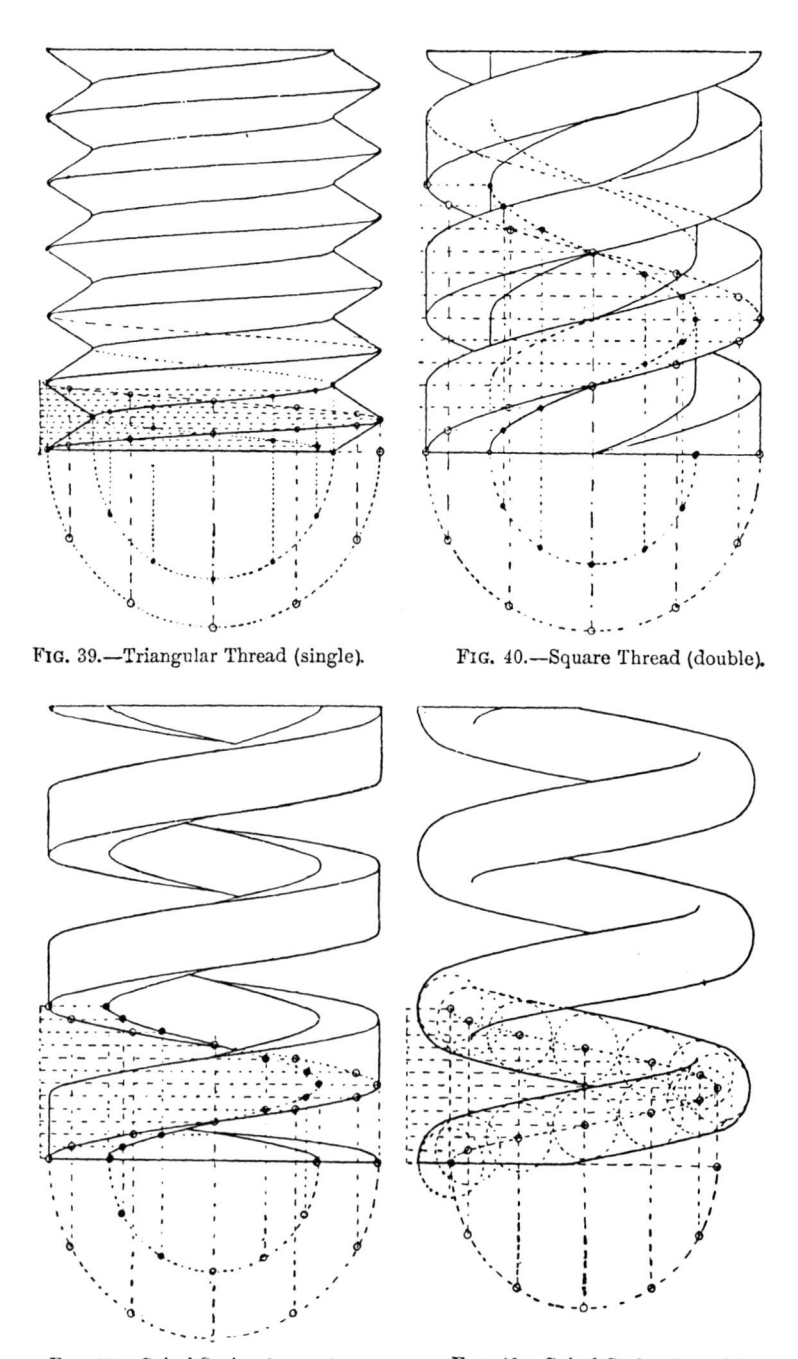

FIG. 40.—Square Thread (double).

FIG. 41.—Spiral Spring (square).

FIG. 42.—Spiral Spring (round.)

16   MACHINE DRAWING AND DESIGN.

equal to the thickness of the round wire from which the spring is made. The outline of the drawing will touch these circles as shown.

When the pitch of a helix is small compared with the diameter of the cylinder upon which it is drawn—that is, when its inclination is small—it may be represented approximately by straight lines. Numerous examples of the application of this approximate method will be found in various parts of this book.

32. Development of the Surface of a Cone.—The development of the surface of a right circular cone is a sector of a circle whose arc is equal in length to the circumference of the base of the cone, and whose radius is equal to the length of the slant side of the cone. Fig. 43 shows how this development is obtained, and also how to draw the development of any line on the surface of the cone.

33. Intersection of Two Cylinders.—Two examples of this problem are shown in Fig. 44. A horizontal and an inclined cylinder are shown intersecting a vertical cylinder. $a'c'$, $a'c'$ are the centre lines of the elevations of the two first-mentioned cylinders. To find points on the curve of intersection, draw lines dab, dab on the plan parallel to the plans of the axes of the horizontal and inclined cylinders. Draw the semicircle ebf. Make $a'b'$ equal to ab. Lines through the points lettered b' parallel to $a'c'$, to meet perpendiculars from the points where dab meets the plan of the vertical cylinder, will determine points on the required intersection. In like manner other points are determined, and a fair curve drawn through them is the intersection required.

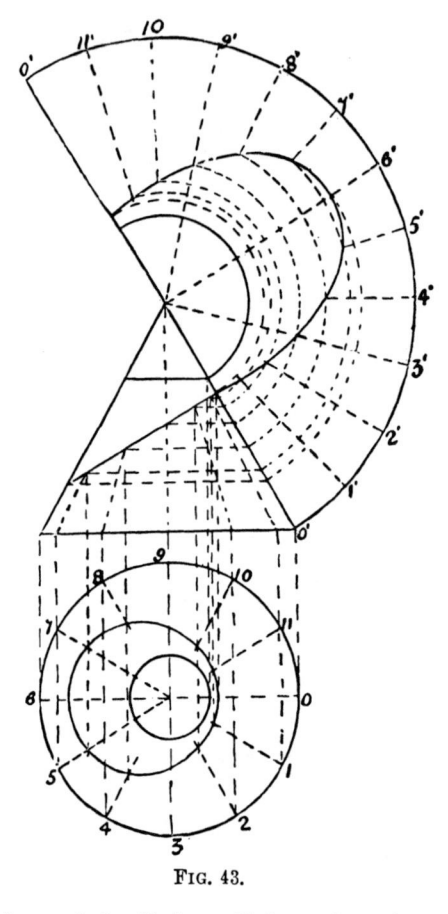

FIG. 43.

34. The Numbers 3·1416 and ·7854.—These numbers occur so frequently in the calculations of the engineer, that we must at this early stage point out what they mean. The first of these numbers is the ratio which the circumference of a circle bears to its diameter, so that the rule for finding the circumference of a circle is—multiply the diameter by the number 3·1416. The number ·7854 is exactly one quarter of the number 3·1416, and is used for finding the area of a circle. The area of a circle is equal to its diameter squared multiplied by ·7854; thus, area

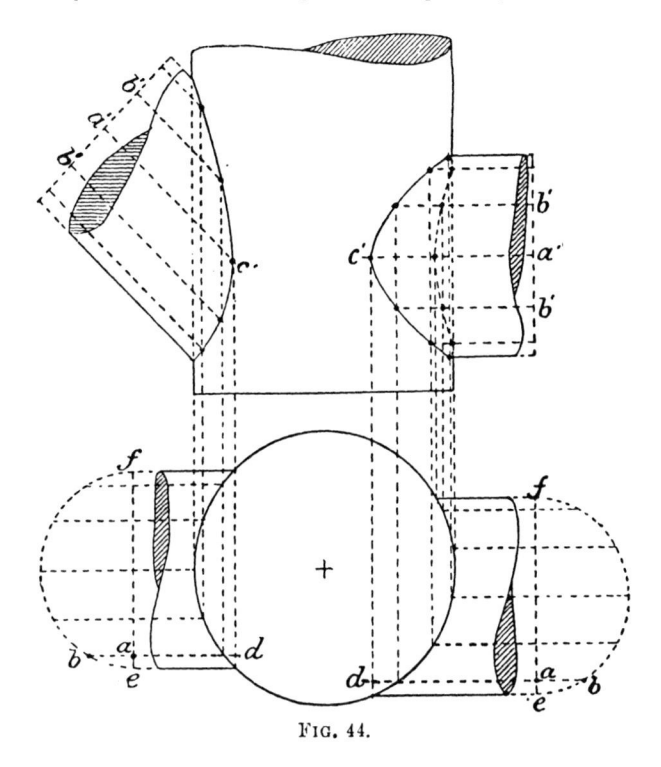

FIG. 44.

$= D^2 \times$ ·7854, where D denotes the diameter. In mathematics the Greek letter π is used to denote the number 3·1416. We might point out that the ratio of the circumference of a circle to its diameter cannot be expressed exactly by any number, so that 3·1416 is only approximate. A closer approximation is 3·14159265.

CHAPTER II.

VARIOUS PRINCIPLES OF MECHANICS.

35. Parallelogram of Forces.—If the straight lines AB and AC (Fig. 45) represent in magnitude and direction two forces acting at the point A, the diagonal AD of the parallelogram ABDC will represent in magnitude and direction the single force which will produce the same effect as the two forces AB, AC acting together, that is, AD will represent the resultant of the forces represented by AB and AC. Conversely, if a parallelogram ABDC be described with AD as a diagonal, AB and AC will represent two forces which will produce the same effect as the single force represented by AD.

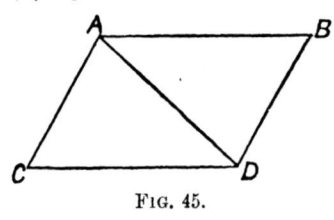

FIG. 45.

36. Triangle of Forces.—If three forces, P, Q, and R (Fig. 46), acting at the same point, be represented in magnitude and direction by the three sides, p, q, r, of a triangle (Fig. 47) taken in order, the three forces P, Q, and R will be in equilibrium, that is, they will balance one another. Conversely, if the forces P, Q, and R balance one another, they

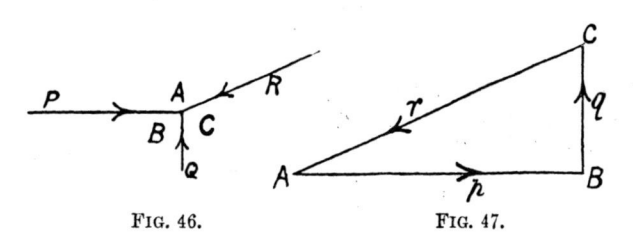

FIG. 46.   FIG. 47.

may be represented in magnitude and direction by the three sides of a triangle taken in order.

The best method of lettering the forces is to place the letters in the angles made by the lines of action of the forces in the one figure, and at the angular points of the triangle in the other. This is known as Bow's notation.

37. Polygon of Forces.—This is an extension of the principle of the triangle of forces. In Fig. 48, four forces, AB, BC, CD, and DA, are shown acting at a point. These forces, if they are in equilibrium, can be represented in magnitude and direction by the sides of the polygon ABCDA, Fig. 49.

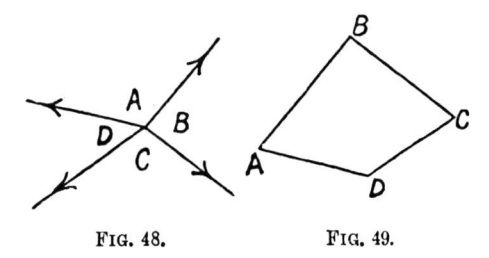

FIG. 48. FIG. 49.

38. Stress Diagrams.—If a system of forces AB, BC, CD, DA, Fig. 50, act at the angular points of a jointed frame, h, k, l, m, and keep that frame at rest, these forces may be represented in magnitude and direction by the sides of the polygon ABCDA, Fig. 51. Now, if through the points A, B, C, and D, Fig. 51, lines be drawn parallel to the sides mh, hk, kl, and lm respectively of the frame shown in Fig. 50, these lines

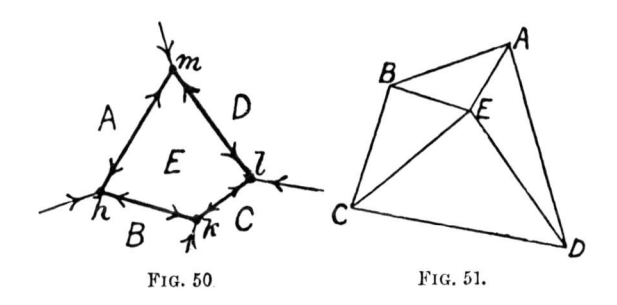

FIG. 50. FIG. 51.

will all meet at the same point, E, and they will represent in magnitude and direction the stresses produced in the bars of the frame by the given external forces acting at its angular points.

If the given external forces are all parallel, then the polygon ABCDA becomes a straight line.

39. Moment of a Force.—The moment of a force with respect to an axis is the product of the magnitude of the force and the perpendicular distance of its line of action from the axis. If the force is measured in pounds and the distance in feet, the moment will be in *foot-pounds*. If the force is in pounds and the distance in inches, the moment will be in *inch-pounds*.

40. Principle of Moments.—When a system of forces acting on a rigid body is in equilibrium, the sum of the moments of those forces which tend to turn the body in one direction about any given axis is equal to the sum of the moments of those forces which tend to turn the body in the opposite direction about the same axis, the moments being taken with respect to the given axis. For example, in Fig. 52 are shown five forces, P, Q, R, S, and T, acting on a rigid body. Three of these forces, namely, P, R, and S, tend to turn the body in one direction about the axis O, while the remaining forces, Q and T, tend to turn the body in the opposite direction about the same axis. If these five forces are in equilibrium, we have from the principle of moments the following relation :—

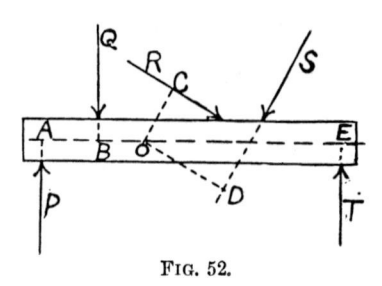

FIG. 52.

$$P \times AO + R \times OC + S \times OD = Q \times BO + T \times EO$$

The principle of moments is very useful in working out questions on levers.

41. Motion in a Circle—Centrifugal Force.—If a body weighing W pounds moves in a circle of radius R feet with a velocity of v feet per second, a radial force is produced equal to $\dfrac{Wv^2}{gR}$ where g is the accelerating effect of gravity or the increase of velocity which a body acquires every second when it falls freely under the action of gravity. The numerical value of g varies at different parts of the earth's surface. For calculations connected with machines g is generally taken at 32·2 feet per second.

42. The Pendulum.—The simple pendulum consists of a small but heavy body attached to one end of a fine thread, the other end of which is attached to a fixed point. If the weight at the end of the thread be moved from its normal position and then be released, it will oscillate backwards and forwards in an arc of a circle under the action of gravity. By an oscillation of a pendulum is meant a movement of the weight from one end of its path to the other. If the length of the arc described by the weight is small compared with the length of the thread the time of an oscillation is given by the formula $t = 3 \cdot 1416 \sqrt{\dfrac{l}{g}}$, where t is the time in seconds, l the length of the pendulum in feet, and g the accelerating effect of gravity (generally taken at 32·2 feet per second).

If the simple pendulum be made to rotate so that the weight describes a horizontal circle, the thread will describe the surface of a cone, and the pendulum becomes a *conical pendulum*. If h is the altitude of the cone in feet, t the time of one revolution in seconds, and n the number of revolutions per minute, then

$$t = 2 \times 3 \cdot 1416 \sqrt{\dfrac{h}{g}} \; ; \text{ and } n = \dfrac{30}{3 \cdot 1416} \sqrt{\dfrac{g}{h}}.$$

43. Work.—A force is said to do work when, acting on a body, it causes that body to move. The amount of work done depends on the magnitude of the force and the distance through which it acts, and is measured by the product of these two quantities. For the purpose of calculating the work done by a force, it is usual to take the magnitude of the force in pounds and the distance through which it acts in feet, then the product, which gives the work done, will be in *foot-pounds*. If the force varies in magnitude, the work done by it is equal to its average magnitude multiplied by the distance through which it acts.

44. Horse-Power.—An engine or working agent is said to be of one horse-power when it can do 33,000 foot-pounds of work in one minute. The horse-power of an engine (or the number of horses to which it is equivalent) is got by dividing the number of foot-pounds of work which it can do in one minute by 33,000.

45. Work in Moving a Body on an Inclined Plane.—The work done in moving a body from A to B on an inclined plane (Fig. 53) is equal to the work done against gravity, together with the work done in overcoming the friction between the body and the plane. The work done against gravity is equal to the weight of the body multiplied by the vertical height BC. The work done on friction is the same as if the body was moved on the horizontal plane from A to C.

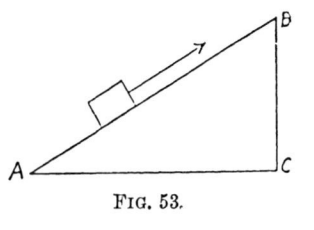

FIG. 53.

The force, P, required to pull the body up the plane is obtained from the equation P × AB = work done on gravity + work done on friction.

46. Accumulated Work.—If a body weighing W pounds be moving with a velocity of v feet per second, it has stored up in it an amount of work equal to $\dfrac{Wv^2}{2g}$ foot-pounds; where g may be taken $= 32 \cdot 2$.

47. Energy of a Rotating Body—The Fly-Wheel.—Those particles of a rotating body which are at different distances from the axis will have different velocities, and, their weights being the same, they will possess different amounts of accumulated work. If all the material of a rotating body could be concentrated at the circumference of a circle of radius R feet, so as to have its accumulated work unchanged, the speed of rotation or angular velocity being the same, then R is called the *radius of gyration* of the rotating body. For a solid cylindrical body, like a grindstone, rotating about its axis, the radius of gyration is equal to the radius of the cylinder divided by $\sqrt{2}$. For a hollow cylinder or hoop rotating about its axis the radius of gyration is equal to $\sqrt{\dfrac{R_1^2 + R_2^2}{2}}$ where R_1 and R_2 are the external and internal radii of the cylinder.

If v is the velocity, in feet per second, of the material of a rotating body at a distance from the axis equal to the radius of gyration, R, then the amount of work accumulated in the body is equal to $\dfrac{Wv^2}{2g}$. If the weight of the body, W, is in pounds, the work accumulated will be given in foot-pounds.

If the rotating body makes N revolutions per minute, then

$$v = \dfrac{2 \times R \times 3 \cdot 1416 \times N}{60}.$$

When the supply of energy to a machine is greater or less than the demand, the speed of the machine must vary; there being an increase of speed when the supply is greater than the demand, and a decrease when the supply is less than the demand. Now, when the velocities of the moving parts of a machine are changed, the amount of work stored up in these parts will also change. This is therefore what becomes of the surplus energy when the supply is greater than the demand; it is stored up in the moving parts of the machine by reason of their increased velocities. Again, when the supply of energy is less than the demand, the deficiency is made up by the accumulated work which is given up when the velocities of the moving parts diminish.

The fly-wheel is made a part of a machine for the purpose of regulating the speed of that machine. It is a wheel with a very heavy rim, and when its speed changes with the speed of the machine, work is stored up in it or is given out by it. If W is the weight of the rim of a fly-wheel and v_1 its velocity, then the energy stored up in it is equal to $\dfrac{Wv_1^2}{2g}$. Let the velocity of the rim change to v_2, the energy now in the rim is equal to $\dfrac{Wv_2^2}{2g}$, and the increase or diminution of the amount of energy stored up is $\dfrac{W}{2g}(v_1^2 - v_2^2)$. From this it is evident that the greater the weight of the rim of the fly-wheel the less will be the change in its speed to store up or give out a given quantity of work. In other words, the heavier a fly-wheel is, the more uniform will the speed of the machine be.

48. Diagrams of Crank Effort and Twisting Moment.—In converting the rectilineal motion of the piston of an engine into the circular motion of the crank-shaft by means of the crank and connecting rod, the turning effort on the crank is continually varying. By means of the

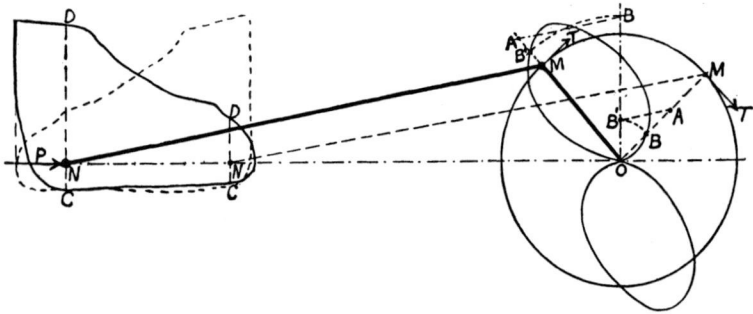

FIG. 54.

following simple construction the turning effort on the crank in any position may be obtained:—Let P be the force on the piston when the crank is in the position OM, Fig. 54. Through O draw OB' at right angles to ON. Make OA = P, and through A draw AB' parallel to MN, meeting OB' at B'. OB' will represent the magnitude of the tangential force T on the crank-pin. If with centre O and radius OB' an arc be described cutting OM, or OM produced at B, and if this construction be repeated for a number of positions of the crank, and the points so obtained be joined, the resulting diagram will show in a graphic manner the variation of the force T as the crank rotates.

The force P on the piston corresponding to any given position of the cross-head N is obtained from the indicator diagram, which is placed as shown, P being equal to CD.

Another form of the diagram of crank effort is shown in Fig. 55. Here the line EF is made equal to the circumference of the crank-pin circle, and it is divided into a number of equal parts. The path of the

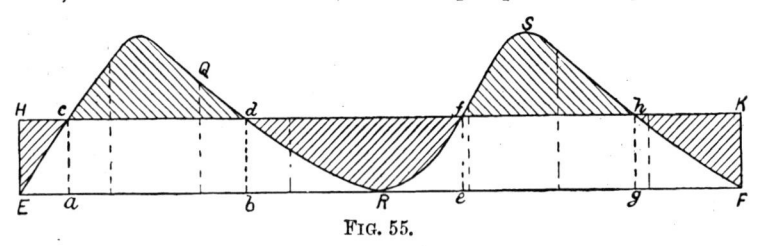

FIG. 55.

crank-pin is divided into the same number of equal parts, and the force T is determined at the end of each of these parts in the manner already explained. Perpendiculars are now drawn from the different points in EF and made equal to the values of T at the corresponding points on crank-pin circle. The tops of these perpendiculars being joined, a diagram is obtained which not only shows the variation of the turning force on the crank, but it has an area proportional to the work done on the crank-pin during one revolution.

A line HK drawn parallel to EF, and at a distance from it equal to the mean height of the diagram EQRSF, will be the *line of mean effort*. If P_m is the mean force on the piston, T_m the mean tangential force on the crank-pin during one revolution, and L the length of the stroke of the piston; then, since the work done on the piston is equal to the work done on the crank-pin in the same time—

$$2P_m L = 3\cdot 1416\, LT_m, \text{ or } T_m = \frac{2P_m}{3\cdot 1416}.$$

If the engine has to overcome a uniform resistance, then the line of mean effort will also be the line of resistance.

Several points of interest can be studied by aid of the diagram in Fig. 55. EF is the path of the crank-pin straightened out, and we will suppose that the crank-pin is moving along the line EF. We will also suppose that the resistance is uniform and equal to EH. While the crank-pin is moving from E to a, the power is less than the resistance, and at a they are equal. In moving from a to b more work is done on the crank-pin than is required by the resistance by an amount represented by the area cQd. This additional work causes the engine to move faster, and the moving parts, including the fly-wheel, if there is one, store up this work. Between b and e the power is less than the resistance, and the speed will diminish and the moving parts will give up an amount of their accumulated work represented by the area dRf. Between e and g the speed will increase, and the surplus energy will be represented by the area fSh. Again, from g to F and E to a the speed will diminish and the deficiency of energy will be represented by the areas hfK and HEc.

Let the areas of the figures cQd, dRf, and fSh, and therefore the work represented by them, be denoted by A_1, A_2, and A_3 respectively. Also let the energy of the moving parts of the engine, when the crank-pin is at a, be denoted by x; then we have—

Energy stored up in moving parts when crank-pin is at $a = x$.
" " " " " $b = x + A_1$.
" " " " " $c = x + A_1 - A_2$.
" " " " " $g = x + A_1 - A_2 + A_3$.

The position of the crank-pin when it has its greatest speed will be that at which the stored-up energy is greatest, namely, either at b or g, according as A_2 is greater or less than A_3. Again, the position of the crank-pin when it has its least speed will be that at which the stored-up energy is least, namely, either at a or c, according as A_1 is greater or less than A_2.

Let the maximum amount of stored-up energy be denoted by X, and the minimum amount by Y. Also let v_1 denote the speed of the fly-wheel rim in feet per second when the stored-up energy is greatest, and v_2 the speed when the stored-up energy is least. Then for a variation of speed $= v_1 - v_2$, the moving parts must give up an amount of energy $= X - Y$. If the total work done during one revolution of the crank shaft be denoted by Z, it will be more convenient to express $X - Y$ in terms of Z, thus $\dfrac{X - Y}{Z} = r$. Approximate values of r for various kinds of engines are as follows:—For single-cylinder condensing engines having connecting rods equal in length to two and a half times the length of the stroke, and working against a uniform resistance—

Point of cut-off = 1 $\tfrac{1}{3}$ $\tfrac{1}{4}$ $\tfrac{1}{5}$ $\tfrac{1}{6}$ $\tfrac{1}{7}$ $\tfrac{1}{8}$
r = ·125   ·163   ·173   ·178   ·184   ·189   ·191

For non-condensing engines—
Point of cut-off = 1 $\tfrac{1}{2}$ $\tfrac{1}{3}$ $\tfrac{1}{4}$ $\tfrac{1}{5}$
r = ·125   ·16   ·186   ·209   ·232

For two similar engines working on cranks at right angles to each other on the same shaft, the value of r is about one-fourth of its value for one engine, and for three similar engines working on cranks at 120° to each other on the same shaft, the value of r is about one-twelfth of its value for one engine.

The value of r for any particular case is best determined by drawing the diagram of crank effort.

When there are two or more cranks on the same crank-shaft, the diagram of effort for each is constructed as already explained, but they must be placed one in advance of the other by an amount depending on the angle between the cranks. The most common case is where there are two cranks at right angles to one another. This case is shown in Fig. 56. EF is, as before, equal to the circumference of the crank-pin circle. The curve of effort for the first crank is moved forward in relation to the curve for the other a distance EE', equal to one-fourth of EF. The curve of total effort is then constructed by adding together the ordinates of the curves for the separate cranks; thus, $ad = ab + ac$.

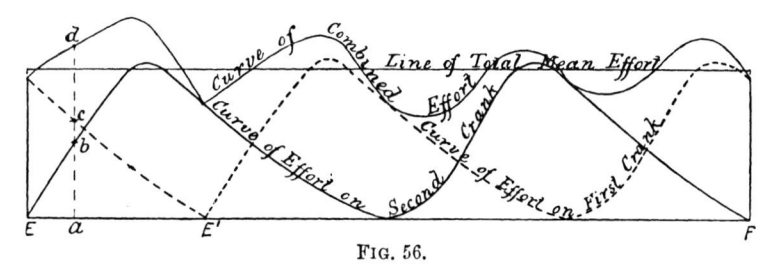

FIG. 56.

As the twisting moment on a crank-shaft is equal to the tangential force or effort multiplied by the radius of the crank-pin circle, and since the radius of the crank-pin circle remains constant, it follows that the diagram of crank effort is also the diagram of twisting moment.

49. Weight of Fly-Wheel.—In the following investigation we will neglect the energy stored up in the arms and boss of the fly-wheel:—

Let W = weight of rim of wheel in pounds.
D = mean diameter of wheel in feet.
N = mean speed of wheel in revolutions per minute.
N_1 = maximum speed of wheel in revolutions per minute.
N_2 = minimum ,, ,, ,,
$\dfrac{N_1 - N_2}{N} = \dfrac{1}{n}$ = co-efficient of fluctuation of speed.
r = ratio of work to be stored up by wheel, in changing its speed from the minimum to the maximum, to the total work done in one revolution.
H = total work done in one revolution.

Work stored up in wheel in changing its speed from N_2 to N_1 revolutions per minute $= \dfrac{W\pi^2 D^2}{2g \times 60^2} (N_1^2 - N_2^2) = \dfrac{W\pi^2 D^2 N^2}{60^2 gn} = rH$.

Therefore $W = \dfrac{60^2 gnrH}{\pi^2 D^2 N^2} = \dfrac{11745 nrH}{D^2 N^2}$.

Values of r are given on page 24. Values of n for various cases are given in the following table:—

Kind of Machinery Driven.	Value of n.
Pumps, and shearing and punching machines	20 to 30
Flour-mills	25 ,, 35
Looms, paper-making machines, and ordinary machine tools	30 ,, 40
Spinning machinery	50 ,, 100
Dynamos	150

The diameter of the fly-wheel is usually from three to five times the stroke of the piston.

CHAPTER III.

STRENGTH AND NATURE OF MATERIALS USED IN MACHINE CONSTRUCTION.

50. Load.—The combination of external forces acting on a piece of a structure is called the *load* on that piece. The *useful load* on a piece is the load arising directly from the purpose for which the piece is designed. Thus, in the lifting chain of a crane the useful load is the weight to be lifted; but this is not the only load on the chain; there is in addition the load due to the weight of the chain itself, and also the load due to the inertia of the weight which increases or diminishes with changes in the velocity of raising or lowering. A *dead load* is one which is applied slowly and remains constant. A *live load* is one which is continually changing.

51. Strain and Stress.—The effect of a load, however small, on a structure is a change of form. This change of form is called *strain*. The internal forces which are called into play in the material of a structure to resist the tendency of the load to produce strain are called *stress*.

In the case of bars which are loaded so as to cause simple lengthening or shortening, the strain is measured as follows—

$$\text{Strain} = \frac{\text{increase or decrease in length}}{\text{length of unloaded bar}}.$$

For example, if a bar 8 feet long is stretched $\frac{1}{16}$th of an inch by the action of a load, the strain is equal to $\frac{\frac{1}{16}}{8 \times 12} = \frac{1}{8 \times 12 \times 16} = \frac{1}{1536}$ = ·065 per cent.

There are three kinds of simple strain and stress, namely: (1) *tensile strain* and *tensile stress*, produced in a piece which is elongated by a load; (2) *compressive strain* and *compressive stress*, produced in a piece which is shortened by a load; (3) *shearing strain* and *shearing stress*, produced in a piece which is distorted by a load which tends to cut it across. Stress is usually measured in pounds or tons per square inch.

52. Elasticity.—The elasticity of a material is the resistance which it offers to change of form on the application of a load, combined with the power of returning to its original shape after the load is removed. A piece of material is said to be perfectly elastic within certain limits of loading, when, the load being removed, the piece returns exactly to its original form. The *limit of elasticity* is reached when the load causes a *permanent set*, that is, when the piece does not recover its original form

MATERIALS USED IN MACHINE CONSTRUCTION.

after the load is removed. When a piece is not loaded beyond its limit of elasticity, the stress produced is directly proportional to the strain, so that the stress divided by the strain is a constant quantity. This constant quantity is called the *modulus of elasticity*. Thus, modulus of elasticity $= \dfrac{\text{stress}}{\text{strain}}$.

If a bar, having a cross section of A square inches, carries a load of W pounds, either in tension or compression, and if the modulus of elasticity of the material of the bar is E pounds per square inch, then the strain produced is given by the formula $S = \dfrac{W}{EA}$.

If the strain S is caused by change of temperature, the force W, which will bring the bar back to its original length, is given by the formula W = EAS, and this is the force which must be applied to prevent the strain or change of length as the temperature changes.

Moduli of Elasticity.

Material.	Modulus of Elasticity in Lbs. per Sq. In.	Material.	Modulus of Elasticity in Lbs. per Sq. In.
Brass, cast	9,150,000	Ash	1,600,000
„ rolled	12,500,000	Box	1,800,000
„ wire	14,200,000	Beech	1,350,000
Bronze (8 copper to 1 tin)	9,000,000	Birch	1,645,000
Copper, cast { from	10,000,000	Chestnut	1,190,000
{ to	15,000,000	Elm	900,000
„ wire, unannealed	17,500,000	Glass, plate	8,000,000
„ „ annealed	14,000,000	Larch	1,200,000
Iron, cast { from	14,000,000	Leather	24,500
{ to	22,900,000	Lignum-vitæ	1,000,000
(average	17,000,000	Mahogany	1,300,000
„ wrought, bars	29,000,000	Oak, English	1,700,000
„ „ plates	26,000,000	Pine, red	1,800,000
„ „ wire	25,300,000	„ yellow	1,600,000
Lead, sheet	720,000	„ white	1,000,000
Steel, mild	30,000,000	Teak, Indian	2,300,000
„ cast, tempered	36,000,000	Willow	1,400,000

The following examples show the use to which the formula $S = \dfrac{W}{EA}$ may be put:—

EXAMPLE 1.—*Find the increase in length of a steel bar 6 feet long and $\frac{3}{4}$ inch in diameter when it is pulled with a force of 4 tons.*

$W = 4 \times 2240$ lbs. $E = 30{,}000{,}000$. $A = \frac{3}{4} \times \frac{3}{4} \times \cdot 7854$.

$$S = \dfrac{\text{increase in length}}{\text{original length}} = \dfrac{W}{EA} = \dfrac{4 \times 2240}{30{,}000{,}000 \times \frac{3}{4} \times \frac{3}{4} \times \cdot 7854};$$

therefore increase in length $= \dfrac{4 \times 2240 \times 6 \times 12}{30{,}000{,}000 \times \frac{3}{4} \times \frac{3}{4} \times \cdot 7854} = \cdot 0486$ inch.

EXAMPLE 2.—*A bar of wrought iron 50 inches long is subjected to a tensile load of 10 tons; what must be the area of its cross section if its increase in length is ·01 inch?*

$$S = \frac{·01}{50} = \frac{W}{EA} = \frac{10 \times 2240}{29,000,000 \ A};$$

therefore $A = \dfrac{10 \times 2240 \times 50}{29,000,000 \times ·01} = 3·86$ square inches.

EXAMPLE 3.—*A bar 10 feet long is made up of two bars, one of copper (E = 15,000,000) and the other of steel (E = 30,000,000), each 1 inch broad and ½ inch thick. This compound bar is stretched by a load of 5 tons. Find the increase of length of the compound bar, and the stress per square inch produced in the steel and copper.*

The portion of the load carried by the steel bar = EAS = 30,000,000 × ½ S = 15,000,000 S.

The portion of the load carried by the copper bar = EAS = 15,000,000 × ½ S = 7,500,000 S.

Therefore the total load = 15,000,000 S + 7,500,000 S = 5 × 2240, or 22,500,000 S = 5 × 2240.

Therefore $S = \dfrac{5 \times 2240}{22,500,000} = \dfrac{\text{increase in length}}{10 \times 12}$, and, increase in length $= \dfrac{5 \times 2240 \times 10 \times 12}{22,500,000} = ·0597$ inch.

Portion of load carried by steel bar = 15,000,000 S =

$$\frac{15,000,000 \times 5 \times 2240}{22,500,000}.$$

Therefore stress in steel bar $= \dfrac{15,000,000 \times 5 \times 2240}{22,500,000 \times \frac{1}{2}} = 14933·33$ lbs. per square inch.

Portion of load carried by copper bar = 7,500,000 S =

$$\frac{7,500,000 \times 5 \times 2240}{22,500,000}.$$

Therefore stress in copper bar $= \dfrac{7,500,000 \times 5 \times 2240}{22,500,000 \times \frac{1}{2}} = 7466·66$ lbs. per square inch.

53. Strength.—The smallest load which will cause the fracture of a piece is called the *ultimate strength* of that piece. The *proof or elastic strength* is the smallest load which will cause permanent set.

54. Effect of Live Loads.—The experiments of Wöhler and others have shown that when a load is applied and withdrawn or reversed a very large number of times, the actual breaking load will be less than the breaking dead load by an amount depending on the *range* of stress produced. For example, a bar of ductile iron or mild steel which would break under a dead load W, would ultimately break under a load W_1 = (about) ⅔ W, if this load W_1 were taken off and applied two or three million times, the load W_1 causing either a tensile or compressive stress. Again, if this bar were subjected alternately to tensile and compressive loads W_2 repeated two or three million times, the bar would ultimately break if W_2 were about equal to $\frac{1}{3}$ W.

55. Factor of Safety.—The ratio of the ultimate strength of a structure to the working load upon it is called the *factor of safety*. This factor varies greatly for different materials and different structures. For materials the quality of which is variable or liable to change, the factor of safety must be larger than for materials the quality of which is more uniform and less affected by exposure to atmospheric and other influences. Also, in structures where the whole load cannot be ascertained with accuracy, the factor of safety must be increased so as to allow for the unknown straining actions. Again, in some structures there is a liability to a sudden increase in the working load; thus, in a crane, if the weight be allowed to descend rapidly and then suddenly stopped, the load may become very much greater than that due to the weight alone. These accidental straining actions must, therefore, be allowed for in the factor of safety.

The following table gives examples of the factor of safety to be met with in practice for various materials under dead and live loads, and when subjected to shocks:—

Factors of Safety.

Material.	Dead Load.	Live Load.	Shocks.
Cast-iron, and brittle metals and alloys	4	6 to 10	10 to 15
Wrought-iron and mild steel	3	5 ,, 8	9 ,, 13
Cast-steel	3	5 ,, 8	10 ,, 15
Copper and other soft metals and alloys	5	6 ,, 9	10 ,, 15
Timber	6	8 ,, 12	14 ,, 18
Masonry and brickwork	10	15 ,, 20	...

56. Resistance to Tension.—If a piece having a cross section of A square inches is subjected to tension by a load W pounds, and if the tensile stress in pounds per square inch is uniformly distributed over the cross section, and is denoted by f, then $W = Af$.

EXAMPLE.—*Find the diameter of a round bar of wrought-iron which will safely carry a tensile load of* 15 *tons. Tensile strength of wrought-iron* 54,000 *lbs. per square inch. Factor of safety* 5.

Let D = diameter of bar in inches.
Safe load = W = 15 × 2240 lbs.
Breaking load = W × 5 = 15 × 2240 × 5 lbs.
Area of cross section of bar = ·7854 D².
·7854 D² × 54000 = 15 × 2240 × 5

$$D^2 = \frac{15 \times 2240 \times 5}{·7854 \times 54000} = 3·9612.$$

$$D = \sqrt{3·9612} = 1·99 \text{ inches.}$$

The following table gives the ultimate tensile strength of various materials:—

Ultimate Tensile Strength of Various Materials.

Material.	Tensile Strength in Lbs. per Sq. In.	Material.	Tensile Strength in Lbs. per Sq. In.
Cast-iron	from 9,000 / to 31,000 / average 17,000	Gun-metal	from 23,000 / to 42,000 / average 32,000
Wrought-iron bars	from 40,000 / to 67,000 / average 54,000	Phosphor-bronze	from 22,000 / to 54,000 / average 35,000
Wrought-iron plates, with fibre	from 45,000 / to 56,000 / average 51,000	Manganese-bronze	from 54,000 / to 72,000
		Brass, yellow	. 26,000
Wrought-iron plates, across fibre	from 40,000 / to 52,000 / average 46,000	Muntz-metal	. 49,000
		Wood, ash	. 15,000
		„ beech	. 12,000
Cast-steel, untempered	from 100,000 / to 150,000	„ elm	. 13,000
		„ fir and pine	from 9,000 / to 14,000
Mild steel	from 60,000 / to 100,000 / average 70,000	„ mahogany	from 8,000 / to 21,000
		„ lignum-vitæ	. 11,000
Copper, cast	from 18,000 / to 27,000 / average 22,000	„ oak	from 9,000 / to 19,000
Copper, sheet	. 30,000	„ teak	from 9,000 / to 19,000
Copper, forged	. 34,000	Leather	. 4,200
Copper bolts	. 36,000	Hempen ropes	. 14,000
Copper wire, unannealed	. 60,000		

57. Resistance to Compression.—The resistance to compression of a piece which is short compared with its transverse dimensions is calculated by the same formula as its resistance to tension, namely, $W = Af$, the stress f being, of course, compressive stress in this case.

The ultimate compressive strength of materials is generally much more difficult to determine than their tensile strength. Especially is this the case with soft or plastic materials which spread out laterally when under compression.

The following table of ultimate compressive strength of various materials has been drawn up after consulting several authorities:—

Ultimate Compressive Strength of Various Materials.

Material.	Compressive Strength in Lbs. per Sq. In.
Cast-iron	60,000 to 140,000
„ average	100,000
Wrought-iron	50,000
Brass, cast	10,500
Wood, ash	8,000
„ box	10,300
„ lignum-vitæ	9,900
„ oak	6,500 to 10,000
„ pine	4,000 to 6,500
Brick, red	550 to 1,100
„ fire	1,700
Granite	6,000 to 11,000
Sandstone	3,300 to 5,500

58. Resistance to Shearing.—The area of the section over which the shearing stress f acts being denoted by A, and the shearing load by W, we have, as before, $W = Af$. In the case of punching a hole of diameter d in a plate of thickness t, the area of the section subjected to shearing is equal to the circumference of the hole multiplied by the thickness of the plate; therefore $A = 3\cdot 1416 dt$.

The ultimate shearing strength of metals is usually from 70 to 100 per cent. of their ultimate tensile strength.

The following table gives examples of the ultimate shearing strength of materials:—

Ultimate Shearing Strength of Various Materials.

Material.	Shearing Strength in Lbs. per Square Inch.
Cast-iron	9,000 to 30,000
Wrought-iron	40,000 ,, 60,000
Mild steel	55,000
Copper	20,000 to 30,000
Wood: ash, across the grain	6,300
,, ,, along ,,	1,400
,, oak, across ,,	4,500
,, ,, along ,,	2,000
,, pine, across ,,	2,500 to 5,500
,, ,, along ,,	500 ,, 800

59. Resistance to Twisting.—This is fully considered in chapter viii.

60. Stresses Induced by Bending.—In a beam which is supported at both ends, the load causes the material in the upper part to be compressed, and that in the lower part to be stretched.

Separating the compressed part of the beam from the stretched part is a horizontal surface, at which there is neither compression nor tension. This surface is called the *neutral surface* of the beam. The line in which the neutral surface intersects a transverse section of the beam is called the *neutral axis* of that section.

61. Bending Moment.—The bending action at any given transverse section of a beam will depend not only on the magnitudes of the forces acting on the beam, but also on the distances of the lines of action of these forces from the given section. Therefore the bending action must be measured by the moments of these forces relatively to the given section.

The resultant moment of the forces acting on the beam on one side of a given section, referred to that section, is called the bending moment on the beam at that section.

For example, in a beam fixed at one end and loaded at the other with a weight W, the bending moment at a cross section at a distance x from the free end is Wx.

As another example, take the case of a beam (Fig. 57) supported at the ends, and carrying a uniformly distributed load of w pounds per unit of length, and a concentrated load of W pounds at a distance a from one end. To determine the bending moment at a section at a distance x from one end, first determine, by the principle of the lever, the reactions R and R' of the points of support. The forces to the left of the given section are, R, W, and wx; the moments of these forces relatively to the section are, Rx, $W(x-a)$, and $\frac{wx^2}{2}$, and the resultant moment is $Rx - W(x-a) - \frac{wx^2}{2}$, and this is the bending moment at the given section.

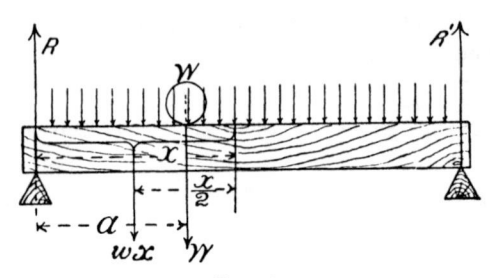

FIG. 57.

Bending moments are generally expressed either in inch-pounds or inch-tons.

62. Shearing Load.—As pointed out in Art. 60, a beam subjected to bending has one part in tension and another in compression; but in general there is also a shearing action, as can be shown by means of a model constructed as in Fig. 58. Here a solid rectangular beam has a portion, ABDC, cut out, and its place supplied by two links, AB and CD, of which AB supports the tension and CD the compression produced by the load W. It will be found that unless an upward force, P, equal to W, be applied at B, the outer part of the beam will drop, as shown by the dotted lines. In the solid beam the part of the upward force P is taken by the shearing resistance of the material of the beam.

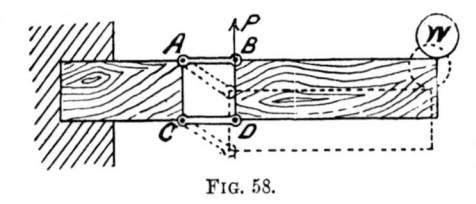

FIG. 58.

The shearing load at any section of a beam is equal to the resultant of all the parallel forces acting on the beam on one side of that section.

63. Moment of Resistance.—The resultant compressive stress on one side of the neutral axis of any transverse section is equal to the resultant tensile stress on the other side of that axis. These two equal and parallel forces form a couple, whose moment is the moment of resistance of the beam to bending at that section.

The moment of resistance is equal to the bending moment.

If f is the greatest tensile or compressive stress at a section of the beam, the moment of resistance may be written in the form fz, where z is a quantity, called the modulus of the section, depending upon the form of the section.

Putting M for the bending moment, the relation between the bending moment and the moment of resistance may be written $M = fz$.

The relation $M = fz$ is only true so long as the beam is not strained beyond the elastic limit, and it is based on the assumption that the stress varies uniformly from a maximum at the top or bottom of a cross section to nothing at the neutral axis of the section.

The following table gives the value of the section modulus z for various sections. The neutral axis of the section is shown in each case by a horizontal chain line :—

Moduli of Sections.

Form of Section.	Modulus of Section z.	Form of Section.	Modulus of Section z.
	$\frac{1}{6} BD^2$		$\frac{1}{6}\left(\frac{BD^3 - bd^3}{D}\right)$
	$\frac{1}{6} B^3$		$\frac{1}{6}\left(\frac{BD^3 - bd^3}{D}\right)$
	$\cdot 118\, B^3$		$\frac{1}{6}\left(\frac{bD^3 + Bd^3}{D}\right)$
	$\frac{3\cdot 1416}{32} D^3$		$\frac{1}{6}\left(\frac{bD^3 + Bd^3}{D}\right)$
	$\frac{3\cdot 1416}{32} BA^2$		$\frac{1}{6}\left(\frac{bD^3 + Bd^3}{D}\right)$
	$\frac{B}{6}\left(\frac{D^3 - d^3}{D}\right)$		$\frac{3\cdot 1416}{32}\left(\frac{D^4 - d^4}{D}\right)$
	$\frac{1}{6}\left(\frac{BD^3 - bd^3}{D}\right)$		$\frac{3\cdot 1416}{32}\left(\frac{BA^3 - ba^3}{A}\right)$

64. Examples of Bending Moments and Shearing Forces on Beams.

CASE I. Fig. 59. Beam fixed at one end and loaded at the other. Greatest bending moment at B, and equal to WL. Shearing force equal to W, and equal from end to end of the beam.

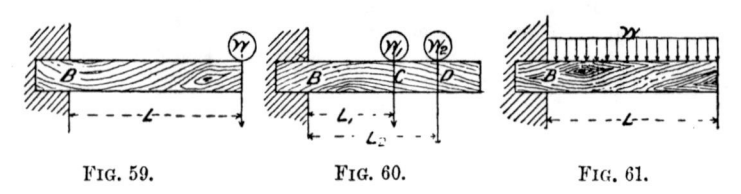

FIG. 59.   FIG. 60.   FIG. 61.

CASE II. Fig. 60. Beam fixed at one end and loaded at more than one point. Greatest bending moment at B, and equal to $W_1L_1 + W_2L_2$. Shearing force, uniform from B to C, and equal to $W_1 + W_2$; also uniform from C to D, and equal to W_2.

CASE III. Fig. 61. Beam fixed at one end and carrying a uniform load, W. Greatest bending moment at B, and equal to $\frac{1}{2}WL$. Greatest shearing force at B, and equal to W.

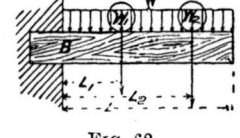

FIG. 62.

CASE IV. Fig. 62. Beam fixed at one end and carrying a uniform load, W, in addition to concentrated loads, W_1 and W_2. Greatest bending moment at B, and equal to $\frac{1}{2}WL + W_1L_1 + W_2L_2$. Greatest shearing force at B, and equal to $W + W_1 + W_2$.

CASE V. Fig. 63. Beam supported at each end and loaded at the

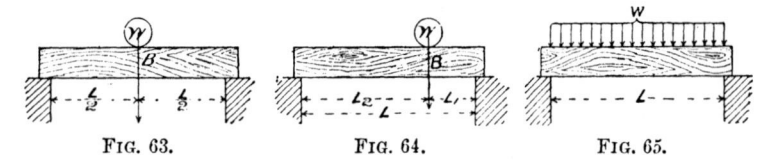

FIG. 63.   FIG. 64.   FIG. 65.

centre. Greatest bending moment at B, and equal to $\frac{1}{4}WL$. Shearing force uniform throughout, and equal to $\frac{1}{2}W$.

CASE VI. Fig. 64. Beam supported at each end and loaded at any point. Greatest bending moment at B, and equal to $\frac{W L_1 L_2}{L}$. Shearing force, uniform from B to the left, and equal to $\frac{WL_1}{L}$; also uniform from B to the right, and equal to $\frac{WL_2}{L}$.

CASE VII. Fig. 65. Beam supported at each end and loaded uniformly. Greatest bending moment at the centre, and equal to $\frac{1}{8}WL$. Greatest shearing force at the ends, and equal to $\frac{1}{2}W$.

CASE VIII. Fig. 66. Beam loaded at the ends and supported at intermediate points. The loads W_1, W_2, and the lengths, L_1, L_2, being such that $W_1L_1 = W_2L_2$. Greatest bending moment equal to either W_1L_1 or

W_2L_2, and is uniform from B to C. Shearing force greatest at B or C, and equal to W_1 at B, and W_2 at C. No shearing force between B and C.

CASE IX. Fig. 67. Beam fixed at each end and loaded at the centre. Greatest bending moment at the centre and at each end, and equal to $\frac{1}{8}WL$. The cross section of the beam is supposed to be uniform. The

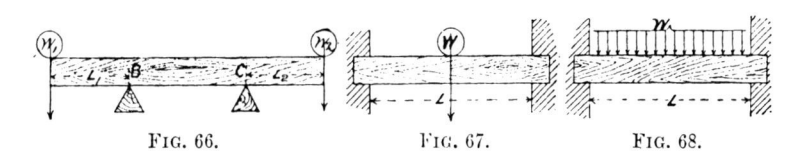

FIG. 66.   FIG. 67.   FIG. 68.

bending moments at the ends are contrary to the bending moment at the centre, that is, at the centre the bottom of the beam will be in tension, while at the ends the bottom will be in compression.

CASE X. Fig. 68. Beam of uniform cross section fixed at each end and loaded uniformly. Bending moment greatest at the ends, and equal to $\frac{1}{12}WL$. The bending moment at the centre is equal to $\frac{1}{24}WL$, and is contrary to the bending moments at the ends.

65. Continuous Beams.—A beam resting on three or more supports is called a *continuous beam*. The bending moments and the reactions at the supports of a continuous beam are determined by means of Clapeyron's theorem of three moments. Let the beam have any number of spans of lengths, L_1, L_2, L_3, etc.; let these

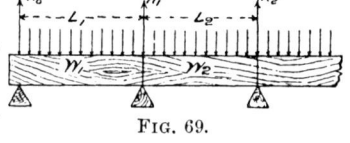

FIG. 69.

spans carry uniform loads, W_1, W_2, W_3, etc.; let the reactions at the supports be R_0, R_1, R_2, etc.; and the bending moments at the supports, M_0, M_1, M_2, etc.

For the first two spans the theorem gives the following equation:—

$$4L_1M_0 + 8(L_1 + L_2)M_1 + 4L_2M_2 = W_1L_1^2 + W_2L_2^2.$$

For the second and third spans the theorem gives the equation:—

$$4L_2M_1 + 8(L_2 + L_3)M_2 + 4L_3M_3 = W_2L_2^2 + W_3L_3^2$$

and so on for every two consecutive spans. Also the bending moments at the ends are each zero. From these equations the bending moments can be calculated, and from the bending moments the reactions can easily be determined.

For example, let the beam consist of two spans each equal to L, and let $W_1 = W_2 = W$. Then the first of the equations above becomes

$$4LM_0 + 16LM_1 + 4LM_2 = 2WL^2.$$

But M_0 and M_2 each equal 0 in this case, therefore $16LM_1 = 2WL^2$, or

$$M_1 = \frac{WL}{8}.$$

But $M_1 = \dfrac{WL}{2} - R_0L$, therefore $R_0L = \dfrac{WL}{2} - \dfrac{WL}{8} = \tfrac{3}{8}WL$, and therefore $R_0 = \tfrac{3}{8}W$. This will evidently also be the value of R_2. Hence $R_1 = 1\tfrac{1}{4}W$.

For a beam of three spans each equal to L, and each carrying a uniform load W, the bending moments at the supports are, $M_0 = M_3 = 0$, $M_1 = M_2 = \dfrac{WL}{10}$; and the reactions at the supports are, $R_0 = R_3 = \tfrac{4}{10}W$, $R_1 = R_2 = \tfrac{11}{10}W$.

66. Cast-Iron.—The product known as *cast-iron* or *pig-iron*, obtained by the smelting of iron ores, is a combination of iron with carbon and various other substances, such as, silicon, manganese, phosphorus, and sulphur.

The total amount of *carbon* in cast-iron varies from 2 per cent. to 5 per cent. of the total weight, but a large proportion of this carbon is generally in the free state, that is, it is simply mixed mechanically with the iron, and not chemically combined with it. The free carbon does not affect the properties of cast-iron to any great extent, but the larger the amount of free carbon there is, the darker is the cast-iron. It is the amount of combined carbon which, as a rule, decides the quality of cast-iron. The greater the amount of carbon there is chemically combined with the iron, the harder and whiter does it become. The harder varieties of cast-iron have usually a high crushing strength, but they are not so suitable for resisting tension as the softer varieties. The softest varieties of cast-iron contain about ·15 per cent. of combined carbon, and the varieties most suitable for resisting tensile and transverse loads contain from ·5 to 1 per cent. of combined carbon.

Mr. Thomas Turner, of the Mason College, Birmingham, has shown that "on gradually adding *silicon* to hard white cast-iron, the metal changes in character to a soft grey iron; by suitable additions of silicon any degree of softness which is desired can be obtained, and it is found that a grey iron, artificially prepared in this manner, is closer in grain, softer, and more sound than metal prepared according to ordinary foundry methods." The amount of silicon should not exceed 3·5 per cent.

The presence of *phosphorus* in cast-iron makes it more fusible and more fluid when melted, but it also makes it more brittle. In good cast-iron the amount of phosphorus does not exceed 1 per cent., and it is usually much less than this. The presence of *sulphur* in cast-iron tends to make the latter white and hard. In good cast-iron the amount of sulphur does not exceed ·15 per cent. The amount of *manganese* in cast-iron should not exceed 1 per cent. An excess of manganese makes cast-iron white, hard, and crystalline.

Cast-iron, after solidifying, contracts in cooling about one-eighth of an inch per foot of length. Patterns for moulding from must therefore be larger than the required casting in this proportion.

67. Strength of Cast-Iron—The ultimate strength of ordinary cast-iron in tension varies from 4 to 14 tons per square inch, the average being about 7½ tons. Cast-iron has, however, been made with a tenacity as high as 18 tons per square inch. The ultimate crushing strength of cast-iron varies from about 25 to 60 tons per square inch, with an average of about 45 tons per square inch. The crushing resistance of cast-iron is about six times its tenacity. The average shearing resistance of cast-iron may be taken at about 12 tons per square inch. The elastic strength of cast-iron is nearly equal to its ultimate strength.

The thickness of the metal in a casting in cast-iron should be as uniform as possible, so that it may cool, and therefore contract, uniformly throughout; otherwise some parts may be in a state of initial strain after the casting has cooled, and will therefore be easier to fracture. Re-entrant angles, unless well rounded out with fillets, are sources of weakness.

68. Chemical Composition of Strong Cast-Iron.—Mr. Thomas Turner tested a number of specimens of cast-iron prepared at Rosebank Foundry, Edinburgh, which had a very high tenacity. The tenacity of four specimens varied from 16·4 tons to 18·2 tons, with an average of 17·1 tons per square inch. The average composition of the four specimens was as follows:—

	Per Cent.
Combined carbon	0·46
Silicon	1·31
Sulphur	0·06
Manganese	0·99
Phosphorus	0·53

69. Chilled Castings.—When grey cast-iron is melted, a portion of the free carbon combines chemically with the iron; this, however, separates out again if the iron is allowed to cool slowly, but if it is suddenly cooled a greater amount of the carbon remains in chemical combination, and a whiter and harder iron is produced. Advantage is taken of this in making *chilled castings*. In this process the whole or a part of the mould is lined with cast-iron, which is protected by a thin coating of loam. The cast-iron lining being a comparatively good conductor of heat, chills a portion of the melted metal next to it, changing it into a hard white iron to a depth varying from one-eighth to half an inch. In a chilled casting the main body of metal is the grey cast-iron, which is the best kind of cast-iron for resisting shocks, while the chilled surface is much harder and better suited to resist wear.

Castings made in ordinary sand moulds are found to have a thin skin of harder iron, for the reason that the outer surface is cooled more quickly than the interior of the casting. This hard skin protects the iron, to a great extent, from oxidising.

70. Malleable Cast-Iron.—This is prepared by imbedding a casting in powdered red hæmatite, which is an oxide of iron, and keeping it at a bright red heat for a length of time varying from several hours to several days, according to the size of the casting. By this process a portion of the carbon in the cast-iron is removed, and the strength and toughness of the casting approach the strength and toughness of wrought-iron or mild steel

71. Wrought-Iron.—Wrought or malleable iron is nearly pure iron, and is made from cast-iron by the puddling process, which consists chiefly of raising the cast-iron to a high temperature in a reverberatory furnace, when the carbon and other impurities are removed by the action of the air, and also by the action of the covering on the bottom of the furnace, called the *fettling*. The iron is removed from the puddling furnace in soft spongy masses, called *blooms*, which are subjected to a process of squeezing or hammering, called *shingling*. These shingled blooms still contain sufficient heat to enable them to be rolled into rough *puddled bars*. These puddled bars are of very inferior quality, having less than half the strength of good wrought-iron. The puddled bars are cut into pieces, which are piled together into *faggots*, re-heated, and again rolled into bars, which are now called *merchant bars*. This process of piling, re-heating, and re-rolling may be repeated several times, depending on the quality of iron required. *Best bar* iron is made from faggots of merchant bars, while *best best* bar iron is made from faggots of best bar. In like manner *best best best* or *treble best* iron is made from faggots of best best iron. The process of piling and rolling gives wrought-iron a fibrous structure.

The presence of a small quantity of phosphorus in wrought-iron produces *cold-shortness*, that is, brittleness when cold. Sulphur causes *red-shortness* or brittleness when hot.

Rivet iron bars are rolled from faggots of selected scrap iron.

Heavy forgings are made from slabs which are formed from faggots of selected scrap iron.

72. Strength of Wrought-Iron.—The tenacity of wrought-iron varies from 18 to 30 tons per square inch. The tenacity and hardness are increased by cold-rolling and by wire-drawing, but if the iron which has been rolled cold or drawn into wire be annealed, it returns to its original strength and softness. The resistance which wrought-iron offers when the tension is in the direction of the fibre is greater than the resistance which it offers when the tension is at right angles to the fibre by an amount which varies from 1 to 4 tons per square inch.

The elastic strength of wrought-iron is generally not less than half its ultimate strength. The elastic strength of good bars and plates may be taken at about 12 tons per square inch for both tension and compression.

In deciding as to the suitability of wrought-iron for a particular purpose, it is generally necessary to know more than its ultimate tensile strength. It is of importance to know how much it elongates before breaking, and also the extent to which the area of the cross section is reduced at the part where fracture occurs. In stating the elongation just before fracture, it is usual to give it as so much per cent. of the length of the specimen. The elongation expressed in terms of the length will be greater for short specimens than for long ones, because the specimen stretches most nearest the fracture. Hence it is usual and necessary in giving the elongation to also give the length of the specimen. The most common length of the specimen for tensile tests is 8 inches.

Iron such as is used for furnace plates in boilers and for difficult forgings, should have a tenacity of about 25 tons per square inch with the fibre, and an elongation of 14 to 18 per cent. in a length of 8 inches. Tested across the fibre, the tenacity should be about 23 tons per square inch, with an elongation of 8 to 12 per cent. in a length of 8 inches.

The strength of a forging is generally less than that of the iron from which it is made. The loss of strength may be taken at from 10 to 30 per cent.

73. Steel.—The difference between wrought-iron and steel is that whereas the former is almost pure iron, the latter is a compound of iron with a small quantity of carbon. Steel is made either by adding carbon to wrought-iron, as in the *cementation* process, or by removing some of the carbon from cast-iron, as in the *Bessemer* and *Siemens-Martin* processes.

In making steel by the cementation process, bars of wrought-iron are embedded in powdered charcoal in a fire-clay trough, and kept at a high temperature in a furnace for several days. The iron combines with a portion of the carbon and forms *blister-steel*, so named because of the blisters which are found on the surface of the bars when they are removed from the furnace. The bars of blister-steel are broken into pieces about 18 inches long, and tied together in bundles by strong steel wire. These bundles are raised to welding heat, and then hammered and rolled into bars of *shear-steel*. *Cast-steel* is produced by melting the broken pieces of blister-steel in a closed crucible, which is then cast into ingots. These ingots are rolled into bars, which also have the name cast-steel, but it is more usual to call these bars *tool-steel*, from the fact that they are chiefly used for making cutting tools.

Bessemer steel is made by pouring melted cast-iron into a vessel called a converter, through which a blast of air is then urged. By this means the carbon is burnt out, and comparatively pure iron remains. To this is added a certain quantity of *spiegeleisen*, which is a compound of iron, carbon, and manganese. The converter is now tilted up, and the molten metal is cast into ingots.

Siemens-Martin steel is produced by melting cast-iron and wrought-iron or cast-iron and certain kinds of iron ore together on the hearth of a reverberatory gas furnace.

Nearly all the steel which is now used by engineers for constructive purposes is made by the Bessemer or by the Siemens-Martin process.

74. Case-Hardening.—Case-hardening is an application of the cementation process for making steel. If after an article, which is made of wrought-iron, is machined and finished, it be heated in contact with substances rich in carbon, such as bone-dust, horn-shavings, or yellow prussiate of potash, the iron at the surface of the article will be converted into steel, which may be hardened by quenching in water. This process is very often applied to the wrought-iron working parts of machines.

75. Strength of Steel.—The strongest kind of steel is cast-steel, such as is used for cutting tools. This steel has a tenacity varying from about 50 to 65 tons per square inch, but it is very hard and brittle, as shown by the fact that its elongation before fracture is only about 5 per cent. This material is therefore not used for constructive purposes. As a rule, high tenacity is accompanied by hardness and want of ductility. Mild steel, having a tenacity of 27 tons per square inch, will elongate as much as 28 per cent. in a length of 8 inches before fracture, whereas steel having a tenacity of 40 tons per square inch would probably not elongate more than 10 per cent. before fracture.

Lloyd's rules for steel boilers state that the minimum tenacity of the material should be 26 tons per square inch, with an elongation of not less than 20 per cent. in a length of 8 inches. The maximum tenacity allowed by these rules varies with the thickness of the plates, being 30 tons per square inch for plates up to 1 inch in thickness, 29 tons for plates between 1 inch and $1\frac{3}{8}$ inches, and for plates having a thickness of $1\frac{3}{8}$ inches and upwards the maximum tenacity allowed is 28 tons per square inch.

The steel used in the tension members of the Forth Bridge has a tenacity of 30 to 33 tons per square inch, with an elongation of not less than 20 per cent. in a length of 8 inches before fracture. For the parts in compression only, the steel used has a tenacity of 34 to 37 tons per square inch, with an elongation of not less than 17 per cent. in a length of 8 inches. The composition of the steel for the tension members is as follows:—Carbon, ·19; silicon, ·02; sulphur, ·024; phosphorus, ·046; manganese, ·69; all being expressed as percentages of the total weight, the remainder being iron. The composition of the stronger steel for the compression members differs from the above in having ·04 per cent. more carbon and ·01 per cent. more silicon. The whole of the steel used for the Forth Bridge was made by the Siemens-Martin process.

Gun-steel, when in the soft or annealed state, has a tenacity of about 30 tons per square inch, and an elongation of not less than 23 per cent. before fracture. When tempered by being heated to a temperature of 1450° F., and plunged into a bath of cool oil, the tenacity is raised to about 45 tons per square inch.

Whitworth's compressed steel is probably superior to any other for constructive purposes, but it is much more expensive. It can be produced having a tenacity of 42 tons per square inch, combined with an elongation of 26 per cent. before fracture; and when the tenacity is lowered to 32 tons per square inch, the elongation before fracture is increased to 34 per cent.

The late Dr. Percy has given an account[1] of some experiments on very strong *steel wire* used for the wire-ropes made by Messrs. John Fowler & Co., Leeds. It was found that the ultimate tensile strength of the wire increased from 90 tons per square inch when the diameter was ·191 inch, to 154 tons per square inch when the diameter was reduced by wire-drawing to ·093 inch. This wire was very hard and broke when bent on itself. The composition of the steel was found to be as follows:—Carbon (total), ·828; manganese, ·587; silicon, ·143; sulphur, ·009; copper, ·03; all in percentages of the total, the remainder being iron. The specific gravity of the wire was 7·81.

76. Copper.—Pure copper has a reddish-brown colour, and is very malleable and ductile. It can be hammered or rolled into thin sheets

---

[1] Journal of the Iron and Steel Institute, 1886.

MATERIALS USED IN MACHINE CONSTRUCTION. 41

or drawn into wire. It can be forged either when hot or cold, but cannot be welded. Slight traces of impurities cause brittleness, although a small quantity of phosphorus increases the tenacity of copper and makes it more fluid when melted.

Hammering, rolling, and wire-drawing increase the strength of copper, but also make it harder and more brittle. Hardness and brittleness induced by working are removed by annealing. While iron and steel are annealed by being heated and then cooled *slowly*, copper is annealed by being heated and then cooled *quickly*.

It is difficult to obtain sound castings with pure copper, but the addition of a small quantity of phosphorus improves it for casting.

Copper is largely used for making pipes for marine and other steam-engines. Copper pipes are generally made from sheets of copper which are rolled and hammered into shape. The joints are brazed scarf or lap joints. Pipes may also be solid drawn; then they have no joints. This process is generally applied to the smaller pipes only, although solid drawn copper pipes up to 10 inches in diameter are made. Flanges are attached to copper pipes by brazing.

A new process for making copper pipes by electro-deposition has been introduced by Mr. W. Elmore, and has produced excellent results. Mr. W. Parker in a paper on "copper steam pipes," read before the Institution of Naval Architects in 1888, thus describes the Elmore process:—
"A mandrel is surrounded by ordinary unrefined Chili bars arranged upon strong supporting frames in a depositing tank of sulphate of copper, and the copper is dissolved or decomposed, and is deposited in the form of pure copper on a revolving mandrel, leaving the copper in the form of a shell or pipe, of any thickness required, fitting closely to the mandrel. When the required thickness has been deposited, the pipe and mandrel are exposed to the action of hot air or steam; then the copper expanding more than iron, admits of the mandrel being drawn, leaving the copper in the form of a pipe, without seam, perfectly round and true, both internally and externally; or the pipe may be expanded by rolling or other mechanical means, and then the mandrel withdrawn." To increase the density, tenacity, and ductility of the deposited copper, "a burnisher or planisher, composed of a small square piece of agate, is supported upon proper arms and levers, and is allowed to press lightly upon the surface of the copper on the revolving mandrel. The burnisher is caused to traverse from end to end of the mandrel by means of a leading screw at any required speed. After it has traversed the whole length of the mandrel, it is automatically reversed and commences its journey backwards. The speed of the revolving mandrel and the speed of the traversing burnisher is so arranged that the whole surface of the copper is acted upon by the burnisher, the result being that every thin film of copper deposited upon the mandrel must be separately acted upon, burnished and compressed into a dense and cohesive sheet of pure copper possessing a great amount of tenacity and ductility."

77. Strength of Copper.—The tenacity of cast copper varies from 8 to 12 tons per square inch. The average tenacity of cast copper may be taken at about 10 tons per square inch. Forged copper has a tenacity of about 15 tons per square inch. When drawn into wire the tenacity may be as high as 28 tons per square inch without annealing. Such wire would have its tenacity lowered to about 18 tons per square inch by being annealed. The fracture of good copper has a fibrous silky appearance.

From experiments made by Mr. W. Parker on sheet copper cut from steam pipes the following results were obtained:—

	A.	B.	C.
Ultimate tenacity in tons per square inch	23¾	20¼	14
Elongation per cent. in length of 8 inches before fracture	13·2	3·37	44
Contraction of area per cent. at fracture	73·6	12·8 / 40·2	45 / 31·2

The results in Column A. were obtained from strips cut from a pipe made by the Elmore depositing process, described in the preceding article. Column B. gives the results of tests with strips cut from a solid drawn pipe. The results given in Column C. were obtained with strips cut from a pipe made of rolled sheet copper, and brazed in the usual way. In each case strips were taken cut circumferentially and also longitudinally from the pipe. There was very little difference in either the tenacity or elongation of the circumferential and longitudinal strips, but in the case of the solid drawn and sheet pipes there was considerable difference between the contraction of area in the circumferential and longitudinal strips. For the solid drawn pipe the contraction of area in the circumferential strip was 12·8, and in the longitudinal strip 40·2. For the rolled sheet copper the contraction in the strip cut in the direction in which the plate was rolled was 45, and in a strip cut across that direction the contraction was 31·2.

Mr. Parker also tested the electro-deposited, solid drawn, and sheet copper at the temperature of steam of 200 lbs. pressure per square inch, when he found the tenacities per square inch to be as follows:— Electro-deposited copper and also the solid drawn copper, about 15 tons; rolled sheet copper about 10½ tons. From these and other experiments Mr. Parker concludes that it is not safe to reckon the tenacity of rolled sheet copper in pipes at more than 10 tons per square inch when these pipes carry high-pressure steam, and this quite apart from the danger and uncertainty of brazed joints.

The addition of from 1 to 3 per cent. of phosphorus to copper increases the tenacity of the latter to about 20 tons per square inch.

78. Alloys.—An alloy is produced by melting together two or more metals. The metals employed in the manufacture of the principal alloys used in machine construction are copper, tin, zinc, and lead. The atomic weights of these metals are copper 63·5, tin 118, zinc 65·2, and lead 207. Some authorities state that the quantities of the metals forming an alloy should bear to one another ratios which are equal to the ratios which multiples of the atomic weights of these metals bear to one another.

For example, an alloy of copper and tin might contain the following parts by weight:—Copper 63·5, and tin 118, or copper 2 × 63·5, and tin 118, or copper 3 × 63·5, and tin 2 × 118, and so on. This rule is, however, not always adhered to by founders.

Alloys of copper and tin are known as *bronze*, and the term *brass* is applied to alloys of copper and zinc. But bronze has often in addition to copper and tin small quantities of zinc and other metals in its composition. Brass also has often other metals in its composition besides copper and zinc. It therefore becomes difficult to say sometimes whether a particular alloy which contains copper, tin, and zinc should be called bronze or brass; the consequence is that in practice the terms bronze and brass are sometimes applied to the same alloy.

79. Bronze or Gun-Metal.—The best gun-metal is composed of about 90 parts of copper to 10 parts of tin. This metal, if well made, has a tenacity of about 17 tons per square inch. It is sufficiently hard for ordinary bearings in machinery. The hardness of bronze is increased by increasing the proportion of tin in the alloy. A very hard bronze suitable for bearings which have to sustain great pressure is composed of 86 parts of copper and 14 parts of tin. Bronze composed of 92 parts of copper and 8 parts of tin is a soft tough bronze suitable for toothed wheels which are liable to shocks.

80. Phosphor-Bronze.—This is the alloy, ordinary bronze, with from 2 to 4 per cent. of phosphorus in its composition. Great attention has been given to the manufacture of this bronze, and it is now often used to take the place of ordinary bronze, and frequently it is employed instead of iron and steel for pump rods, propellers, &c. When in the form of wire, phosphor-bronze has a tenacity, in the hard or unannealed state, as high as 70 tons per square inch, the average tenacity of such wire being about 56 tons per square inch. The same wire when annealed has a tenacity of about 40 per cent. of the tenacity of the unannealed wire.

The softer varieties of phosphor-bronze have a tenacity of about 20 tons per square inch, with an elongation of about 30 per cent. before fracture.

The composition of some very heavy phosphor-bronze castings (over 10 tons) made for the stern-framing of several wood-sheathed and copper-bottomed cruisers was, copper 90 per cent., tin 9·5 per cent., and phosphorus ·5 per cent.

81. Manganese-Bronze.—White bronze or manganese-bronze is an alloy of ordinary bronze and ferro-manganese. It can be forged into bolts, nuts, pump rods, &c. It can also be rolled into bars, plates, or sheets. This alloy is remarkable for its strength and toughness, being equal to mild steel in these respects. Manganese-bronze should be forged at a cherry-red heat. Plates and sheets may be worked either cold or hot. Owing to the fact that manganese-bronze resists the corroding action of sea-water, it is better suited for the blades of screw propellers than steel or iron.

82. Brass.—The composition of yellow brass is about 2 of copper to 1 of zinc. This alloy has a tenacity of about 12 tons per square inch. The composition of the metal for the brass condenser tubes used in the

English navy is, according to the specification of the Admiralty, 70 parts of best selected copper to 30 parts of Silesian zinc. For brass boiler tubes the Admiralty specify 68 parts of copper to 32 parts of zinc.

Muntz metal is a form of brass, its most usual composition being 3 parts of copper to 2 parts of zinc. This alloy when rolled into sheets is used for the sheathing of wooden ships. It may also be rolled or forged when hot into bars, bolts, and ship fastenings, which are liable to rust when made of iron or steel. The tenacity of Muntz metal is about 22 tons per square inch. It is very ductile, and elongates considerably before fracture. *Naval brass* is a modification of Muntz metal, being that alloy with about 1 per cent. of tin added to it to give it greater strength.

83. Babbitt's Metal.—This is a soft white alloy of copper, tin, and antimony, which has been largely used as a lining for bearings. Authorities differ in giving the composition of this alloy, but we believe that the following is the correct recipe for the real Babbitt's metal:—First, an alloy of 4 parts of copper, 8 parts of antimony, and 24 parts of tin is made, which is called "hardening." This alloy is then melted with twice its weight of tin, and cast into the recesses formed in the bearing to receive it. The true composition of the white metal lining is therefore 4 parts of copper, 8 parts of antimony, and 96 parts of tin.

84. Wood.—The principal woods used in machine construction are *ash, beech, boxwood, elm, fir, hornbeam, lignum-vitæ, mahogany, oak, pine,* and *teak*.

Ash, a straight-grained, tough, and elastic wood, is largely used where sudden shocks have to be resisted, as in the handles of tools, shafts of carriages, and the framing and other portions of agricultural machinery, when such are not made of metal. This wood is very durable if it is protected from the weather.

Beech takes a smooth surface, and is very compact in its grain. It is largely used for the cogs of mortise-wheels and for joiners' tools.

Boxwood is very hard and heavy, and takes a very smooth surface. It is of a bright yellow colour, and is used for sheaves of pulley-blocks, bearings in machinery, small rollers, &c.

Elm is valuable on account of its durability when constantly wet, and is used for piles, floats of paddle-wheels, &c.

Fir and *pine* are largely used for various purposes, because they are cheap, easy to work, and possess considerable strength. White or yellow pine is much used for pattern-making.

Hornbeam is chiefly used for cogs of mortise-wheels.

Lignum-vitæ is a very hard wood of very high specific gravity, being $1\frac{1}{3}$rd times the weight of the same volume of water. It is very valuable for bearings of machinery which are under water. It is also used for sheaves of pulley blocks, and for other purposes where great hardness and strength are required.

Mahogany is a durable, strong, straight-grained wood, and is less liable to crack or twist in seasoning than almost any other wood. It is used to a considerable extent by pattern-makers for light patterns.

Oak is one of the strongest and most durable of woods. It is tough

MATERIALS USED IN MACHINE CONSTRUCTION.

and straight-grained, and is durable in either a wet or dry situation. It is good for framing or for bearings of machinery. English oak is considered the best.

Teak is also a very strong and durable wood, possessing considerable toughness. It is also valuable for many purposes, on account of the small amount of shrinkage which takes place with seasoning. The oil which this wood contains prevents the rusting of bolts or other iron parts which may be used in framing it.

85. Weight of Materials.—The following table gives the weight of one cubic foot of various materials:—

Weight of Materials.

Material.		Weight in Lbs. of One Cubic Foot.	Weight in Lbs. of One Cubic Inch.	Material.		Weight in Lbs. of One Cubic Foot.
Cast-iron	from	430	·249	Brick	from	125
	to	460	·266		to	135
	average	445	·258	Brickwork		112
Wrought-iron	from	474	·274	Granite	from	150
	to	487	·282		to	190
	average	480	·278	Sandstone	from	130
Steel	from	487	·282		to	157
	to	493	·285	Masonry	from	115
	average	490	·284		to	144
Copper, cast		537	·311	Timber—		
„ sheet		549	·318	Ash		46
„ hammered		556	·322	Beech		44
„ wire		554	·321	Birch		45
Brass, cast	from	490	·284	Boxwood		62
	to	525	·304	Chestnut		38
	average	505	·292	Elm		34
„ wire		533	·308	Larch		34
„ sheet		527	·305	Lignum-vitæ		80
Muntz metal		512	·296	Mahogany, Honduras		35
Gun-metal, 84 copper, 16 tin		534	·309	„ Spanish		53
„ „ 79 „ 21 „		544	·315	Oak, English		54
„ „ 90 „ 10 „		561	·325	Pine, red	from	30
Lead, cast		708	·410		to	44
„ sheet		712	·412	„ yellow	from	29
Tin, cast		462	·267		to	41
Zinc, cast		428	·248	„ white		31
„ sheet		449	·260	Teak	from	41
Aluminium, cast		160	·093		to	55
„ sheet		167	·097	Water distilled at 39·1° F.		62·425
Mercury at 32° F.		848·75	·491	„ sea, ordinary		64
				Oils, various, about		57

CHAPTER IV.

SCREWS, BOLTS, AND NUTS.

86. Forms of Screw Threads.—Screw threads are generally either triangular or square in section, but besides these there are other forms in use. Of the triangular threads the most important is the *Whitworth* V thread. This form of thread, which is shown in Figs. 70 and 71, is the standard triangular thread in use amongst English engineers. The angle of the V, as shown in Fig. 71, is 55 degs., and an amount, C, equal to one-sixth of the total height, B, is rounded off at the top and bottom as shown.

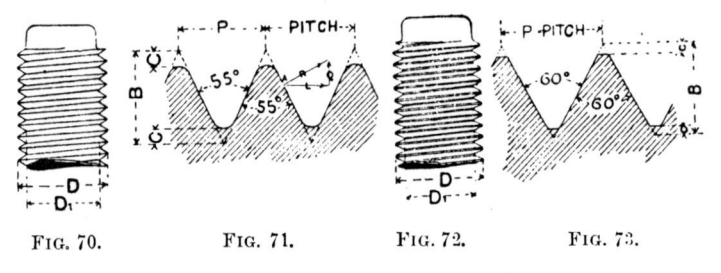

Fig. 70.   Fig. 71.   Fig. 72.   Fig. 73.

From the proportions of the Whitworth triangular thread which have just been stated, it is easy to prove that the diameter of the screw at the bottom of the thread is given by the following formula:—

$$D_1 = D - \frac{1 \cdot 28065}{N}$$

where D_1 = diameter at bottom of thread, D = diameter over the thread, and N = number of threads per inch. It is obvious that if P = pitch,

$$P = \frac{1}{N}, \text{ and also, } N = \frac{1}{P}.$$

The following table of dimensions of Whitworth screw threads will be found useful. The diameter at the bottom of the thread D_1 is also the diameter of the hole in a nut previous to tapping it. The area A at the bottom of the thread is required in calculating the strength of a bolt or screwed bar.

The number of threads per inch is given approximately by the formula:—

$$N = \frac{9D + 67}{8D + 2}$$

Dimensions of Whitworth Screws.

Diameter of Screw D.	Number of Threads per Inch N.	Diameter at Bottom of Thread D_1.	Area at Bottom of Thread A.	Diameter of Screw D.	Number of Threads per Inch N.	Diameter at Bottom of Thread D_1.	Area at Bottom of Thread A.
Inches.	Threads.	Inches.	Sq. Ins.	Inches.	Threads.	Inches.	Sq. Ins.
$\frac{1}{8}$	40	·093	·0068	2	$4\frac{1}{2}$	1·715	2·3101
$\frac{3}{16}$	24	·134	·0141	$2\frac{1}{4}$	4	1·930	2·9255
$\frac{1}{4}$	20	·186	·0272	$2\frac{1}{2}$	4	2·180	3·7325
$\frac{5}{16}$	18	·241	·0456	$2\frac{3}{4}$	$3\frac{1}{2}$	2·384	4·4637
$\frac{3}{8}$	16	·295	·0683	3	$3\frac{1}{2}$	2·634	5·4490
$\frac{7}{16}$	14	·346	·0940	$3\frac{1}{4}$	$3\frac{1}{4}$	2·856	6·4063
$\frac{1}{2}$	12	·393	·1213	$3\frac{1}{2}$	$3\frac{1}{4}$	3·106	7·5769
$\frac{5}{8}$	11	·509	·2035	$3\frac{3}{4}$	3	3·323	8·6726
$\frac{3}{4}$	10	·622	·3038	4	3	3·573	10·027
$\frac{7}{8}$	9	·733	·4219	$4\frac{1}{4}$	$2\frac{7}{8}$	3·805	11·371
1	8	·840	·5542	$4\frac{1}{2}$	$2\frac{7}{8}$	4·055	12·914
$1\frac{1}{8}$	7	·942	·6969	$4\frac{3}{4}$	$2\frac{3}{4}$	4·284	14·414
$1\frac{1}{4}$	7	1·067	·8942	5	$2\frac{3}{4}$	4·534	16·146
$1\frac{3}{8}$	6	1·162	1·0597	$5\frac{1}{4}$	$2\frac{5}{8}$	4·762	17·810
$1\frac{1}{2}$	6	1·287	1·3009	$5\frac{1}{2}$	$2\frac{5}{8}$	5·012	19·729
$1\frac{5}{8}$	5	1·369	1·4719	$5\frac{3}{4}$	$2\frac{1}{2}$	5·238	21·549
$1\frac{3}{4}$	5	1·494	1·7530	6	$2\frac{1}{2}$	5·488	23·654
$1\frac{7}{8}$	$4\frac{1}{2}$	1·590	1·9855				

The *Sellers* screw thread, which is the standard triangular thread used in America, is shown in Figs. 72 and 73. The enlarged section of the thread given in Fig. 73 shows that the angle between the sides of the thread is 60 degs., and that a part is cut square off at the top and bottom. The amount, C, cut off is one-eighth of the total height, B. From these proportions we get the diameter of the screw at the bottom of the thread by the following formula:—

$$D_1 = D - \frac{1{\cdot}29904}{N}$$

where D_1 = diameter at bottom of thread, D = diameter over the thread, and N = number of threads per inch.

The relation between the pitch P and the diameter D of the Sellers screw thread is given very approximately by the formula

$$P = 0{\cdot}24\sqrt{D + 0{\cdot}625} - 0{\cdot}175.$$

The number of threads per inch is

$$N = \frac{1}{P} = \frac{1}{0{\cdot}24\sqrt{D + 0{\cdot}625} - 0{\cdot}175}.$$

The following table gives the standard dimensions of Sellers screw threads:—

Dimensions of Sellers Screws.

Diameter of Screw D.	Number of Threads per Inch N.	Diameter at Bottom of Thread D_1.	Area at Bottom of Thread A.	Diameter of Screw D.	Number of Threads per Inch N.	Diameter at Bottom of Thread D_1.	Area at Bottom of Thread A.
Inches.	Threads.	Inches.	Sq. Ins.	Inches.	Threads.	Inches.	Sq. Ins.
1/4	20	·185	·0269	2	4 1/2	1·711	2·3001
5/16	18	·240	·0454	2 1/4	4 1/2	1·961	3·0212
3/8	16	·294	·0678	2 1/2	4	2·175	3·7161
7/16	14	·345	·0933	2 3/4	4	2·425	4·6194
1/2	13	·400	·1257	3	3 1/2	2·629	5·4276
9/16	12	·454	·1620	3 1/4	3 1/2	2·879	6·5090
5/8	11	·507	·2018	3 1/2	3 1/4	3·100	7·5492
3/4	10	·620	·3020	3 3/4	3	3·317	8·6414
7/8	9	·731	·4193	4	3	3·567	9·9930
1	8	·838	·5510	4 1/4	2 7/8	3·798	11·3304
1 1/8	7	·939	·6931	4 1/2	2 3/4	4·028	12·7404
1 1/4	7	1·064	·8898	4 3/4	2 5/8	4·255	14·2204
1 3/8	6	1·158	1·0541	5	2 1/2	4·480	15·7661
1 1/2	6	1·283	1·2938	5 1/4	2 1/2	4·730	17·5746
1 5/8	5 1/2	1·389	1·5151	5 1/2	2 3/8	4·953	19·2676
1 3/4	5	1·490	1·7441	5 3/4	2 3/8	5·203	21·2617
1 7/8	5	1·615	2·0490	6	2 1/4	5·423	23·0944

The *square screw thread* is represented in Figs. 74, 75, 76. The first of these figures is an elevation of a single-threaded right-handed screw, the second is an enlarged section of the thread, and the third is an elevation of a double-threaded left-handed screw. As its name implies, the cross section of a square screw thread is a square, but in practice the angles of this square are generally more or less rounded, so that the edges of the thread shall not be so easily injured. When this rounding of the angles is carried to excess we get the form of thread shown in

FIG. 74.

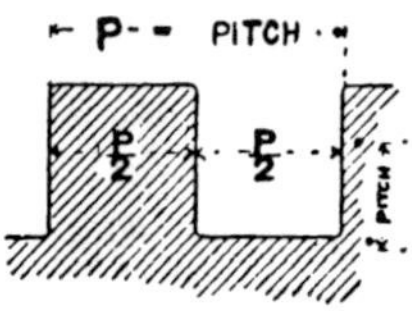

FIG. 75.

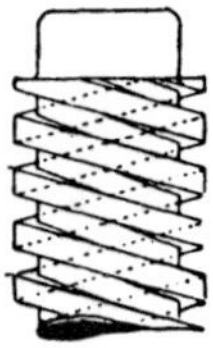

FIG. 76.

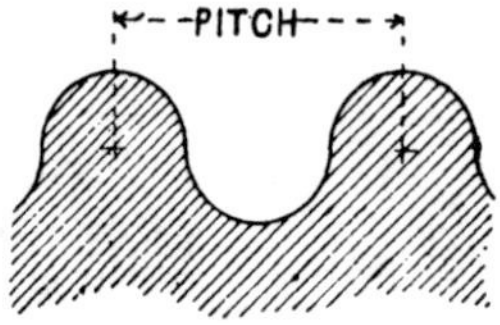

FIG. 77.

Fig. 77; this form is sometimes called a *knuckle thread.* The screw thread shown in Figs. 78 and 79 is a modification of the square thread, and is used for the leading screws of lathes. It will be observed that this thread in section tapers slightly from the root to the point. This is for the purpose of allowing the nut, which is divided into two parts, to readily disengage or engage with the screw. The number of threads per inch in a square-threaded screw is half the number for a triangular-threaded screw of the same diameter.

The *buttress screw thread* is shown in Figs. 80 and 81. From the enlarged section of the thread shown in Fig. 81, it will be seen that the angle between the sides is 45 deg., and that one of these sides is perpendicular to the axis of the screw. An amount C is cut off at the top

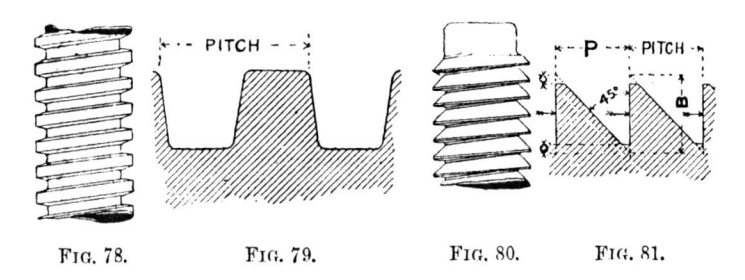

FIG. 78.   FIG. 79.   FIG. 80.   FIG. 81.

and bottom, varying from one-eighth to one-sixth of the total depth, B. In Fig. 81 the point and root of the thread are shown cut off square, but they may be rounded instead. There are no standard proportions for the buttress thread, but those which have been given are proportions generally adopted.

87. Uses and Relative Advantages of Various Screw Threads.—The greatest load on a bolt or screwed rod is generally in the direction of its axis. This load is supported by the reaction between the surface of the thread on the bolt, and that in contact with it on the nut. Referring to Fig. 71, R is the force of reaction at the point A on the surface of the thread, that is, the normal pressure between the bolt and nut at that point. This force R may be resolved into two others, one L, parallel to the axis of the bolt, and the other Q, perpendicular to it. L will represent the portion of the load on the bolt carried by the surface of the thread at A. The force Q will not assist in carrying the load, but will tend to burst the nut. It is evident that for a given load, the magnitude of R and also of Q will be greater the greater the angle of the V. Now the friction between two surfaces is proportional to the normal pressure between them; therefore *the friction between a bolt and a nut is greater the greater the angle of the V of the screw thread. The bursting action on the nut is also greater the greater the angle of the V.*

In the case of the square thread the load acts at right angles to the surface of the thread, and the reaction R is therefore very nearly equal to the load L, so that the friction is less with a square thread than with a V thread. The square thread is therefore preferred for transmitting motion. Another advantage which the square thread has over the V thread is that in the former there is no bursting action on the nut. The V thread is, however, about double the strength of the square thread, because for a given height of nut the amount of material at the root of the thread, to resist the shearing action of the load, is about double in the V thread what it is in the square thread.

The buttress thread is designed to combine the advantages of the V and square threads without their disadvantages. The load must, how-

D

ever, only act on that side of the thread which is perpendicular to the axis of the screw, as shown by the arrows in Fig. 81. If the load be transferred to the other side of the thread, the friction and also the bursting action on the nut will be greater than in the case of the V thread, because of the steeper inclination of the slant side of the buttress thread to the direction of the load.

Comparing the "Whitworth" and "Sellers" screw threads, the former is stronger than the latter because of the rounding at the root. The point of the Whitworth thread is also less liable to injury than the Sellers. The form of the Sellers thread is, however, one which is more easily produced with accuracy, in the first place, because it is easier to get with certainty an angle of 60 degs. than an angle of 55 degs.; and in the second place, because it is easier to make the point and root perfectly parallel to the axis than to ensure a truly circular point and root. The Sellers thread has also a slight advantage in that the normal pressure, and therefore the friction, at every part of the acting surface is the same; while in the Whitworth thread the normal pressure, and therefore the friction, is greater at the rounded parts. The surface of the Sellers thread will therefore wear more uniformly than the surface of the Whitworth thread. The total friction, and also the bursting action on the nut, are slightly greater in the Sellers thread than in the Whitworth, because of the greater angle of the V. From a comparison of the tables of dimensions of Whitworth and Sellers screws already given, it will be seen that for a given diameter of screw the diameter at the bottom of the thread is greater in the case of the Whitworth than in the Sellers. A bolt screwed with a Sellers thread is therefore weaker than the same size of bolt screwed with a Whitworth thread. The strength of the Sellers screw is still further reduced on account of the sudden change of the cross section of the bolt at the bottom of the thread.

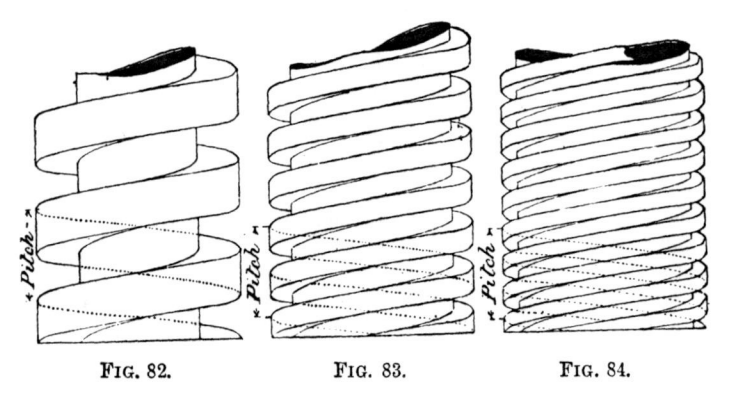

FIG. 82.   FIG. 83.   FIG. 84.

The "knuckle" thread is one capable of withstanding a large amount of rough usage, but otherwise it is not a good design.

88. Multiple Threaded Screws.—The distance which a nut advances for one revolution of its screw is equal to the pitch of the thread of that screw. Now it is evident that if, with a screw of given diameter, it is required that the nut shall move a considerable distance for one revolution of the screw, the pitch of the latter must be large, and if the thread is square, its depth will be half the pitch, and therefore may be so large that the diameter of the screw at the bottom of the thread would be so much reduced as to make the screw too weak. This difficulty is surmounted by making the screw with two or more threads, each having the same pitch and running parallel to one another. Fig. 82 shows a single threaded screw, Fig. 83 is a double-threaded, and Fig. 84 a treble-threaded screw, the pitch being the same in each case. It will be observed that the diameter of the screw at the bottom of the thread in Fig. 83 is greater than the corresponding diameter in Fig. 82, and still greater in Fig. 84. The total working surface of the screw is also increased by making it double or treble-threaded. The pitch of a screw divided by the number of threads is called the *divided pitch*.

89. Forms and Proportions of Bolt-Heads.—A number of forms of bolt-heads are shown in Figs. 85 to 90. The *hexagonal head* is shown in Fig. 85, and the *square head* in Fig. 86. These are two very common forms. The usual proportions for hexagonal and square heads are, width across the flats equal to $1\frac{1}{2}$ D + $\frac{1}{8}$, height from $\frac{2}{3}$ D to D. Fig. 87 shows a *hexagonal head* with a *collar* or *flange* formed on it to give it a larger bearing surface. Such a head would have the same proportions as an ordinary hexagonal head, the height including the thickness of the collar. The *cylindrical* or *cheese head* is represented in Fig. 88. If the bolt is parallel, the diameter of the cheese head is usually from 1·3 D to 1·4 D and its height from ·5 D to ·8 D. Fig. 89 shows a *spherical head*, the diameter of which may be 1·5 D and its height ·75 D. In all the above examples the bearing surface of the head of the bolt is flat and at right angles to the axis of the bolt. In Fig. 90 the bearing surface of the head is spherical, and the seat upon which it rests is of the same shape. This permits of the bolt leaning over to one side, while the head remains

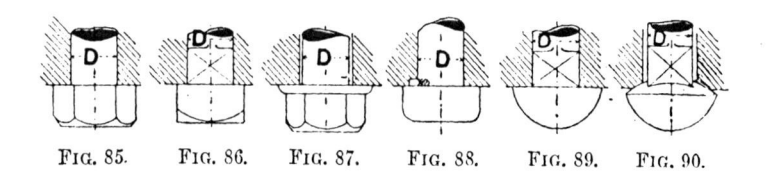

FIG. 85.   FIG. 86.   FIG. 87.   FIG. 88.   FIG. 89.   FIG. 90.

in contact with its seat all round, instead of on one side only, as would be the case where the bearing surface of the head is flat and perpendicular to the axis.

The head of an *eye-bolt* is shown in Fig. 91. The thickness A varies from D to 1·2 D, and the thickness B should be such that $A \times B = ·5 D^2$. The diameter D_2 of the pin which passes through the eye may be from ·8 D to D when it is supported at both sides of the bolt, as in Fig. 91, but if the pin is overhung, its diameter should be larger, say, 1·2 D. The head of a *hook-bolt* is shown in Fig. 92. This form of bolt is useful in connecting hangers for shafts to the iron joists or flanged beams in a building without having to make holes in their flanges, which would weaken the beams considerably. The proportions of this form of bolt-head are marked on the figure.

The *countersunk-head* is shown in Fig. 93. E, the largest diameter

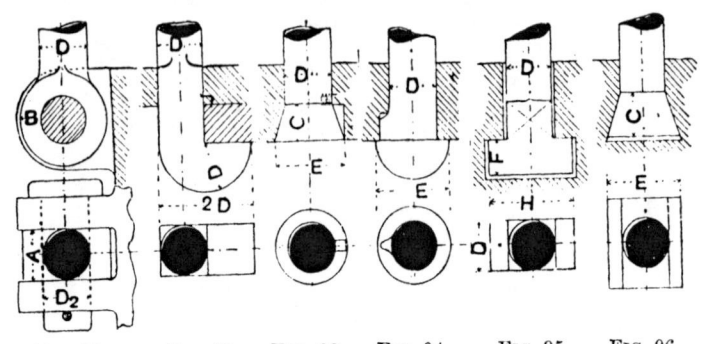

FIG. 91.   FIG. 92.   FIG. 93.   FIG. 94.   FIG. 95.   FIG. 96.

of the head, may be $1\frac{1}{2}$ D, and C, its height, may be $\frac{3}{4}$ D. The *T-head* is shown in Fig. 95. The proportions of this head may be as follows :—
F = from $\frac{3}{4}$ D to D, and H = $1\frac{1}{2}$ D + $\frac{1}{8}$. A wedge-shaped head is shown in Fig. 96. The proportions of this head may be as follows :— C = $\frac{3}{4}$ D, E = $1\frac{1}{2}$ D.

Three forms of *foundation bolt-heads* are shown in Figs. 97, 98, and

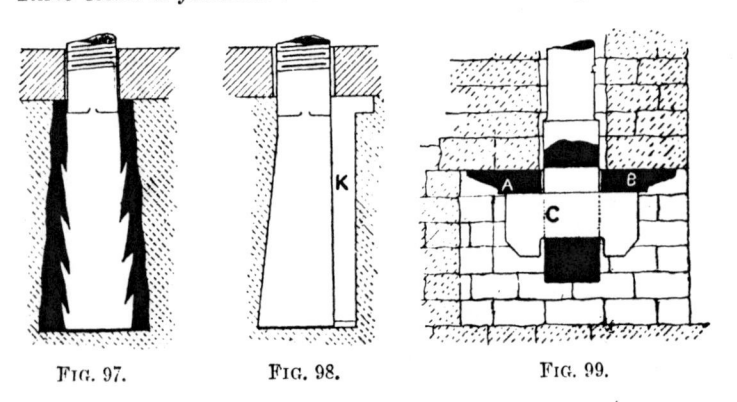

FIG. 97.   FIG. 98.   FIG. 99.

99. In the first the head is jagged, and is let into a hole in a large stone, this hole being wider at the bottom than at the top. After the bolt-head is placed in the hole, the space between it and the sides of the hole is filled with molten lead or sulphur. Such a bolt as this is difficult to remove. The second form is an improvement on the first, in that it can be removed by first taking out the key K. The third form has

a cotter or key C, which rests against a square or circular cast-iron plate AB. The diameter of the cast-iron plate should be such that the area of the surface of the plate pressing on the foundation multiplied by the safe compressive stress of the material of the formation is equal to the area of the bolt multiplied by its safe tensile stress.

90. **Methods of Preventing Bolts from Rotating.**—We may here notice a few of the methods of preventing a bolt from rotating when the nut is being rotated. A very common device is to make the neck of the bolt next the head square, and also the hole through which the bolt passes, or part of it, at the end next the head of the bolt. This arrangement is shown in Figs. 86, 89, 90. Another design for the same purpose is shown in Figs. 88 and 93. Here a hole is drilled into the neck of the bolt close to the head, and a short pin is driven tight into it, or the hole may be tapped, and the pin screwed. The projecting part of this pin fits into a recess formed to receive it. In Fig. 94 a snug is shown forged on the neck of the bolt to take the place of the pin in the last arrangement. Sometimes a set screw is screwed through one of the pieces connected by the bolt, so that the point of the set screw presses on the bolt. This prevents the bolt from rotating, and also prevents it from dropping out when the nut is removed.

91. **Bolts of Uniform Strength.**—When a bolt is in tension only, it is made lighter and stronger by making the cross section of the unscrewed part equal in area to that of the screwed part at the bottom of the thread. This may be effected by forging or turning down the unscrewed part to the same diameter as the screwed part at the bottom of the thread, but a part at each end should be left of the full diameter, as

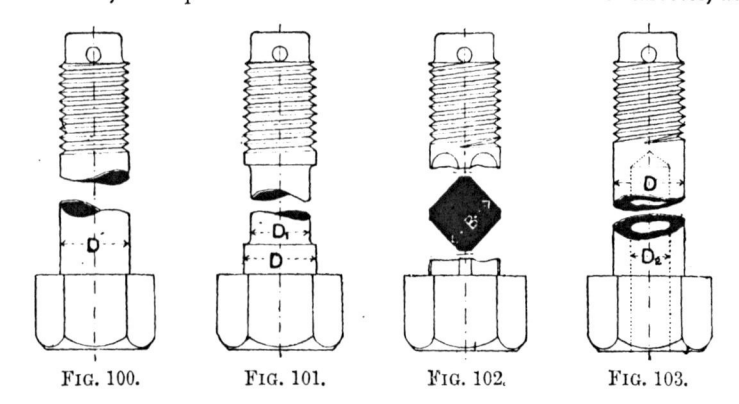

FIG. 100.   FIG. 101.   FIG. 102.   FIG. 103.

shown in Fig. 101, so that the bolt may not shake in the hole through which it has to pass. A second method is to forge or cut flats on the unscrewed part, as shown in Fig. 102, so that the section of the unscrewed part is a square with the corners rounded off. For bolts above half-inch in diameter the breadth across the flats, to secure uniformity of sectional area, is given approximately by the following formula:—

$$B = 1.41D - \sqrt{D^2 - A}$$

where B is the breadth across the flats, D the diameter of the screw over the threads, and A the area of the section of the bolt at the bottom of the screw thread. Values of A for different diameters of Whitworth screws are given in the table on page 47. A third method is to drill a hole from the head of the bolt up to near where the screw ends, as shown in Fig. 103, the axis of the hole coinciding with the axis of the bolt. The diameter of the hole is found by the formula—

$$D_2 = \sqrt{D^2 - D_1^2}$$

where D_2 is the diameter of the hole, D the diameter of the unscrewed part, and D_1 the diameter of the bolt at the bottom of the screw-thread.

The following table gives the values of B and D_2 for various sizes of bolts with Whitworth screw threads:—

D	1	1½	2	2½	3	3½	4	4½	5	5½	6
B	·74	1·14	1·52	1·94	2·35	2·77	3·20	3·64	4·07	4·51	4·95
D_2	·54	·77	1·03	1·22	1·44	1·61	1·80	1·95	2·11	2·26	2·43

92. Forms and Proportions of Nuts.—The nuts in common use are generally either hexagonal or square. For the same strength the hexagonal nut is lighter than the square, and also requires less room. The spanner used for rotating the nut takes a better hold of a square nut than of a hexagonal one. When a nut is in a confined situation, the smallest angle through which it can be turned by a straight spanner before the latter can be shifted round on the nut so as to turn through the same angle again, is 90 degrees, but in the case of the hexagonal nut this angle is 60 degrees. Hence a hexagonal nut is easier screwed up in a confined situation than a square nut.

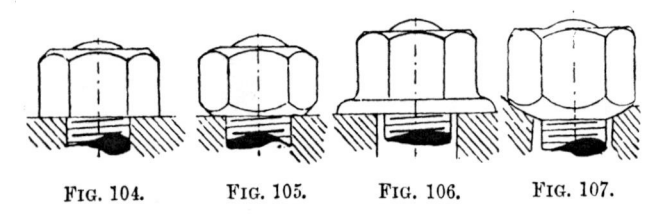

FIG. 104. FIG. 105. FIG. 106. FIG. 107.

Various forms of hexagonal nuts are shown in Figs. 104 to 109. In Fig. 104 the nut is chamfered off on one end only, while in Fig. 105 it is chamfered on both ends. A *flanged nut* is shown in Fig. 106. The flange gives a larger bearing surface for the nut, and is useful when the hole through which the bolt passes is larger than the bolt. In Fig. 107 the bearing surface of the nut is spherical, and the surface against which it fits is of the same shape. This permits of the bolt canting over as shown. This arrangement is often applied to the tool-holders of the slide-rests of lathes. *Cap nuts* are shown in Figs. 108 and 109. These

SCREWS, BOLTS, AND NUTS.

are used to prevent leakage of a liquid or gas past the screw threads. Leakage at the bearing surface of the nut is prevented in the arrangement shown in Fig. 109 by a thin, soft, copper washer.

The usual rules for proportioning hexagonal nuts are, width across

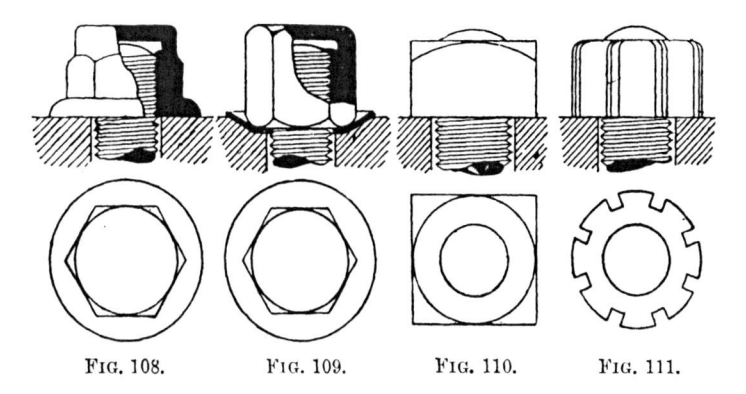

FIG. 108.   FIG. 109.   FIG. 110.   FIG. 111.

the flats $= 1\frac{1}{2}$ D $+ \frac{1}{8}$, and height $=$ D. The width across the angles is got by multiplying the width across the flats by 1·155.

In drawing small nuts or large nuts to a small scale, it is near enough to take the diameter of the hexagonal nut across the angles as twice the diameter of the screw.

Dimensions of Whitworth Standard Hexagonal Nuts and Bolt-Heads.

Diameter of Bolt.	Width of Nut or Bolt-Head across Flats.	Height of Bolt-Head.	Diameter of Bolt.	Width of Nut or Bolt-Head across Flats.	Height of Bolt-Head.
$\frac{1}{8}$	·338	·109	$1\frac{1}{4}$	2·048	1·094
$\frac{3}{16}$	·448	·164	$1\frac{3}{8}$	2·215	1·203
$\frac{1}{4}$	·525	·219	$1\frac{1}{2}$	2·413	1·312
$\frac{5}{16}$	·601	·273	$1\frac{5}{8}$	2·576	1·422
$\frac{3}{8}$	·709	·328	$1\frac{3}{4}$	2·758	1·531
$\frac{7}{16}$	·820	·383	$1\frac{7}{8}$	3·018	1·641
$\frac{1}{2}$	·919	·437	2	3·149	1·750
$\frac{9}{16}$	1·011	·492	$2\frac{1}{8}$	3·337	1·859
$\frac{5}{8}$	1·101	·547	$2\frac{1}{4}$	3·546	1·969
$\frac{11}{16}$	1·201	·601	$2\frac{3}{8}$	3·750	2·078
$\frac{3}{4}$	1·301	·656	$2\frac{1}{2}$	3·894	2·187
$1\frac{3}{16}$	1·390	·711	$2\frac{5}{8}$	4·049	2·297
$\frac{7}{8}$	1·479	·766	$2\frac{3}{4}$	4·181	2·406
$1\frac{5}{16}$	1·574	·820	$2\frac{7}{8}$	4·346	2·516
1	1·670	·875	3	4·531	2·625
$1\frac{1}{8}$	1·860	·984	...	...	...

The height of the nut is in each case equal to the diameter of the bolt.

The proportions adopted for the *Sellers* or *American standard nuts* and *bolt-heads* are as follows :—

Width across the flats $= 1\frac{1}{2} D + \frac{1}{8}$ (rough).
,,   ,,   ,, $= 1\frac{1}{2} D + \frac{1}{16}$ (finished).
Height of nut $= D$ (rough).
,,   ,, $= D - \frac{1}{16}$ (finished).
Height of head $= \frac{3}{4} D + \frac{1}{16}$ (rough).
,,   ,, $= D - \frac{1}{16}$ (finished).

A *square nut* is shown in Fig. 110. The width across the flats is the same as in the hexagonal nut for the same size of bolt. The height of

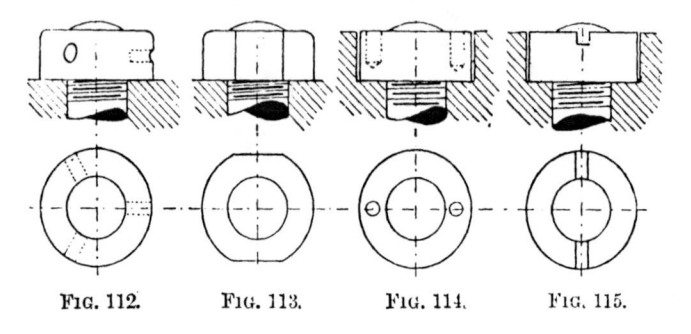

FIG. 112.   FIG. 113.   FIG. 114.   FIG. 115.

the square nut is also the same as the height of the hexagonal nut, namely, the diameter of the bolt. The width of a square nut across the angles is equal to the width across the flats multiplied by 1·414.

The nuts shown in Figs. 111 to 115 are *circular nuts;* that shown in Fig. 111 is a *fluted nut*, the longitudinal grooves being for the reception of the teeth on a specially-formed spanner. The nut in Fig. 112 is rotated by means of a bar known in the workshops as a "tommy," which is inserted in one of the holes shown.

93. Necessity for Locking Nuts.—To permit of a nut turning freely on a bolt, there is a very small clearance space between the screw threads of the nut and those of the bolt. In Fig. 116 the clearance space between the threads of a nut C and bolt E is shown exaggerated. When a nut is loose upon a bolt, that is, when the nut and bolt are not pulled or pushed in opposite directions, the vibration or shaking of the bolt may cause the nut to rotate slightly. If the bolt E is stretched by a force T and the nut F be screwed up, the upper surfaces of the projecting threads of the nut will press on the under surfaces of the threads of the bolt by the reaction R between the nut and the piece H upon which it rests, as shown in Fig. 116. When in this condition the friction between the screw threads of the nut and bolt and between the nut and the supporting piece H, due to the forces T and R, prevents any vibration of the bolt from rotating the nut, so that there is no danger of the nut slackening back. If, however, the tension T be momentarily removed or diminished, as it would be in the case of a marine connecting rod, the friction which opposes the turning of the nut might be so much diminished that a vibration might cause it to slacken back through a small angle; and if this is repeated a great many times, the nut might slacken back so far as to become useless.

From the foregoing remarks it will appear that in many cases it will be very desirable, if not absolutely necessary, to secure the nut in some way so as to prevent it from working loose.

94. Common Lock Nut.—A common arrangement for locking a nut is shown in Fig. 117, where C is an ordinary nut and F one of half the thickness. F is first screwed up so as to act on the bolt in the manner shown in Fig. 116. C is then screwed down on F, and when C is almost as tight as it can be made, it is held by one spanner while F is turned back with another as far as it will go, which will be through a small angle only. The action of the nuts upon one another and upon the bolt will now be as shown in Fig. 117, from which it will be seen that the nuts are wedged tightly on the bolt, and that this action is independent

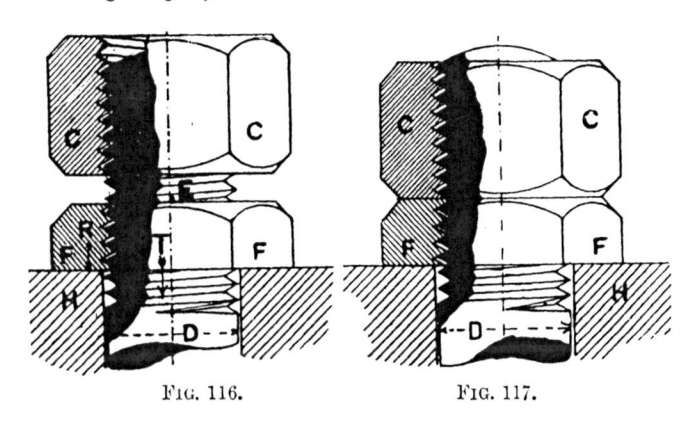

FIG. 116.   FIG. 117.

of the tension in the bolt. In this arrangement the outer nut C is called a lock-nut.

An examination of Fig. 117 will show that it is the outer nut C which carries the load on the bolt, and it should, therefore, be the thicker of the two nuts. In practice, however, the thin nut is often on the outside, for the reason that ordinary spanners are generally too thick to act on the thin nut when placed under the other. Sometimes a compromise is made by having each nut of a thickness equal to three-fourths of the thickness of an ordinary nut, that is, equal to three-fourths of the diameter of the bolt.

95. Locking Nuts by Set-Screws.—A nut may be very effectively locked to a bolt by tapping a hole through the side of the nut and inserting a set-screw, the point of which is screwed up tight on the thread of the bolt, as shown in Fig. 118. This, however, is a clumsy method, as the screw thread of the bolt is injured by the point of the set-screw, and it becomes difficult to remove the nut when required. An improvement on this method is shown in Fig. 119. Here a portion of the bolt has the screw thread removed, and the point of the set-screw presses on this portion. For the sake of appearance and lightness, the upper part of the nut, through which the set-screw passes, is made

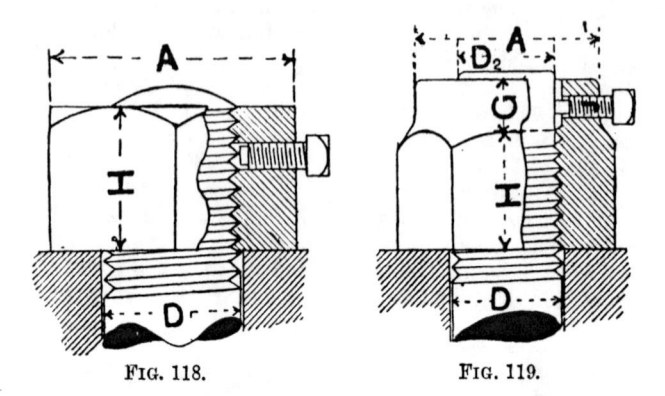

Fig. 118.   Fig. 119.

circular. A modification of this design is shown in Fig. 120, where the part corresponding to the circular part of the nut in Fig. 119 is made separate, and is made to rotate with the nut by means of a pin, P. Fig. 121 shows an improvement on the arrangement shown in Fig. 118,

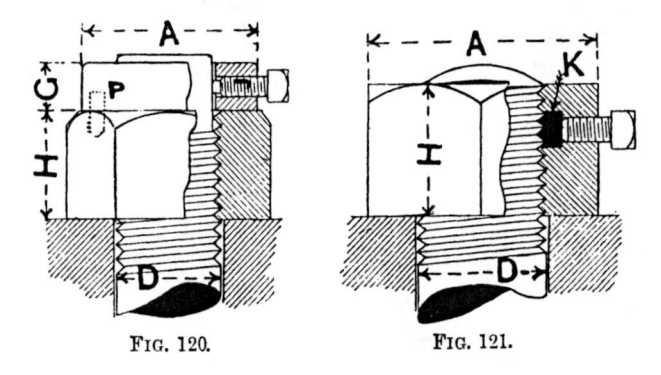

Fig. 120.   Fig. 121.

which might be adopted for large nuts. The inner end of the hole for the set-screw is enlarged to receive a piece K which is screwed along with the nut.

The following rules may be used in proportioning the nuts and set-screws shown above :—

$$A = 1\tfrac{1}{2} D + \tfrac{1}{8}. \qquad H = D.$$
$$J = \tfrac{1}{8} D + \tfrac{1}{8}. \qquad G = 2 J = \tfrac{1}{4} D + \tfrac{1}{4}.$$

The diameter of the stop-pin P may be equal to the diameter of the set-screw, and D_2 may be $\tfrac{1}{16}$th of an inch less than the diameter of the screw of the bolt at the bottom of the thread.

SCREWS, BOLTS, AND NUTS.

Two very neat and very much used methods of locking a nut are shown in Figs. 122 and 123. The lower part of the nut is turned circular, and has a groove turned on it as shown. The circular part of the nut fits into a collar, as in Fig. 122, or into one of the pieces connected by the bolt, as in Fig. 123. Through this collar, or through this

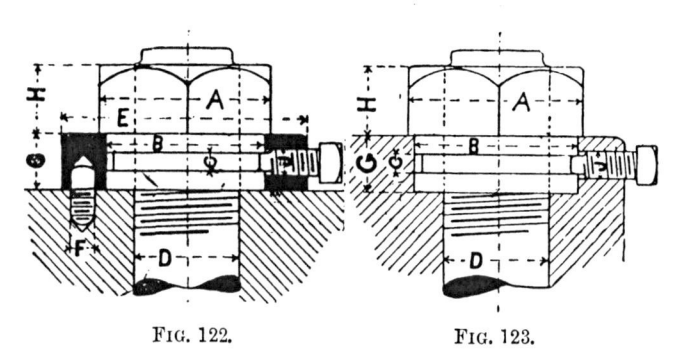

FIG. 122.   FIG. 123.

piece passes a set-screw, the point of which is made to press on the nut at the bottom of the groove. The use of the groove is to prevent the bur, raised by the point of the set-screw, interfering with the rotation of the nut. The following proportions may be adopted in designing the arrangements shown in Figs. 122 and 123:—

$A = 1\frac{1}{2} D + \frac{1}{8}$.
C = diameter of set-screw at bottom of thread.
$J = \frac{1}{8} D + \frac{1}{8}$.
$F = \frac{1}{8} D + \frac{1}{16}$.

$B = 1\frac{1}{2} D - \frac{1}{16}$.
$E = 1\frac{7}{8} D + \frac{1}{8}$.
$G = 2 J = \frac{1}{4} D + \frac{1}{4}$.
Depth of groove $= \frac{1}{6} J$.
$H = $ from $\frac{3}{4} D - \frac{1}{4}$ to $D - \frac{1}{4}$.

96. Stop-Plates for Nuts.—Two arrangements for locking nuts by means of stop-plates are shown in Figs. 124 and 125. The stop-plate AB in Fig. 124 is fixed by the tap-bolt C, and is shaped so as to fit the nut when a face of the latter is perpendicular to the line CD, and also when a face is parallel to that line, so that the nut may be locked at intervals of one-twelfth of a revolution.

The second arrangement is an improvement on the first, as it allows of the nut being locked in any position. It will be observed that the stop-plate in this example (Fig. 125) has a slot, through which passes the stud-bolt which holds down the plate. Should the nut on the stud-bolt slacken, the stop-plate and the main nut cannot slacken back more than one-sixth of a revolution. To facilitate the screwing in of the stud-bolt, and to prevent its slackening back, the part of it within the slot in the stop-plate is square.

97. Miscellaneous Locking Arrangements for Nuts.—Grover's spring-washer, shown in Fig. 126, serves to keep a nut tight on its bolt when

60   MACHINE DRAWING AND DESIGN.

the tension on the latter is diminished. The washer is made of steel, and has a cross section, which is shown by the dotted section lines in

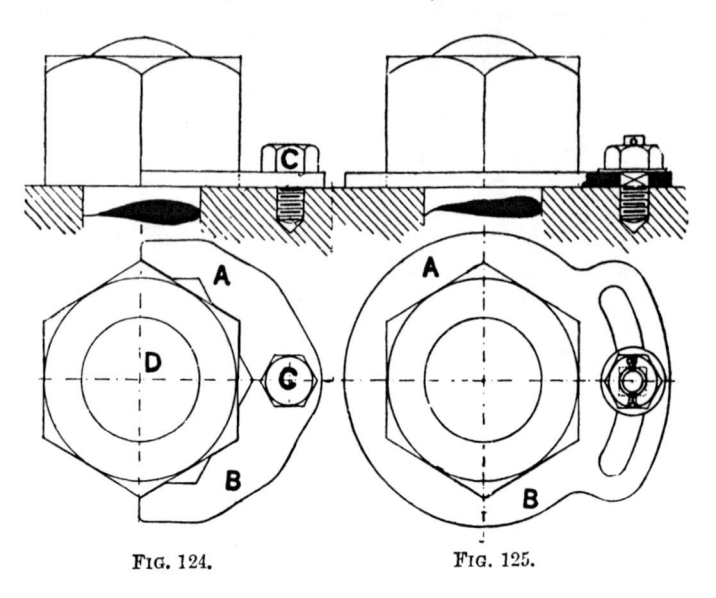

FIG. 124.   FIG. 125.

the upper part of the figure. The shape of this washer, when unloaded, is shown in the lower part of Fig. 126.

The arrangement shown in Fig. 127 consists of a plate AB, having a square hole in the centre, which passes over the point of the bolt, which is shaped to fit it. The ends of the plate AB are bent over to catch the sides of the nut, as shown. The plate is prevented from coming off by the pin C, which may be a split pin, or a tapered pin driven in tight.

The lock-washer, shown in Fig. 128, is simple and effective. The washer is stamped out of manganese sheet steel, and has on its under side two projections, which fit into two short grooves in the piece supporting it, and thus prevent the rotation of the washer. When first put on the bolt,

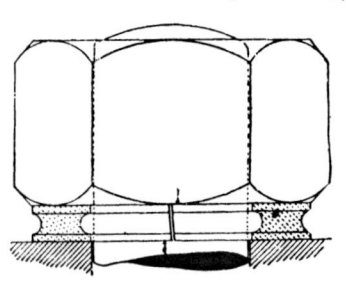

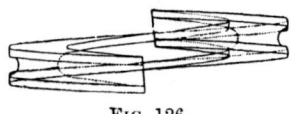

FIG. 126.

the washer is quite flat, with the exception of the projecting pieces on the bottom, but after the nut is screwed up tight, one of the teeth, A, on its edge is bent up to hold the nut as shown. When it is required to slacken the nut, the bent tooth on the washer is straightened out again.

The nuts shown in Figs. 129 and 130 have grooves cut across their

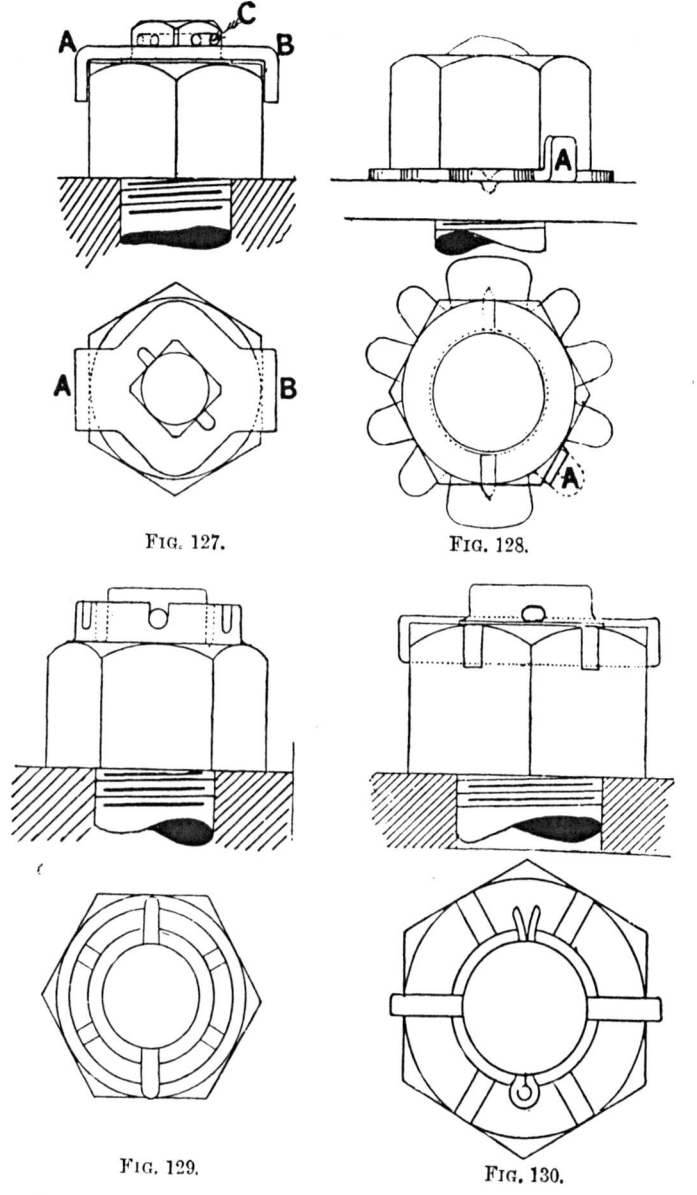

FIG. 127.

FIG. 128.

FIG. 129.

FIG. 130.

upper faces, one of these grooves being opposite a hole in the point of the bolt through which passes a pin or cotter.

The nut shown in Fig. 131 is cut half way through, and after the nut is screwed up the semi-detached portion is sprung closer to the other by a screw, which has the effect of tightening the screw of the nut on that of the bolt.

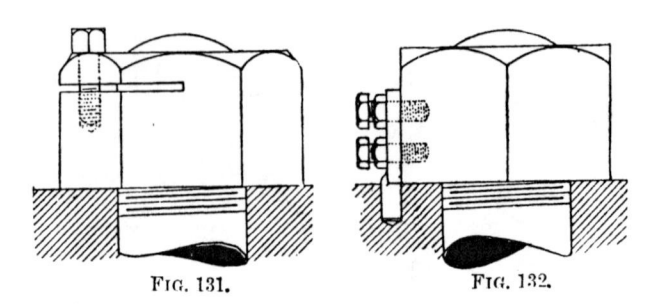

FIG. 131.　　　　　FIG. 132.

In Fig. 132 the nut shown has a plate bolted to one of its faces, this plate having a tail-pin which fits into a corresponding hole in the piece supporting the nut.

98. Various Forms of Bolts.—A *stud* or *stud-bolt* is one which is screwed at both ends, one end being screwed into one of the pieces to be connected, while the other end carries an ordinary nut. Fig. 133 shows a plain stud. Fig. 134 shows a stud with a collar, which may be circular or square. This collar serves as a shoulder, against which the stud may be screwed up tight. The collar, if square, is also useful for receiving a spanner to screw up the stud. In Fig. 135 the stud has a

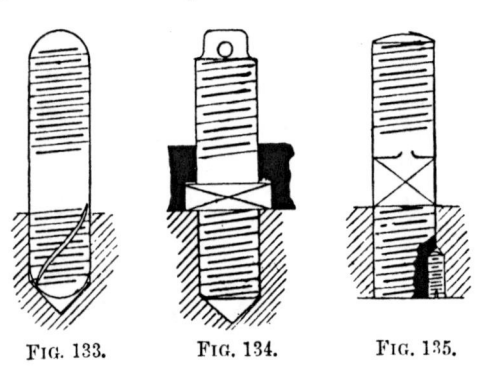

FIG. 133.　　　FIG. 134.　　　FIG. 135.

square neck to receive a spanner for screwing it in. This figure also shows a method of locking the stud so that it cannot slacken back. If the stud is required to go to the bottom of the hole, it is sometimes desirable to cut a spiral groove on it as shown in Fig. 133, so as to allow the air and any water or oil which may be in the hole to escape. This groove may be made with the edge of a file.

A *tap-bolt* is a bolt with a head, and is screwed into one of the pieces to be connected while the head presses on the other piece as shown in Fig. 136.

Tap and stud bolts combined are shown in Figs. 137 and 138. Bolts

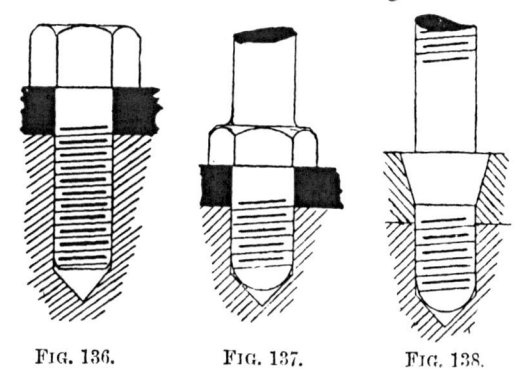

FIG. 136.   FIG. 137.   FIG. 138.

having nuts at each end are shown in Figs. 139, 140, and 141. Fig. 142 shows bolts or screws with heads slotted to receive a screwdriver.

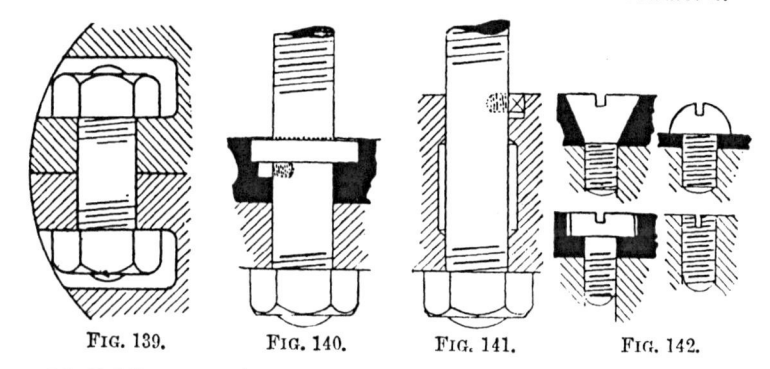

FIG. 139.   FIG. 140.   FIG. 141.   FIG. 142.

99. Set-Screws.—A set-screw is a screw or bolt which presses on a piece so as to prevent the rotation or sliding of that piece. Examples

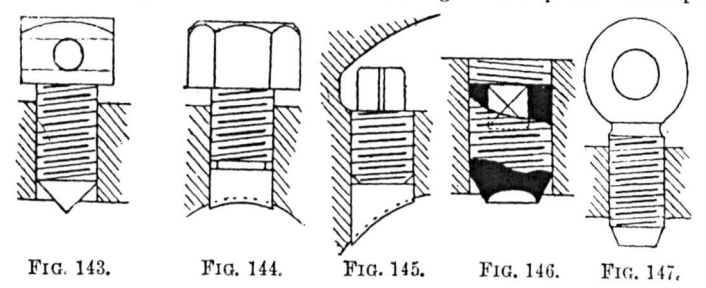

FIG. 143.   FIG. 144.   FIG. 145.   FIG. 146.   FIG. 147.

of set-screws are shown in Figs. 143 to 147. In Fig. 143 the set-screw is shown with a sharp, conical point, which enables it to penetrate into the piece which it is required to set or lock. A good holding form of point is that shown in Fig. 146. In cases where the damaging action of the point of the set-screw would be objectionable, a metal pad may be used, as shown in Figs. 144 and 145. It is a good practice to let the point of the set-screw press on the bottom of a shallow groove in the piece to be locked, in cases where the piece is not locked permanently in one position, as in the locking arrangements for a nut shown in Fig. 122, and in the cotters for connecting rod ends. By this artifice the bur raised by the point of the set-screw does not interefere with the rotation or sliding of the piece when the latter has to be shifted. Set-screws are generally made of steel, and their points should be hardened.

100. Backlash in Screws—Divided Nuts.—The amount of rotary motion which can be given to a nut without causing any movement of the screw on which it works is called *backlash*, and is due either to imperfection in the manufacture or to wear. Backlash may be overcome by using a *divided nut*. For a square-threaded screw the nut should be divided at right angles to its axis; but for a triangular-threaded screw

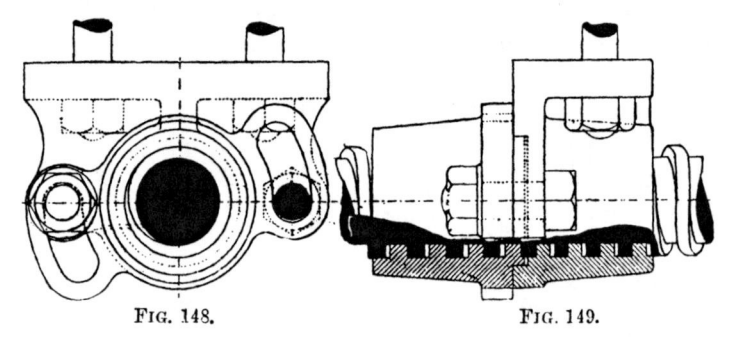

FIG. 148.   FIG. 149.

the nut may be divided at right angles to the axis, or in a plane containing the axis. In each case the two halves of the nut are held together by bolts and nuts. In the design shown in Figs. 148 and 149, the two halves of the divided nut are provided with flanges which are bolted firmly together. The backlash is taken up by rotating one half of the nut on the other previous to tightening up the bolts. To permit of this rotation the bolt-holes are in the form of slots, as shown in Fig. 148.

CHAPTER V.

KEYS.

101. Keys.—Keys are wedges, generally rectangular in cross section, which are used for securing wheels, pulleys, cranks, etc., to shafts. Steel is the best material for ordinary keys, but wrought-iron is often used instead.

102. Saddle-Key.—The *hollow* or *saddle key* is shown in Figs. 150, 151, and 152. The first of these figures is an enlarged section of the key, while the others show it in position, securing a wheel or pulley to

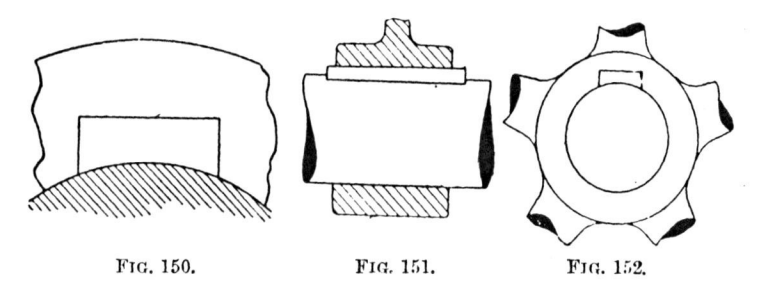

FIG. 150.   FIG. 151.   FIG. 152.

a shaft. With this form of key it is not necessary to cut the shaft in any way, but its holding power is small, and it is therefore only used for light work.

103. Flat Key.—The *flat key* receives its name from the fact that it rests on a *flat*, which is formed on the shaft to receive it. Fig. 153

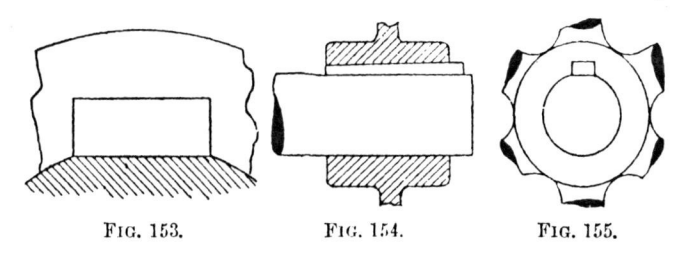

FIG. 153.   FIG. 154.   FIG. 155.

is an enlarged cross section of a flat key, while Figs. 154 and 155 show it in position. The holding power of this key is much greater than that of the saddle-key.

104. Sunk Key.—The *sunk key* is a very secure form of key, and is much used. It derives its name from the fact that it is sunk into the shaft. Figs. 156, 157, and 158 show this form of key. The enlarged

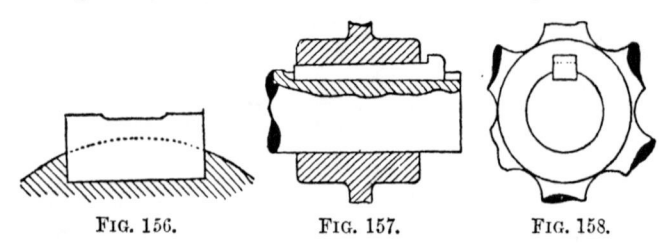

FIG. 156.   FIG. 157.   FIG. 158.

section of the key, shown in Fig. 156, shows a shallow flat-bottomed groove on the top. This is often put on large keys, to facilitate fitting them into their key-ways. Large keys are made in the following manner:—A wooden pattern is first fitted into the key-way, and this is given to the man in charge of a planing or shaping machine, who planes or shapes the forging for the key to the same dimensions. In fitting the key itself, it should then only be necessary to file the top face, and this is easier done if it has been previously grooved in the machine. It must not be supposed that the grooving of keys, as described above, is confined to sunk keys, or that sunk keys are always grooved. Keys are oftener found without the grooves than with them.

FIG. 159.

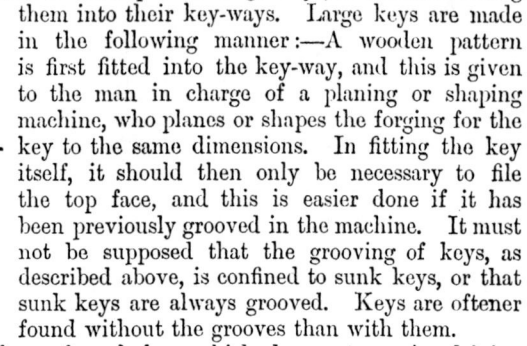

Fig. 159 shows a form of sunk key which does not require driving in. This key fits the key-way at the sides, but there is a small clearance

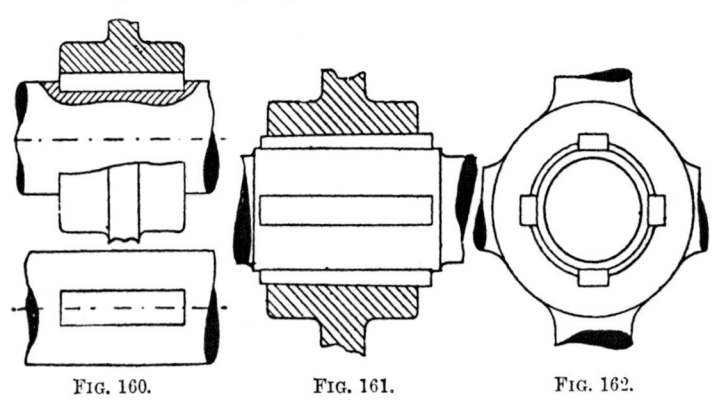

FIG. 160.   FIG. 161.   FIG. 162.

space on the top. The key is put in position, and then locked or secured by one or more set-screws.

In general, when a piece is keyed to a shaft, the piece is placed in position and the key driven in, and if a sunk key is used, the key-way

KEYS.

on the shaft must be cut for some distance outwards to allow of this; but in many cases this prolongation of the key-way is objectionable, and the key is bedded in, as shown in Fig. 160, and the wheel or piece to be keyed to the shaft is then driven on to the key.

Other examples of the use of the three forms of keys which have been described are shown in Figs. 161, 162, and 163. Figs. 161 and 162 show a common method of fixing heavy wheels on shafts. The eye of the wheel in this case is larger in diameter than the shaft, for one or both of two reasons. First, the wheel may have to pass over a larger part of the shaft than that upon which it has to be fixed; second, the wheel and shaft may be in an awkward position, where it would be difficult to disconnect them if the eye of the wheel was bored to fit the shaft. Four keys are used, which generally rest on flats on the shaft. These keys must be carefully fitted, so as to make the wheel run true with the shaft, and after being fitted they should be numbered, and marks put upon them, showing how far each has to be driven in.

Fig. 163 shows the combination of a saddle-key with the sunk key, which is useful where the hole in the piece to be keyed to the shaft is not a good fit on the latter. The sunk key serves to transmit the turning force, while the saddle-key, placed at right angles to the sunk key as shown, prevents the piece from rocking on the shaft. The same figure

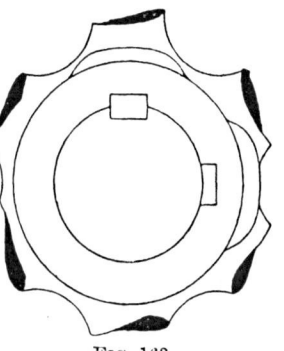

FIG. 163.

also illustrates what is known as a *key-boss*, which is simply a thickening up of the boss of a wheel or pulley where it is weakened by the key-way.

105. Round or Pin Keys.—When a piece such as a crank is *shrunk on* to a shaft, a practice which is sometimes adopted is to drill a hole half into the shaft and half into the crank, into which a round key or pin is driven, as shown in Figs. 164 and 165. If the hole does not go

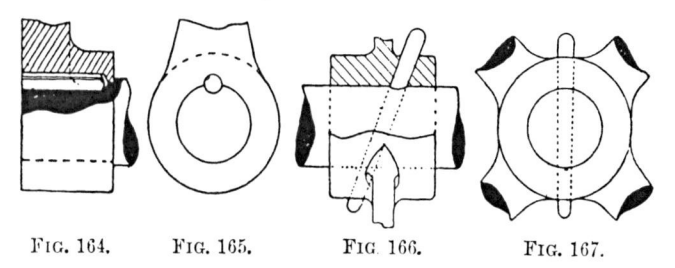

FIG. 164. FIG. 165. FIG. 166. FIG. 167.

right through, a small groove should be cut on the key to allow the air and any superfluous oil to escape.

Figs. 166 and 167 show a method sometimes adopted to secure a piece to a shaft. Here a hole is drilled through the piece and shaft into which a round pin is driven. The hole may be at right angles to the axis of the shaft, or, to suit convenience in drilling, the hole may be inclined as shown in Fig. 166. A pin used in this way is, however, really a *cotter* and not a key.

106. Gib-Heads on Keys.—When the point of a key cannot be conveniently reached for the purpose of driving it out, the head should be formed as shown in Fig. 168. This is known as a *gib-head*.

107. Sliding or Feather Keys.—The function of a sliding or feather key is to secure a piece to a shaft, so far as to prevent the one from rotating without the other, but at the same time allow of a relative motion in the direction of the axis of the shaft. This form of key has no taper, and it is generally secured to the piece carried by the shaft, in which case it is made a sliding fit in the key-way of the shaft. But the feather-key may be fixed to the shaft and made a sliding fit in the piece carried by it.

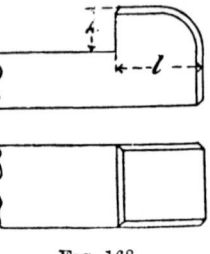

FIG. 168.

Examples of feather-keys which slide in the key-way of the shaft are shown in Figs. 169 to 176. In Fig. 169 the key has a projecting pin, A, formed on it, which enters a corresponding hole in the piece B carried by the shaft. In Figs. 170 and 171 gib-heads, C, are formed on the ends of the key, which serve the same purpose as the pin A in Fig. 169, namely, to ensure that the key and the piece B remain or move together.

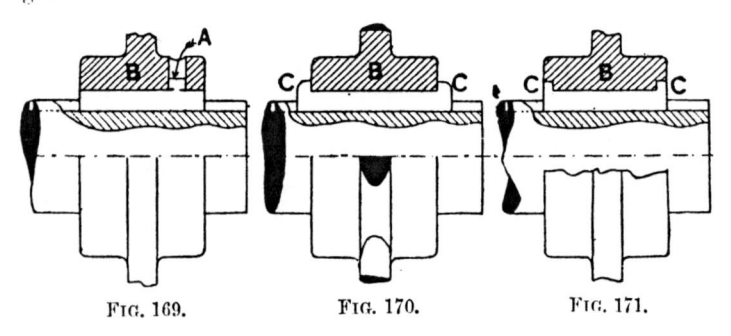

FIG. 169.   FIG. 170.   FIG. 171.

Figs. 172 and 173 show a feather-key which is secured to the piece carried by the shaft by a screw, which passes through a tail-piece formed on the key. One or more screws passing into the body of the key, as shown in Fig. 174, may be used for the same purpose.

The key shown in Figs. 175 and 176 is dovetailed in cross section, and it is made a driving fit in the piece B. In cases where the resistance to rotation is small, the point of a screw may be used to take the place of a feather-key. Fig. 177 shows two screws used in this way. In all the above examples the key-way in the shaft must be long enough to permit of the necessary longitudinal motion.

KEYS.

Figs. 178 and 179 show two examples of feather-keys which are fixed to the shaft. In Fig. 178 the key is sunk into the shaft, and in Fig. 179 the key is either a saddle-key or a key on a flat, and is secured to

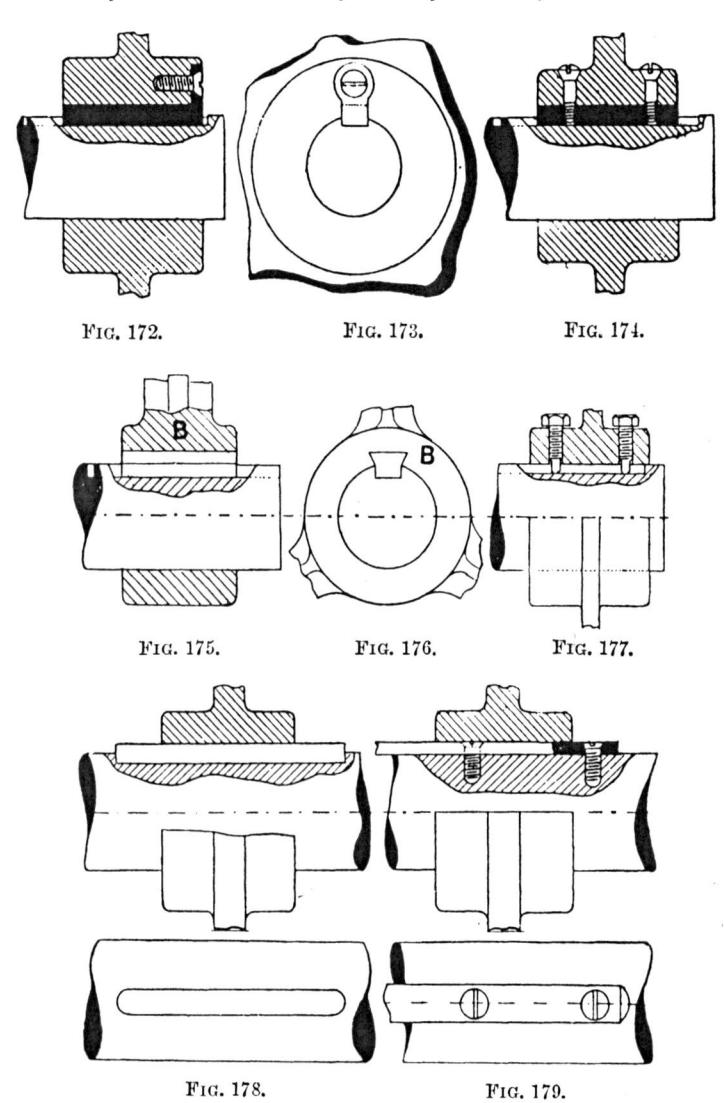

FIG. 172. FIG. 173. FIG. 174.

FIG. 175. FIG. 176. FIG. 177.

FIG. 178. FIG. 179.

the shaft by screws as shown. In these cases the key must be long enough to permit of the necessary sliding motion.

108. Cone-Keys.—These are shown in Figs. 180 and 181. The piece B to be keyed to the shaft is in this case bored out larger than the shaft and the hole is tapered. The space between the shaft and the piece B is filled with three *saddle* or *cone-keys*, C. These keys are made of cast-iron and are all cast together in one piece, and before being divided the casting is bored to fit the shaft and turned to fit B. By this arrangement of keys the same piece B may be fixed on shafts of different diameters by using keys of different thicknesses; also the piece B may be bored out large enough to pass over any enlargement on the shaft.

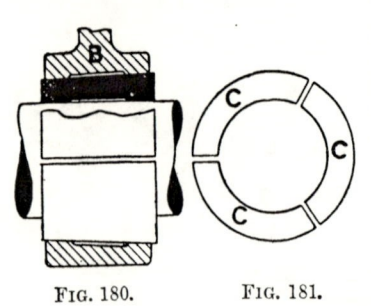

FIG. 180.   FIG. 181.

Smith's screwed conical bush may be used for the same purpose as the cone-keys just described. This bush is made in four pieces, which are fastened to a flexible lining.

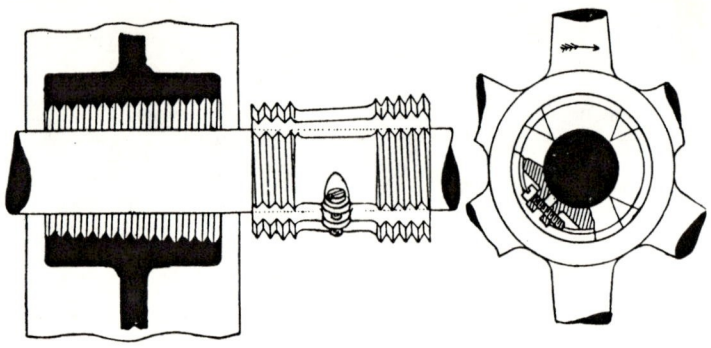

FIG. 182.   FIG. 183.

Externally the bush is tapered and screwed to fit the inside of the boss of the pulley or other piece to be fixed to the shaft, while internally it fits the shaft. Fig. 184 shows the bush open before it is put on the shaft. Fig. 182 shows the bush on the shaft ready to receive the pulley, which must be screwed on to the bush. The bush must be so placed that the driving force on the pulley tends to screw it tighter on the bush.

109. Strength of Keys.—The straining actions to which an ordinary sunk key is subjected are shearing and crushing. Let the length of the key be denoted by L and its breadth by B, and let the shearing stress on the key be denoted by f_s. Then the shearing load on the key $= \text{LB}f_s$. If this shearing load is produced by a force, P, acting on a lever at a perpendicular distance, R, from the axis of the shaft, and in a plane at right angles to that axis, then $2\,\text{PR} = \text{LB}f_s\text{D}$, where D is the diameter of the shaft.

FIG. 184.

If f_s'' is the shearing stress at the outside of the shaft, where the stress is greatest, then if the key and the shaft are strained to the same extent—

$$LBf_s\frac{D}{2} = \frac{3\cdot1416}{16} D^3 f_s'' \text{ or } LBf_s = \frac{3\cdot1416}{8} D^2 f_s''.$$

Assuming that the key is sunk to half its thickness in the shaft, then the area subjected to crushing is equal to $\frac{1}{2}$ TL, T being the thickness of the key. If f_c is the crushing stress, the crushing load on the key and on the side of the key-way is equal to $\frac{1}{2}$ TLf_c.

If the shearing resistance of a key is to be equal to its crushing resistance, then $LBf_s = \frac{1}{2} TLf_c$ or $Bf_s = \frac{1}{2} Tf_c$. But the last relation only holds in the case of a feather-key or a key which is not driven tight into the key-way. In a key which is wedged in tight, the crushing action on the sides is much less than in the case of keys which are made a sliding fit.

Taking the safe crushing stress at twice the safe shearing stress, the equation $Bf_s = \frac{1}{2} Tf_c$ gives B = T, or the key is square in section, which would be the case with sliding or feather keys, but in other cases T would be less than this.

110. Proportions of Keys.—In applying the formulæ which were proved in the preceding article to the calculation of the dimensions of keys, the safe shearing stress f_s or f_s'' may be taken at 9000 lbs. per square inch for wrought-iron and 11,000 for steel.

For a steel key the formula $2PR = LBf_sD$ gives the result $LB = \dfrac{PR}{5500\,D}$; and for a wrought-iron key $LB = \dfrac{PR}{4500\,D}$.

If, instead of the twisting moment PR on the piece carried by the shaft, we have the horse-power H transmitted by it at N revolutions per minute, then $LB = \dfrac{11\cdot459\,H}{DN}$ for a steel key, and $LB = \dfrac{14\cdot005\,H}{DN}$ for a wrought-iron key.

If the full power of the shaft is transmitted by the key, we have from the preceding article, $LBf_s = \dfrac{3\cdot1416}{8} D^2 f_s''$. For a steel key and a wrought-iron shaft this gives $LB = \cdot32\,D^2$. If the key and shaft are both of wrought-iron or both of steel, then $LB = \cdot39\,D^2$. These formulæ are still further simplified if L is expressed in terms of D. Thus if $L = nD$, then we have $B = \dfrac{\cdot32}{n} D$ for the first case, and $B = \dfrac{\cdot39}{n} D$ for the second case. The following table gives the value of B in terms of D for various values of n:—

	L =	·9 D.	D.	1·2 D.	1·4 D.	1·6 D.
B =	Steel key and wrought-iron shaft	·36 D	·32 D	·27 D	·23 D	·20 D
	Key and shaft of same material	·43 D	·39 D	·33 D	·28 D	·24 D

Keys are very often proportioned according to empirical rules such as the following:—

Diameter of eye of wheel or boss of shaft = D.
Width of key = B = $\frac{1}{4}$ D + $\frac{1}{8}$.
Mean thickness of sunk key = $\frac{1}{2}$ B.
,, ,, ,, key on flat = $\frac{1}{3}$ B.

The above empirical rules are taken from "Unwin's Machine Design." Mr. Thomas Box in his book on "Mill Gearing" gives the following rules for sunk keys:—

Diameter of shaft = D.
Breadth of key = $\frac{1}{4}$ D + $\frac{1}{8}$.
Thickness of key = $\frac{1}{11}$ D + ·16.
Depth of key-way in shaft = $\frac{1}{40}$ D + ·075.

The depth of the key-way by the last rule is measured at the *side* of the key.

The taper of keys varies from $\frac{1}{8}$-inch to $\frac{3}{16}$-inch per foot of length, that is, from 1 in 96 to 1 in 64.

The following table, taken from "Low's Introduction to Machine Drawing and Design," gives dimensions of keys agreeing with average practice:—

Dimensions of Keys.

D = diameter of shaft.
B = breadth of key.
T = thickness of sunk key.
T_1 = thickness of flat key, also = thickness of saddle-key. Taper of key $\frac{1}{8}$-inch per foot of length, *i.e.*, 1 in 96.

D.	$\frac{3}{4}$	1	$1\frac{1}{4}$	$1\frac{1}{2}$	$1\frac{3}{4}$	2	$2\frac{1}{4}$	$2\frac{1}{2}$	$2\frac{3}{4}$	3	$3\frac{1}{2}$
B	$\frac{5}{16}$	$\frac{3}{8}$	$\frac{7}{16}$	$\frac{1}{2}$	$\frac{9}{16}$	$\frac{5}{8}$	$\frac{11}{16}$	$\frac{3}{4}$	$\frac{13}{16}$	$\frac{7}{8}$	1
T	$\frac{1}{4}$	$\frac{1}{4}$	$\frac{1}{4}$	$\frac{5}{16}$	$\frac{5}{16}$	$\frac{5}{16}$	$\frac{3}{8}$	$\frac{3}{8}$	$\frac{7}{16}$	$\frac{7}{16}$	$\frac{1}{2}$
T_1	$\frac{3}{16}$	$\frac{3}{16}$	$\frac{3}{16}$	$\frac{3}{16}$	$\frac{1}{4}$	$\frac{1}{4}$	$\frac{1}{4}$	$\frac{5}{16}$	$\frac{5}{16}$	$\frac{5}{16}$	$\frac{3}{8}$

D.	4	$4\frac{1}{2}$	5	$5\frac{1}{2}$	6	7	8	9	10	11	12
B	$1\frac{1}{8}$	$1\frac{1}{4}$	$1\frac{3}{8}$	$1\frac{1}{2}$	$1\frac{5}{8}$	$1\frac{7}{8}$	$2\frac{1}{8}$	$2\frac{3}{8}$	$2\frac{5}{8}$	$2\frac{7}{8}$	$3\frac{1}{8}$
T	$\frac{1}{2}$	$\frac{9}{16}$	$\frac{5}{8}$	$\frac{11}{16}$	$\frac{3}{4}$	$\frac{13}{16}$	$\frac{15}{16}$	1	$1\frac{1}{16}$	$1\frac{3}{16}$	$1\frac{1}{4}$
T_1	$\frac{7}{16}$	$\frac{1}{2}$	$\frac{1}{2}$	$\frac{9}{16}$	$\frac{5}{8}$	$\frac{11}{16}$	$\frac{3}{4}$	$\frac{7}{8}$	$\frac{15}{16}$	$1\frac{1}{16}$	$1\frac{1}{8}$

The proportions of the gib-head for a key, Fig. 168, may be h = thickness of key, $l = 1\frac{1}{2} h$.

CHAPTER VI.

COTTERS.

111. Cotters.—Cotters, like keys, are wedges usually rectangular in cross section. The cross section is often rounded at the ends, as shown in Figs. 190, 191, and 192. With this form of section the cotter is not so easily fitted, but the hole to receive the cotter is easier made, and does not weaken the piece in which it is made quite so much as a rectangular hole. A key has to resist a shearing action over a *longitudinal section*, while a cotter has to resist a shearing action over two *transverse sections*. Cotters are generally made of steel or wrought-iron.

112. Gibs and Cotters Combined.—When a cotter, AC, Fig. 185, is used to pull a strap, EFGH, along a rod, HK, the friction between the cotter and the strap causes the latter to open out, as shown by the dotted lines, when the cotter is driven in. This may be prevented by the use of a gib, LM, Fig. 186. The gib also gives a larger surface for the cotter to slide on. To make the sliding surface on each side of the cotter the same, two gibs are used, as shown in Fig. 187. When a cotter is used by itself as in Fig. 185, the hole through which it passes

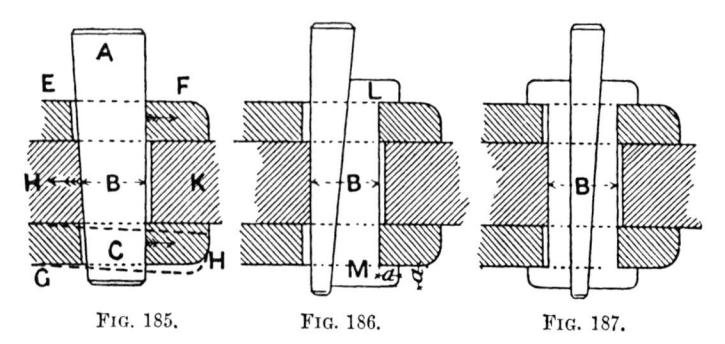

FIG. 185.   FIG. 186.   FIG. 187.

must be tapered the same as the cotter; but when a gib is used with the cotter, the hole is made parallel, and the gib and cotter have equal and opposite tapers, as shown in Fig. 186. When two gibs are used, one only need be tapered as shown in Fig. 187, or both may be tapered.

The thickness of the cotter is usually about one-fourth of the breadth B, and the total breadth B for the gib and cotter is the same as for a cotter used by itself for the same purpose.

The depth and length, a, of the gib-head may be equal to the thickness of the cotter.

113. Strength and Proportions of Cotters—Suppose a cotter of breadth B and thickness t to pass through a bar of diameter D, as shown in Figs. 188 and 189. The load on the bar being T, the principal straining actions to which this cotter is subjected are shearing and crushing.

The shearing resistance of the cotter is equal to $2Btf_s$ where f_s is the shearing stress. The crushing resistance is Dtf_c where f_c is the crushing stress. If these resistances are equal, we have therefore

$$2Btf_s = Dtf_c, \text{ or } B = \frac{Df_c}{2f_s}.$$

The bar through which the cotter passes is in tension, and its weakest section is at the cotter. The area of the cross section of the bar at the cotter is very nearly equal to $\cdot 7854\ D^2 - Dt$, and therefore the resistance of the bar to tension is $(\cdot 7854\ D^2 - Dt)f_t$, where f_t is the tensile stress.

If the bar and cotter are to be of equal strength, we must have $2Btf_s = Dtf_c = (\cdot 7854\ D^2 - Dt)f_t$.

If the bar and cotter are both made of wrought-iron or both of steel, the compressive stress f_c may be taken as equal to twice the tensile stress f_t, and the tensile strength may be taken at one and a quarter times the shearing stress f_s. That is, $f_c = 2f_t$, and $f_t = 1\frac{1}{4}f_s$.

Solving the above equations we then get $B = 1\cdot 25\ D$, and $t = \cdot 26\ D$, or $t = \dfrac{D}{4}$ nearly. These rules agree very well with actual practice, although the breadth B is sometimes as small as D.

If the cotter is of steel and the bar of wrought-iron, and if the shearing resistance of steel be taken as equal to the tenacity of wrought-iron,

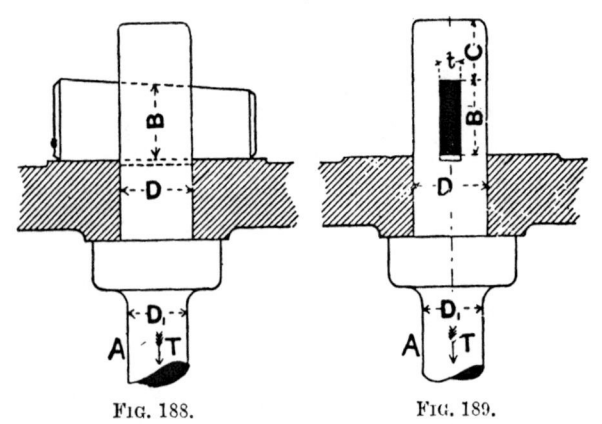

FIG. 188.   FIG. 189.

we get from the above equations $B = D$, and $t = \cdot 26\ D$. It should be noticed that although the crushing resistance of steel is greater than that of wrought-iron, no allowance can be made for this when a steel cotter is used in a wrought-iron bar. A steel cotter has therefore the same thickness as a wrought-iron one, and consequently weakens the bar through which it passes quite as much. A steel cotter will, however, withstand hammering better than a wrought-iron one, being harder.

The distance C of the cotter from the end of the bar cannot be determined properly on account of the uncertainty of the longitudinal shearing resistance of the material of the bar. But C should not be less than $\frac{D}{2}$, and if made equal to D this part will have ample strength.

Unless there is some special reason for making the bar of uniform diameter throughout the part, A should be reduced to a diameter D_1 such that $\cdot 7854 D_1^2 = \cdot 7854 D^2 - Dt$. If $t = \cdot 26D$, then $D_1 = \cdot 82D$. Or if D_1 is given, then $D = 1\cdot 22 D_1$. If the part A is screwed, D_1 must be its diameter at the bottom of the thread. The diameter of the collar on the rod may be 1·5D, and its thickness ·6C.

114. Sockets for Cottered Joints.—When a bar is cottered into a boss as shown in Fig. 190, or into a socket as shown in Figs. 191 and 192, the proportions of the boss or socket are determined in the following manner. The bar being in tension, there are three straining actions on the boss or socket. First, the tensile straining action, the resistance to which is $\cdot 7854(D_2^2 - D^2)f_t - (D_2 - D)tf_t$, this being the resistance at the section through the cotter holes. Second, the straining action which

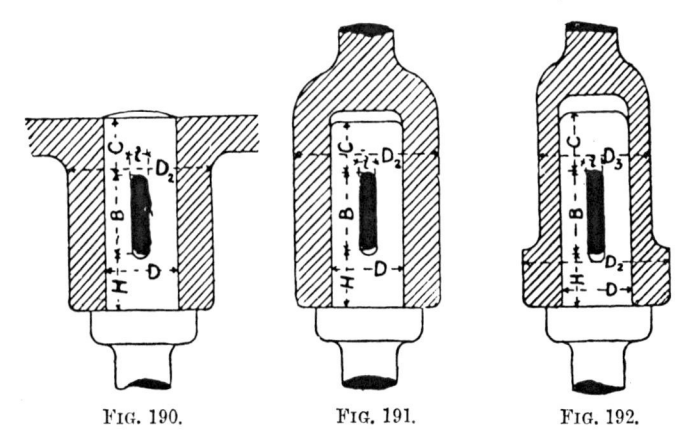

FIG. 190.   FIG. 191.   FIG. 192.

tends to shear out the pieces in front of the cotter, the resistance to which is $2(D_2 - D)Hf_s$. Third, the crushing action on the part supporting the cotter, the resistance to which is $(D_2 - D)tf_c$.

If the crushing resistance of the material of the boss or socket be not less than that of the bar or that of the cotter, the bearing surface of the cotter on the boss or socket should be equal to its bearing surface on the bar. This would require D_2 to be equal to 2 D, which would generally make the resistance of the boss or socket to tension greater than, or at least equal to, the resistance of the bar to tension. This rule, $D_2 = 2D$, is therefore generally used in proportioning joints such as are shown in Figs. 190 and 191.

The socket may be lightened as shown in Fig. 192 by making the part which is in tension only of a diameter D_3 sufficient for the tension. In determining the rule for D_3 we will assume that the rod and socket are both made of the same material, from which assumption it follows that the area of the cross section of the socket through the cotter holes must be equal to the area of the cross section of the bar through its cotter hole, hence $\cdot 7854(D_3^2 - D^2) - (D_3 - D)t = \cdot 7854 D^2 - Dt$; putting $t = \frac{1}{4}D$, the above equation gives $D_3 = 1\cdot 34 D$. In practice we find, however, that D_3 is not less than $1\cdot 5 D$.

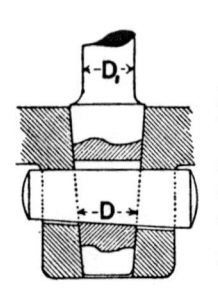

The remarks already made with reference to the distance, C, of the cotter from the end of the bar apply also to the distance, H, of the cotter from the end of the boss or socket. It will be sufficient if H is made equal to C, that is, not less than $\dfrac{D}{2}$ and not greater than D.

FIG. 193.

Instead of putting a collar on the rod to resist the thrust, the end of the rod may be tapered, as shown in Fig. 193. The diameters D and D_1 are related in the same way as in the other cases, namely, $D_1 = \cdot 82 D$, or $D = 1\cdot 22 D_1$. The total taper on the tapered part of the rod, that is, the variation in diameter per unit of length, may vary from $\frac{3}{8}$ inch to $\frac{3}{4}$ inch per foot of length—that is, from 1 in 32 to 1 in 16.

115. Taper of Cotters.—The greatest taper that can be given to a cotter, while it does not slacken back by the longitudinal force transmitted by the rod through which it passes, depends on the co-efficient of friction between the sliding surfaces, but will generally be about 1 in 7. It is not usual to make the taper more than 1 in 24 unless special means are adopted for locking the cotter. With a taper greater than 1 in 7 the cotter should be provided with a screw and nut to pull it into position, as in Fig. 200.

116. Tightening and Locking Arrangements for Cotters.—In general, cotters are driven in by hammering, but frequently they are designed so that they may be moved by a screw. Fig. 194 shows a separate bolt provided with nuts for moving the cotter out or in. Examples of cotters and gibs with screwed tail-pieces are shown in Figs. 195 to 200. When the axis of the screw is not parallel to the side of the cotter in contact with the gib, the eye through which the screw passes must have a slotted hole as shown in the upper part of Fig. 197. These screws and nuts are, however, of more use in locking the cotters than for tightening or slackening them. The tightening and slackening are more rapidly and effectively performed by hammering the end of the cotter. The diameter of the screw on the gib or cotter should not exceed, but may be a little less than the thickness of the gib or cotter. A common way of locking a cotter is by one or two set-screws as shown in Figs. 201 and 202. The

COTTERS.

point of the set-screw presses on the bottom of a shallow groove, so that the bur raised by the screw does not interfere with the motion of the

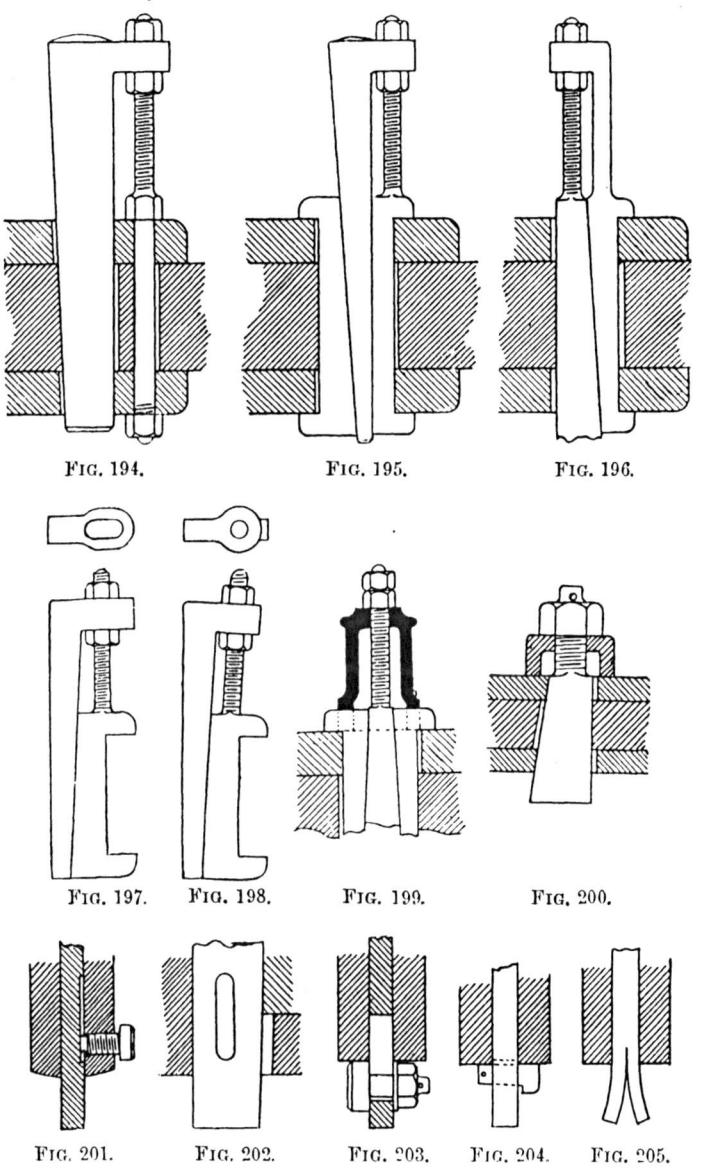

FIG. 194.   FIG. 195.   FIG. 196.

FIG. 197.   FIG. 198.   FIG. 199.   FIG. 200.

FIG. 201.   FIG. 202.   FIG. 203.   FIG. 204.   FIG. 205.

cotter. Fig. 203 shows a method of locking a cotter by a bolt passing through a slotted hole near the point of the cotter. A cotter is sometimes secured by a smaller cotter passing through it at right angles to its broadest face, as shown in Fig. 204, this smaller cotter being in turn secured by a split pin, or by being split itself at the point. The method of locking the smaller cotter just mentioned by splitting it at the point may be applied to the main cotter, as shown in Fig. 205.

117. Split Pins.—We have frequently mentioned split pins. These are used not so often for locking one piece firmly to another, as for preventing the two pieces from separating altogether, while allowing a

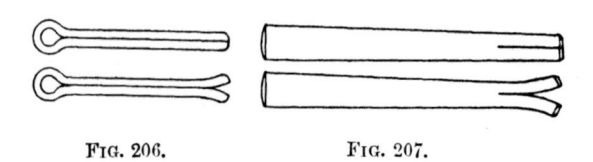

FIG. 206.   FIG. 207.

certain amount of "play" or relative motion. Small split pins are made as shown in Fig. 206 from wire of semicircular cross section. Large split pins are generally made solid and tapered, as shown in Fig. 207.

118. Carver's Circlip.—This is a device designed to take the place of split pins on bolts, shafts, and pins of knuckle-joints. It consists of a split steel ring which is sprung into a groove turned on the bolt, shaft, or pin. The ring is of conical form, and the smaller end presses on one side of the groove just mentioned, while the other end presses on the piece which it is desired to secure.

CHAPTER VII.

PIPES AND PIPE-JOINTS.

119. Thickness of Pipes subjected to Internal Pressure.—Let D be the internal diameter of a pipe in inches, and t its thickness also in inches, and let P be the internal pressure in pounds per square inch, and f the stress in pounds per square inch, produced in the material of the pipe by the pressure P. If the thickness t is small compared with the diameter D, then $t = \dfrac{PD}{2f}$. In actual practice the thickness given by the above formula is generally increased by an amount, c, depending on various circumstances. The rule for the thickness adopted in practice therefore takes the form—

$$t = \frac{PD}{k} + c.$$

The following table gives various values of c and k:—

	k.	c.
Cast-iron steam or water pipes	4,000	·3
Cast-iron steam-engine cylinders	3,500	·5
Lap-welded wrought-iron tubes	17,000	·06
Solid-drawn steel tubes	40,000	·0
Copper steam-pipes	7,000	·1
Lead pipes	450	·3

For thick pipes the thickness t may be determined from the formula

$$t = \frac{D}{2}\left\{ \sqrt{\frac{3f + 2P}{3f - 4P}} - 1 \right\}$$

where f is the safe stress on the material and P the excess of internal over external pressure in pounds per square inch. Taking $f = 2400$ for cast-iron

$$t = \frac{D}{2}\left\{ \sqrt{\frac{3600 + P}{3600 - 2P}} - 1 \right\}$$

120. Flanges for Cast-Iron Pipes.—The following rules may be used in proportioning the flanges for cast-iron pipes of the form shown in

Fig. 208. The rule for the thickness of the pipe has already been given, namely, $t = \dfrac{PD}{4000} + \cdot 3$. The other proportions are as follows :—

$T = 1\cdot 4\, t + \cdot 15.$ $\qquad n =$ number of bolts $= \cdot 6\, D + 2.$
$d = \cdot 83\, t + \cdot 3.$ $\qquad B = 2\tfrac{1}{4}\, d.$

EXAMPLE.—*Required the dimensions of a flanged cast-iron pipe* 10 *inches in diameter to carry a pressure of* 100 *lbs. per square inch.*

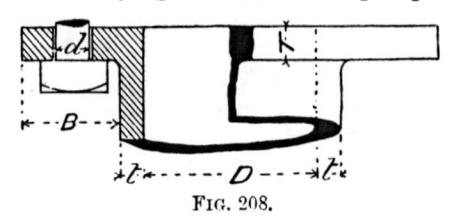

FIG. 208.

$P = 100$ and $D = 10$, therefore $t = \dfrac{100 \times 10}{4000} + \cdot 3 = \cdot 55$ inch, or very nearly $\tfrac{9}{16}$ inch.

$T = 1\cdot 4 \times \cdot 55 + \cdot 15 = \cdot 92$ inch, say $1\tfrac{5}{16}$ inch.
Number of bolts $= \cdot 6 \times 10 + 2 = 8.$
$d = \cdot 83 \times \cdot 55 + \cdot 3 = \cdot 7565$ inch, say $\tfrac{3}{4}$ inch.
$B = 2\tfrac{1}{4} \times \tfrac{3}{4} = 1\tfrac{11}{16}$ inches.

External diameter of pipe $= D + 2\, t = 10 + \dfrac{2 \times 9}{16} = 11\tfrac{1}{8}$ inches.

Diameter of flange $= D + 2\, t + 2\, B = 10 + \dfrac{2 \times 9}{16} + 2 \times 1\tfrac{11}{16} = 14\tfrac{1}{2}$ inches.

The breadth B of the flange is often much greater than that stated above, being as much as $3\tfrac{1}{2}\, d$ in some cases.

Pipes are frequently strengthened by being made thicker near the flanges, as shown in Fig. 209. This thicker part may be equal in thickness to $\dfrac{T + t}{2}$, and of a length equal to $2\, t + \cdot 6.$

Flanges are sometimes stiffened by brackets, as shown in Fig. 210,

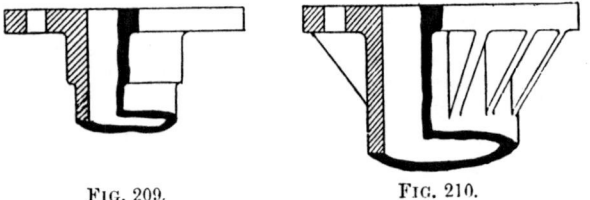

FIG. 209.   FIG. 210.

one between each pair of bolts. The thickness of these brackets may be equal to $\dfrac{T + t}{2}$, but in this case the thickness of the flange may be reduced so as to be equal to the thickness of the pipe.

Cast-iron pipes of 3 inches in diameter and upwards are generally made in lengths of 9 feet.

121. Spigot and Socket Joints for Cast-Iron Pipes.—Cast-iron gas or water pipes which have to be embedded in the earth are usually connected by spigot and socket joints such as are shown in Figs. 211 and 212. The joint shown in Fig. 211 is made tight by first partly filling the space between the spigot and the socket with gasket, and then filling the remaining space with lead, which is run in in the molten state. After the lead has solidified it is stemmed in tight with hammer and

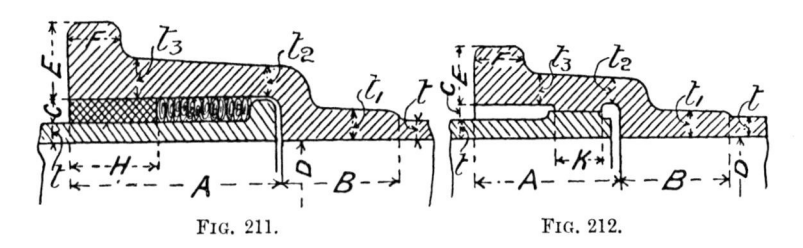

FIG. 211.   FIG. 212.

drift. In the joint shown in Fig. 212 a strip on the spigot is turned to a slight taper, and a strip in the socket is bored to fit the turned part of the spigot. Before being put together the turned and bored parts are painted with red-lead or Portland cement.

The thickness t of the pipe is determined by the formula already given, namely, $t = \dfrac{PD}{4000} + \cdot 3$, where D is the internal diameter of the pipe in inches and P the pressure of the water in pounds per square inch. The other dimensions may be determined by the following rules, all dimensions being in inches :—

$t_1 = 1 \cdot 1\, t + \cdot 07.$ $C = \frac{1}{4}$ to $\frac{1}{2}.$
$t_2 = \cdot 025\, D + \cdot 375,$ $E = \cdot 045\, D + \cdot 8.$
   but not less than t_1. $F = \cdot 04\, D + \cdot 7.$
$t_3 = t_2 + \frac{1}{8}.$ $H = \cdot 06\, D + 1.$
$A = \cdot 1\, D + 2 \cdot 75.$ $K = \cdot 05\, D + 1.$
$B = \cdot 09\, D + 2.$ Slope of turned part 1 in 32.

As illustrations of recent practice in the construction of large cast-iron pipes, we show in Figs. 213 and 214 spigot and socket joints used for pipes for the Manchester waterworks.[1] The joints used are of three kinds: (1) deep sockets (Fig. 213), (2) shallow sockets, and (3) collars (Fig. 214). The shallow sockets are similar in construction to that shown in Fig. 213, but are only $3\frac{3}{4}$ inches deep instead of $5\frac{3}{4}$ inches, and are used on sloping ground where there is no danger from settlement. Collars 6 inches wide (Fig. 214) have been used on the more level ground, but in the coal districts the collars are from 8 inches to 12

[1] Engineering, vol. lii. p. 615.

inches wide. The joints were run full with lead, no yarn being used. To prevent the lead running into the pipe, a steel spring ring was

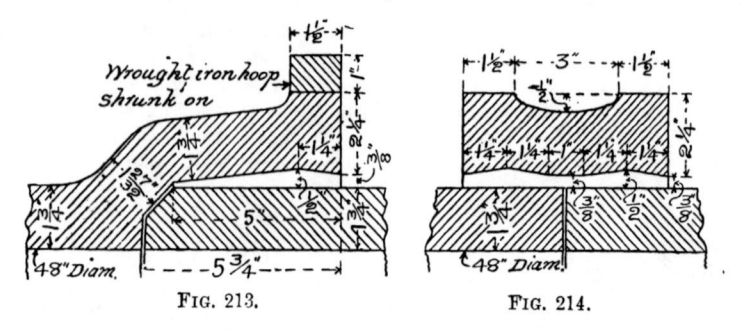

FIG. 213.   FIG. 214.

placed temporarily inside the pipe. The pipes were made in lengths of 12 feet.

122. Flanges for Copper Pipes.—Copper pipes are either "solid-drawn" or made from sheet copper. In the latter case there is a longitudinal brazed joint. The flanges for these pipes are made of tough brass, and are secured to the pipes by brazing, or by riveting and brazing. Fig. 215 shows the form of the flange for brazing only, which may

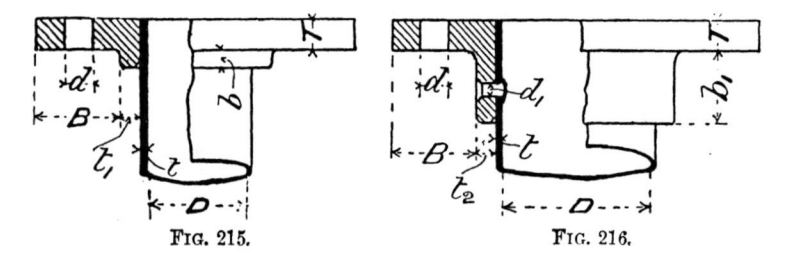

FIG. 215.   FIG. 216.

be used for pipes up to 12 inches in diameter. For pipes above 12 inches in diameter the flange should be constructed as shown in Fig. 216, and riveted to the pipe, in addition to being brazed to it.

The following rules may be used in designing the above, all the dimensions being in inches, and the pressure P of the steam in pounds per square inch :—

$$t = \frac{PD}{7000} + \cdot 1.$$

$$d = 1\tfrac{1}{8} T.$$

$$B = 2\tfrac{1}{4} d.$$

$$t_2 = t + \tfrac{1}{4}.$$

$$b_1 = 3 d_1.$$

$$T = 2\cdot 7\, t + \cdot 14$$

$$t_1 = \frac{T}{2}.$$

$$b = \frac{T}{2} - \tfrac{1}{16}.$$

$$d_1 = 1\tfrac{1}{4} t.$$

PIPES AND PIPE JOINTS.

The number of bolts in the flange must be such as to make their distance apart, measured from centre to centre, about $4\frac{1}{2}$ times their diameter. When the flanges are riveted to the pipe, as shown in Fig. 216, copper rivets are used, their distance apart being about 3 to $3\frac{1}{2}$ times their diameter.

123. Pope's Flanges for Pipe Joints.—Mr. R. B. Pope, of the Engine Works, Dumbarton, has patented the pipe joint shown in Figs.

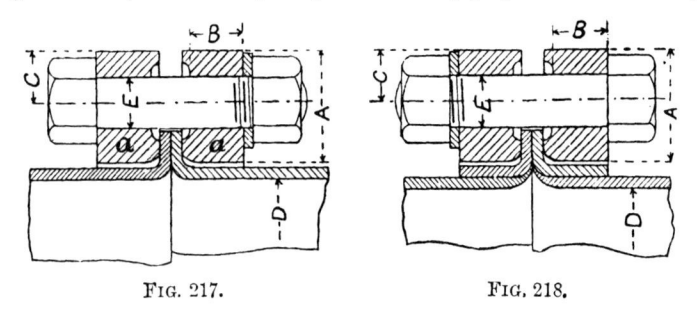

FIG. 217.   FIG. 218.

217 and 218. The following description and table of dimensions are based on those given by the inventor:—

A copper pipe, if solid-drawn, can have the flange formed on it as shown in Fig. 217; but it is preferred to form the flange separately, as shown in Fig. 218. The joint rings aa are made preferably of cast-steel, but cast-iron, wrought-iron, and steel have been used.

When the rings are made in one piece, they have to be strung on to the copper pipe before both ends are flanged out. They may, however, be made in halves, half checked at the meeting ends, or each ring may consist of two thicknesses, each thickness being in halves, and the four parts arranged so as to break joint.

These joints have been used for steam, feed, and exhaust pipes, from $1\frac{1}{2}$ inches diameter up to 36 inches diameter.

Dimensions of Pope's Flanges for Pipe Joints.

D.	A.	B.	C.	E.	No. of Bolts.	D.	A.	B.	C.	E.	No. of Bolts.
$1\frac{1}{2}$	$1\frac{3}{4}$	$\frac{3}{4}$	$\frac{7}{8}$	$\frac{3}{4}$	5	5	$2\frac{1}{8}$	1	1	1	8
2	$1\frac{7}{8}$	$\frac{13}{16}$	$\frac{7}{8}$	$\frac{3}{4}$	5	6	$2\frac{3}{16}$	$1\frac{1}{16}$	1	1	9
$2\frac{1}{2}$	$1\frac{7}{8}$	$\frac{13}{16}$	$\frac{7}{8}$	$\frac{13}{16}$	6	7	$2\frac{5}{16}$	$1\frac{1}{8}$	$1\frac{1}{16}$	$1\frac{1}{8}$	9
3	$1\frac{15}{16}$	$\frac{7}{8}$	$\frac{15}{16}$	$\frac{7}{8}$	6	8	$2\frac{5}{16}$	$1\frac{1}{8}$	$1\frac{1}{16}$	$1\frac{1}{8}$	10
$3\frac{1}{2}$	$1\frac{15}{16}$	$\frac{7}{8}$	$\frac{15}{16}$	$\frac{7}{8}$	6	9	$2\frac{5}{16}$	$1\frac{3}{16}$	$1\frac{1}{16}$	$1\frac{1}{8}$	10
4	2	$\frac{15}{16}$	$\frac{15}{16}$	$\frac{7}{8}$	7	10	$2\frac{5}{16}$	$1\frac{1}{4}$	$1\frac{1}{16}$	$1\frac{1}{8}$	11
$4\frac{1}{2}$	2	$\frac{15}{16}$	$\frac{15}{16}$	$\frac{7}{8}$	8	11	$2\frac{5}{16}$	$1\frac{1}{4}$	$1\frac{1}{16}$	$1\frac{1}{8}$	11

All the above dimensions are in inches.

124. Armstrong's Pipe Joint.—The form of pipe joint introduced by Sir William Armstrong & Co. for cast-iron pipes carrying water under

84   MACHINE DRAWING AND DESIGN.

great pressure is shown in Figs. 219 and 220. The joint is kept water-tight by a gutta-percha ring ¼ inch thick, which is forced into the recess *a*, Fig. 219.

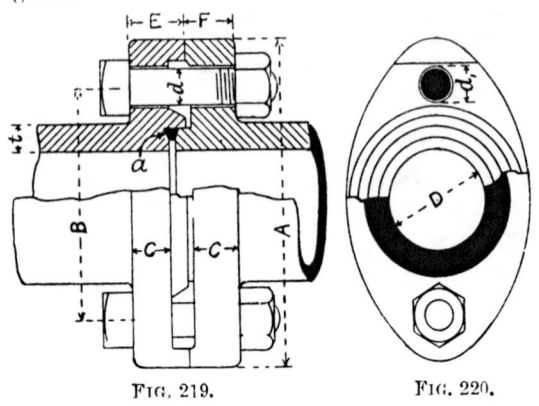

FIG. 219.   FIG. 220.

The following rules may be used in proportioning the parts of the above joint:—

D = internal diameter of pipe.
t = thickness of pipe.
A = D + 7 t + 1·3.
B = D + 4·5 t + ·3.
C = 2 t.

$d = 1·25\ t.$
$d_1 = d + ·125.$
E = 2 t + ·3.
F = 2 t + ·125.

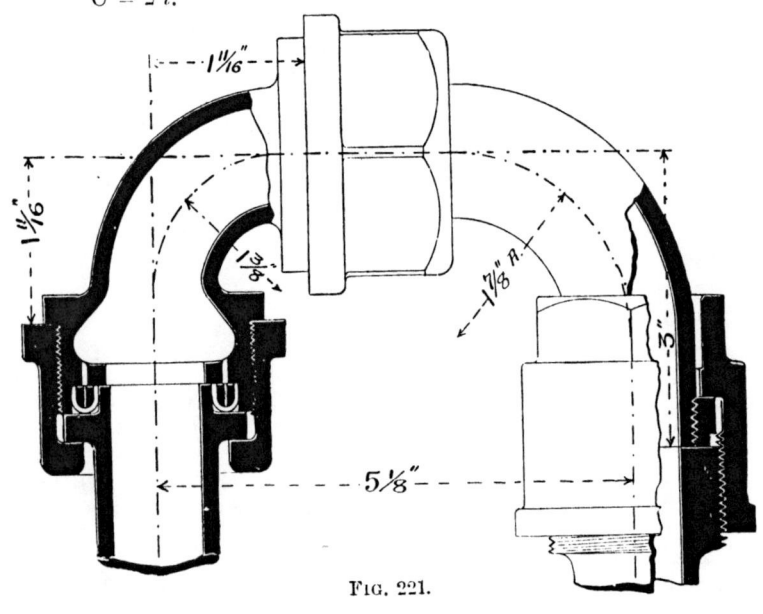

FIG. 221.

125. Swivel Joints for Hydraulic Pipes.—For a certain class of portable hydraulic machines, such as the hydraulic jiggers largely used about docks, the pipes which connect the machines with the hydraulic main must be capable of being swung round into various positions.

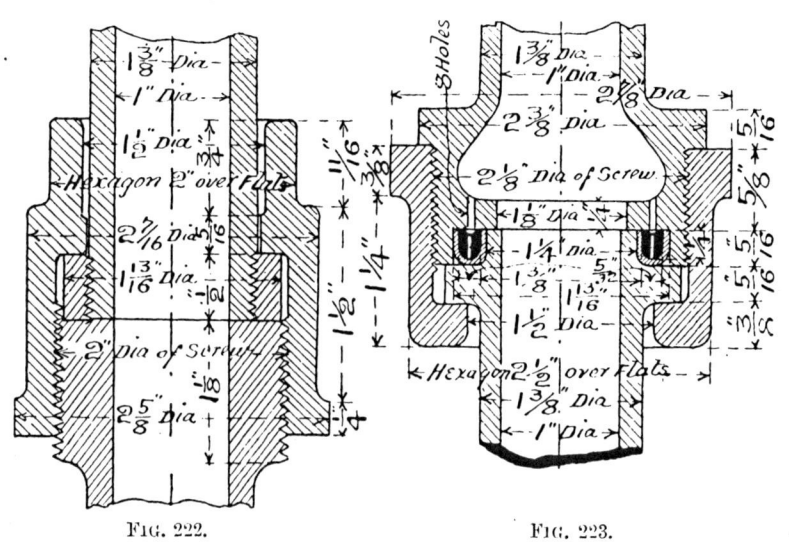

FIG. 222.   FIG. 223.

This is provided for by fitting the pipes with one or more swivel joints such as are shown in Fig. 221.

Detailed dimensioned illustrations of these joints are shown in Figs. 222 and 223.

126. Details Connected with Wrought-Iron and Steel Tubes.—Wrought-iron and steel tubes are made either butt-welded, lap-welded, or solid drawn. They are usually made in lengths of 10 feet to 15 feet, but they can also be made up to 20 feet long.

Lengths of tube in the same line are generally connected by a socket,

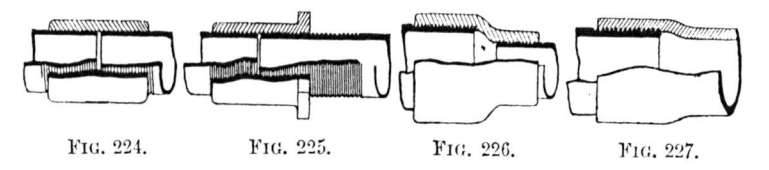

FIG. 224.   FIG. 225.   FIG. 226.   FIG. 227.

Fig. 224, which is first screwed half way on to the end of one length, and the other length is then screwed into the part of the socket which projects beyond the end of the first length. When it is not possible or convenient to rotate the lengths of tube so as to screw them into the socket, a "long screw" is made on one, Fig. 225. The socket is first screwed on to the long screw, and after the lengths of tube are butted together, the socket is screwed back until it is half on each length as shown. Tubes of different diameters are connected by a reducing socket, Fig. 226. A socket may be formed on the tube by swelling one end as shown in Fig. 227.

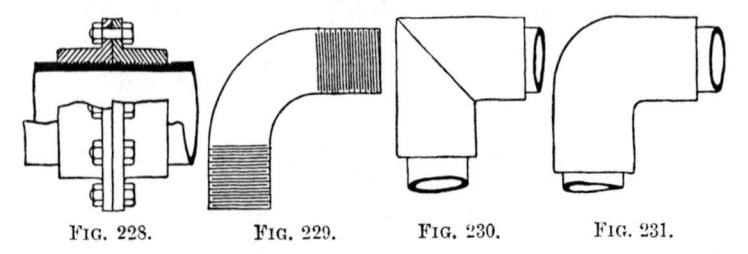

FIG. 228.   FIG. 229.   FIG. 230.   FIG. 231.

Cast-iron flanges screwed on to the tubes and bolted together as shown in Fig. 228 are sometimes used instead of a screwed socket. Tubes at right angles are connected by bends, elbows, tees, or crosses, Figs. 229, 230, 231, 232, and 233.

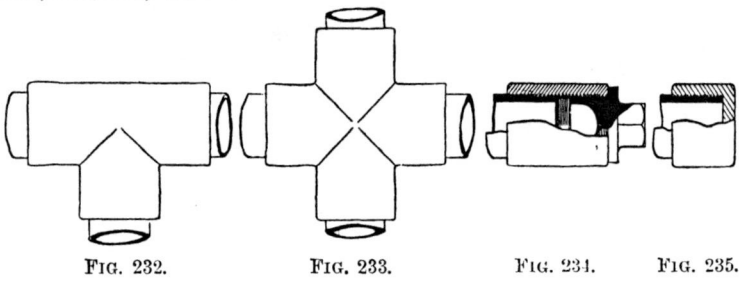

FIG. 232.   FIG. 233.   FIG. 234.   FIG. 235.

The end of a tube may be closed by a socket and plug, Fig. 234, or by a cap, Fig. 235.

127. Joints for Light Wrought-Iron and Steel Pipes.—Wrought-iron and steel pipes frequently take the place of cast-iron pipes for the transmission of water. For waterworks in the interior of a new country, where the cost of transit is great, the wrought-iron or steel pipes are preferred because they are much lighter than cast-iron pipes of the same strength. These wrought-iron or steel pipes are either riveted or lap-welded. The riveted pipes are usually made in lengths up to 25 feet, but they have been made as long as 40 feet. The welded pipes are made in lengths up to 18 feet.

Figs. 236, 237, and 238 [1] show three forms of joints used for connecting different lengths of wrought-iron or steel pipes. The joint shown in Fig. 236 is a flanged joint. The flanges are made from angle-irons bent and welded into rings which are riveted to the pipes. Fig. 237 shows an ordinary spigot and socket joint. The socket is rolled and welded into a ring which is riveted to the pipe. The "Kimberley joint," Fig. 238, is made with a collar containing two sockets.

A flanged joint in which the rings are stamped in dies from solid

[1] These illustrations were prepared from drawings supplied by Messrs. Thomas Piggott & Co., Birmingham.

plates, and are therefore without welds, is shown in Fig. 239.[1] These rings are truly circular and require no facing.

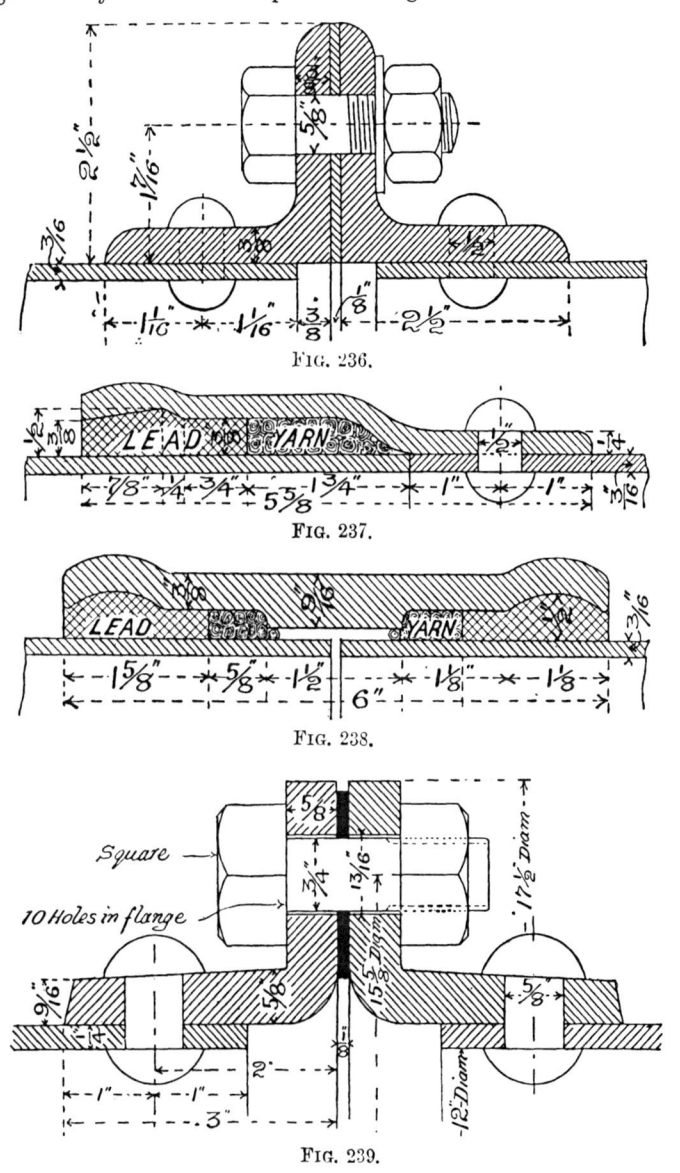

FIG. 236.

FIG. 237.

FIG. 238.

FIG. 239.

[1] Figs. 239 to 243 were prepared from working drawings supplied by the Steel Pipe Co., Kirkcaldy.

88   MACHINE DRAWING AND DESIGN.

A light and strong form of socket and spigot joint is shown in Fig. 240. This is known as Riley's stamped steel socket.

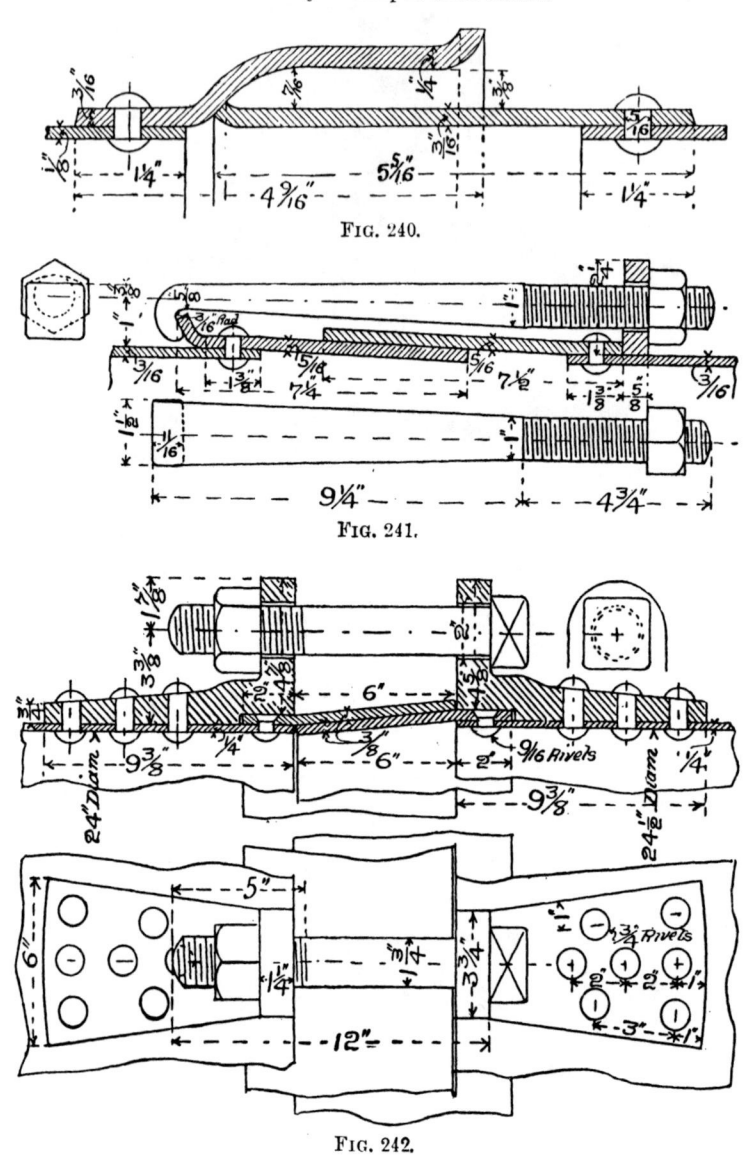

FIG. 240.

FIG. 241.

FIG. 242.

Brown's socket and spigot joint, shown in Fig. 241, is used for drainage pipes of large diameter. The socket and spigot are of conical form, and

PIPES AND PIPE JOINTS.

are pulled together by hook bolts. The screwed ends of these bolts pass through a loose ring, which is threaded on to the pipe before the socket piece is riveted on, and the hooked ends of the bolts hook on to a flange formed on the tail-end of the spigot piece as shown.

Another form of tight-fitting socket and spigot joint is shown in Fig. 242, and is known as the "Bournemouth" joint, from the fact that it was first used for the storm outfall (3 feet diameter) at the town of that name. The socket and spigot are drawn together by bolts which pass through wrought-iron lugs riveted to the pipes.

The riveted joints in wrought-iron and steel pipes are generally lap-joints. The longitudinal joints may be single or double riveted. For the circumferential seams single riveting is sufficient. Fig. 243 shows the best form of circumferential seam. This design not only maintains uniformity in the internal diameter of the pipe, but greater accuracy in riveting. The plates are punched in multiple punching machines, bent into cylinders, and the ends required to overlap are expanded by suitable machinery.

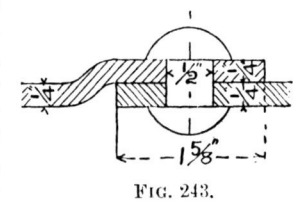

FIG. 243.

128. The Converse Pipe Joint.—Figs. 244 and 245 show an excellent design of pipe joint which has been extensively used in America, and is known as the "Converse" lock-joint, from the name of the inventor, Mr. Converse, of the National Tube Works Co., McKeesport, Pa. This joint, which is chiefly used for wrought-iron and steel pipes, consists of

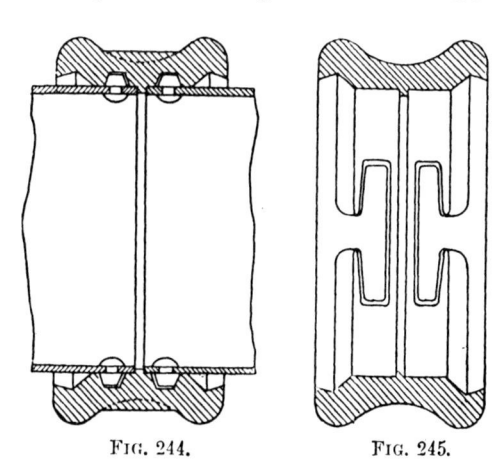

FIG. 244.   FIG. 245.

a cast-iron sleeve or collar, with a lead space at each end. The sleeve has recesses, clearly shown in Fig. 245, into which projecting rivet-heads or studs fixed on the ends of the pipes are inserted and locked by turning the pipe through a small angle to the right or left. One side of each of the recesses just mentioned slopes slightly, so that when the pipe is twisted round it is at the same time forced tight against the projecting ring on the inside of the sleeve. The lead, when poured in, not only fills the space specially made for it, but also penetrates into the recesses round the rivet-heads. After the lead has solidified, it is stemmed in or caulked in the usual way. No yarn is used in this joint.

129. Screwed Cast-Iron Flanges for Wrought-Iron Pipes.—In a paper read before the American Society of Mechanical Engineers in 1889, Mr. E. F. C. Davis described the joint used for the steam-pipes in the collieries of the Philadelphia and Reading Coal and Iron Company. The pipes are of wrought-iron, and carry cast-iron flanges of the form shown in Figs. 246 and 247. The flange is screwed tightly on the pipe, and the end

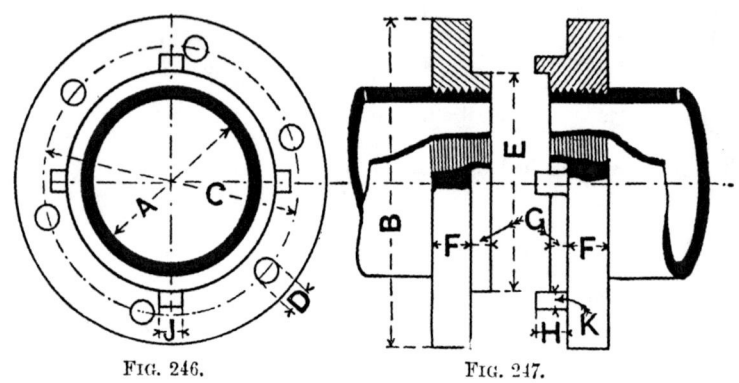

FIG. 246.   FIG. 247.

of the latter is faced off flush with the facing piece, which is cast on the former. One of the two flanges which are to be bolted together has cast on its face a number of lugs, which have their inner faces bored to fit the facing piece on the other flange, and thus ensure the pipes being in line. The joint is made steam-tight by a "gum-joint" ring which is placed between the abutting ends of the pipes and within the circle of the lugs, so that the latter keep the ring central.

The following table gives the dimensions (in inches) adopted for this pipe joint:—

A.	B.	N_1.	C.	D.	E.	F.	G.	N_2.	H.	J.	K.
3	$7\frac{3}{4}$	4	6	$\frac{3}{4}$	5	$\frac{3}{4}$	$\frac{1}{2}$	4	$\frac{7}{8}$	$\frac{3}{4}$	$\frac{1}{2}$
$3\frac{1}{2}$	$8\frac{1}{4}$	4	$6\frac{3}{4}$	$\frac{3}{4}$	$5\frac{1}{2}$	$\frac{3}{4}$	$\frac{1}{2}$	4	$\frac{7}{8}$	$\frac{3}{4}$	$\frac{1}{2}$
4	$9\frac{1}{2}$	4	$7\frac{1}{2}$	$\frac{7}{8}$	6	$\frac{7}{8}$	$\frac{1}{2}$	4	$\frac{7}{8}$	$\frac{3}{4}$	$\frac{1}{2}$
5	$10\frac{1}{2}$	4	$8\frac{1}{2}$	$\frac{7}{8}$	7	$\frac{7}{8}$	$\frac{1}{2}$	4	$\frac{7}{8}$	$\frac{3}{4}$	$\frac{1}{2}$
6	12	6	10	$\frac{7}{8}$	8	1	$\frac{1}{2}$	4	$\frac{7}{8}$	$\frac{3}{4}$	$\frac{1}{2}$
7	13	6	11	$\frac{7}{8}$	9	1	$\frac{1}{2}$	4	$\frac{7}{8}$	$\frac{3}{4}$	$\frac{1}{2}$
8	14	6	12	$\frac{7}{8}$	10	$1\frac{1}{4}$	$\frac{1}{2}$	6	$\frac{7}{8}$	1	$\frac{5}{8}$
10	$16\frac{1}{4}$	8	14	1	12	$1\frac{1}{4}$	$\frac{1}{2}$	6	$\frac{7}{8}$	1	$\frac{5}{8}$

N_1 = number of bolts. N_2 = number of lugs.

The pipes are put together in lengths of from 16 to 20 feet.

130. Specification for Mild Steel Pipes 49 Inches Diameter:[1]

Material.—All the mild steel used in plates and bars shall be of the best quality, and of approved brand and manufacture, the brand to be plainly visible on each piece. The plates shall be capable of withstanding a tensile load of not less than 27 tons per square inch of original section, with an extension of not less than 20 per cent. of the tested length. The plates for sockets shall be capable of withstanding a tensile load of not less than 24 tons per square inch of original section, with an extension of not less than 15 per cent. of the tested length.

The rivets shall be made from mild steel of approved brand and manufacture, and shall be capable of withstanding a tensile load of not less than 24 tons per square inch of original section, with an extension of 15 per cent. of the tested length. The rivets shall also bear being bent double when cold without showing sign of fracture.

The above-named tensile tests are to be made on carefully prepared pieces, each with a sectional area of not less than ·75 square inch, and a length of not more than 8 inches between the shoulders, all pieces to be cut and tested the lengthway of the grain.

Pipes.—The pipes shall be 25 feet in length, exclusive of length of socket, and ¼ inch thick. They shall be constructed of plates not less than 6 feet 3 inches wide, and of such length as to be equal to the circumference of the pipe, plus the necessary lap. Plates to be bent end on, the longitudinal lap-joints to be double riveted zigzag, and the circumferential lap-joints to be expanded and single riveted. Proportions of laps, pitches, etc., to be as follows:—

Thickness of plates	¼ inch
Breadth of lap, longitudinal joints	3¼ ,,
Breadth of lap, circumferential joints	1⅝ ,,
Diameter of rivets	½ ,,
Pitch of rivets in line, longitudinal joints	1¾ ,,
Distance apart of pitch lines, longitudinal joints	1½ ,,
Pitch of rivets, circumferential joints	1¼ ,,

Each pipe to be fitted with a welded and rolled socket and spigot of the sections shown in Fig. 248, shrunk on to the pipe, and riveted and caulked in an approved manner.

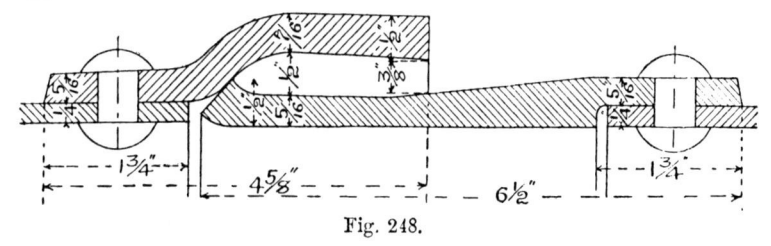

Fig. 248.

Riveting.—The rivet-holes to be regularly and evenly spaced to the given pitch, and carefully punched opposite to each other, so as to require the least amount of rimering. Drifting will on no account be permitted in any portion of the work. The punching shall be so performed that the smaller ends of the holes shall come together, and the larger ends against the rivet heads, and all burrs shall be removed. When closed, all the rivets must completely fill their holes, and they must be long enough and of suitable shape for making good snap-heads both inside and outside. All riveting shall be performed with approved hydraulic machinery, and the plates shall be punched with multiple machines so as to ensure accuracy. The edges of all plates and straps that have to be caulked shall be planed or ground with a bevel edge for caulking. All seams to be caulked in an approved and workmanlike manner.

[1] By the Steel Pipe Company, Kirkcaldy.

Hydrostatic Testing.—The whole of the mild steel pipes shall be proved to a hydrostatic pressure of 120 lbs. per square inch. During the time the pipe is undergoing the proof, it shall be repeatedly struck in various parts of its length and circumference with a suitable hammer, and be kept externally dry, so as to exhibit more perceptibly any escape of water resulting from the pressure applied. All hydrostatic testing is to be done before the pipes are coated.

Coating.—Every pipe must be perfectly clean, dry, and free from rust on all surfaces when the coating is applied. The mixture for coating shall be as follows:—

Best crude Trinidad natural asphalt or bitumen	.	about 44 per cent.
Coal tar, distilled until free from naphtha	. .	,, 55 ,,
Resin		,, 1 ,,

The ingredients must be of the very best quality, and free from all impurities, to be carefully heated in a suitable vessel to 400° Fah., and kept at this temperature during the whole time of dipping. Two dipping tanks must be used, the first to be charged with the above specified mixture, and the second with a larger proportion of natural asphalt, (say) at least 25 per cent. Each pipe must be left at least half-an-hour in the boiling liquid in the first tank, so as to acquire its temperature, unless previously heated by hot air; it shall then be slowly removed and laid upon skids to drip, and when the coating is well set the pipe must be again dipped into the second tank and lifted out to dry. The object of the first dipping is to get the coating well attached to the iron, and of the second dipping to thicken it. The coating when finished must be at least $\frac{1}{32}$ of an inch in thickness. When the coating material shall have partially set, the pipe shall be placed in a frame in a horizontal position, in which it shall be made to revolve on its axis, during which process dry sand is to be allowed to fall on it uniformly, externally, until a sufficient quantity of sand shall have become incorporated with the coating material so as to render it less liable to run in hot weather, and to protect it (the coating material) from injury in transit or otherwise. As the mixture will deteriorate after a number of pipes have been dipped, fresh materials in the right proportions must be frequently added, and the vessels must occasionally be entirely emptied and replenished with new materials.

131. Joint Rings for Flange Joints.—Flange joints of pipes are made steam or water-tight by introducing a packing ring between the flanges. This packing ring may be a flat ring of lead or a ring of lead or copper wire. Wrought-iron rings covered with canvas or gasket are also used. For water-pipes 18 inches in diameter and upwards a hoop of American elm has proved very efficient. The section of the hoop is a rectangle $\frac{3}{4}$ inch × $\frac{1}{2}$ inch, and the joint of the hoop is a copper-riveted lap-joint. Rings of india-rubber, asbestos, and gasket are also largely used for flange joints.

132. Provision for Expansion in Pipes.—In a long line of pipes exposed to an atmosphere of varying temperature, and in a comparatively short line of pipes carrying steam or hot water, provision has to be made for the expansion and contraction of the pipes due to changes in temperature. Three methods of allowing for this expansion and contraction are shown in Figs. 249, 250, and 251. In the first of these illustrations a copper pipe, *abc*, in the form of a loop is shown forming a part of a line of iron or copper pipes. The expansion and contraction of the pipes will cause the loop portion to bend, which from its shape, and being made of copper, it can readily do, to a certain extent, without injury. Messrs. Hopkinson of Huddersfield have introduced the copper corrugated expansion piece for pipes shown in Fig. 250. A common and effective arrangement is a gland and stuffing-box as shown in Fig. 251. On account of the liability of cast iron to corrode in the presence of moisture, especially when the skin has been removed by boring or

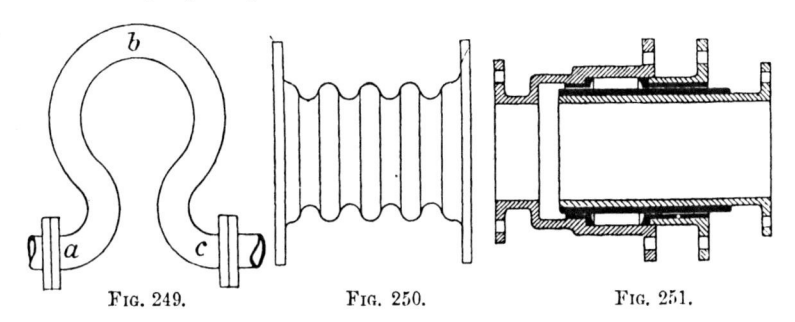

FIG. 249.   FIG. 250.   FIG. 251.

turning, the stuffing-box and the gland should be fitted with brass bushes, and the portion of the pipe which slides inside the gland and stuffing box should be sheathed with brass as shown.

CHAPTER VIII.

SHAFTING AND SHAFT COUPLINGS.

133. Twisting Moment.—If a shaft carries a lever, Fig. 252, or a wheel, Fig. 253, and a force P acts at the outer end of the radius, or if the shaft carries a pulley, Fig. 254, over which passes a band subjected to tensions T_1 and T_2, so that the effective turning force at the rim of the pulley is $T_1 - T_2 = P$, then the force P in each case produces a

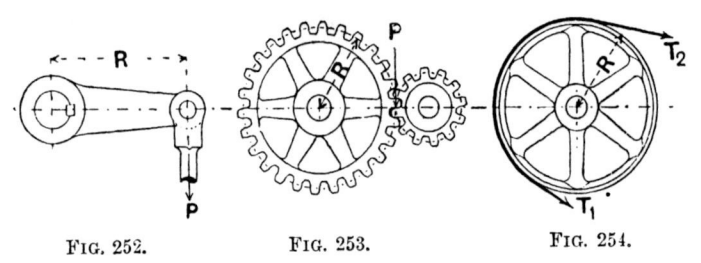

FIG. 252.   FIG. 253.   FIG. 254.

twisting action on the shaft which will be proportional to the radius R and to the force P, and is measured by the product PR. This product is called the *twisting moment* on the shaft. If the force P is in pounds and the radius R in feet, the twisting moment is in *foot-pounds*. If P is in pounds and R in inches, the twisting moment is in *inch-pounds*. By expressing the force P in tons and the radius R in feet or in inches, we get the twisting moment in *foot-tons* or in *inch-tons*. In practice it is generally most convenient to express the twisting moment in inch-pounds.

Let N = speed of shaft in revolutions per minute.
    H = horse-power which is being transmitted.
    T = twisting moment on the shaft in inch-pounds.
    P = force in pounds acting at a perpendicular distance of R inches from the centre of the shaft.

Then, T = PR.

$$H = \frac{2 \times 3\cdot1416\ TN}{12 \times 33000} = \cdot00001587\ TN$$

$$T = \frac{12 \times 33000\ H}{2 \times 3\cdot1416\ N} = 63025\cdot2\ \frac{H}{N}$$

$$N = \frac{12 \times 33000\ H}{2 \times 3\cdot1416\ T} = 63025\cdot2\ \frac{H}{T}$$

SHAFTING AND SHAFT COUPLINGS.

EXAMPLE 1.—*The diameter of a spur wheel is 50 inches, and the total pressure on the teeth at the pitch line is 2000 lbs. Required the twisting moment on the shaft due to the pressure on the wheel teeth and the horse-power transmitted by the wheel at a speed of 150 revolutions per minute.*

Here, R = 25 inches, and P = 2000 lbs.; therefore, T = PR = 2000 × 25 = 50,000 inch-pounds.

Horse-power = H = ·00001587 TN = ·00001587 × 50,000 × 150 = 119·025, say 119.

EXAMPLE 2.—*Required the twisting moment on a shaft which transmits 20 horse-power at a speed of 90 revolutions per minute.*

$$\text{Twisting moment} = \text{T} = 63025\cdot2 \frac{\text{H}}{\text{N}} = 63025\cdot2 \times \frac{20}{90} = 14005\cdot6$$

inch-pounds.

EXAMPLE 3.—*Required the speed of a shaft which transmits 15 horse-power under a twisting moment of 10,000 inch-pounds.*

$$\text{Speed} = \text{N} = 63025\cdot2 \frac{\text{H}}{\text{T}} = 63025\cdot2 \times \frac{15}{10000} = 94\cdot5 \text{ revolutions per}$$

minute.

134. Resistance of a Shaft to Twisting.—When a shaft is subjected to twisting, the stress induced is a shearing stress, which varies uniformly from nothing at the centre to a maximum at the circumference.

Let D = diameter of shaft in inches.

f = maximum shearing stress in pounds per square inch.

T = twisting moment on shaft in inch pounds.

The moment of resistance of the shaft to twisting is $\frac{3\cdot1416}{16}\text{D}^3 f$; and as this must balance the twisting moment we have

$$\text{T} = \frac{3\cdot1416}{16}\text{D}^3 f = \cdot19635 \text{ D}^3 f.$$

$$\text{D} = \sqrt[3]{\frac{16\,\text{T}}{3\cdot1416 f}} = 1\cdot72\sqrt[3]{\frac{\text{T}}{f}}.$$

The moment of resistance of a *square* shaft to twisting is $\cdot208 s^3 f$, where s is the length of the side of the square. Rankine and other authorities give the moment of resistance of a square shaft to twisting as $\cdot281 s^3 f$; but Professor Cotterill, in his "Applied Mechanics," has pointed out that it ought to be that which we have given, namely, $\cdot208 s^3 f$.

EXAMPLE.—*Required the diameter of a steel shaft to transmit 20 horse-power at a speed of 100 revolutions per minute, f to be taken = 11,000.*

$$\text{T} = 63025\cdot2\,\frac{\text{H}}{\text{N}} = \frac{63025\cdot2 \times 20}{100} = 12605 \text{ inch-lbs.}$$

$$\text{D} = 1\cdot72\sqrt[3]{\frac{\text{T}}{f}} = 1\cdot72\sqrt[3]{\frac{12605}{11000}} = 1\cdot72 \times 1\cdot05$$

$$= 1\cdot8 \text{ inches} = 1\tfrac{13}{16} \text{ inches nearly.}$$

Strength of Shafts Calculated by the Formula $T = \cdot 19635 \, D^3 f$.

Diameter of Shaft in Inches.	Twisting Moment in Inch-Pounds when the Stress in Pounds per Square Inch is—					
	8,000	9,000	10,000	11,000	12,000	13,000
1	1,571	1,767	1,964	2,160	2,356	2,553
$1\frac{1}{8}$	2,237	2,516	2,796	3,075	3,355	3,634
$1\frac{1}{4}$	3,068	3,451	3,835	4,218	4,602	4,985
$1\frac{3}{8}$	4,083	4,594	5,104	5,615	6,125	6,636
$1\frac{1}{2}$	5,301	5,964	6,627	7,289	7,952	8,615
$1\frac{5}{8}$	6,740	7,583	8,425	9,268	10,110	10,953
$1\frac{3}{4}$	8,419	9,471	10,523	11,575	12,628	13,680
$1\frac{7}{8}$	10,354	11,649	12,943	14,237	15,532	16,826
2	12,566	14,137	15,708	17,279	18,850	20,420
$2\frac{1}{4}$	17,892	20,129	22,365	24,602	26,839	29,075
$2\frac{1}{2}$	24,544	27,612	30,680	33,748	36,816	39,884
$2\frac{3}{4}$	32,668	36,751	40,835	44,918	49,002	53,085
3	42,412	47,713	53,015	58,316	63,617	68,919
$3\frac{1}{2}$	67,348	75,767	84,185	92,604	101,022	109,441
4	100,531	113,098	125,664	138,230	150,797	163,363
$4\frac{1}{2}$	143,139	161,032	178,924	196,816	214,709	232,601
5	196,350	220,894	245,438	269,981	294,525	319,069
$5\frac{1}{2}$	261,342	294,010	326,677	359,345	392,013	424,681
6	339,293	381,704	424,116	466,528	508,939	551,351
7	538,784	606,132	673,481	721,194	808,177	881,525
8	804,250	904,781	1,005,312	1,105,843	1,206,374	1,306,506
9	1,145,113	1,288,252	1,431,392	1,574,531	1,717,670	1,860,809
10	1,570,800	1,767,150	1,963,500	2,159,850	2,356,200	2,552,550
11	2,090,735	2,352,077	2,613,418	2,874,760	3,136,102	3,397,444
12	2,714,342	3,053,635	3,392,928	3,732,221	4,071,514	4,410,806
13	3,451,048	3,882,429	4,313,810	4,745,190	5,176,571	5,607,952
14	4,310,275	4,849,060	5,387,840	5,926,628	6,465,413	7,004,197
15	5,301,450	5,964,131	6,626,813	7,289,494	7,952,175	8,614,563
16	6,433,997	7,238,246	8,042,496	8,846,746	9,650,995	10,455,245
17	7,717,340	8,682,008	9,646,676	10,611,343	11,576,011	12,540,678
18	9,160,906	10,306,019	11,451,132	12,596,245	13,741,358	14,886,472
19	10,774,117	12,120,882	13,467,647	14,814,411	16,161,176	17,507,940
20	12,566,400	14,137,200	15,708,000	17,278,800	18,849,600	20,420,400

For cast-iron shafts the working stress may be taken at from 4000 to 6000. For wrought-iron the stress should not exceed 8000 or 9000; and for steel such as is generally used for shafts the stress may be taken at from 10,000 to 13,000.

135. Resistance of a Shaft to Combined Twisting and Bending.—If a shaft is subjected to a twisting moment T, and has to support at the same time a bending moment B, these two together will produce the same effect as a twisting moment T_1 given by the formula,

$$T_1 = B + \sqrt{(B^2 + T^2)}.$$

The twisting moment T_1 is called the *equivalent twisting moment*, and this must be used in determining the diameter of the shaft.

EXAMPLE.—*Required the diameter of a wrought-iron shaft which has to*

SHAFTING AND SHAFT COUPLINGS.

support a twisting moment of 480,000 *inch-pounds and a bending moment of* 360,000 *inch-pounds; the working stress to be* 9000 *lbs. per square inch.*

Here T = 480,000 and B = 360,000.

$$T_1 = B + \sqrt{B^2 + T^2} = 360,000 + \sqrt{360,000^2 + 480,000^2}$$
$$= 360,000 + 600,000 = 960,000 \text{ inch-pounds.}$$

If D is the diameter of the shaft, then

$$D = 1\cdot72 \sqrt[3]{\frac{T_1}{f}} = 1\cdot72 \sqrt[3]{\frac{960,000}{9000}} = 8\cdot16 \text{ inches.}$$

If the size of the shaft was calculated for the twisting moment only, we would get its diameter to be $1\cdot72 \sqrt[3]{\frac{480,000}{9000}} = 6\cdot47$ inches. This shows the importance of taking into account the bending action on the shaft.

The formula $T_1 = B + \sqrt{(B^2 + T^2)}$ may be put in another form thus: Let $\frac{B}{T} = p$, then $T_1 = pT + \sqrt{p^2T^2 + T^2}$; that is, $T_1 = T(p + \sqrt{p^2 + 1})$. If $q = p + \sqrt{p^2 + 1}$, then $T_1 = qT$. Again, if D is the diameter of a shaft suitable for the twisting moment, T and D_1 is the diameter required for the twisting moment T_1, then $D_1 = D \sqrt[3]{q}$.

The following table gives the values of q and $\sqrt[3]{q}$ for various values of p:—

p.	q.	$\sqrt[3]{q}$.	p.	q.	$\sqrt[3]{q}$.	p.	q.	$\sqrt[3]{q}$.
·2	1·220	1·068	1·2	2·762	1·403	2·2	4·617	1·665
·4	1·477	1·139	1·4	3·120	1·461	2·4	5·000	1·710
·6	1·766	1·209	1·6	3·487	1·516	2·6	5·386	1·753
·8	2·081	1·277	1·8	3·859	1·569	2·8	5·773	1·794
1·0	2·414	1·342	2·0	4·236	1·618	3·0	6·162	1·833

136. Hollow Shafts.—It has already been pointed out that the stress induced by the twisting moment on a shaft is nothing at the centre, and varies uniformly to a maximum at the circumference. It follows that the material of a shaft which is near the circumference is much more effective than that which is near the centre, and therefore that the shaft will be much stronger for the same weight of material by making it hollow. Let D_1 be the external and D_2 the internal diameter of a hollow shaft. The moment of resistance of a solid shaft of diameter D_1 is $\frac{3\cdot1416}{16} D_1^3 f$. Now, if a portion of diameter D_2 be removed from the centre of this shaft, the moment of resistance of the remainder will be equal to the moment of resistance of the solid shaft of diameter D_1 diminished by the moment of resistance of the part of diameter D_2. But the moment of resistance of the part of diameter D_2 is not the same when this part is in the interior of another shaft, as it is when it forms a

G

separate solid shaft, because the stress at its circumference is less in the former case than in the latter. If f is the stress at the outer circumference of the hollow shaft, then the stress at the inner circumference is $f \dfrac{D_2}{D_1}$, and therefore the moment of resistance of the inner part of diameter D_2 *before it is removed* is $\dfrac{3 \cdot 1416}{16} D_2^3 f \dfrac{D_2}{D_1}$; that is, $\dfrac{3 \cdot 1416 D_2^4}{16 D_1} f$. Hence the moment of resistance of the hollow shaft is

$$\frac{3 \cdot 1416}{16} D_1^3 f - \frac{3 \cdot 1416 D_2^4}{16 D_1} f = \frac{3 \cdot 1416}{16} \left(\frac{D_1^4 - D_2^4}{D_1} \right) f.$$

If D is the diameter of a solid shaft, and D_1 and D_2 the diameters of a hollow shaft, then the relation between the weights of these two shafts, if they are of the same material, is as follows:—

$$\frac{\text{Weight of hollow shaft}}{\text{Weight of solid shaft}} = \frac{D_1^2 - D_2^2}{D^2},$$

and the relation between their strengths is

$$\frac{\text{Strength of hollow shaft}}{\text{Strength of solid shaft}} = \frac{D_1^4 - D_2^4}{D_1 D^3}.$$

If the solid and hollow shafts are of equal weight, then $D^2 = D_1^2 - D_2^2$.

If the shafts are of equal strength, then $D^3 = \dfrac{D_1^4 - D_2^4}{D_1}$.

Large hollow shafts are forged hollow, but the smaller sizes are first forged solid and then bored out. Not only is the weight of a solid shaft reduced by removing the central part without reducing the strength to the same extent, but the central part of a solid shaft is liable to be defective, as in forging the hammer acts more on the outside than on the centre. The removal of the central portion tends also to remove the initial stress and strain which there may be in a shaft through unequal cooling or through unequal working in forging.

A hollow shaft has greater stiffness, and yields less to bending action, than a solid shaft having the same resistance to twisting. This is an objection in the case of the propeller shaft of a steamship, where the bearings are liable to get slightly out of line by the straining of the ship in rough weather.

137. Angle of Torsion.—If a shaft of diameter D and length L is subjected throughout its entire length to a twisting moment T, one end will rotate in advance of the other by an angle θ given by the formula—

$$\theta = \frac{32\, TL}{3 \cdot 1416\, CD^4} = 10 \cdot 186 \frac{TL}{CD^4}$$

where C is the modulus of transverse elasticity of the material of the shaft. The above formula gives the angle of torsion in circular measure. If n is the number of degrees in the angle of torsion, then—

$$n = \frac{180 \times 32\, TL}{3 \cdot 1416^2\, CD^4} = 583 \cdot 61 \frac{TL}{CD^4}.$$

The value of C may be taken at from 9,000,000 to 10,500,000 for wrought-iron, and at 11,000,000 for steel.

Mr. D. K. Clark gives as the working limit of the angle of torsion one degree for a length equal to twenty diameters. Applying this to the formula above, we get the following relations—

$$CD^3 = 11672 \cdot 2 \, T$$

$$T = \frac{CD^3}{11672 \cdot 2}$$

$$D = \sqrt[3]{\frac{11672 \cdot 2 \, T}{C}}.$$

If in the formula $n = 583 \cdot 61 \dfrac{TL}{CD^4}$ we substitute the value T from the equation $T = \dfrac{3 \cdot 1416}{16} D^3 f$ already given, we get the formula $n = 114 \cdot 59 \dfrac{fL}{CD}$, and if we make $L = 20\,D$ and substitute the ordinary values of f which are generally taken in calculating the diameter of a shaft, we find that the angle of twist is about double that given by Mr. Clark as the working limit of twist.

For a hollow shaft having an external diameter D_1 and an internal diameter D_2 the angle of torsion is—

$$\theta = 10 \cdot 186 \, \frac{TL}{C(D_1^4 - D_2^4)} \text{ in circular measure};$$

$$n = 583 \cdot 61 \, \frac{TL}{C(D_1^4 - D_2^4)} \text{ in degrees.}$$

It should be remembered that the degree of stiffness which a shaft requires depends on the class of work which it has to do, and if it is known that a shaft of diameter D and length L has sufficient stiffness when transmitting H horse-power at a given speed, then another shaft of the same material having a diameter d and length l will have the same stiffness when transmitting h horse-power at the same speed if the following relation between the different quantities is satisfied, viz. :—

$$\frac{HL}{D^4} = \frac{hl}{d^4}, \text{ or } d = D \sqrt[4]{\frac{hl}{HL}}.$$

138. Span between Bearings of Shafts.—Let D be the diameter of the shaft in inches and S the span between the bearings in feet. Then

$$S = C \sqrt[3]{D^2}$$

where C is equal to from 5 to 6 for shafts which carry their own weight only, and from 4·5 to 5 for shafts which give off power by an ordinary number of pulleys or wheels between the supports.

Diameter in Inches.	1	1½	2	2½	3	4	5	6	7	8
Span in feet C=6	6	7·86	9·52	11·05	12·48	15·12	17·54	19·81	21·95	24
Span in feet C=4·5	4·5	5·10	7·14	8·29	9·36	11·34	13·16	14·86	16·47	18

139. Box or Muff Couplings.—The coupling illustrated in Figs. 255 and 256 consists of a solid box or muff, made of cast-iron, bored out to fit the shafts, whose ends are made to butt together inside the box. The box may be secured to the shafts by means of a sunk key which extends the whole length of the box. A better arrangement is to use two keys, both driven from the same end of the box, as shown in Fig. 255. With two keys it is not so important that the keyways in the shafts have exactly the same depth. Two keys may be more tightly driven in, and

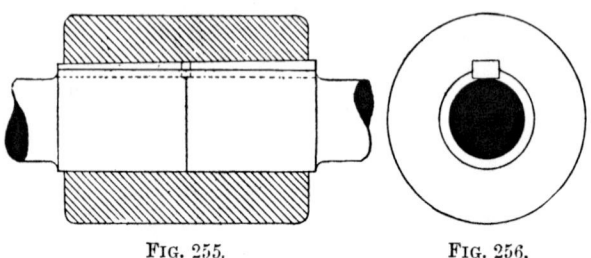

FIG. 255.   FIG. 256.

are easier to drive back than a single key of the same total length. There must, however, be a clearance space between the head of the forward key, and the point of the hind one, so as to ensure that the latter is driven in with the same degree of tightness as the former. In driving the keys back also, they can be started separately, and therefore more easily, when this clearance space is allowed. Fig. 255 shows the ends of the shafts enlarged where the keyways are cut, so that the latter do not weaken the shafts, the amount of the enlargement being such that the bottom of the keyway touches the outside of the shaft.

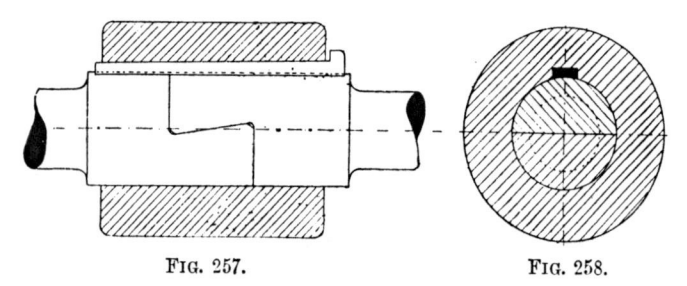

FIG. 257.   FIG. 258.

The half-lap coupling, first introduced by Sir William Fairbairn, is shown in Figs. 257 and 258. The lap-joint slopes slightly, so as to prevent the shafts being separated so long as the box covers the lap. The box is kept in position by means of a saddle-key. The portion of the shaft within the box should be enlarged so as to make up for the loss of strength due to the cutting away of the material to form the lap-joint. If D is the diameter of the shaft, and D_1 the diameter of the enlarged part, then $D_1 = D \sqrt[3]{2}$ very nearly, or $D_1 = 1\cdot 26\ D$.

SHAFTING AND SHAFT COUPLINGS.

The half-lap coupling is expensive when properly made, and it is now seldom used.

Dimensions of Solid Box Couplings.

D = diameter of shaft.
D_1 = diameter of shaft at lap.
l = length of lap.
T = thickness of metal in box.
L = length of box for butt coupling.
L_1 = length of box for lap coupling.

D.	$1\frac{1}{2}$	2	$2\frac{1}{2}$	3	$3\frac{1}{2}$	4	$4\frac{1}{2}$	5	$5\frac{1}{2}$	6
D_1	$2\frac{5}{16}$	3	$3\frac{11}{16}$	$4\frac{3}{8}$	$5\frac{1}{16}$	$5\frac{3}{4}$	$6\frac{7}{16}$	$7\frac{1}{8}$	...	...
l	$1\frac{7}{16}$	$1\frac{7}{8}$	$2\frac{5}{16}$	$2\frac{3}{4}$	$3\frac{3}{16}$	$3\frac{5}{8}$	$4\frac{1}{16}$	$4\frac{1}{2}$	...	...
T	$1\frac{1}{8}$	$1\frac{5}{16}$	$1\frac{1}{2}$	$1\frac{11}{16}$	$1\frac{15}{16}$	$2\frac{1}{8}$	$2\frac{5}{16}$	$2\frac{1}{2}$	$2\frac{3}{4}$	$2\frac{15}{16}$
L	$5\frac{3}{4}$	7	$8\frac{1}{4}$	$9\frac{1}{2}$	$10\frac{3}{4}$	12	$13\frac{1}{4}$	$14\frac{1}{2}$	$15\frac{3}{4}$	17
L_1	$4\frac{5}{8}$	$5\frac{1}{4}$	$6\frac{3}{8}$	$7\frac{1}{2}$	$8\frac{5}{8}$	$9\frac{3}{4}$	$10\frac{7}{8}$	12	...	...

140. Split Muff Couplings.—This form of coupling, which is shown in Figs. 259 and 260, is very easily put on or taken off; it has no projecting parts, the bolts being completely covered, and, like the solid muff couplings, it may be used as a driving pulley, or a driving pulley may be placed on it. In making this coupling the faces for the joint

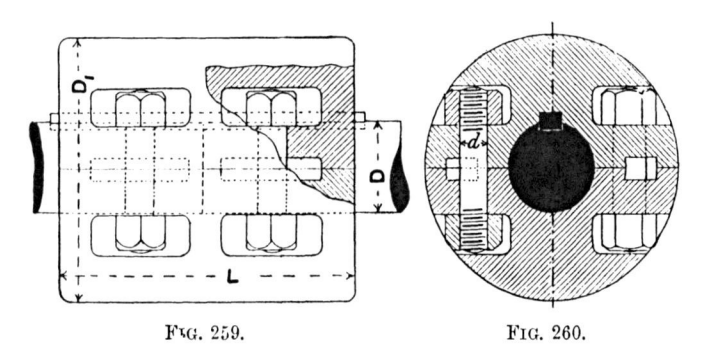

FIG. 259. FIG. 260.

between the two halves of the box are first planed. The bolt-holes are then drilled and the two halves bolted together with pieces of paper between them; then the muff is bored out to the exact size of the shaft. When the paper is removed and the box put on the shaft and bolted up, the box grips the shaft firmly. The key has no taper and should fit on the sides only.

Dimensions of Split Muff Couplings.

Diameter of shaft, D . .	$1\frac{1}{2}$	$1\frac{3}{4}$	2	$2\frac{1}{4}$	$2\frac{1}{2}$	$2\frac{3}{4}$	3	$3\frac{1}{2}$	4
Diameter of Box, D_1 . .	$4\frac{1}{2}$	$5\frac{1}{4}$	$5\frac{3}{4}$	6	$6\frac{1}{2}$	$7\frac{1}{4}$	$7\frac{7}{8}$	$9\frac{3}{4}$	$10\frac{1}{4}$
Length of Box, L . .	6	7	8	9	10	11	12	14	16
Diameter of Bolts, d . .	$\frac{1}{2}$	$\frac{5}{8}$	$\frac{5}{8}$	$\frac{5}{8}$	$\frac{3}{4}$	$\frac{7}{8}$	$\frac{7}{8}$	$\frac{7}{8}$	$\frac{7}{8}$
Number of Bolts . . .	4	4	4	4	4	4	4	6	6

141. Cast-Iron Flange Coupling.—In this form of coupling, which is shown in Fig. 261, there are cast-iron centre pieces or bosses, provided with flanges which are keyed to the ends of the shafts to be connected. These flanges are fastened together by means of bolts and nuts as shown. Sometimes the shaft is enlarged where it enters the coupling, so as to allow for the weakening effect of the keyway, but more frequently it is parallel throughout, or very slightly reduced, as shown in Fig. 261, so as to form a shoulder, which prevents the shaft going farther into the coupling. The key, being driven in from the inside, prevents the shaft coming out of the coupling if it is pulled in the direction of its length.

To ensure that the face of each flange is exactly perpendicular to the axis of the shaft upon which it is keyed, it should be "faced" in the lathe after being keyed to the shaft. As a precaution against the shafts being out of line, the end of one may enter into the flange of the other.

If the bolts are as strong as the shaft, and they are made of the same material, then, if D is the diameter of the shaft, d the diameter of the bolts, C the diameter of the bolt circle, and n the number of bolts—

$$nd^2 C = \tfrac{1}{2} D^3, \text{ and } d = \sqrt{\frac{D^3}{2nC}}.$$

As C is about 2·8D, the above may be put in the simpler form:

$$d = \frac{\cdot 423 D}{\sqrt{n}}.$$

Comparing the results of this formula with actual practice, we find that the bolts have generally a diameter about $\frac{5}{16}$ inch greater than that given by the formula. Thus in practice

$$d = \frac{\cdot 423 D}{\sqrt{n}} + \cdot 3.$$

As the screwed part of the bolt has only to resist the tension due to screwing up, this part may be made smaller in diameter than the unscrewed part.

To guard against the nuts and bolt-heads catching the clothes of workmen, or an idle driving band which might be near the coupling, the flanges may be made thicker and have recesses for the nuts and bolt-heads.

Dimensions of Cast-Iron Flange Couplings.

Diameter of Shaft D.	Diameter of Flange F.	Thickness of Flange T.	Diameter of Boss B.	Depth at Boss L.	Number of Bolts.	Diameter of Bolts. d.	Diameter of Bolt Circle C.
1	$5\frac{5}{8}$	$\frac{3}{4}$	$2\frac{5}{8}$	$2\frac{1}{16}$	3	$\frac{1}{2}$	$4\frac{1}{4}$
$1\frac{1}{4}$	$6\frac{3}{16}$	$\frac{13}{16}$	$3\frac{1}{16}$	$2\frac{5}{16}$	3	$\frac{5}{8}$	$5\frac{1}{16}$
$1\frac{1}{2}$	$7\frac{1}{4}$	$\frac{7}{8}$	$3\frac{1}{2}$	$2\frac{5}{8}$	3	$\frac{5}{8}$	$5\frac{1}{2}$
$1\frac{3}{4}$	$7\frac{11}{16}$	1	$3\frac{15}{16}$	$2\frac{7}{8}$	4	$\frac{5}{8}$	$5\frac{13}{16}$
2	$8\frac{5}{8}$	$1\frac{1}{16}$	$4\frac{3}{8}$	$3\frac{3}{16}$	4	$\frac{3}{4}$	$6\frac{3}{4}$
$2\frac{1}{4}$	$9\frac{1}{16}$	$1\frac{1}{8}$	$4\frac{3}{4}$	$3\frac{7}{16}$	4	$\frac{3}{4}$	$7\frac{1}{8}$
$2\frac{1}{2}$	$10\frac{9}{16}$	$1\frac{1}{4}$	$5\frac{1}{16}$	$3\frac{3}{4}$	4	$\frac{7}{8}$	$8\frac{1}{8}$
$2\frac{3}{4}$	11	$1\frac{5}{16}$	$5\frac{3}{4}$	$4\frac{1}{16}$	4	$\frac{7}{8}$	$8\frac{7}{16}$
3	$12\frac{3}{8}$	$1\frac{7}{16}$	$6\frac{1}{4}$	$4\frac{7}{16}$	4	1	$9\frac{1}{2}$
$3\frac{1}{4}$	$12\frac{5}{8}$	$1\frac{1}{2}$	$6\frac{5}{8}$	$4\frac{5}{8}$	4	1	$9\frac{13}{16}$
$3\frac{1}{2}$	$13\frac{1}{4}$	$1\frac{5}{8}$	$7\frac{5}{8}$	$4\frac{7}{8}$	4	1	$10\frac{5}{16}$
$3\frac{3}{4}$	$13\frac{9}{16}$	$1\frac{11}{16}$	$7\frac{9}{16}$	$5\frac{3}{16}$	4	1	$10\frac{3}{4}$
4	14	$1\frac{3}{4}$	8	$5\frac{3}{16}$	6	1	$11\frac{1}{4}$
$4\frac{1}{4}$	$14\frac{7}{16}$	$1\frac{7}{8}$	$8\frac{7}{16}$	$5\frac{3}{4}$	6	1	$11\frac{5}{8}$
$4\frac{1}{2}$	$15\frac{5}{8}$	2	$8\frac{7}{8}$	6	6	$1\frac{1}{8}$	$12\frac{1}{2}$
$4\frac{3}{4}$	$16\frac{1}{8}$	$2\frac{1}{16}$	$9\frac{3}{8}$	$6\frac{5}{16}$	6	$1\frac{1}{8}$	13
5	$17\frac{5}{16}$	$2\frac{1}{8}$	$9\frac{13}{16}$	$6\frac{5}{8}$	6	$1\frac{1}{4}$	$13\frac{13}{16}$
$5\frac{1}{4}$	$17\frac{3}{4}$	$2\frac{1}{4}$	$10\frac{1}{4}$	$6\frac{7}{8}$	6	$1\frac{1}{4}$	$14\frac{1}{16}$
$5\frac{1}{2}$	$18\frac{3}{16}$	$2\frac{5}{16}$	$10\frac{3}{4}$	$7\frac{1}{4}$	6	$1\frac{1}{4}$	$14\frac{11}{16}$
$5\frac{3}{4}$	$19\frac{1}{2}$	$2\frac{7}{16}$	$11\frac{1}{4}$	$7\frac{7}{16}$	6	$1\frac{3}{8}$	$15\frac{5}{8}$
6	$19\frac{7}{8}$	$2\frac{1}{2}$	$11\frac{5}{8}$	$7\frac{3}{4}$	6	$1\frac{3}{8}$	16

For the dimensions of the keys for the above see p. 72.

The projection of one shaft into the flange on the other varies from $\frac{1}{4}$ inch in small shafts to $\frac{1}{2}$ inch in large ones.

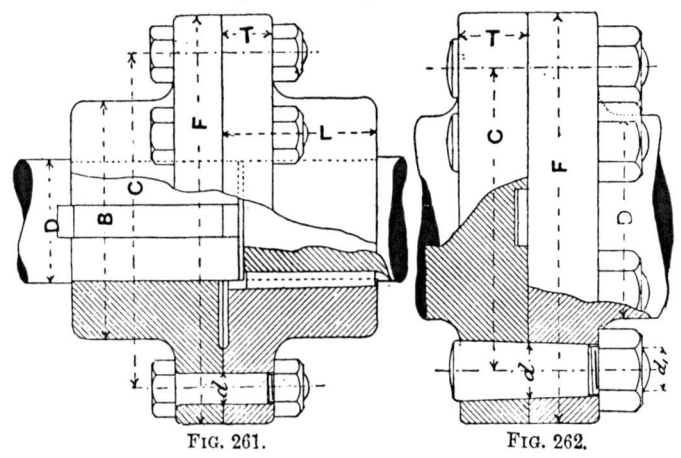

FIG. 261.   FIG. 262.

142. Marine or Solid Flange Coupling.—The form of coupling shown in Fig. 262, in which flanges are forged on the ends of the shaft, is called a marine coupling, from the fact that it is practically the only form of coupling used for the shafts of marine engines. The bolts for connecting the flanges are usually tapered in couplings for large shafts. Tapered bolts are sometimes made with, and often without heads. To ensure that the different lengths of shaft are in line, there is sometimes a projecting piece formed on the centre of one flange which fits into a corresponding recess in the centre of the other, as shown in Fig. 262. This is, however, a precaution of less importance in marine couplings, because the bolts are fitted tightly into the bolt-holes, and the flanges being solid with the shaft, have no chance of shifting. Another method of putting the lengths of shafting in line is shown in Fig. 263. Here the flanges are recessed to an equal depth to receive a steel disc. In a steamship of the P. & O. Company with a shaft 19 inches in diameter this disc is 6 inches in diameter and $1\frac{1}{2}$ inch thick. Many engineers prefer to recess the flanges as shown in Fig. 264, so that they come in contact at the outer portions of their faces only.

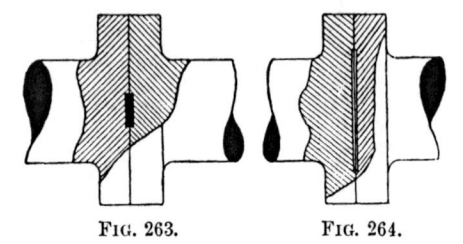

FIG. 263. FIG. 264.

Occasionally a cross-key is fitted in between the flanges, being sunk half into each for the purpose of diminishing the shearing action on the bolts. The keyway for this key should enter a little way into at least one of the bolt-holes, and at this point the key must be hollowed out on one side to allow the bolt to pass through. This prevents the key coming out after the coupling is bolted up.

If the bolts and the shaft are made of the same material, then the former will be of equal strength to the latter when $nd^2C = \frac{1}{2}D^3$, where n is the number of bolts, d their diameter, C the diameter of the bolt circle, and D the diameter of the shaft. As a first approximation C may be taken at $1 \cdot 6D$, then

$$1 \cdot 6nd^2D = \frac{D^3}{2}, \text{ and } d = \frac{D}{\sqrt{3 \cdot 2n}}.$$

But C is more nearly equal to $D + 2d$. Using this value of C and the above approximate value of d, we get a nearer approximation to the value of d in the equation.

$$nd^2\left(D + \frac{2D}{\sqrt{3 \cdot 2n}}\right) = \frac{D^3}{2},$$

and $d = D \sqrt{\dfrac{\sqrt{3 \cdot 2n}}{2n(\sqrt{3 \cdot 2n} + 2)}}$ or $d = cD$.

The following table gives values of c for various values of n:—

n	3	4	5	6	7	8	9	10
c	·318	·283	·258	·236	·224	·207	·201	·192

SHAFTING AND SHAFT COUPLINGS.

The other parts of the marine coupling may be proportioned by the following rules:—

Number of bolts $= n = \frac{1}{3}D + 2$.
Diameter of bolt circle $= C = D + 1\frac{1}{2}d + \frac{5}{8}$.
Diameter of flange $= F = D + 3d + 1\frac{3}{8}$.
Thickness of flange $= T = \dfrac{2D + 1}{7}$.
Diameter of screwed part of bolt $= d_1 = \dfrac{7d + 1}{8}$.

Taper of bolts $= \frac{3}{8}$ inch (on the diameter) per foot of length.

143. Sellers' Cone Coupling.—This form of shaft coupling, which is shown in Figs. 265 and 266, consists of a box or muff which is cylindrical externally, but internally it is of a double conical form. This muff receives two sleeves which are turned externally to fit the inside of the muff, and bored out to fit the shaft. These sleeves are pulled together by means of three bolts, and are thus bound firmly to the muff. To permit of the sleeves gripping the shaft tightly, they are split at one of the bolt holes as shown. The bolts are square in cross section, and

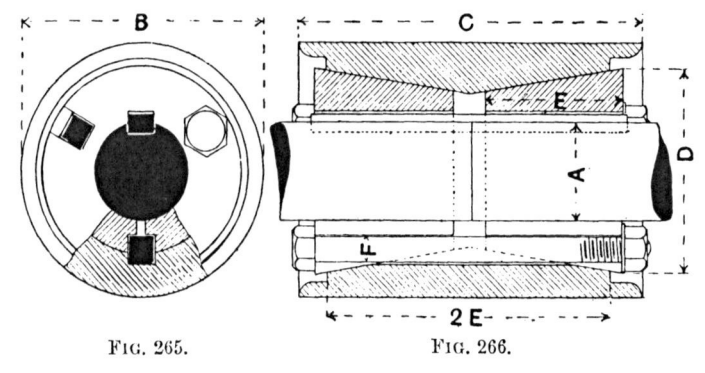

FIG. 265.   FIG. 266.

pass through holes which are slotted in the sleeves and in the muff. The friction between the sleeves and the shaft may be quite sufficient to prevent slipping, but as an additional security the sleeves are generally keyed to the shaft. In Figs. 265 and 266 the key shown has no taper, and it fits on the sides only.

Dimensions of Sellers' Cone Couplings.

A	$1\frac{1}{2}$	$1\frac{3}{4}$	2	$2\frac{1}{4}$	$2\frac{1}{2}$	$2\frac{3}{4}$	3	$3\frac{1}{2}$	4	5	6
B	$4\frac{1}{2}$	$5\frac{1}{4}$	$6\frac{1}{8}$	$6\frac{1}{2}$	$7\frac{1}{4}$	$7\frac{5}{8}$	$8\frac{1}{2}$	$9\frac{3}{4}$	11	$12\frac{7}{8}$	$14\frac{1}{2}$
C	$5\frac{5}{8}$	$6\frac{5}{8}$	$7\frac{3}{4}$	$8\frac{1}{4}$	$9\frac{1}{4}$	$10\frac{1}{4}$	$11\frac{3}{4}$	$13\frac{1}{4}$	$14\frac{7}{8}$	$18\frac{1}{4}$	$21\frac{1}{4}$
D	$3\frac{1}{2}$	4	$4\frac{3}{4}$	5	$5\frac{5}{8}$	6	$6\frac{3}{4}$	$7\frac{5}{8}$	$8\frac{1}{4}$	$10\frac{1}{4}$	$11\frac{1}{4}$
E	$2\frac{1}{4}$	$2\frac{5}{8}$	3	$3\frac{3}{8}$	$3\frac{3}{4}$	$4\frac{1}{4}$	$4\frac{1}{2}$	$5\frac{1}{4}$	6	$7\frac{1}{2}$	9
F	$\frac{7}{16}$	$\frac{1}{2}$	$\frac{5}{8}$	$\frac{5}{8}$	$\frac{3}{4}$	$\frac{3}{4}$	$\frac{7}{8}$	1	$1\frac{1}{8}$	$1\frac{1}{4}$	$1\frac{1}{4}$

Taper of conical sleeves 3 inches on the diameter per foot of length. Hence, if D = larger diameter of sleeve, d = smaller diameter, and E = length, then $d = D - \tfrac{1}{4}E$.

144. Butler's Frictional Coupling.—In Butler's frictional coupling, which is shown in Figs. 267 and 268, there is a box or muff, A, which is bored out to a double conical form to receive two conical split bushes, B B, as in Sellers' coupling; but in Butler's coupling the bushes are driven in tight like ordinary keys, and are then kept in position or locked by nuts, C C. The bushes, B B, have their split sides at opposite sides of the shaft, so that an ordinary steel set or key-driver may be intro-

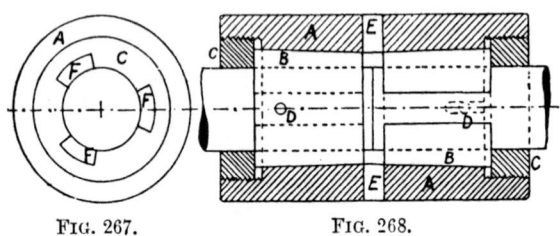

FIG. 267. FIG. 268.

duced through one of them to drive out the other after the nuts C C have been removed. To ensure that the bushes are placed in their proper positions, each is provided with a pin, D, which enters a groove on the inside of the muff. The muff has two holes, E E, in it, and when the workman sees through these that the shafts meet, he knows that the coupling is in its correct position on the shaft. The locknuts are rotated by a special form of spanner, which has projecting pieces which fit into two of the recesses F F F. There are no keyways cut in the shafts. The driving-power of this coupling depends on the tightness with which the bushes, which are really saddle-keys, are driven in.

The proportions for this coupling are: diameter of muff equal to two and a quarter times the diameter of the shaft, length of muff equal to four times the diameter of the shaft.

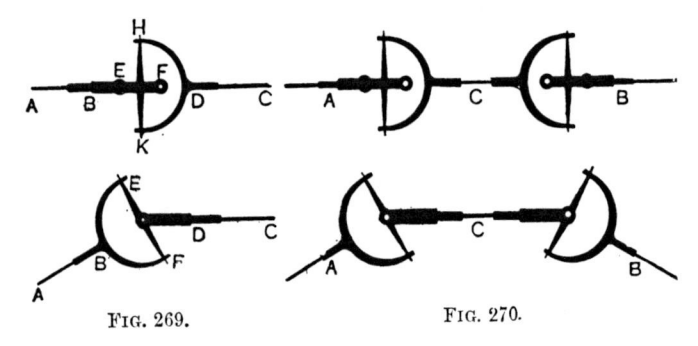

FIG. 269. FIG. 270.

145. Hooke's Joint or Universal Coupling.—This form of coupling is used to connect two shafts whose axes intersect, and it has the advantage that the angle between the shafts may be varied while they are in motion.

The diagram Fig. 269 shows the main features of this coupling. AB and CD, the shafts to be coupled, are forked at their ends. These forked ends carry between them a cross, EKFH, the arms of which are at right angles to one another. The arms of the cross are jointed to the forks so that they may turn freely about their axes.

It can be shown that the angular velocities of the shafts AB and CD will be unequal except at every quarter of a revolution. This inequality of angular velocities is greater the greater the acute angle between the shafts.

By using a double Hooke's joint, as shown in Fig. 270, the shafts A and B will have the same angular velocities, provided they make equal angles with the intermediate shaft C and are in the same plane with it.

The details of the construction of a Hooke's coupling are shown in Fig. 271, the shafts being placed for convenience in the same line.

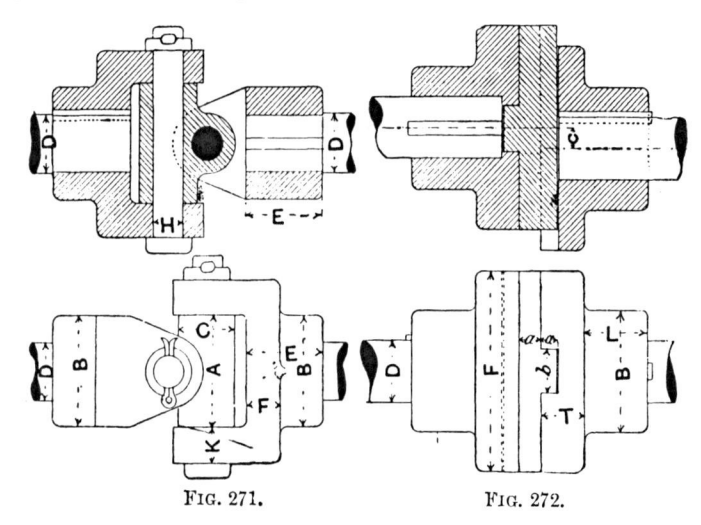

FIG. 271.   FIG. 272.

The following rules may be used in designing Hooke's coupling, all the parts being of wrought-iron :—

$A = 2\ D.$ $E = 1\cdot6\ D.$
$B = 1\cdot8\ D.$ $F = H = \cdot5\ D.$
$C = D.$ $K = \cdot6\ D.$

146. Oldham's Coupling.—This coupling may be used for connecting two shafts whose axes are parallel, but not in the same straight line. It consists of two pieces, one fixed on each shaft, with a piece between them which has on its opposite faces two rectangular projecting parts, at right angles to one another, which fit into corresponding recesses in the two former pieces. The three pieces mentioned are generally in the form of discs, as shown in Fig. 272. In this coupling the motion is transmitted by means of "sliding contact," and the angular velocities of the two shafts and the intermediate disc are equal at every instant.

The following rules may be used in designing this coupling, the discs being of cast-iron :—

Diameter of shaft = D.
Distance between centre lines of shafts = C.
Diameter of discs or flanges = F = 3D + C.
Diameter of bosses = B = 1·8D + ·8.
Length of boss = L = ·75D + ·5.
Breadth of grooves = $b = \dfrac{F}{6} = \dfrac{D}{2} + \dfrac{C}{6}$.
Depth of grooves = $a = \dfrac{b}{2} = \dfrac{D}{4} + \dfrac{C}{12}$.
Thickness of central disc = a.
Thickness of flanges = $T = 2\frac{1}{2}a$.

147. Claw Coupling.—For large and slow moving shafts, the cast-iron claw coupling shown in Figs. 273 and 274 is very simple and effective. In Fig. 273 one half of the coupling is shown fitted with a feather key,

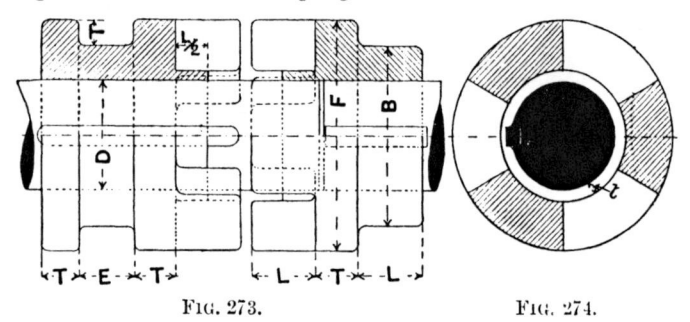

FIG. 273.   FIG. 274.

so that this half may be disconnected from the other or geared with it at pleasure. If the claws are shaped as shown in Fig. 275, the two

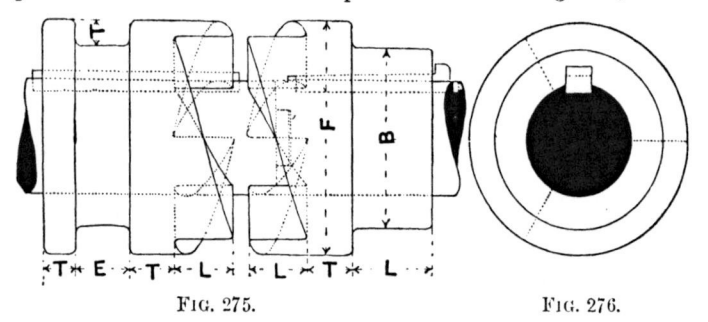

FIG. 275.   FIG. 276.

halves of the coupling are more easily put into gear, but in this case the motion must always take place in the same direction.

The following rules may be used in designing cast-iron claw couplings:—

$F = 2D + 2$ $\qquad T = \cdot 3D + \cdot 3$
$B = 1\cdot 7D + 1$ $\qquad E = \cdot 4D + \cdot 4$
$L = \cdot 5D + \cdot 5$ $\qquad t = \cdot 1D + \cdot 1$

148. Shaw's "Transmitter."—A very neat and ingenious friction coupling, being one of the forms of Shaw's power transmitter, is shown in Figs. 277 and 278. To the shaft A, a bush, B, is keyed, and encircling this bush is a coil which is a sliding fit on the bush. The head end, C, of this coil is attached to the end, E, of the casing, the latter being keyed to the shaft F. The tail end of the coil is fixed to a ring, H, which may be made to rotate through a small angle on the bush B by moving the lever K. The rotation of the ring H causes the coil to gripe the bush B, so that the coil and the bush, and therefore the casing and shafts A and F, rotate as one piece. A small motion of the ring H is sufficient to cause the coil to gripe the bush B so firmly that the full power of the shaft A may be transmitted to the shaft F.

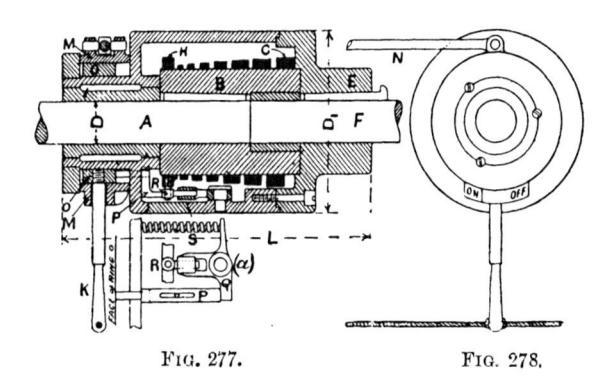

FIG. 277. FIG. 278.

The manner in which the rotation of the ring H on the bush B is obtained is as follows :—The lever K passes through an oblique slot in a ring M, which is prevented from rotating by being anchored to some fixed point by the rod N. The inner end of K is attached to a ring O, which therefore rotates when K is moved, but since K moves in a helical slot, the rotation of O is accompanied by a longitudinal motion, which is communicated to the slipper P, which, acting on one end of the bell-crank lever Q, causes a movement of the pin R, which is attached to the ring H. The bell-crank lever Q is kept against the slipper P by the spring S, as shown in detail plan (*a*).

The lubrication of the moving parts is ensured by having a quantity of a semi-fluid lubricant in the interior of the casing.

It should be noticed that this form of coupling will not drive backwards, also there is no end thrust on the shafting.

The following table gives the over all dimensions, in inches, of Shaw's transmitter: —

D	$1\frac{1}{2}$ to $2\frac{1}{4}$	$1\frac{3}{4}$ to $2\frac{3}{4}$	$2\frac{1}{4}$ to 3	$2\frac{1}{2}$ to $3\frac{1}{2}$	$2\frac{3}{4}$ to 4	3 to $4\frac{1}{2}$	$3\frac{1}{4}$ to 5	$3\frac{3}{4}$ to $5\frac{1}{2}$	$4\frac{1}{2}$ to $6\frac{1}{4}$	5 to 7	$5\frac{1}{2}$ to $7\frac{1}{2}$	6 to 8
D_1	$6\frac{5}{8}$	$7\frac{1}{2}$	$8\frac{1}{4}$	9	$9\frac{7}{8}$	$10\frac{5}{8}$	$11\frac{3}{4}$	$12\frac{3}{8}$	$13\frac{7}{8}$	$15\frac{1}{4}$	$16\frac{5}{8}$	$18\frac{1}{4}$
L	$13\frac{1}{2}$	$15\frac{5}{8}$	$17\frac{7}{16}$	$18\frac{1}{4}$	$19\frac{9}{16}$	$20\frac{1}{16}$	$22\frac{5}{16}$	$23\frac{1}{4}$	$25\frac{7}{8}$	$27\frac{3}{16}$	29	$31\frac{1}{16}$

149. Edmeston's Friction Clutch.—Figs. 279 and 280 show very fully the construction of a simple and effective friction clutch or coupling, designed and made by Messrs. Edmeston & Sons, Salford. The piece A is in the form of a ring of unequal thickness, and it is divided at its thinnest part. From the thickest part of the ring A a strong arm proceeds to a central boss which is keyed to the shaft. This ring, arm, and boss are all cast in one piece. An outer ring or shell B is bored or turned to fit easily over the ring A. On the back of the piece B is a boss which serves to carry a wheel or pulley. When it is desired to use the arrangement as a shaft coupling, the boss on the back of B is keyed to one of the shafts, the boss of A being keyed to the other. When used to take the place of fast and loose pulleys, as shown in our illustrations, the piece B rides loose on the shaft, except when it is bound to the ring A. The pieces A and B are bound together by the expansion of the former caused by the rotation of right and left-handed screws working in two nuts which fit into sockets in the ring A, one on each side of the line of division. The screw is rotated by pushing the sliding boss C along the shaft. This movement of C is communicated to the screw by means of the link D and lever E. The lever E is secured to the middle portion of the screw by a grooved key, which is held by a set-screw. At the back of this key there is a clearance space sufficient to allow of the key being withdrawn clear of the grooves, so that the lever may be turned on the screw into another position, and so take up the wear of the screw threads.

It will be observed that the pressure between the two main parts of this clutch acts in such a way that there is an entire absence of end thrust on the shaft. The makers guarantee that this clutch will transmit at least half the full working power of the shaft.

150. Mather and Platt's Friction Clutch.—In the friction clutch of Messrs. Mather & Platt, of Manchester, and which is shown in Figs. 281 and 282, there is a sleeve, A, which is keyed to the shaft and also attached to the friction band, B, by means of bolts, as shown. The friction band, B, encircles the boss, C, of the wheel or pulley, the latter being loose on the sleeve, A. The band, B, may be bound to, or released from, the boss, C, by the action of right and left-handed screws controlled by the sliding sleeve, D, and levers, E. It is evident that the band, B, moves with the shaft, and that the wheel or pulley will only rotate when the band, B, is tight on the boss, C. When used as a shaft coupling the

SHAFTING AND SHAFT COUPLINGS.   111

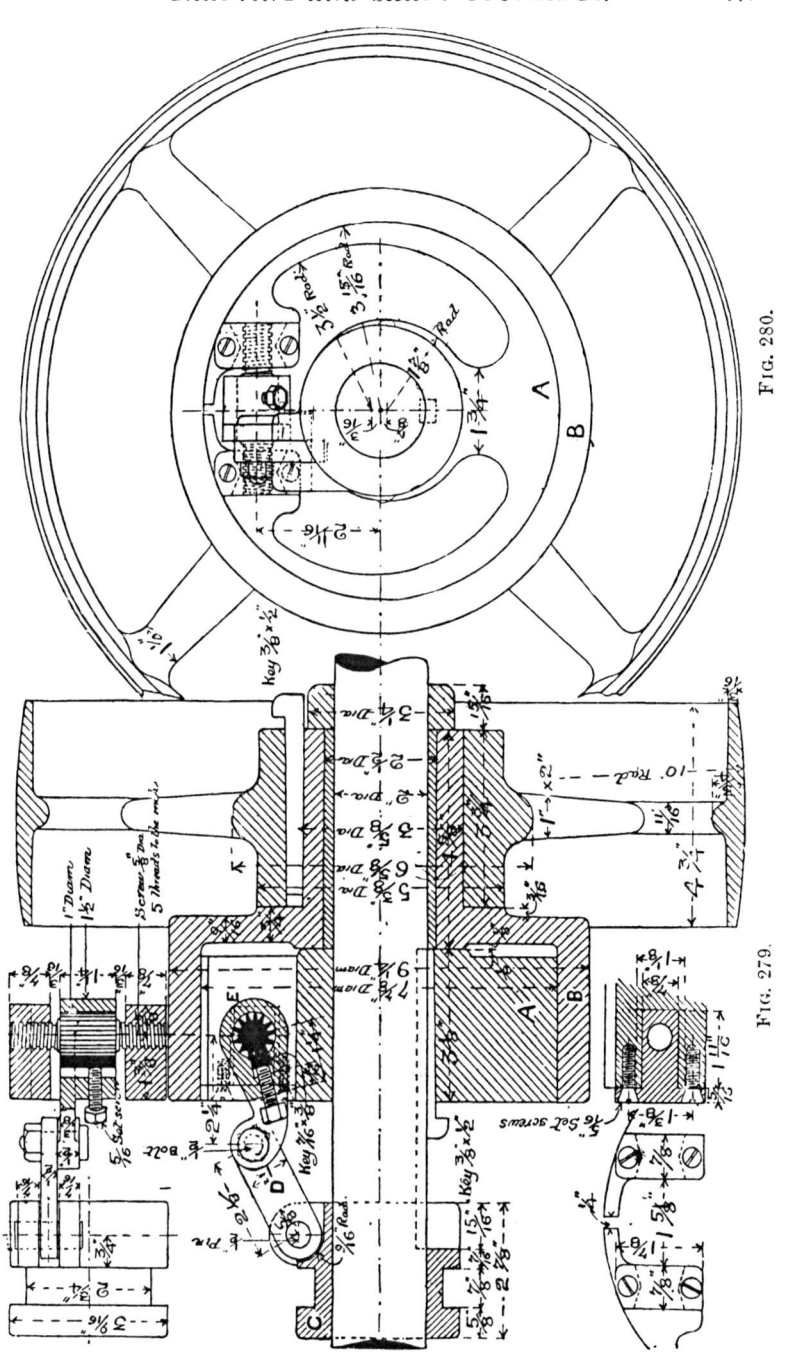

FIG. 280.

FIG. 279.

boss, C, is prolonged and keyed to one of the shafts, the sleeve, A, being keyed to the other.

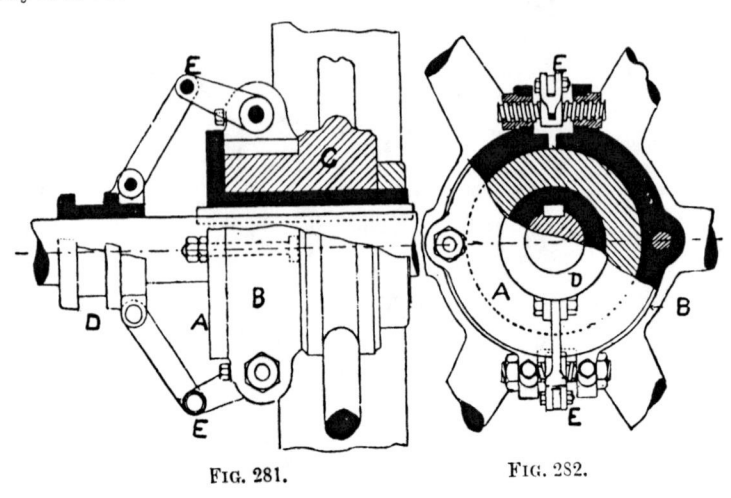

FIG. 281.

FIG. 282.

151. Conical Friction Coupling.—The friction coupling shown in Fig. 283, although simpler in construction than those which have already been described, is open to the objection that it requires a much greater force to put it into gear, and also this force, acting parallel to the axis of the shaft, causes an objectionable end thrust. The mean diameter of the conical part may be from four to eight times the diameter of the shaft, being larger the greater the amount of power which the coupling has to transmit. The inclination α of the slant side of the cone to its axis may vary from four degrees to ten degrees. The other proportions may be as follows:—

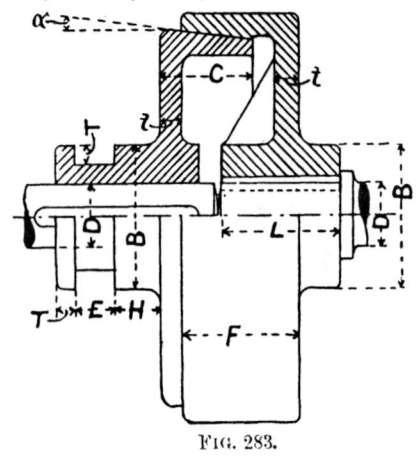

FIG. 283.

$B = 2 D + 1.$
$C = 1\cdot 5 D.$
$F = 1\cdot 8 D.$
$H = \cdot 5 D.$

$E = \cdot 4 D + \cdot 4.$
$T = \cdot 3 D + \cdot 3.$
$t = \cdot 2 D + \cdot 1.$
$L = 2 D.$

152. Shifting Gear for Clutch Couplings.—The most common way of putting clutch couplings in or out of gear is by means of a forked

SHAFTING AND SHAFT COUPLINGS.

lever, the prongs of which enter the groove on the sliding piece of the coupling, as shown in Fig. 284. The lever is generally worked directly by hand, but a screw and hand wheel are sometimes used, the screw working in a nut on the end of a lever. To increase the wearing surface the forked end of the lever is fitted with blocks made of brass or cast-iron, as shown in Fig. 285. To still further increase the wearing surface a strap such as is shown in Fig. 286 is often used. This strap is gener-

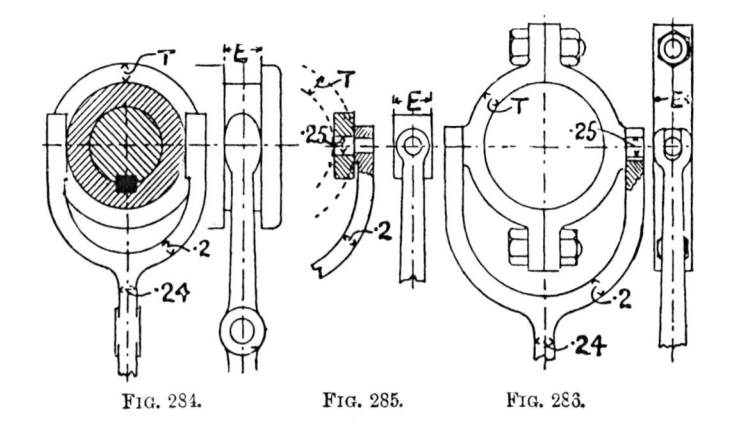

FIG. 284.　　FIG. 285.　　FIG. 286.

ally made of brass, and completely fills the groove on the sliding portion of the coupling.

In designing the arrangements shown in Figs. 284, 285, and 286, take as a unit the diameter of the shaft in inches plus half an inch, and multiply this unit by the dimensions marked on the figures. The dimensions E and T must be made to suit the groove on the coupling, rules for which have already been given.

153. Alley's Flexible Coupling.—As is well known, the shafts of screw and paddle steamers are often subjected to great strain through the bearings getting slightly out of line by the straining of the ship's hull in rough weather, and in designing these shafts it is clear that an allowance must be made for this. It is also evident that if the bearings of a rigid shaft are not kept in a straight line, or if they are kept in a straight line by the rigidity of the shaft, a greater amount of power will be required to overcome the increased friction. The shaft coupling of Mr. Stephen Alley is designed to allow of the shaft bearings getting slightly out of line without throwing any extra strain on the shaft or causing any increase in the journal friction. In this coupling, shown in Fig. 287, the shaft A has, projecting from the centre of its flange, a piece, B, which is part of a sphere. The face of the piece B has turned on it a blunt point, C, which rests against the centre of the flange on the end of the shaft D, and transmits the thrust when the engines are going ahead. Surrounding the piece B is a ring, E, turned to fit the spherical part of B as shown. This ring E is cut in halves

H

along a diameter, and it is bolted to the flange on the shaft D by bolts with tapered or countersunk heads. The concave part of the ring E takes up the pull when the engines are going astern. The heads of the bolts are prolonged to form barrel-shaped pins, F, which enter into holes in the flange on the shaft A, and thus act as drivers to transmit the twisting moment on the shaft. The holes in the flange on A are lined with hard steel bushes, and the pins F are case-hardened and made an easy fit in the bushes. Between the ring E and the flange on A there is a space, which, in a shaft 17 in. in diameter, amounts to one-eighth of an inch. This would allow of a deflection of 2 in. in a shaft 20 ft. long. The pins F have a diameter 1·58 times the diameter of the ordinary coupling bolts.

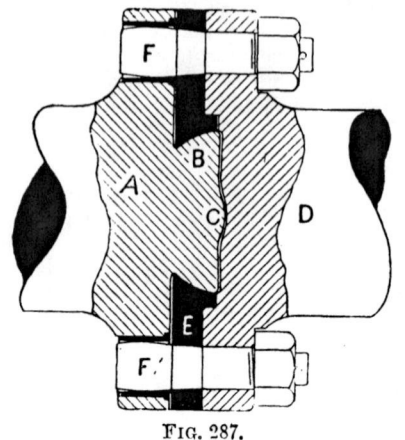

FIG. 287.

CHAPTER IX.

SUPPORTS FOR SHAFTS

154. Journal, Pivot, and Collar Bearings.—The bearings for a shaft or other rotating piece must be of such a form as to permit of the free rotation of the shaft or rotating piece. If the shaft has a rotary motion only, it is clear that the surface of the bearing and the corresponding surface of the shaft must be surfaces of revolution. In this chapter we will consider bearings for shafts which have rotary motion only.

The principal consideration which decides the form of a shaft bearing is the direction of the pressure on it. If the pressure on the bearing is perpendicular to the axis of the shaft, the bearing is a *journal bearing*. If the pressure is parallel to the axis of the shaft, and the shaft terminates at the bearing surface, the bearing is a *pivot bearing*. If the pressure is parallel to the axis of the shaft and the shaft is continued through the bearing, the latter is a *collar bearing*. Sketches showing these various

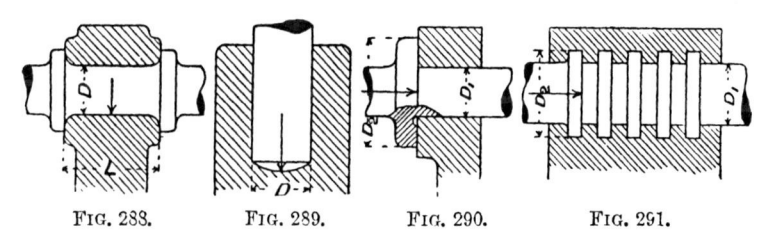

FIG. 288.   FIG. 289.   FIG. 290.   FIG. 291.

forms of bearings are given above. Fig. 288 represents a journal bearing, Fig. 289 a pivot bearing, Fig. 290 a collar bearing in which there is only one collar on the shaft, and Fig. 291 a collar bearing in which there are several collars on the shaft. The arrows show the direction of the pressure on the bearing in each case.

155. Area of a Bearing.—By the area of the surface of a bearing is meant the area of its projection on a plane perpendicular to the direction of the pressure. This is also called the "projected area."

For a cylindrical journal bearing of diameter D and length L the projected area is DL.

For a pivot bearing of diameter D the projected area is $.7854\, D^2$.

For a collar bearing having N collars of inside diameter D_1 and outside diameter D_2 the projected area is $.7854(D_2^2 - D_1^2)N$.

In all calculations on the area of the surface of a bearing the projected area only is considered.

156. Intensity of Pressure on Bearings.—If R is the total load on a journal bearing of diameter D and length L, and p is the mean intensity of pressure, then $p = \dfrac{R}{DL}$.

For a pivot bearing of diameter D and carrying a total load R, $p = \dfrac{R}{\cdot 7854 D^2}$.

For a collar bearing with N collars of diameters D_1 and D_2, and carrying a total load R, $p = \dfrac{R}{\cdot 7854 (D_2^2 - D_1^2) N}$.

The magnitude of p varies greatly in different cases in practice. It is generally smaller the greater the speed; and, in cases where the load is intermittent or changes from one side of the bearing to the other during each revolution, the intensity of the pressure may be greater than in cases where the load is a steady load. With a changing load the lubricant has generally a better chance of getting in between the shaft and the bearing

The pressure on a lubricated journal bearing is not uniform over the circumference, but varies from a maximum near the point where the line of action of the load cuts the surface, to nothing at points 90 degrees from this, that is, at the points where the bearing is usually divided. Neither is the pressure uniform over the length of the bearing, but it does not vary so much in this direction as in the circumferential direction. The pressure is greatest at the middle of the length.

The maximum intensity of pressure p on the main journal bearings of steam-engines, in pounds per square inch, is 600 for slow-going engines and 400 for fast engines, but where space will permit it is desirable to make the bearings of such a length that the pressure is from 200 to 300 lbs. per square inch.

For railway axle bearings p varies from 160 to 300 lbs. per square inch. According to Mr. Joseph Tomlinson the pressure on railway axle bearings should not exceed 280 lbs. per square inch.

The rule used by Sellers in designing cast-iron journal bearings for factory shafts is $p = 15$ lbs. per square inch.

For collar bearings such as are used for taking the thrust in propeller shafts, $p = 50$ to 70 lbs. per square inch.

For pivot bearings running at moderate speeds Reuleaux gives the following values of p in pounds per square inch:—

Wrought-iron pivot on gun-metal bearing	$p = 700$
Cast-iron pivot on gun-metal bearing	$p = 470$
Wrought-iron pivot on lignum vitæ bearing	$p = 1400$

The information available on the safe pressure for pivot bearings is, however, of a very conflicting nature, and there can be no doubt that it is often overstated. The experiments carried out by a committee of the Institution of Mechanical Engineers seem to prove that for pivots which have to run continuously the pressure should not exceed 250 lbs. per square inch.

SUPPORTS FOR SHAFTS.

157. Solid Journal Bearings.—The simplest form of journal bearing, for an axle or shaft of a machine, is made by boring a hole through the framing of the machine to receive the shaft as shown in Fig. 292. End motion of the shaft is prevented by two collars, one of which may be forged on the shaft, but the other must be made separate and secured to the shaft by a pin or set-screw. To give the necessary length of bearing, the framing is increased in thickness by casting bosses on it as shown. This form of bearing has no adjustment for wear, and it cannot be renewed without renewing part of the framing. An improvement on this design is to line the bearing with a solid bush of brass, or other metal, as shown in Fig. 293. The bush is kept in place by a screw as shown. To renew this bearing it is only necessary to renew the bush.

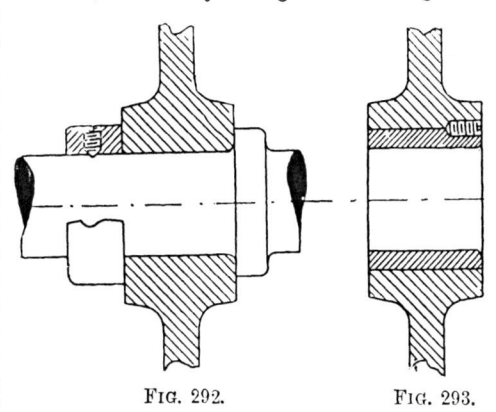

FIG. 292.   FIG. 293.

It will be observed that the solid bearing can only be used when there are no enlargements on the shaft which would prevent it passing into the bearing.

158. Divided Journal Bearings.—When it is not possible or not desirable to place a shaft in its bearings by introducing it endwise, the bearing is divided, and the parts are held together by bolts or other fastenings. The dividing of a bearing into two or more parts also permits of adjustment for wear. The division of a bearing is generally made in a direction perpendicular to the line of the load on the bearing. Examples of the ways in which a bearing may be divided are given in subsequent articles in this chapter. These examples also show various methods of taking up the wear.

159. Brasses or Steps.—The "brasses" or steps for bearings are generally made of gun-metal, but phosphor-bronze and other alloys are

Unit $= t = \cdot 09d + \cdot 15$.

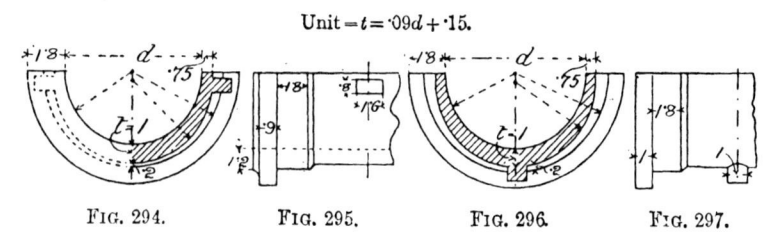

FIG. 294.   FIG. 295.   FIG. 296.   FIG. 297.

also used for steps. Several common forms of steps are shown in Figs. 294 to 301. The proportions marked on these illustrations are in terms of the thickness of the bush at the bottom, which may be obtained from the formula $t = \cdot 09d + \cdot 15$.

In the steps shown in Figs. 294 to 297 the fitting-strips may be turned in the lathe, and the bed which receives them may be bored. Hence these

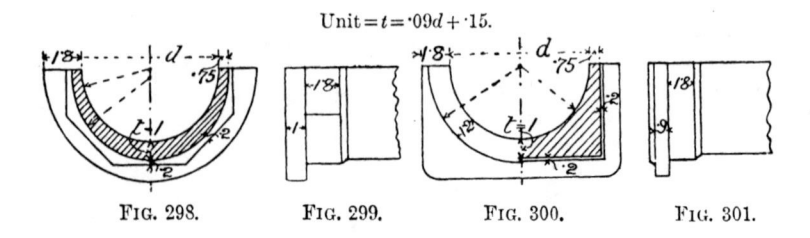

Unit $= t = \cdot 09d + \cdot 15$.

FIG. 298.   FIG. 299.   FIG. 300.   FIG. 301.

forms of steps are easily fitted. To prevent the step from rotating with the shaft, it may be provided with lugs as in Figs. 294 and 295, or with a projecting pin as in Figs. 296 and 297. The lugs or pin fit into corresponding recesses in the supports for the steps.

Steps for long bearings have fitting strips in the middle as well as at the ends, and for heavy work it is a common practice to bed the steps all over their length, thus dispensing with fitting strips.

160. Linings for Steps.—Steps, especially those for large bearings, are frequently lined with white metal. It has been found that it is easier to keep large bearings cool when the steps are lined with white metal, but this is probably due not so much to the anti-friction property of the white metal as to the fact that from its softness it accommodates itself to the journal better than the rigid gun-metal. For instance, if the heavy crank shaft of a large steam-engine or the bearings for that shaft should get slightly out of their original line, the pressure on one bearing may be thrown towards one end, with the result that if the step is quite rigid the bearing surface will be much diminished, and there will be rapid wear and consequent heating. But if the step is lined with soft white metal, the latter will yield where the pressure is greatest until the journal bears uniformly on the supporting surface.

Figs. 302 and 303 show a bush lined with white metal in such a way that the white metal covers nearly the whole of the bearing surface of

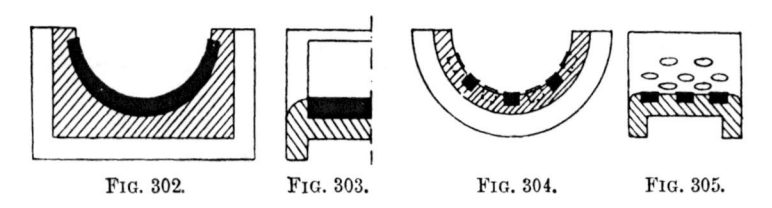

FIG. 302.   FIG. 303.   FIG. 304.   FIG. 305.

the bush. The bush shown in Figs. 304 and 305 has the white metal cast into a number of round holes. In Fig. 306 the white metal is shown filling a series of spiral grooves in the bush, while in Figs. 307 and 308 it is shown filling a number of rectangular recesses. In all of these examples the white metal is supported all round by the metal of the bush.

Fig. 309 shows a bearing in which the step is entirely of white metal run into the cast-iron framing which supports the shaft. In this case the white metal extends the whole length of the bearing.

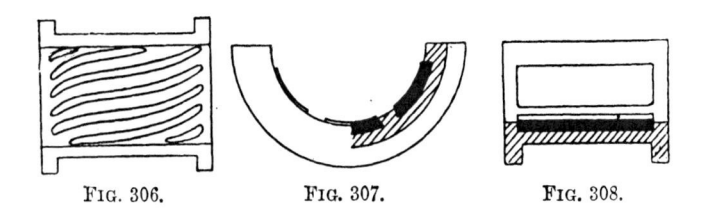

FIG. 306.   FIG. 307.   FIG. 308.

The white metal in all of these examples is poured, when in a molten state, into the recesses formed to receive it after a mandril has been placed into the position which the journal of the shaft will afterwards occupy. Care must be taken to heat the parts with which the white metal will come in contact before the lining is poured in. For large and important bearings, the white metal, after it has solidified, is

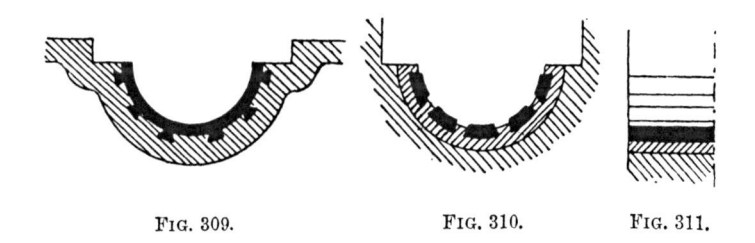

FIG. 309.   FIG. 310.   FIG. 311.

hammered in, and the bearing is then bored out to the exact size of the journal. But for small and less important bearings the operations of hammering and boring are dispensed with.

Some engineers recommend, for large bearings, that the white metal be fitted in separate solid strips, as shown in Figs. 310 and 311, each strip being fitted and driven in like a key.

161. Ordinary Plummer Block.—Fig. 312 shows an ordinary *plummer block*, *pillow block*, or *pedestal* for supporting a shaft. A is the block proper, B the base through which passes the holding-down bolts, and C is the cap. The shape of the interior of the block and cap must conform to that of the steps used. In the example shown the steps are of the form shown in Figs. 296 and 297, but any of the forms shown in Figs. 294 to 301 may be used.

In actual practice the proportions of plummer blocks by different makers vary considerably. The following rules represent average practice for shafts up to 8 inches in diameter :—

Diameter of journal	$= d$.
Length of journal	$= l$.
Height to centre	$= 1\cdot05\ d + \cdot5$.
Length of base	$= 3\cdot6\ d + 5$.
Width of base	$= \cdot8\ l$.
,, block	$= \cdot7\ l$.
Thickness of base	$= \cdot3\ d + \cdot3$.
,, cap	$= \cdot3\ d + \cdot4$.
Diameter of bolts	$= \cdot25\ d + \cdot25$.
Distance between centres of cap bolts	$= 1\cdot6\ d + 1\cdot5$.
,, ,, base bolts	$= 2\cdot7\ d + 4\cdot2$.
Thickness of step at bottom	$= t = \cdot09\ d + \cdot15$.
,, ,, sides	$= \tfrac{3}{4}\ t$.

The length of the journal varies very much in different cases, and depends upon the speed of the shaft, the load which it carries, the workmanship of the journal and bearing, and the method of lubrication. For ordinary shafting one rule is to make $l = d + 1$. Some makers use the rule $l = 1\cdot5 d$; others make $l = 2d$.

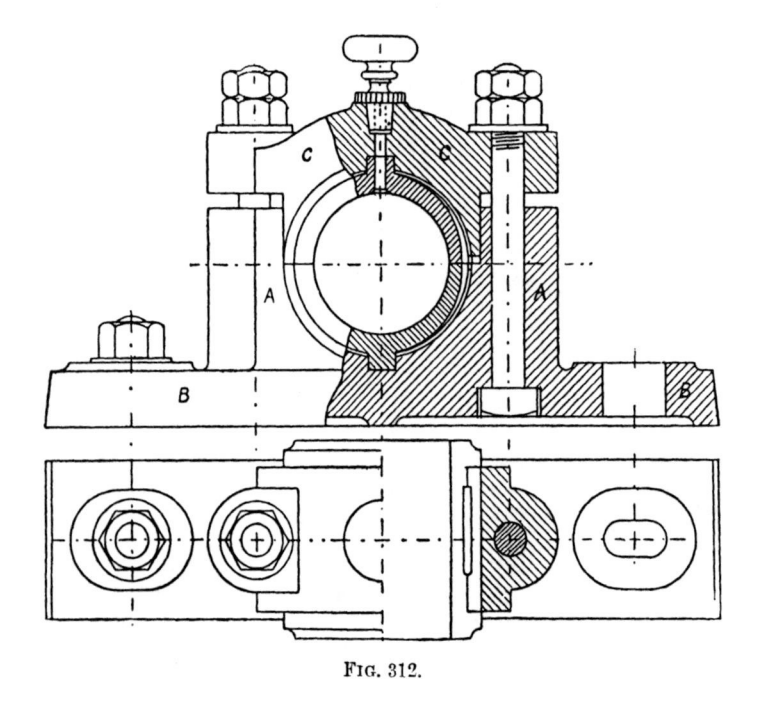

A pillow block for a large shaft has generally four bolts in the cap and

FIG. 312.

four in the base, as shown in Fig. 313. The design shown in Fig. 313 may be used for shafts from 8 inches to 12 inches in diameter. The unit for the proportions marked on the Fig. is $1\cdot 15d + \cdot 4$.

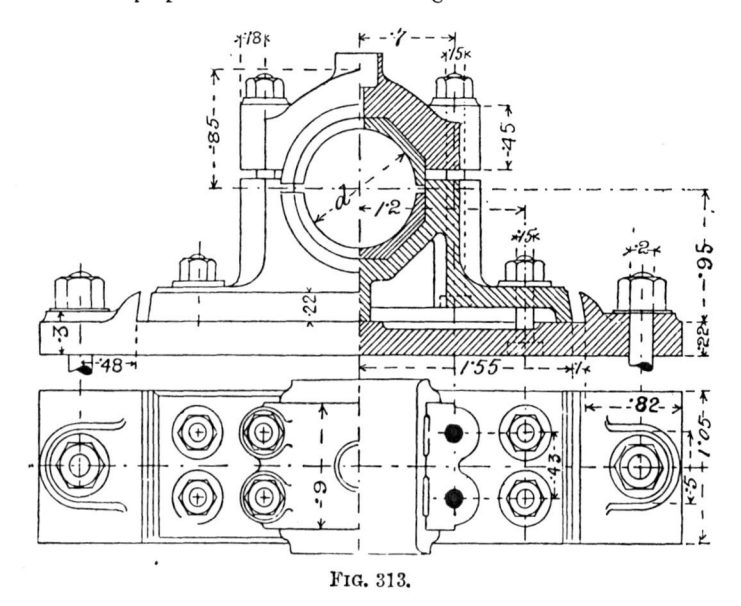

FIG. 313.

162. Sellers Pillow Block.—The American firm, Messrs. Sellers & Co., introduced the form of pillow block shown in Figs. 314 and 315. The steps are made of cast-iron, and are exceptionally long. The steps have

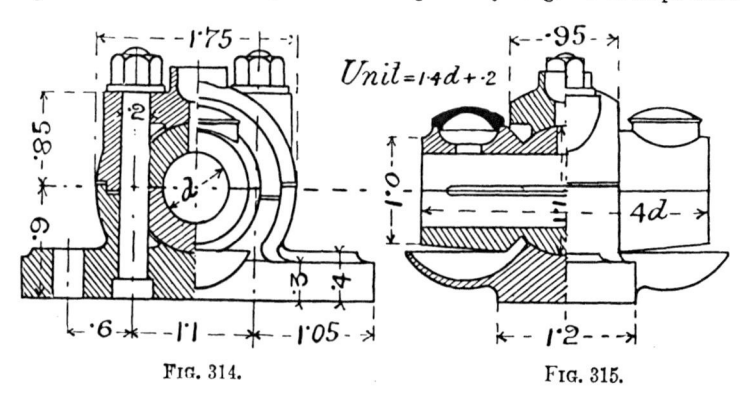

FIG. 314.   FIG. 315.

a spherical enlargement at the centre which fits in corresponding recesses in the block and cap. This arrangement permits of the steps adjusting themselves to the journal of the shaft. The steps are prevented from rotating by the cap bolts which pass through recesses in the central portions of the steps. The ordinary lubrication is through the centre of the cap, but the upper step has two cups containing a mixture of oil and tallow, which is solid at ordinary temperatures, but melts when the bearing becomes heated to about 100° F. The proportions marked on the Figs. are those given by Reuleaux, the unit being $1\cdot 4d + \cdot 2$.

163. Crank Shaft Bearings.—The load on the bearings for the crank shaft of an engine is made up of the weight of the shaft and the alternate thrust and pull along the connecting-rod. The shaft being horizontal, its weight is supported by the lower parts of the bearings, and in vertical engines the force acting along the connecting-rod will act alternately on the lower and upper parts of the bearings. Hence for vertical engines adjustment for wear is obtained by making the bearings in two parts, the plane of division being horizontal. But in horizontal engines the force acting along the connecting-rod acts alternately on opposite sides of the bearings, while the weight of the shaft acts on the lower parts. Hence for horizontal engines proper adjustment for wear can only be obtained by dividing the bearings into three, or four parts. Approximate adjustment may, however, be obtained by making the bearings in two parts, the plane of division being inclined to the horizontal.

Fig. 316[1] shows an example of a three-part bearing suitable for the crank shaft of a horizontal engine. The block for the bearing is here

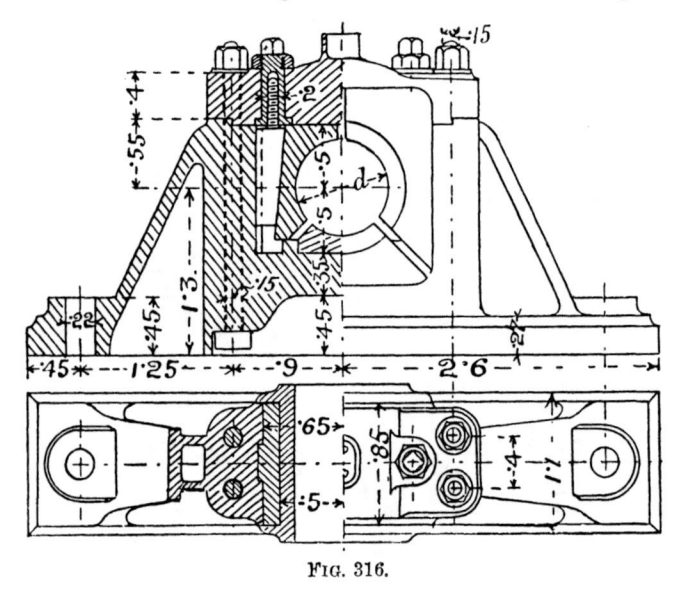

FIG. 316.

shown complete in itself, but generally it would form part of the framing of the engine. Adjustment for wear in a horizontal direction is made by means of wedges having screwed pieces on their upper ends which

---

[1] Copied from Reuleaux' "Der Konstrukteur."

SUPPORTS FOR SHAFTS. 123

enter nuts contained by the cap. These nuts are flanged at their lower ends, and are circular except at their upper ends, where they are square or hexagonal, and these are locked with hexagonal nuts as shown. Adjustment for wear on the lower part of the bearing in the example shown can only be made by placing packing-pieces under the lower brass. In other designs the lower brass is adjusted by means of horizontal wedges placed underneath it. The unit for the proportions marked on the Fig. is $1\cdot15d + \cdot4$.

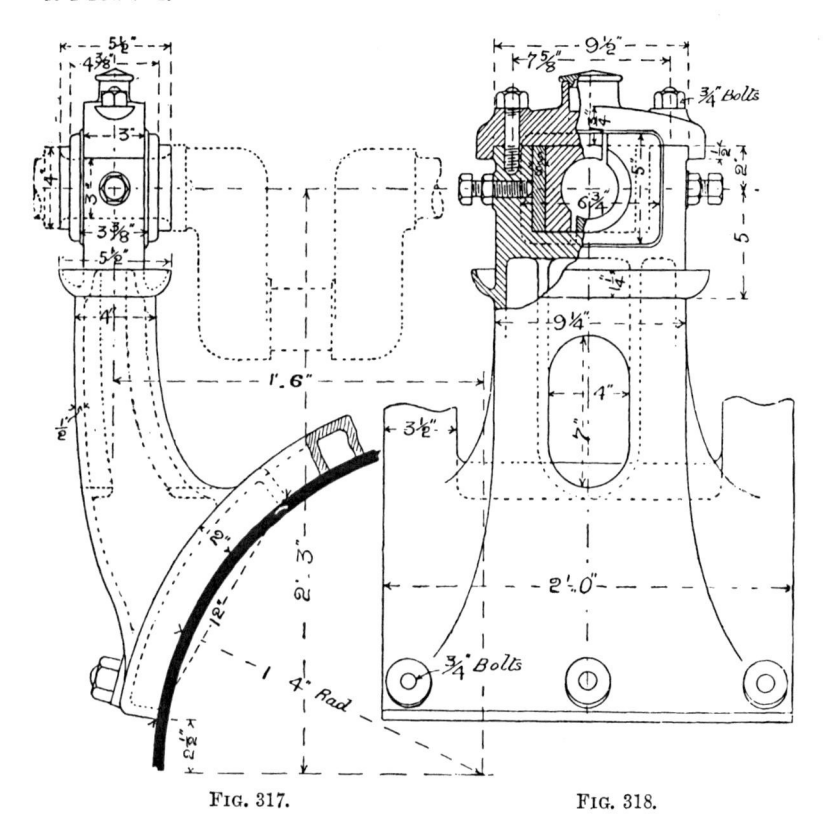

FIG. 317.   FIG. 318.

An example of a bearing for the crank shaft of a portable engine is shown in Figs. 317 and 318. There are two bearings for the shaft, one at each end, and they are carried by pedestals which are cast with a base in the form of a saddle which is bolted to the barrel of the boiler. The steps are divided in a zigzag direction, and are adjusted horizontally by set-screws, the points of which press on steel plates at the back of each step.

164. Brackets for Shaft Bearings.—When a shaft has to be carried on a support fixed to the side of a wall or to a pillar, a bracket is generally

124   MACHINE DRAWING AND DESIGN.

used. Figs. 319 and 320 show a form of bracket suitable for supporting an ordinary pillow block.

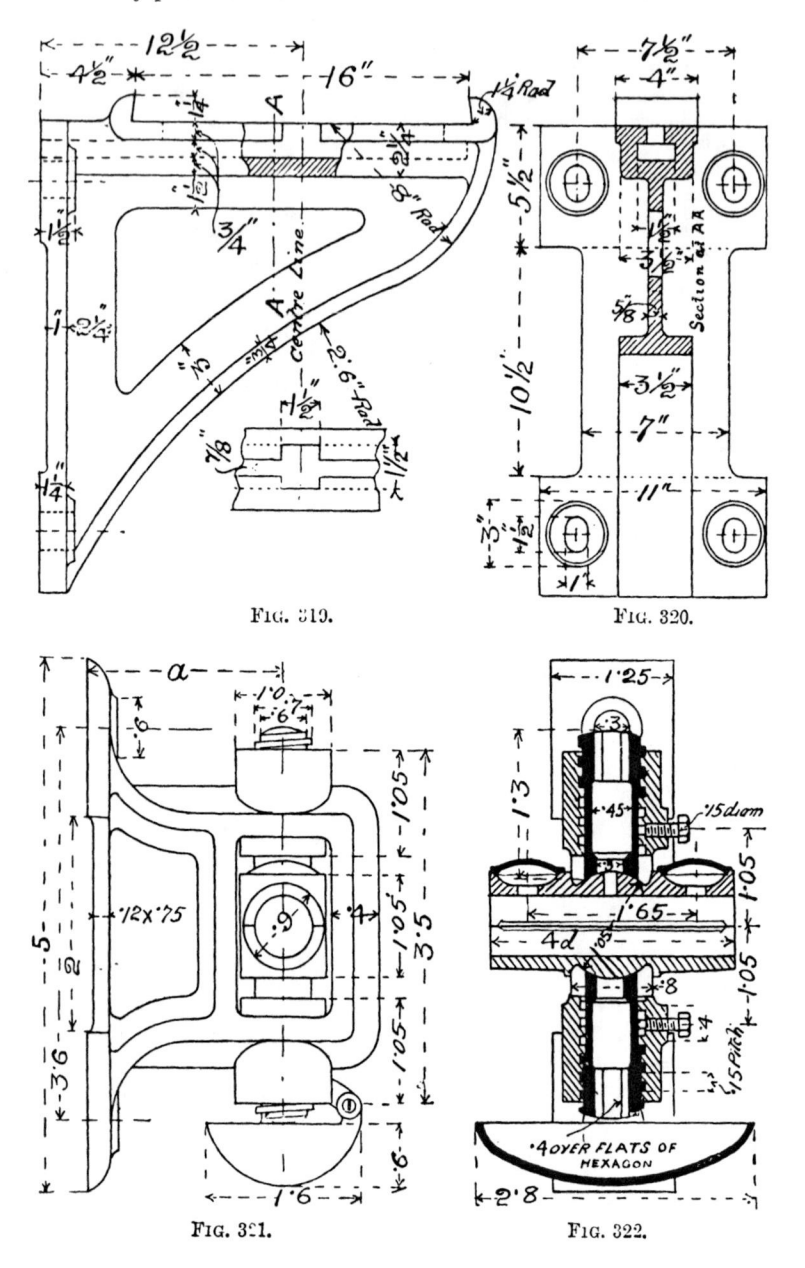

Fig. 319.

Fig. 320.

Fig. 321.

Fig. 322.

SUPPORTS FOR SHAFTS. 125

An elegant design of wall-bearing, due to Sellers, which has been copied and modified by many makers, is shown in Figs. 321 and 322. The bearing is of the same design as in the Sellers pillow block. The spherical centre of the bearing-box is held between the ends of two hollow

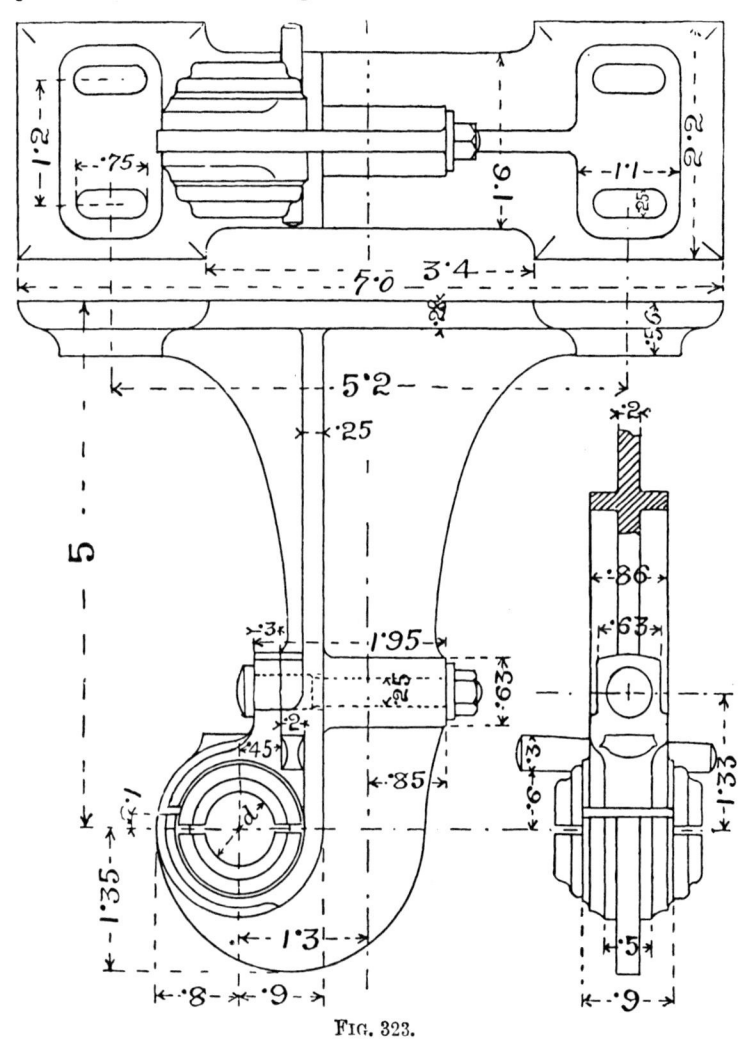

FIG. 323.

cast-iron stems. These stems are screwed at their outer ends with shallow rectangular threads, which fit into screwed bosses on the bracket. The screw threads on the stems and bosses are cast on them. The holes in the stems at their outer ends are of hexagonal form to receive a key, by means of which they may be rotated so as to adjust the height of the bearing. The stems are locked in position by set screws. The distance a from the wall to the centre of the bearing is usually 6 inches for shafts from 2 inches to 6 inches in diameter. The proportions marked on the Figs. are those given by Reuleaux, the unit being $1·4d + ·2$.

165. Hangers.—When a shaft bearing is suspended from a ceiling a hanger is used. A common form of hanger is shown in Fig. 323. The cap for the bearing is secured by a bolt and a key. The unit for the proportions given is $1·15d + ·4$, where d is the diameter of the journal.

In the hanger, Figs. 324 and 325, designed by Sellers, the bearing is of the same form, and is held and adjusted in the same way as in the Sellers wall bracket already described. The unit for the proportions given is $1·4d + ·2$, where d is the diameter of the journal.

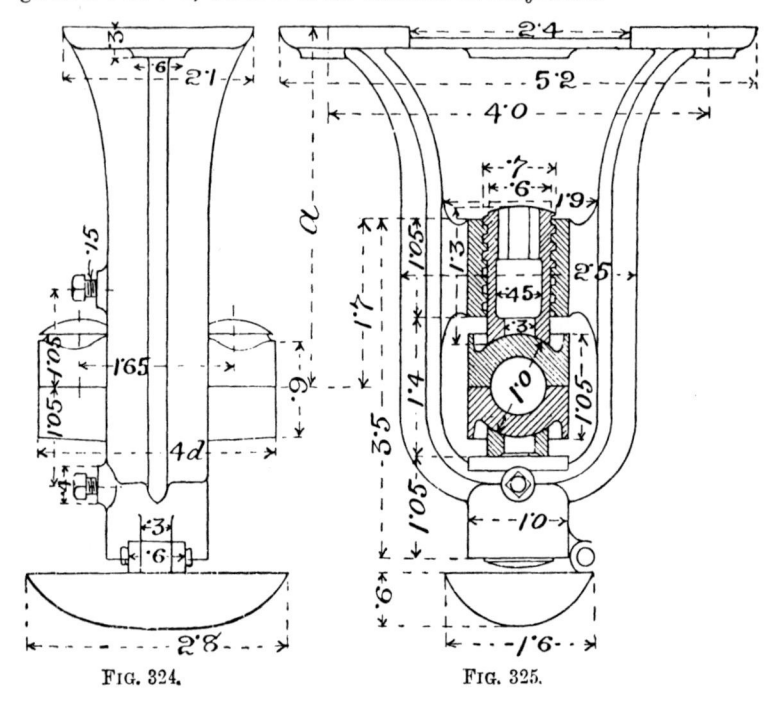

FIG. 324.   FIG. 325.

166. Footstep or Pivot Bearings.—The bearing which supports the weight of a vertical shaft is either a collar bearing or a footstep bearing. Fig. 326 shows one form of footstep bearing. The unit for the proportions given is $1·15d + ·4$. Unless this bearing is supported on a stool or pedestal which may be removed after the shaft has been propped up, the shaft must be raised a distance equal to the depth of the bush before the bearing can be removed for examination or repair. The latter operation is often attended with great trouble and inconvenience, and to avoid this the footstep bearing may be divided vertically as shown in Fig. 327   The unit for the proportions of this design is $1·15d + ·4$.

SUPPORTS FOR SHAFTS.

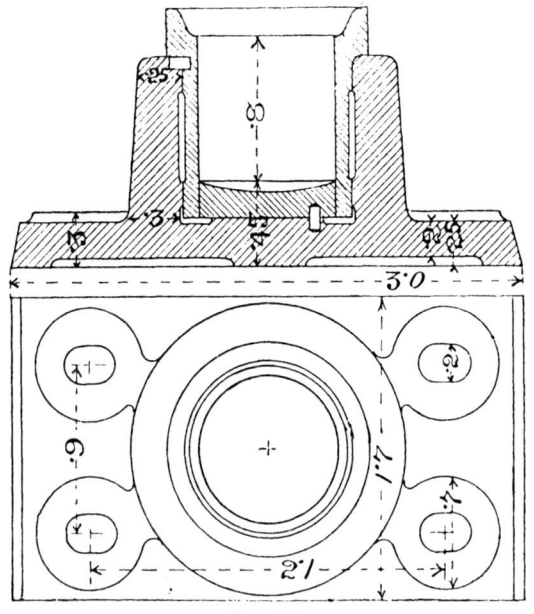

FIG. 326.

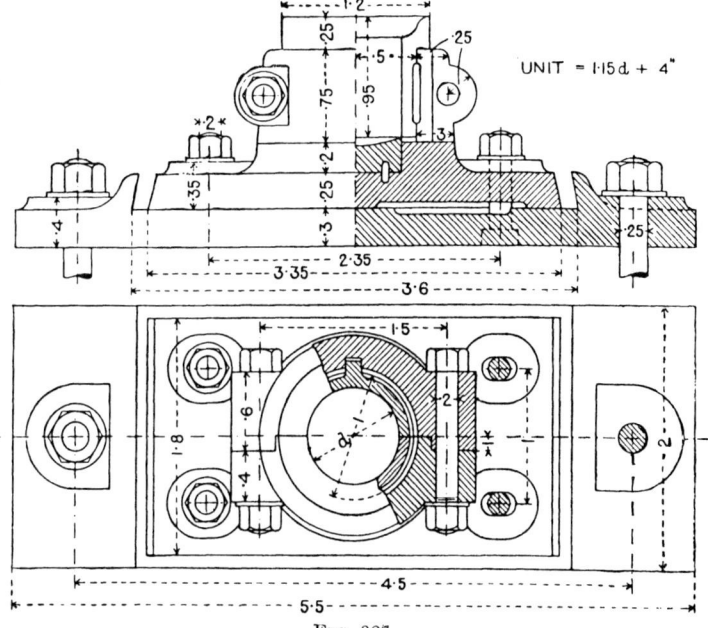

UNIT = 1·15 d + 4"

FIG. 327.

128   MACHINE DRAWING AND DESIGN.

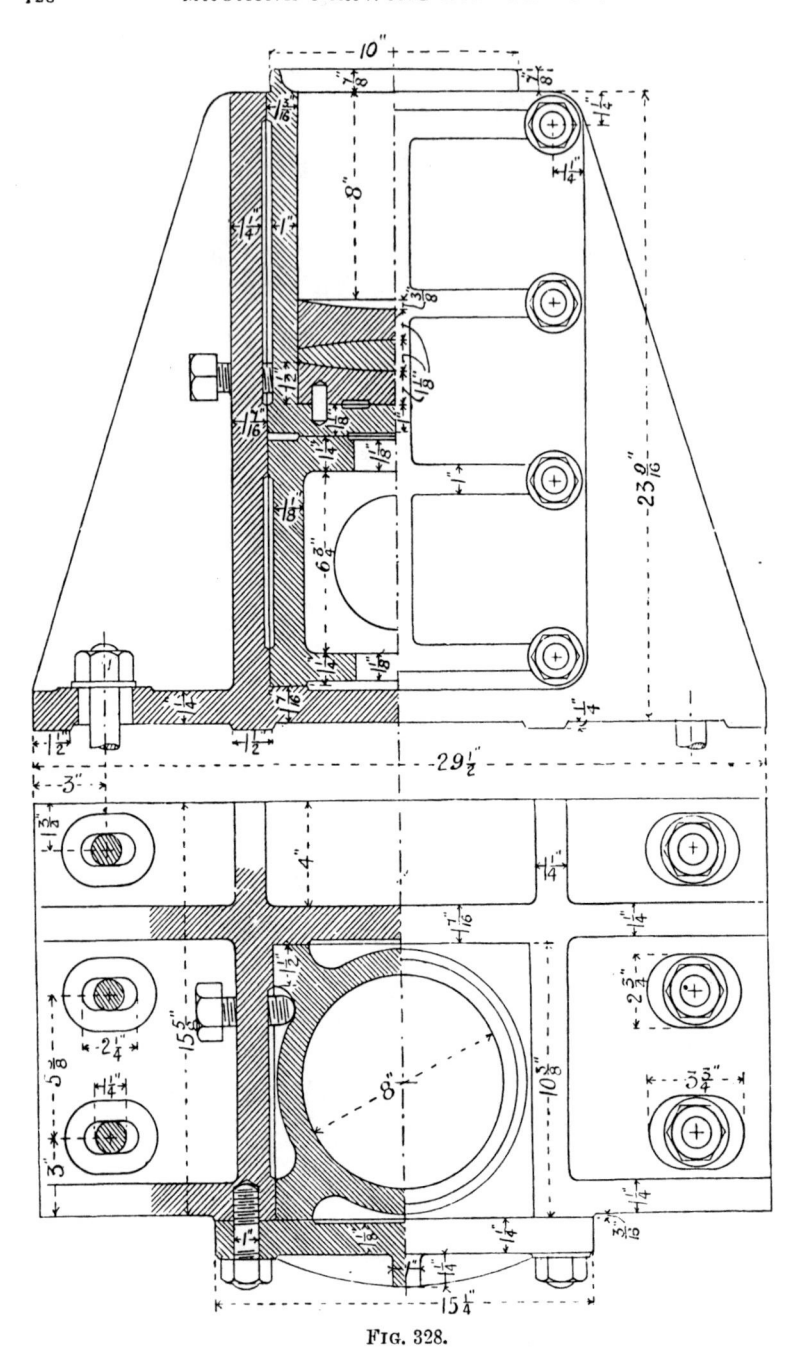

Fig. 328.

The foot-step end of a shaft should be of steel, and it may be quite flat on the end, or slightly cup-shaped. The weight of the shaft is supported on a disc of steel, brass, or bronze. It is a common practice to have several of these discs, as shown in Fig. 328, so that if one should heat up and seize, the next will come into action and give the first time to cool.

The footstep bearing shown in Fig. 328 is for a very heavy upright shaft. The bottom disc is of tough brass, and is not allowed to rotate; the other discs are loose. The middle disc is of steel and the upper one of brass. To remove the step, the shaft is raised slightly and propped up. The front cover is then removed, and the stool under the step is taken out. While the stool is being removed the step is prevented from dropping by the set-screw shown. After the stool is removed the set-screw is slackened, and the step is lowered clear of the shaft and then withdrawn.

167. Bearings for Turbine Shaft.—For the vertical shaft of a turbine the ordinary form of foot-step bearing is unsuitable, and bearings of special design have been introduced to support the weight of the shaft and its attachments and the axial pressure of the water on the wheel. Generally, the principal bearing for supporting the weight of the shaft and the axial water pressure is placed at the upper end of the shaft, so that the shaft is suspended from the top. An example of a well-designed bearing of this kind is shown in Fig. 329. This is from the practice of Messrs. Ganz & Co., Engineers, Buda-Pesth.

The upper end of the vertical shaft A is fitted with a bell-shaped cast-steel cap, B, which is made to rotate with the shaft by the feather-key C. The position of the cap on the shaft is adjusted by the nuts D and E. The cap B runs on a phosphor bronze ring, F, which contains two circular grooves into which oil is forced through the pipe H by a small force pump. The pressure of the oil is sufficient to lift the cap slightly, so that the latter really runs on a layer of oil. The oil after passing through the bearing is filtered and used over again. The casting which supports the bronze ring F has within it a space, J, of a zigzag shape in plan, as shown at (a), through which water may be circulated to cool the bearing if necessary. Any oil which runs down the shaft past the neck bush K is deflected by the piece L into the annular-shaped drip-box M.

The total load on the bearing described above is about $17\frac{1}{4}$ tons, and the bearing area of the ring F, after deducting the area of the grooves, is 172·8 square inches, so that the load is about 231 lbs. per square inch of bearing surface. The speed of the shaft is 134 revolutions per minute, and the mean velocity of the rubbing surfaces is 528 feet per minute. The horse-power of the turbine is 772.

The load on the top bearing is reduced to about one half of that stated above by means of a hydrostatic foot-step designed by Professor Radinger, Vienna. This foot-step is shown in Fig. 330. A steel plunger, N, is connected to the lower end of the shaft A by an Oldham coupling, P, and rotates inside the cylinder Q, which is supplied with water under pressure. The use of the Oldham coupling is to provide for the possibility of the plunger N being eccentric to the shaft A through error

I

130   MACHINE DRAWING AND DESIGN.

in erecting. The lower part of the plunger is surrounded by a brass

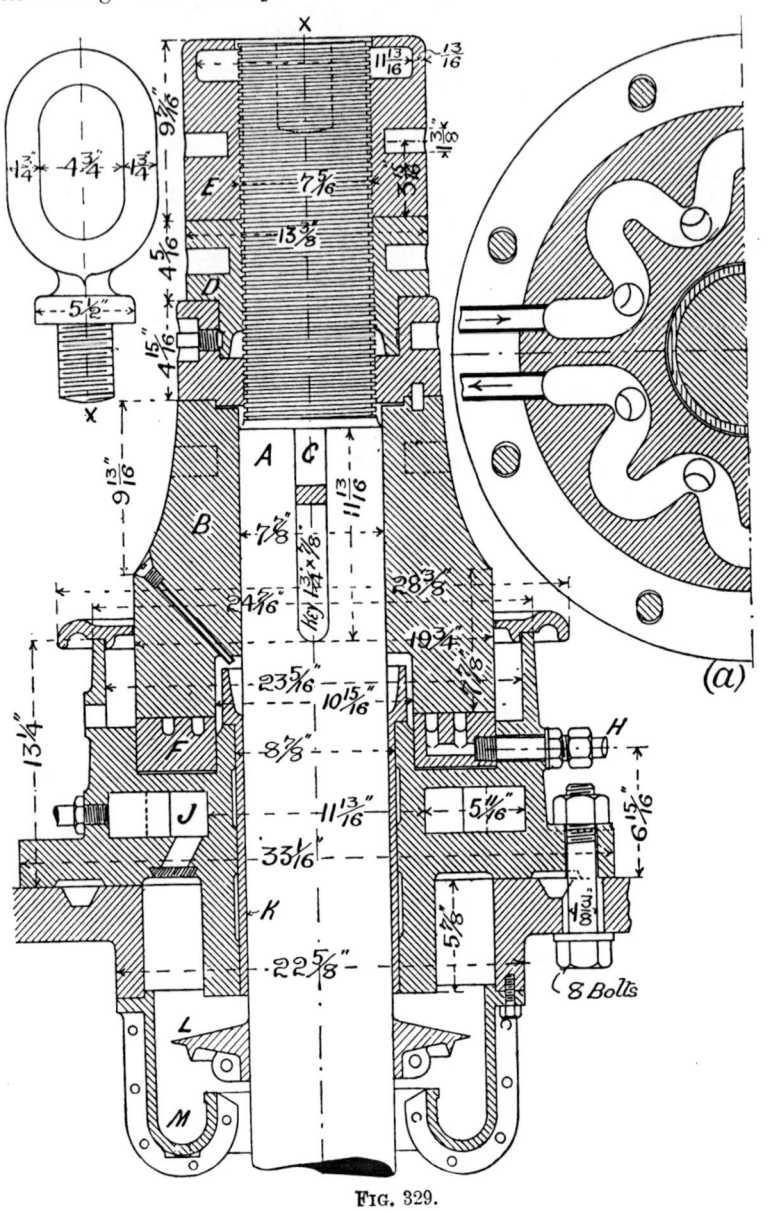

FIG. 329.

bush R. Above this bush there is a series of split rings wedge-shaped in section. These rings are made of a special alloy. The main gland is made water-tight with an india-rubber ring, S. Inside the main gland is a stuffing-box packed with leather rings placed obliquely as shown. The packing arrangement is designed so as to keep up the pressure of water in the cylinder, and at the same time permit a leakage of about 13 gallons per hour. This leakage takes place through the pipe T, and serves to keep the plunger cool. The pipe T is led up to the engineer's stand, where it terminates in a pressure gauge and a cock, through which the water flows at a regulated rate.[1]

168. Thrust Bearings for Screw Propeller Shafts.—The thrust along the shaft of a steamship propelled by a screw is taken up by a collar bearing, and through it transmitted to the ship. Small shafts up to eight inches in diameter have generally only one thrust collar, but sometimes comparatively small shafts have several small thrust collars. In the latter case the bearing may have a brass bush in halves containing grooves

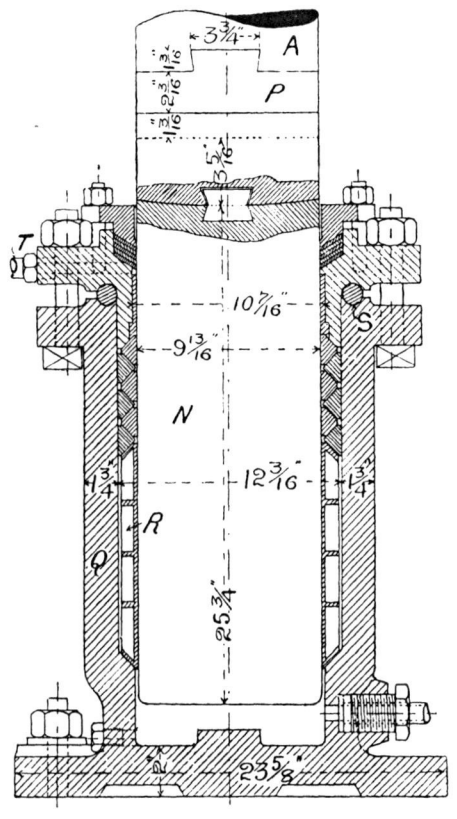

FIG. 330.

to receive the collars on the shaft, or the bearing may contain a number of rings in halves which fit between the collars on the shaft. The general practice, however, now is to fit between the collars cast-iron or cast-steel horse-shoe shaped pieces faced with brass or white metal.

An example of a thrust bearing for a shaft $11\frac{1}{2}$ inches in diameter is shown in Fig. 331. The shaft has four thrust collars, which run against four cast-iron horse-shoe shaped pieces, which in this case are faced with brass. At each end of the thrust bearing there is a bearing lined with white metal to support the weight of the shaft. Where the shaft passes through the bulk-head into the engine-room there is a gland and stuffing-box, so that in the event of the tunnel through which the propeller

---

[1] Further information on turbine shaft bearings by Messrs. Ganz & Co. will be found in *Engineering*, vol. lii. p. 307, and vol. liii. p. 228, also in the *Engineer*, vol. lxxii. p. 464.

132   MACHINE DRAWING AND DESIGN.

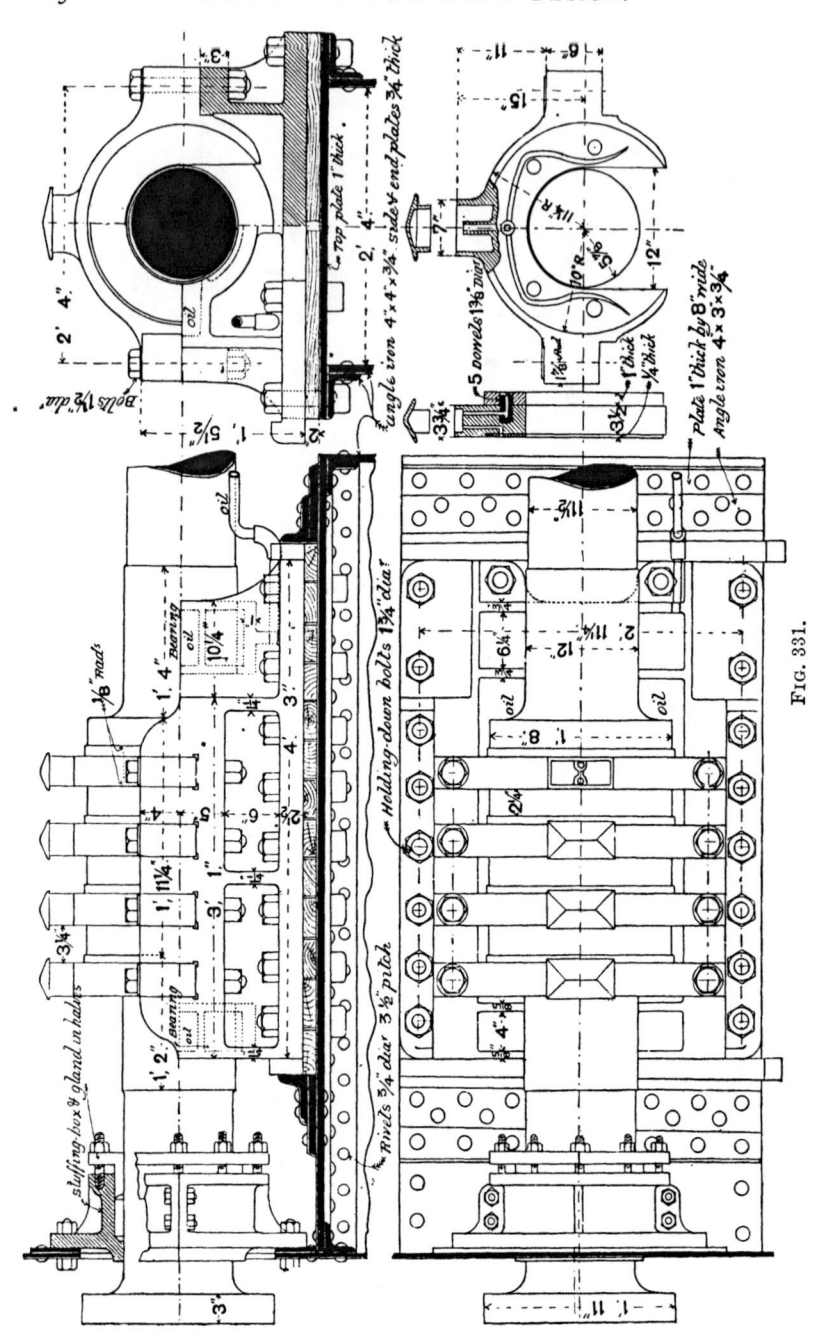

FIG. 331.

SUPPORTS FOR SHAFTS. 133

shaft passes becoming flooded the water cannot reach the engine-room. The thrust block is here shown in the tunnel, but the usual practice is now to fix the thrust block in the engine-room to a prolongation of the bed of the engine. The collars are lubricated by running in a bath of oil, or of oil and soapy water. In this example the horse-shoe pieces are firmly secured between jaws in the main casting of the block. There is here no simple adjustment to ensure that each collar will exert the

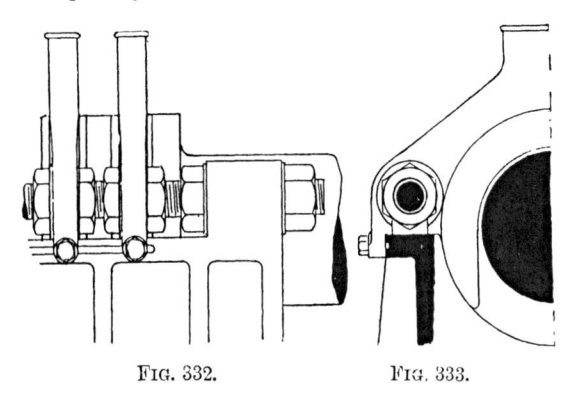

FIG. 332.   FIG. 333.

same pressure on its bearing. Modern practice for large thrust blocks is to clamp each of the horse-shoe pieces at each side between two nuts which are threaded on a screwed steel bar, supported at its ends by solid bearings cast on the block, as shown in Figs. 332 and 333. In this design each horse-shoe piece may be adjusted separately by the nuts on each side, or they may be all moved together by means of the nuts at the ends of the bars.

169. Bearing Surface of Thrust Collars.—The total area of the bearing surface of thrust collars for propeller shafts should be such that the pressure does not exceed 70 lbs. per square inch. If D_1 and D_2 are the larger and smaller diameters of the collars respectively and n the number of collars, then

$$\text{total bearing surface} = \cdot 7854 n (D_1^2 - D_2^2).$$

If p denotes the pressure per square inch on the bearing surface, and T the total thrust, then

$$T = \cdot 7854 n p (D_1^2 - D_2^2), \text{ and } D_1 = \sqrt{\left(\frac{T}{\cdot 7854 n p} + D_2^2\right)}.$$

The number of collars may be $n = \dfrac{D}{2} - 2$, where D is the diameter of the shaft.

CHAPTER X.

BELT GEARING.

170. Leather Belting.—The most common material used for the bands or belts in belt gearing is leather tanned with oak bark from ox-hides. The best leather is made from that part of the hide which covered the back of the animal. The thickness of the leather is about $\frac{3}{16}$ths inch, and it can be obtained in strips up to about 5 feet long. By joining these strips together belts of any length may be made. *Single* belts are made of one thickness, and *double* belts of two thicknesses of leather.

In the system of making belts introduced by Messrs. Sampson & Co., of Stroud, a large disc of leather is cut from the hide, and after being cured and dressed in the usual way it is cut in a spiral direction to within about 9 inches of the centre. In this way a very long strip is obtained, the length depending of course on the width. From a disc 4 feet 9 inches in diameter a strip 2 inches wide and over 100 feet long would be obtained. The strips obtained in this way are stretched, and when in a stretched condition they are rubbed and thereby made straight. Any required width of belt is made by sewing these strips together side by side. One double belt made on this system was 75 inches wide and 153 feet 6 inches long.

The outside of the leather, called the grain or hair side, is smoother than the inside or flesh side, and a considerable difference of opinion exists as to which side should be placed next the pulley. Many authorities prefer the grain or smooth side next the pulley, because of the greater driving power. Mr. John Tullis, of Glasgow, states that the band is more durable if the flesh side is placed next the pulley, and if it receives one coating of curriers' dubbin and three coatings of boiled linseed oil every year, the flesh side becomes as smooth as the grain side, and the driving power fully as good.

171. Cotton and other Belting.—In damp situations *cotton belting* is superior to leather; it is also cheaper, and since it can be made of great strength and of great length, without the numerous joints found in leather belts, it is much used. Cotton belts are made of widths up to 5 feet. The necessary thickness is obtained by sewing together from four to ten plies of cotton duck.

Belting known as *Scandinavian cotton belting* is woven of the required thickness in a loom at one operation, and not by sewing together several thicknesses of cotton.

India-rubber belting is made by cementing several plies of cotton duck together with india-rubber. This kind of belting is considered the best in wet situations.

BELT GEARING.

Paper belting has been used successfully in America. It is said to possess great strength and durability when not exposed to moisture. It is also recommended for not stretching when at work.

172. Leather Link Belting.—A form of belt which has been largely used with great success is shown in Fig. 334. This consists of links made of leather, connected by iron or steel pins as shown. A belt of this design can be made of any width, and it works freely on pulleys of small diameter, and can be driven at very high speed. These advantages, combined with its great strength, make this form of belt very suitable for driving dynamos. When the link belt has to run between guides or on flanged pulleys, the rivet heads on the edge of the band are objectionable, but this may be got over by facing the outside links with leather after the belt is riveted up.

When a link belt of considerable width works on a curved pulley,

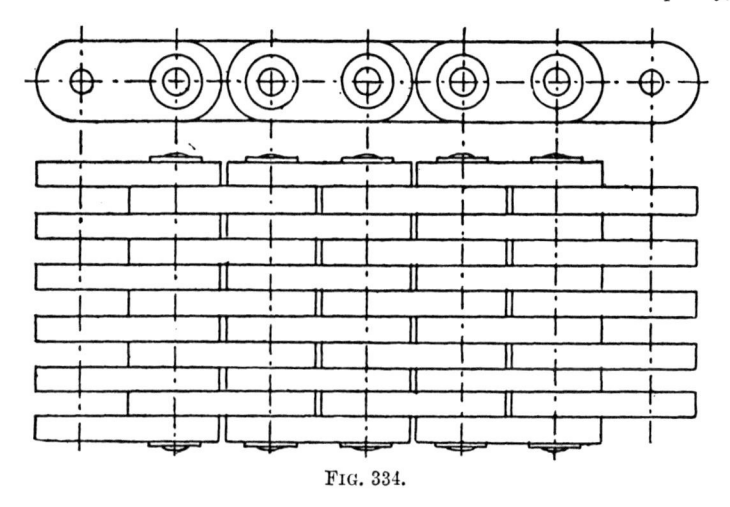

FIG. 334.

either there is contact at the centre only, or the pins are bent where the band is in contact with the pulley. This difficulty is got over by making the band in two or more longitudinal strips, which are hinged together as shown in Fig. 335. In the case illustrated, where the band is hinged longitudinally at the centre, the rim of the pulley is best made of a double conical form, so that the section consists of two straight portions instead of the usual curve. Messrs. Tullis, of Glasgow, get over the same difficulty by making the section of the band to suit the curved section of the pulley, as shown in Fig. 336.

173. Belt Joints and Belt Fasteners.—As already stated, the strips of leather obtained from hides in the ordinary way do not exceed five feet in length, and to form a belt these strips are joined together. The joints used in belts may be divided into two classes—permanent joints, which cannot readily be broken, and temporary joints, which can easily be made or broken. The first or permanent class of joints is used for con-

136   MACHINE DRAWING AND DESIGN.

necting the comparatively short strips of which the belt is made up, and the other class is used for uniting the two ends of the completed belt.

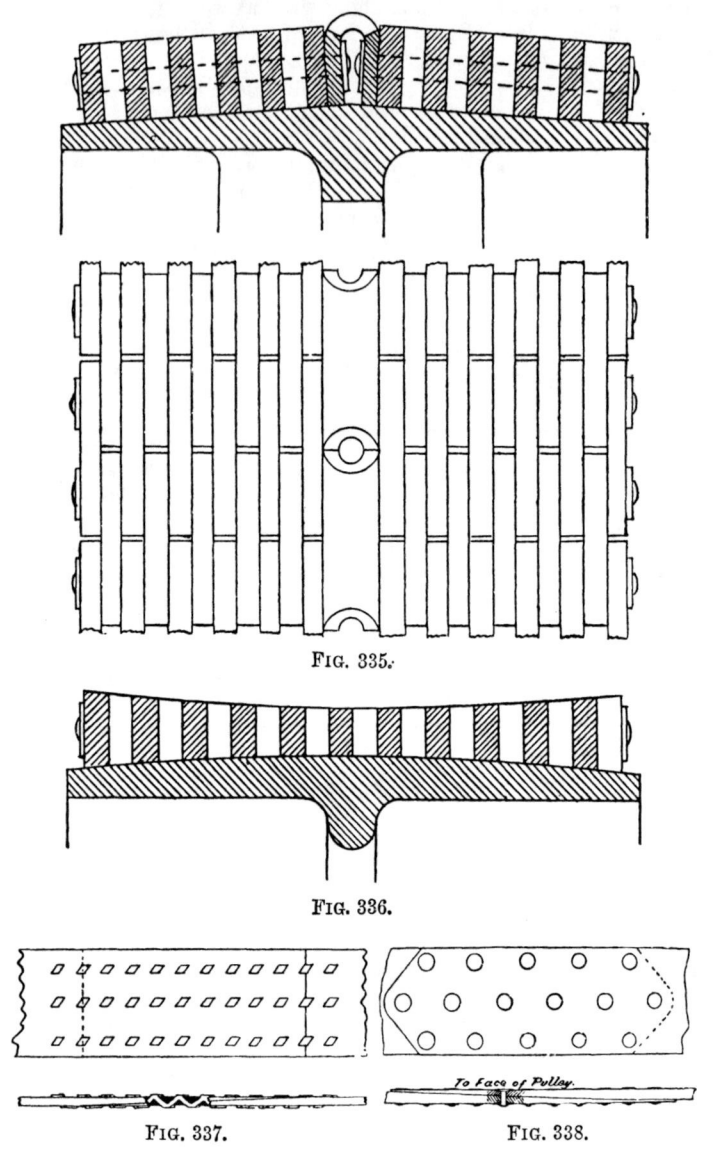

FIG. 335.

FIG. 336.

FIG. 337.   FIG. 338.

The two most common forms of permanent belt joints are shown in Figs. 337 and 338. In both of these examples the ends of the strips to be connected are pared down and cemented together, forming a lap-joint, which is then laced, as in Fig. 337, or riveted, as in Fig. 338, with copper rivets and washers. In Fig. 337 the holes for the lace are made in a diagonal direction, and the lace is sunk into the belt.

Of the joints which are easily made and easily undone, the simple laced joint, shown in Fig. 339, is the most common. The ends of the belt are cut square and butted together, and the lace is threaded through

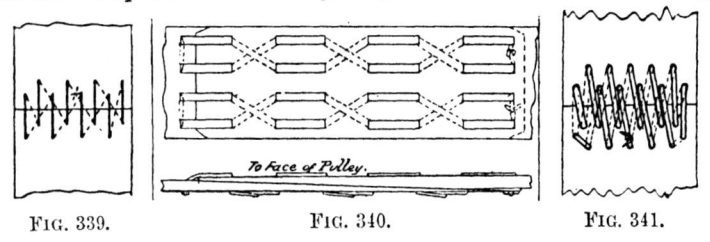

FIG. 339.   FIG. 340.   FIG. 341.

round or oval holes made with a hand-tool known as a belt-punch. When the holes are oval the larger diameter should lie along the belt, and not across it. The belt is weakened less with oval holes than with round ones. On the face of the belt, next the pulley, the lace should be parallel to the edge of the belt. This ensures smoother and more even running. Fig. 340 shows a laced lap-joint. Fig. 341 shows a laced butt-joint in which there is no crossing of the lace on itself.

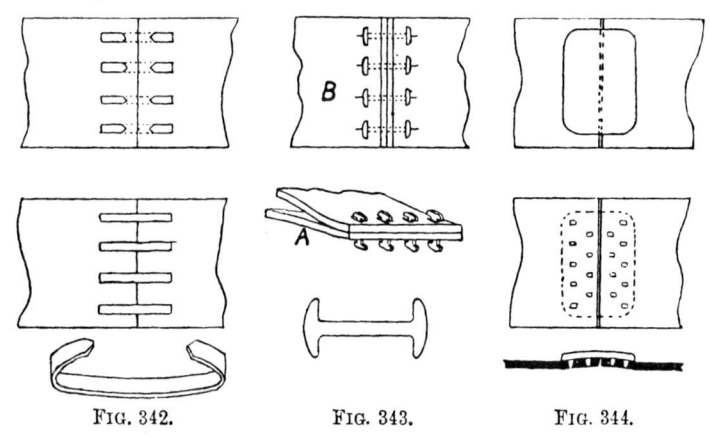

FIG. 342.   FIG. 343.   FIG. 344.

There are in the market a large number of metallic connecting pieces for belts, called "belt fasteners," many of them being very simple, and more easily applied than lacing.

The fastener shown in Fig. 342 is very easily applied. Holes are first punched near the ends of the belt with an ordinary belt-punch. The ends of the fasteners are then passed through these holes and flattened down with a hammer. After raising their ends, these fasteners may be taken out and used again. Fig. 343 shows Blake's or Greene's belt fastener or belt stud. The ends of the belt are placed together as shown at A, and the holes, which are in the form of slits, are punched through the two simultaneously. The fasteners are then put in as shown at A, and then twisted round through 90 deg. The band is next pulled straight and flattened with a mallet or hammer. The completed joint is then as shown at B. Harris's belt fastener, Fig. 344, consists of a malleable cast-iron plate which is slightly curved and has teeth cast on its hollow side. The ends of the belt, *one at a time*, are driven on to the teeth of the fastener. It will be observed that the teeth, being at right angles to the curved plate, are inclined to the band, so that the tension in the latter causes it to keep in contact with the plate.

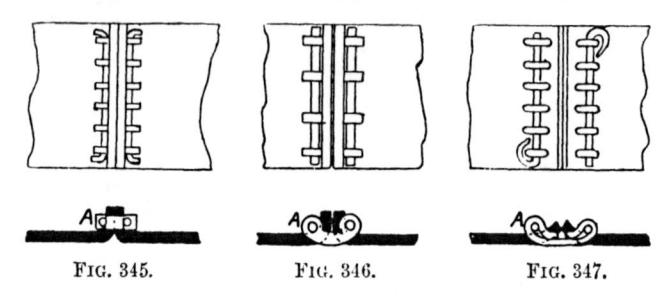

FIG. 345. FIG. 346. FIG. 347.

Three very similar forms of belt fasteners are shown in Figs. 345, 346, and 347. In each case the ends of the belt are laid together, and the holes punched in the manner explained for Greene's fastener. The pieces A are then passed through the holes in the belt, and through the holes in the ends of these pieces pins are passed and secured. The joint being now complete, the belt is straightened. These fasteners have the great advantage that the strain on the ends of the belt is better distributed over the width, and there is not the same tendency to cut through in front of the holes in the belt as there is in most other forms of fasteners.

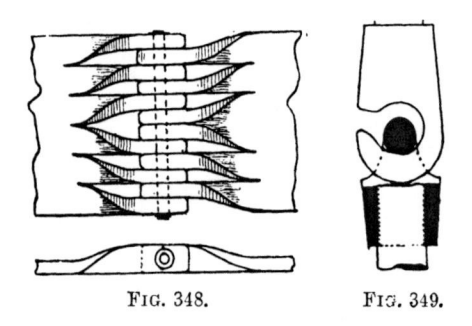

FIG. 348. FIG. 349.

In Moxon's belt fastener, Fig. 348, the ends of the belt are slit up into fingers, and each finger has a hole punched in it near the outer end. These fingers are now twisted through 90 degs., and the ends of

BELT GEARING.

the one set are placed between the ends of the other, and connected to them by a pin as shown. This belt joint, it will be seen, resembles a portion of a link belt. Fig. 349 shows a form of hook and eye used for connecting the ends of small round leather or catgut bands. The hook and eye are made of cast steel, and are screwed internally as shown, so that they may be screwed on to the ends of the band.

174. Transmission of Motion by Bands.—If motion be transmitted from one pulley to another by means of a thin inextensible band, every part of the latter will have the same velocity, and the outer surface

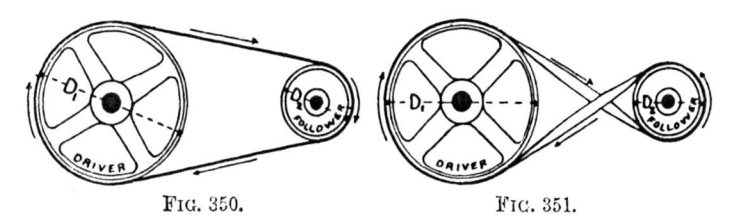

FIG. 350.   FIG. 351.

of the rim of each pulley will have the same velocity as the band. This fact enables us to prove the formula connecting the speeds of two pulleys with their diameters. Let D_1 and D_2 be the diameters of the driver and follower respectively, and let N_1 and N_2 denote their speeds in revolutions per minute. The speed of the outer surface of the rim of the driver = $D_1 \times 3\cdot1416 \times N_1$, and the speed of the outer surface of the rim of the follower = $D_2 \times 3\cdot1416 \times N_2$; but each of these is equal to the speed of the band; therefore, $D_1 \times 3\cdot1416 \times N_1 = D_2 \times 3\cdot1416 \times N_2$; that is, $D_1 N_1 = D_2 N_2$ or $\dfrac{N_2}{N_1} = \dfrac{D_1}{D_2}$. This formula is true whether the band is open as in Fig. 350, or crossed as in Fig. 351; but the direction of the motion will not be the same in each of these cases, as is shown by the arrows. With an open band the pulleys rotate in the same direction, while with a cross band the pulleys rotate in opposite directions.

EXAMPLE 1.—*The driving pulley is 3 ft. 6 in. in diameter, and it makes* 100 *revolutions per minute. The following pulley is* 2 *ft.* 6 *in. in diameter. To find its speed.*

$$D_1 N_1 = D_2 N_2$$
$$42 \times 100 = 30 \times N_2$$
$$\text{therefore } N_2 = \frac{42 \times 100}{30} = 140 \text{ rev. per minute.}$$

EXAMPLE 2.—*A band moving with a velocity of* 1000 *feet per minute passes over a pulley* 3 *ft.* 3 *in. in diameter. To find the number of revolutions made by the pulley in one minute.*

Circumferential speed of pulley = $\dfrac{39 \times 3\cdot1416 \times N}{12} = 1000$; therefore $N = \dfrac{1000 \times 12}{39 \times 3\cdot1416} = 97\cdot94$ revolutions per minute.

EXAMPLE 3.—*The speeds of the driver and follower are 120 and 90 revolutions per minute respectively, and the sum of their diameters is 49 in. To find the diameters of the driver and follower*

$$\frac{D_1}{D_2} = \frac{N_2}{N_1} = \frac{90}{120} = \frac{3}{4} \text{ or } D_1 = \frac{3}{4} D_2$$

also $D_1 + D_2 = 49$; therefore $\frac{3}{4} D_2 + D_2 = 49$

that is, $1\frac{3}{4} D_2 = 49$; therefore $D_2 = \frac{49}{1\frac{3}{4}} = 28$ in.

and $D_1 = \frac{3}{4} D_2$; therefore $D_1 = \frac{3}{4} \times 28 = 21$ in.

The diameters of the driver and follower are therefore 21 in. and 28 in. respectively.

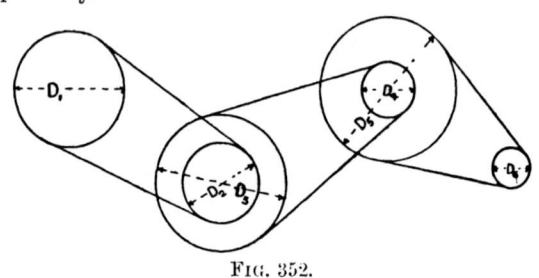

FIG. 352.

If motion is transmitted from one shaft to another through one or more intermediate shafts by means of belt gearing, as shown in Fig. 352, it is easy to prove that $\frac{N_6}{N_1} = \frac{D_1}{D_2} \times \frac{D_3}{D_4} \times \frac{D_5}{D_6}$, where D_1, D_2, etc., are the diameters of the pulleys taken in order, and N_1 and N_6 the speeds of the first and last pulleys respectively.

EXAMPLE 4.—*At the People's Palace Technical Schools, London, the following example occurs:* A, *Fig.* 353, *the crank-shaft of a steam-engine,*

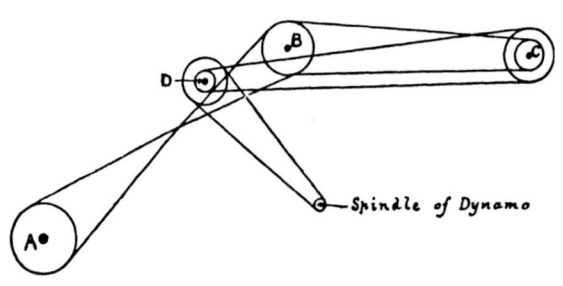

FIG. 353.

carries a pulley 26 in. in diameter, from which passes a belt to a pulley 20 in. in diameter on one end of a shaft, B. *This shaft* B *passes through a workshop and gives power to various machines, and at its end farthest from the engine carries another pulley 20 in. in diameter, which drives a*

BELT GEARING.

pulley 12 *in. in diameter on one end of a shaft* C, *which passes along a corridor and carries at its other end a pulley* 20 *in. in diameter, which drives a pulley* 8 *in. in diameter on a countershaft* D. *The countershaft* D *carries a pulley* 18 *in. in diameter, which drives a pulley* 6 *in. in diameter, which is fixed to the spindle of a dynamo in the electrical laboratory. The speed of the engine is* 95 *revolutions per minute. To find the speed of the dynamo.*

$$\frac{\text{Speed of dynamo}}{\text{Speed of engine}} = \frac{26}{20} \times \frac{20}{12} \times \frac{20}{8} \times \frac{18}{6};$$

Therefore speed of dynamo $= \dfrac{26}{20} \times \dfrac{20}{12} \times \dfrac{20}{8} \times \dfrac{18}{6} \times 95 = 1543{\cdot}75$ revolutions per minute.

175. Effect of Thickness of Band on Velocity Ratio of Pulleys.—When a thick band is bent over a pulley its inner surface is compressed and its outer surface is stretched, but the surface mid-way between these two remains of the same length. It follows, therefore, that the velocity of the inner surface of the band in contact with the pulley must be less than the velocity of the middle surface of the band, and it is only the middle surface of the band which has the same velocity at every point. Hence the velocity ratio of two pulleys connected by a thick band will be the same as if the band is indefinitely thin, and the diameters of the pulleys are increased by an amount equal to the thickness of the band. If D_1 and D_2 are the diameters of the pulleys, N_1 and N_2 their speeds, and t the thickness of the band, then

$$\frac{D_1 + t}{D_2 + t} = \frac{N_2}{N_1}.$$

EXAMPLE.—*To find the speed of the dynamo mentioned in Example* 4 *of the preceding article, when the thickness of the bands is taken into account; the thickness of each band being* $\tfrac{3}{16}$*th inch.*

The effective diameters of the pulleys will be those already given, increased by the thickness of the band; hence

$$\frac{\text{Speed of dynamo}}{\text{Speed of engine}} = \frac{26\tfrac{3}{16}}{20\tfrac{3}{16}} \times \frac{20\tfrac{3}{16}}{12\tfrac{3}{16}} \times \frac{20\tfrac{3}{16}}{8\tfrac{3}{16}} \times \frac{18\tfrac{3}{16}}{6\tfrac{3}{16}}$$

$$= \frac{419}{323} \times \frac{323}{195} \times \frac{323}{131} \times \frac{291}{99}$$

Therefore speed of dynamo $= \dfrac{419}{323} \times \dfrac{323}{195} \times \dfrac{323}{131} \times \dfrac{291}{99} \times 95 = 1479{\cdot}42$ revolutions per minute.

When the thickness of the band was neglected, we saw that the speed of the dynamo was 1543·75 revolutions per minute. The thickness of the band in this example, therefore, affects the speed of the dynamo by $1543{\cdot}75 - 1479{\cdot}42 = 64{\cdot}33$ revolutions per minute, or $\dfrac{64{\cdot}33 \times 100}{1543{\cdot}75} = 4{\cdot}16$ per cent.

176. Strength of Belting.—The ultimate tenacity of the leather used for belting is generally taken as 3000 lbs. per square inch of section, but some qualities have a tenacity as high as 5000 lbs. per square inch. The strength at a laced joint is only about one-third of the strength of the solid leather. The safe working stress may be taken at from one-fourth to one-third of the ultimate strength of the weakest part of the belt. If, therefore, the belt has a laced joint, the safe working tension may be taken at from 250 to 350 lbs. per square inch of belt section.

It is more convenient, however, to give the strength of a belt *per inch of width*, as shown in the following table.

Safe Working Tension for Leather Belts in Lbs. per Inch of Width.

Thickness of belt in inches		$\frac{3}{16}$	$\frac{7}{32}$	$\frac{1}{4}$	$\frac{5}{16}$
Safe tension in lbs. per inch of width when the safe stress in lbs. per square inch is	250	47	55	62	78
	300	56	66	75	94
	350	66	77	87	109

Information on the strength of cotton belting is not so plentiful as that on leather belting, but it is generally assumed that a four-ply cotton belt is equivalent to a single leather belt, and an eight-ply cotton belt to a double leather belt; but a test made by Mr. Kirkaldy showed that an eight-ply cotton belt had an ultimate tenacity of 1135 lbs. per inch of width, the test being made on a belt 6 inches wide.

177. Transmission of Power by Bands.—When the driving pulley AB, Figs. 354 and 355, begins to rotate in the direction of the arrow, the tension in the portion BD of the band will be increased, while the tension in the portion AC will be diminished, and while the motion continues the tension in BD will be greater than the tension in AC. It is evident that if the tensions in BD and AC were equal, the pulley CD

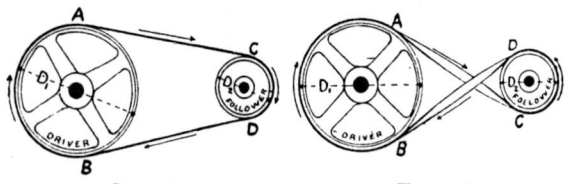

FIG. 354.   FIG. 355.

would not rotate, because it would then be pulled round in opposite directions by equal forces acting at the same leverage. Let T_1 be the tension in BD, the driving side of the band, and T_2 the tension in AC, the slack side; then $T_1 - T_2 = P$, is the effective force acting at the circumference of the pulley, tending to turn it. This force, P, is called the driving force or driving tension. If H is the horse-power which is being transmitted by a band which has a linear velocity of V feet per minute, and if P is the driving force, in lbs., then

$$H = \frac{V P}{33000}.$$

BELT GEARING.

If the band passes over a pulley which makes N revolutions per minute, and which has a diameter D (in feet), then

$$H = \frac{3 \cdot 1416\, DNP}{33000}.$$

Mr. Alfred Towler, of Leeds, gives the following simple formulæ for determining the horse-power transmitted by belts:—

$$H = \frac{DWAR}{2000} \text{ for single belts.}$$

$$H = \frac{1 \cdot 75\, DWAR}{2000} \text{ for double belts.}$$

D = diameter of smaller pulley in inches.
W = width of belt in inches.
A = ratio of arc covered by belt to circumference.
R = revolutions of smaller pulley per minute.

For rough calculations on the power of single leather belts the following rule may be used:—$H = \frac{BV}{800}$, where H = horse-power, B = width of belt in inches, and V = velocity of belt in feet per minute.

178. Friction of a Band on a Pulley.—In Fig. 356 is shown a band passing over a pulley, the arc of contact being ABC. Let T_1 be the tension on the driving side of the band, and T_2 the tension on the slack side. As already stated, T_1 will be greater than T_2. When the ratio of T_1 to T_2 is greatest, that is, when the band is on the point of slipping on the pulley, it will be equal to $e^{f\theta}$, that is, $\frac{T_1}{T_2} = e^{f\theta}$ where f is the coefficient of friction between the band and the pulley, θ the circular measure of the angle of contact AOC, so that $\theta = \frac{\text{arc ABC}}{\text{radius OA}}$, and e is the base of the Naperian system of logarithms.

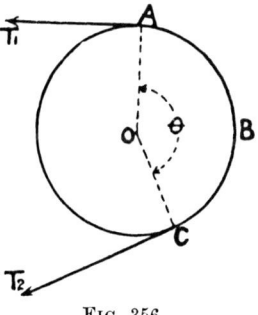

FIG. 356.

The above formula can only be used by aid of logarithms, thus—

$$\log. \frac{T_1}{T_2} = f\theta \log.e = \cdot 4343\, f\theta.$$

If the angle AOC contains n degrees, then—

$$\log. \frac{T_1}{T_2} = \cdot 00758\, fn.$$

Having calculated the value of $\log. \frac{T_1}{T_2}$, the value of $\frac{T_1}{T_2}$ may be obtained from a table of logarithms.

The following table gives values of $\frac{T_1}{T_2}$ corresponding to various values of f and n.

Values of the Ratio of the Tensions on the Tight and Slack Sides of a Band when on the Point of Slipping.

Angle of Contact in Degrees $= n$.	Ratio of Tensions $= \dfrac{T_1}{T_2}$.			
	$f = \cdot 2$.	$f = \cdot 3$.	$f = \cdot 4$.	$f = \cdot 5$.
20	1·072	1·110	1·150	1·191
40	1·150	1·233	1·322	1·418
60	1·233	1·369	1·520	1·688
80	1·322	1·520	1·748	2·010
100	1·418	1·688	2·010	2·393
120	1·520	1·874	2·311	2·850
140	1·630	2·081	2·658	3·393
160	1·748	2·311	3·056	4·040
180	1·874	2·566	3·514	4·811
200	2·010	2·850	4·040	5·728
220	2·155	3·164	4·646	6·820
240	2·311	3·514	5·342	8·121
270	2·566	4·111	6·586	10·551
300	2·850	4·811	8·121	13·709

For leather belting on iron pulleys the coefficient of friction may be taken at ·4.

179. Speed of Belts.—For main driving belts the lineal velocity should be from 3000 to 4000 feet per minute. In America a speed as high as 6000 feet per minute has been attained in a main driving band, but such a high velocity as this is unusual.

180. Examples.—The following are examples of the calculations on belting which have very frequently to be made by engineers:—

EXAMPLE 1.—*A belt running at a speed of* 2000 *feet per minute transmits* 20 *horse-power. Required: first, the driving tension, that is, the difference between the tensions on the tight and slack sides; second, the tension on the tight side, assuming that it is twice the tension on the slack side; third, the width of the belt, assuming that the safe tension per inch of width is* 90 *lbs.*

Taking the formula $H = \dfrac{VP}{33000}$

we have $20 = \dfrac{2000 \times P}{33000}$

or P, the driving tension $= \dfrac{20 \times 33000}{2000} = 330$ lbs.

But $P = T_1 - T_2$; therefore $T_1 - T_2 = 330$
and $T_2 = \tfrac{1}{2} T_1$; therefore $T_1 - \tfrac{1}{2} T_1 = 330$
or $\tfrac{1}{2} T_1 = 330$; therefore $T_1 = 660$ lbs.,

and the required width of belt is $\dfrac{660}{90} = 7\cdot 33$ in.

EXAMPLE 2.—*To find the horse-power which may be transmitted by a belt* 6 *inches wide, which passes over a pulley* 5 *feet in diameter, making*

130 *revolutions per minute*. *The arc of contact between the belt and the pulley subtends an angle of* 120 *degrees at the centre, and the coefficient of friction is known to be* ·4, *and the tension on the tight side of the belt is not to exceed* 80 *lbs. per inch of width.*

Referring to the table already given, we find that with an arc of contact of 120 degrees and a coefficient of friction of ·4, the ratio of the tension on the tight side to that on the slack side is 2·311, that is,

$T_2 = \dfrac{T_1}{2·311}$. But $T_1 = 6 \times 80 = 480$ lbs. Therefore $T_2 = \dfrac{480}{2·311} = 207·7$ lbs.

The driving tension $P = T_1 - T_2$ is therefore $= 480 - 207·7 = 272·3$ lbs.
The velocity of the belt is $= 5 \times 3·1416 \times 130 = 2042$ feet per minute.

Therefore the horse-power $= H = \dfrac{VP}{33000} = \dfrac{2042 \times 272·3}{33000} = 16·85$.

To obtain the above result it would be necessary to stretch the belt over the pulleys in the first instance under a tension at least $= \dfrac{T_1 + T_2}{2} = \dfrac{480 + 207·7}{2} = 343·85$ lbs.

181. "Creep" in Belts.—When a belt is placed under tension it stretches, and the greater the tension the greater must be the stretching. Now it has been shown that when power is transmitted from one pulley to another by a belt, the driving side is subjected to a greater tension than the other or slack side. It follows therefore that, paradoxical as it may seem, for every foot of belt which goes on to the driving pulley, less than a foot comes off and goes on to the following pulley. In other words, the speed of the slack side is slightly less than the speed of the tight or driving side. But the speed of the rim of a pulley is equal to the speed of the band which comes on to it; therefore the rim of the driver must have a higher speed than the rim of the follower, and therefore the speed of the follower will be less than that given by the formula

$$\dfrac{N_2}{N_1} = \dfrac{D_1}{D_2}.$$

For example, if the pulleys were of exactly the same diameter, the driver would make a larger number of revolutions per minute than the follower. The loss of speed due to the above may be taken at from 1 to 3 per cent., depending on the elasticity of the material of the belt.

182. Rims of Pulleys for Flat Bands.—The rims of pulleys for flat

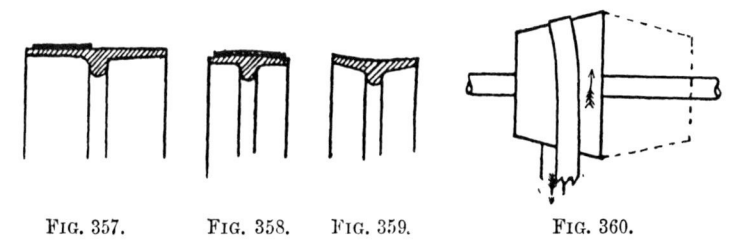

Fig. 357.   Fig. 358.   Fig. 359.   Fig. 360.

bands are made either flat or straight on the outside of their cross sections, as in Fig. 357, or curved, as in Fig. 358 At first sight it

K

would seem as if the band would remain on the straight rim more readily than on the curved one, and that it would be still more secure from falling off if the section of the rim were concave, as in Fig. 359, instead of convex, as in Fig. 358, but experiment soon shows that the band tends to run on the highest part of the pulley. This action of the band in climbing to the highest part of the rim is easily explained. If a band runs on a conical pulley, as shown in Fig. 360, each part of the band as it approaches the pulley receives a set towards the base or larger end of the cone, so that each part of the band as it passes on to the pulley is a little nearer the base than the part in front of it, but once in contact with the pulley, it remains in contact, without slipping, until it leaves on the other side. The result is, therefore, that the band ultimately reaches the highest part, and to prevent it falling off a similar pulley is placed, as shown by the dotted lines. To prevent the sharp edge where the bases of the two cones meet from cutting the band, this part is rounded, and we get the ordinary convex form of rim.

The flat or straight rim is used where it is necessary to move the band from one side of the rim to the other, as in the case where a pulley drives a pair of fast and loose pulleys.

It must be borne in mind, however, that the convex section of rim only tends to keep the band on the highest part so long as the band and

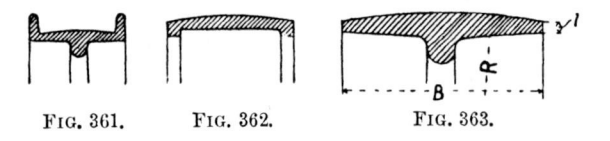

FIG. 361.   FIG. 362.   FIG. 363.

pulley move together. If the band should slip, through the resistance being too great, it will then fall off more readily if the pulley is convex than if it is straight.

When there is a frequent danger of a belt slipping through a temporary increase in the resistance, the pulley may have flanges, as shown in Fig. 361.

The rims of cast-iron pulleys are strengthened by a circumferential rib on the inside, as in Figs. 357, 358, 359, 361, and 363, or by two ribs, as in Fig. 362.

The amount of curvature in the section of the rim may be greater the greater the speed at which it runs. The curve may be struck with a radius R, equal to from 3 B to 5 B, where B is the breadth of the pulley. The breadth of the pulley is generally from $\frac{1}{4}$ inch to $\frac{1}{2}$ inch greater than the width of the belt which it carries.

The thickness t of the rim at the edge in inches may be $\frac{D}{200} + \frac{1}{8}$ for a single belt, and $\frac{D}{200} + \frac{1}{4}$ for a double belt, where D is the diameter of the pulley in inches. The inside of the section should slope slightly, so that the pattern may leave the mould easily.

BELT GEARING.

183. Arms of Pulleys.—In cross section the arms of cast-iron pulleys are generally oval shaped, and of the proportions shown in Fig. 364. In the direction of their length they may be straight, as in Fig. 365, or curved, as in Figs. 366 and 367. For the same strength the curved arm is heavier than the straight one, but owing to its shape it will yield more readily in a radial direction, and therefore allow for the unequal contraction of the rim and boss in cooling. With care, however, the founder can cast a straight-armed pulley free from initial strain, and this form of arm is preferable to the curved one. The curved arms may be drawn by the methods shown in Figs. 366 and 367.

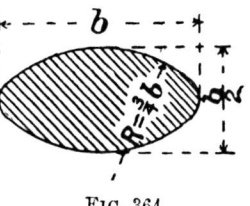

FIG. 364.

There is no recognised rule for the number of arms in a pulley, but in making the preliminary calculations for the dimensions of the arms, it will be sufficient to assume that pulleys up to 18 inches in diameter have

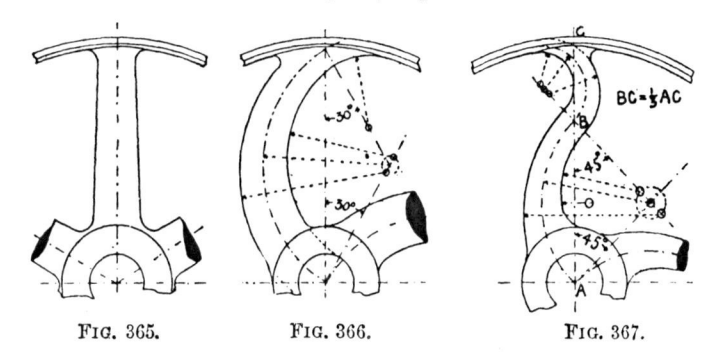

FIG. 365.   FIG. 366.   FIG. 367.

four arms, and pulleys of larger diameter six. If it should then be found that the number of arms assumed, causes them, in the opinion of the designer, to be too heavy or too light for the rest of the pulley, the number may be increased or diminished.

Let P be the driving force in pounds at the circumference of the pulley, that is, the difference between the tensions in the tight and slack sides of the belt, D the diameter of the pulley in inches, and n the number of arms. The greatest bending moment on one arm is approximately $\frac{PD}{2n}$, and the moment of resistance of the arm to bending, if it has a section of the proportions shown in Fig. 364, is approximately $\cdot 05 b^3 f$, where f is the greatest stress in the arm due to the bending action.

Taking f at 2000 lbs. per square inch, so as to allow for the weakening effect of the unequal contraction of the pulley in cooling in the mould, and also to allow of the bending action not being distributed equally to all the arms, we get the equation $\frac{PD}{2n} = \cdot 05 b^3 \times 2000$, or $b = \sqrt[3]{\frac{PD}{200n}}$.

If the driving force P is not definitely known, we may take its maximum value at 50 B for single belts and 100 B for double belts, where B is the width of the pulley. We then have

$$b = \sqrt[3]{\frac{BD}{4n}} \text{ for single belts,}$$

$$\text{and } b = \sqrt[3]{\frac{BD}{2n}} \text{ for double belts.}$$

The breadth b given in the above equations is the breadth of the arm measured at the centre of the pulley, supposing the arm to be produced through the boss to that point. The breadth and thickness of the arm at the rim of the pulley may be two-thirds of the breadth and thickness respectively at the centre.

When a pulley is very wide and the arms are required to be light, two sets of arms may be put in as shown in Fig. 368.

184. Naves of Pulleys.—The central part of a pulley is called the "nave," "eye," or "boss." Its thickness varies considerably in practice. In giving the thickness some authorities state it in terms of the diameter of the pulley only, others take into account the diameter of the pulley and also the diameter of the shaft, while others again express the thickness in terms of the diameter of the pulley and its breadth. The following is a slight modification of a rule recommended by Box in his "Mill Gearing":—

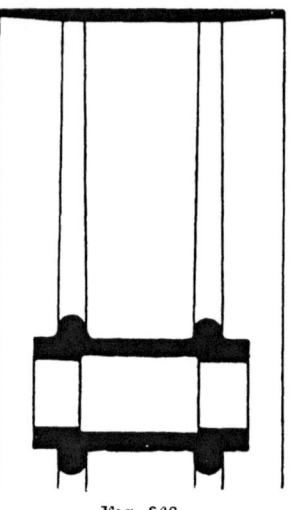

FIG. 368.

$$t = \frac{D}{96} + \frac{d}{8} + \frac{5}{8}$$

where t is the thickness of the nave, D the diameter of the pulley, and d the diameter of the shaft, all the dimensions being in inches.

Unwin, in his "Machine Design," gives the following:—

$$t = \cdot 14 \sqrt[3]{BD} + \tfrac{1}{4} \text{ for a single belt,}$$
$$\text{and } t = \cdot 18 \sqrt[3]{BD} + \tfrac{1}{4} \text{ for a double belt,}$$

t being the thickness of the nave as before, B the width of the pulley, and D its diameter, all in inches.

The length of the nave may be from $\tfrac{2}{3}$ B to B.

The above rules apply to "fast" pulleys, or pulleys which have to be keyed to their shafts. Loose pulleys have generally longer naves, and they need not be so thick, and they are frequently lined with brass bushes.

Fig. 369 shows the forms of bosses or naves for fast and loose pulleys made by Messrs Geo. Richards & Co., Manchester. It will be observed

BELT GEARING.

that the boss of the loose pulley is of considerable length, so as to give a large wearing surface.

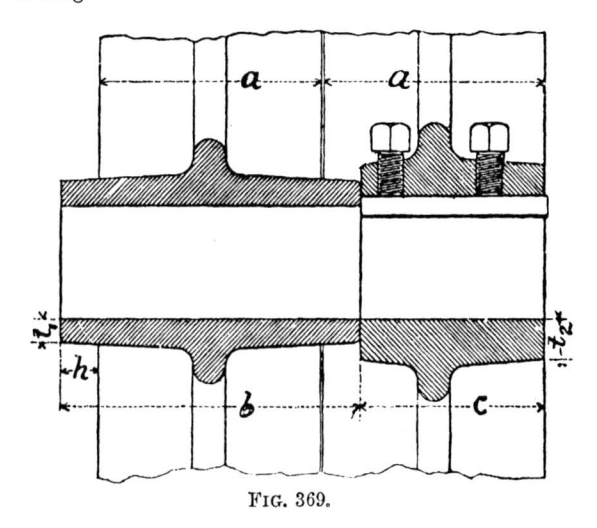

FIG. 369.

The proportions are as follows, all dimensions being in inches:—
Breadth of pulley = a.
Length of boss of loose pulley = $b = a + 1\frac{1}{2}$.

Length of boss of fast pulley = $c = \dfrac{a}{2} + 1\frac{3}{8}$.

Projection of boss of loose pulley = $h = \frac{3}{4}$ inch.

t_1, the thickness of the boss of the loose pulley, and t_2, the thickness of the boss of the fast pulley, are shown in the following table:—

Diameter of Pulley.	Width of Pulley.										
	$3\frac{1}{4}$		$4\frac{1}{4}$		$5\frac{1}{4}$		$6\frac{1}{4}$		$7\frac{1}{4}$		$8\frac{1}{4}$	
	t_1	t_2	t_1	t_2	t_1	t_2	t_1	t_2	t_1	t_2	t_1	t_2
9	$\frac{3}{8}$	$\frac{5}{8}$	$\frac{7}{16}$	$\frac{11}{16}$	$\frac{7}{16}$	$\frac{3}{4}$	...	...	...	...	...	...
10	$\frac{3}{8}$	$\frac{5}{8}$	$\frac{7}{16}$	$\frac{11}{16}$	$\frac{7}{16}$	$\frac{3}{4}$	$\frac{1}{2}$	$\frac{13}{16}$	...	...	...	...
12	$\frac{3}{8}$	$\frac{11}{16}$	$\frac{7}{16}$	$\frac{3}{4}$	$\frac{1}{2}$	$\frac{13}{16}$	$\frac{1}{2}$	$\frac{7}{8}$	...	...	...	...
14	$\frac{7}{16}$	$\frac{3}{4}$	$\frac{7}{16}$	$\frac{13}{16}$	$\frac{1}{2}$	$\frac{7}{8}$	$\frac{1}{2}$	$\frac{9}{16}$	$\frac{15}{16}$	...	...	...
16	$\frac{1}{2}$	$\frac{13}{16}$	$\frac{1}{2}$	$\frac{7}{8}$	$\frac{9}{16}$	$\frac{7}{8}$	$\frac{9}{16}$	$\frac{15}{16}$	...	...	...	...
18	...	...	$\frac{9}{16}$	$\frac{7}{8}$	$\frac{9}{16}$	$\frac{7}{8}$	$\frac{5}{8}$	$\frac{15}{16}$	$\frac{5}{8}$	1	$\frac{5}{8}$	1

Further details of pulleys made by the same firm are given in Art. 187.

185. Tension in a Pulley Rim Due to Centrifugal Force.—As the rim of a pulley rotates, each portion tends to fly outwards and produces a tension in the rim, and if the speed is high enough the rim may burst. The relation between the velocity of the pulley rim and the tension in it is determined as follows :—

Let a be the area of the cross section of the rim in square inches; w the weight of a cubic inch of the material, in pounds; v the lineal velocity of the rim, in feet per second; and D its mean diameter, in inches.

The weight of a portion of the rim 1 in. long is equal to aw lbs., and the centrifugal force of this portion is, by the formula given in Art. 41, p. 20, equal to $\dfrac{awv^2}{gR}$; where g is the accelerating effect of gravity, and may be taken at 32 ft. per second, and R is the radius of the circle described by the body *in feet*, so that R is equal to $\dfrac{D}{24}$. We have therefore the centrifugal force of 1 in. of the rim equal to $\dfrac{awv^2}{\frac{32D}{24}} = \dfrac{3awv^2}{4D} = q$. Each inch of the rim will have the same amount of centrifugal force acting on it, so that the result is a uniformly distributed force acting on the rim as shown by the small radial arrows q in Fig. 370. Now by a well-known theorem in mechanics the resultant force Q of all the forces qq, etc., on one side of a diameter AB is equal to qD. Therefore $Q = \dfrac{3awv^2 D}{4D} = \dfrac{3awv^2}{4}$. Examining Fig. 370, we see that the tendency of the resultant forces QQ is to tear the rim asunder at the points A and B, and if f is the stress produced, we have $2af = Q = \dfrac{3awv^2}{4}$, or

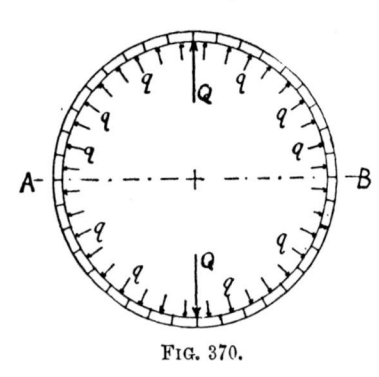

FIG. 370.

$f = \dfrac{3wv^2}{8}$. From this result we see that the area of the section of the rim has no influence on the stress produced by the centrifugal force, so that to add to the section does not add to the strength to resist bursting, because although an increase in the section of the rim produces a corresponding increase in the resistance to bursting, the bursting force is increased in the same proportion.

The value of w may be taken at ·26 for cast-iron, and ·28 for wrought-iron. The following table gives values of the stress f corresponding to various velocities of the rim calculated from the formula $f = \dfrac{3wv^2}{8}$:—

BELT GEARING.

Velocity of rim in feet per second	60	80	100	150	200
Velocity of rim in feet per minute	3,600	4,800	6,000	9,000	12,000
Stress in rim in lbs. per square inch — Cast-iron	351	624	975	2,194	3,900
Stress in rim in lbs. per square inch — Wrought-iron	378	672	1,050	2,362	4,200

In addition to the stress due to the centrifugal force there is the stress due to the unequal contraction in cooling, and also that due to the driving force of the belt. In practice it is not considered safe to run a cast-iron pulley rim at a higher speed than 6000 ft. per minute.

186. Split Pulleys.—For convenience in fixing pulleys on shafts it is a common practice to make them in halves, which are bolted together as shown in Figs. 371 and 372, which illustrate an example of a cast-

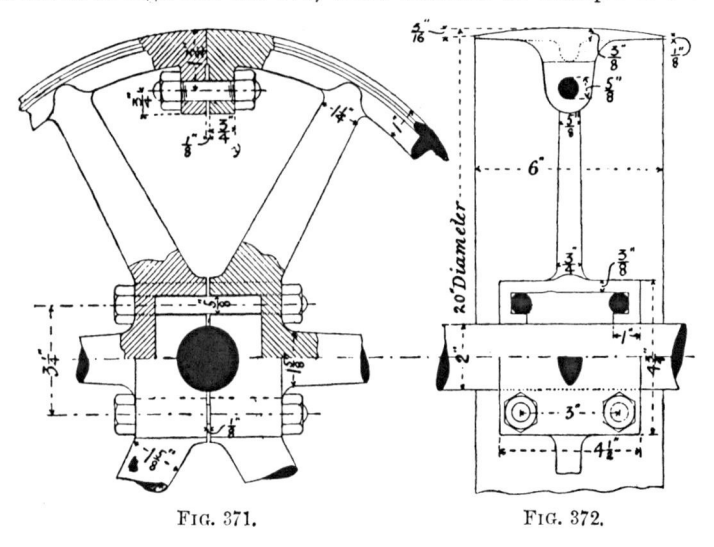

FIG. 371. FIG. 372.

iron split pulley 20 in. by 6 in. for a 2-in. shaft. By leaving a small clearance space between the two halves of the pulley at the nave, the latter may be tightened on to the shaft so firmly by means of the bolts that no key is necessary.

187. Wrought-Iron Pulleys.—An enormous number of wrought-iron pulleys are now made, and they possess several important advantages over cast-iron ones. They are much lighter, and, being built up, they are free from the initial strains which exist in cast-iron pulleys due to unequal contraction in cooling. They are also safer at high speeds, not only because wrought-iron has a greater tenacity than cast-iron, but because if a wrought-iron pulley should burst, it will not fly to pieces like a cast-iron one, since cast-iron is so much more brittle than wrought-iron. Wrought-iron pulleys can be made of any diameter required, but it is when they are of large diameter that their superiority over cast-iron pulleys is most decided.

In a wrought-iron pulley the rim is made of sheet-iron, and riveted to the arms, which are made from bar-iron. The boss is sometimes also made of wrought-iron, but generally it is made of cast-iron.

The differences in the design of wrought-iron pulleys are chiefly in the form of the arms. A very common form is that shown in Fig. 373. The arms in this case are round bars fixed in the boss at one end, and riveted to the rim at the other, as shown. In Macbeth's pulley the arms

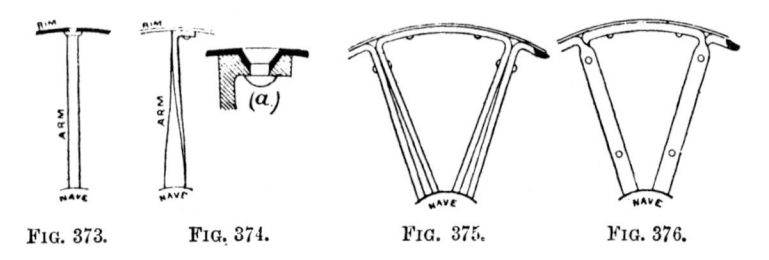

FIG. 373.   FIG. 374.   FIG. 375.   FIG. 376.

are made from flat bars, which are twisted and kneed, as shown in Fig. 374. The outer end of the arm is shown in the detail drawing at (a), from which it will be seen that the bed for the countersunk head of the rivet is not formed by drilling, but by forcing the thin rim into the countersunk hole in the arm. By this means a larger rivet head can be used, and the rim is more firmly secured to the arm. Fig. 375 shows the form of the arms used in Mackie's pulleys. They are made from bars having a cross section which is a segment of a circle. In the pulley named by its makers the "universal" pulley, the arms are constructed as shown in Fig. 376. The cross section of the wrought-iron bars from which the arms are made is a segment of a circle.

In the wrought-iron pulleys as made by Messrs. Croft & Perkins, Bradford, the rim is rolled of a special section, being thicker at the

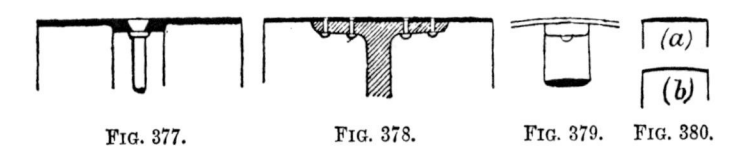

FIG. 377.   FIG. 378.   FIG. 379.   FIG. 380.

centre, as shown in Fig. 377. This gives greater strength to the rim, and allows of a larger rivet head being formed on the end of the arm.

In Medart's pulley, as made by Messrs. Geo. Richards & Co., Manchester, the arms and boss are made of cast-iron in one piece, and the arms are riveted to a wrought-iron rim, as shown in Figs. 378 and 379.

The rims of wrought-iron pulleys are generally cylindrical, that is, straight or flat in their cross section. If the rim is required to be barrel-shaped, it may be rolled so as to have the section shown at (*a*) Fig. 380, or it may be bulged out by rolling to the form shown at (*b*).

Steel often takes the place of wrought-iron in the construction of pulleys.

When wheels and pulleys run at a high speed, there is a loss of power due to the churning of the air by the arms. This loss is diminished by substituting for the arms a central disc of sheet-iron, as in Sherwin's disc pulleys.

In Hall's corrugated disc pulley there is a central disc of sheet-iron which has radial corrugations like those in a lady's fan. These corrugations add greatly to the strength and stiffness of the disc.

When a broad flat band runs on a pulley at a high speed, a cushion of air is drawn in between the band and the pulley, and diminishes the resistance to slipping. This cushion of air is got rid of in Shepherd's perforated pulley. In this pulley the rim has a large number of small holes in it, which allow the air between the band and the pulley to escape.

CHAPTER XI.

ROPE GEARING.

188. The System of Rope Gearing.—The credit of the introduction of the system of rope gearing, now so largely used in mills, is due to Mr. James Combe, of Belfast, who first applied it about the year 1863, but the extension of the system is largely due to Messrs. Pearce Brothers, engineers, Dundee. In this system the prime mover has a large fly-wheel, the circumference of which is grooved to receive a number of ropes, which pass direct to pulleys similarly grooved in the various rooms of the mill or factory. It follows, therefore, that a breakdown in the connection between the fly-wheel and any one room does not interfere with the transmission of the power to any other room.

Mr. D. K. Clark, in his exhaustive treatise on the steam-engine, gives an excellent example of the transmission of power by rope gearing, and to that source we are indebted for the following particulars. In

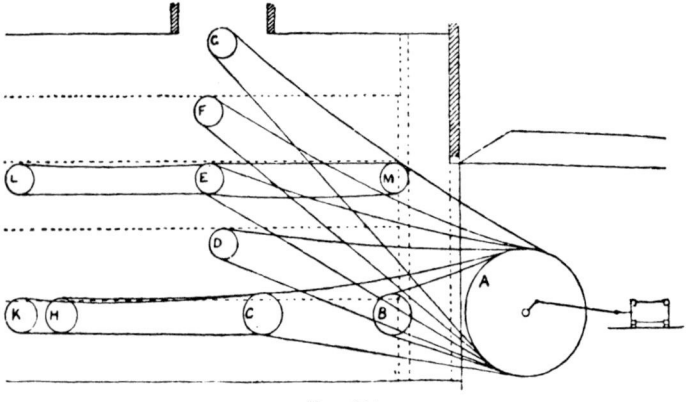

FIG. 381.

this example, which was designed by Messrs. Hick, Hargreaves, & Co., Bolton, A, Fig. 381, is the fly-wheel of a steam-engine. This fly-wheel is 26 ft. in diameter, and is grooved to receive thirty-five ropes, each 5 in. in circumference, or nearly $1\frac{5}{8}$ in. in diameter. Four of the thirty-five ropes pass from the fly-wheel to the pulley B, which is 8 ft. 3 in. in diameter. Five ropes pass to the pulley C, which is also 8 ft. 3 in. in diameter. Five ropes pass to the pulley D, which is 5 ft. 8¾ in. in diameter. Seven ropes pass to each of the pulleys E, F, and G, which have the same diameter as D. On the same shaft as the pulley C there is a pulley 8 ft. 3 in. in diameter, from which passes four ropes, two to each of the pulleys H and K, which are each 6 ft. 4 in. in diameter. The pulleys L and M are driven from the pulley E by two ropes to each. All the above pulleys are in a chamber called the rope chamber, which is built against the mill. The power required for the various machines in any one room is taken by belts directly or indirectly from the main shaft in that room, this main shaft being driven direct from the engines as already described.

189. Ropes for Rope Gearing.—The material used for the ropes in the rope gearing of mills is either hemp or cotton. Formerly hemp was much preferred, but cotton is now largely used as the material for these ropes. The ropes are white or untarred, and are "hawser laid," that is, they are formed with three "strands" twisted together right-handed. A strand is made by twisting "yarns" together left-handed. The yarns are spun from the fibres of the hemp or cotton with a right-handed twist.

The ropes most commonly used are from 1½ inches to 2 inches in diameter. The size of a rope is very often given by its circumference or *girth*. The net section of a rope is about 9-10ths of the area of the circumscribing circle, and its girth is a little less than the circumference of that circle.

The breaking weight of the ropes for rope gearing varies greatly, being as low as 7000 lbs. per square inch in some cases, and as high as 12,000 lbs. per square inch in others. 1200 lbs. per square inch is considered a fair working stress for these ropes, but in rope gearing the practice is to make the stress on the tight or driving side about 140 lbs. per square inch, so that the working stress is only about 1-60th of the average breaking load. This ensures great durability.

The weight of ropes when dry is given approximately by the formula $W = \cdot 3 D^2$, where W is the weight per foot of length, and D the diameter of the rope.

To make a rope into an endless one, its ends are connected by splicing. The splice should be carefully made, so as to preserve a uniformity in the rope, and it should have a length about sixty times the diameter of the rope.

According to Mr. James Durie, the life of a rope in rope gearing is from three to five years, but some ropes have lasted ten and a half years.

The speed of the rope is from 3000 to 6000 feet per minute. The average which we have found from a great many examples in practice is a speed of 4500 feet per minute.

In the following table particulars are given which will facilitate calculations on rope gearing. The girth of the rope is taken as the circumference of the circumscribing circle. The gross section is the area of the circumscribing circle, and the net section is taken as 9-10ths of the gross section. The weight of the rope per foot of length is taken $= \cdot 3 D^2$. The breaking weight is calculated at 8500 lbs. per square inch of gross section. The tension on the driving side of the rope is taken at 140 lbs. per square inch, and the driving force at 105 lbs. per square inch of gross section.

Diameter of Rope in Inches.	Girth of Rope in Inches.	Gross Section of Rope in Sq. Inches.	Net Section of Rope in Sq. Inches.	Weight of Rope per Foot in Lbs.	Breaking Load in Lbs.	Tension on Driving Side in Lbs.	Driving Force in Lbs.
1	3·14	·785	·706	·30	6,672	109·9	82·4
1⅛	3·53	·994	·895	·38	8,449	139·2	104·4
1¼	3·93	1·227	1·104	·47	10,429	171·8	128·8
1⅜	4·32	1·484	1·336	·57	12,614	207·8	155·8
1½	4·71	1·767	1·590	·67	15,019	247·4	185·5
1⅝	5·10	2·073	1·866	·79	17,620	290·2	217·7
1¾	5·50	2·405	2·164	·92	20,442	336·7	252·5
1⅞	5·89	2·761	2·485	1·05	23,468	386·5	289·9
2	6·28	3·141	2·827	1·20	26,698	439·7	329·8

190. Pulleys for Rope Gearing.—The pulleys for rope gearing are generally made entirely of cast-iron, but occasionally the arms are made of wrought-iron. The rim is grooved to receive the ropes, as shown in Fig

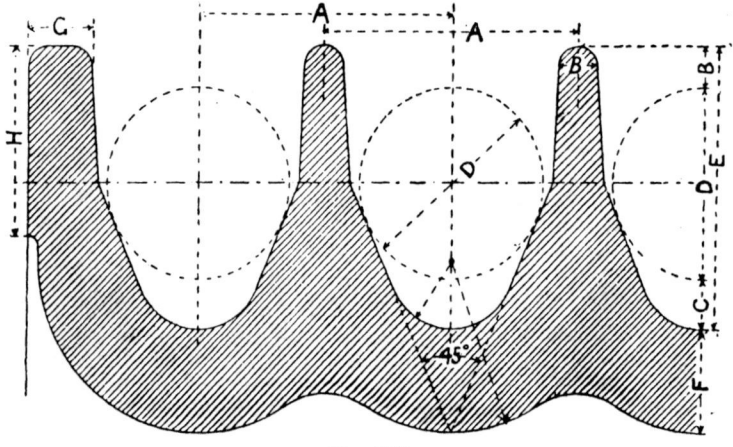

FIG. 382.

382, which is a cross section of a portion of a wheel rim, as made by several good makers. The angle of the groove for the rope is usually 45 degrees, but it is sometimes as small as 30 degrees. The following are the proportions of the different parts, when the angle of the groove is 45 degrees:—

D = diameter of rope.
A = $1\frac{1}{4}$D + $\frac{1}{4}$.
B = $\frac{1}{8}$D + $\frac{3}{16}$.
C = $\frac{1}{4}$D.

E = $1\frac{3}{8}$D + $\frac{3}{16}$.
F = $\frac{1}{2}$D + $\frac{1}{16}$.
G = $\frac{1}{4}$D + $\frac{3}{16}$.
H = D.

ROPE GEARING.

It will be observed that the rope does not rest on the bottom of the groove, but on the sides only, so that it is wedged in, causing the resistance to slipping to be much greater.

To diminish the chance of a rope leaving its groove some engineers make the ribs between the grooves higher than is shown in Fig. 382, and more like that shown in Fig. 383.

When a pulley is used as a guide pulley only, the rope rests on the bottom of the groove, which has the form shown in Fig. 384.

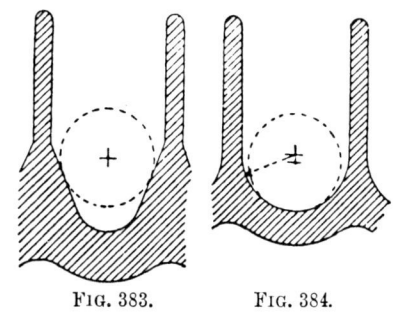

FIG. 383.   FIG. 384.

The diameter of a rope pulley is measured to the centre of the rope. The diameter of a rope pulley should not be less than thirty times the diameter of the rope. The larger the pulley is the less work is consumed in bending the rope as it goes on, and in straightening it as it comes off the pulley.

The horizontal distance between the centres of two pulleys connected by ropes is not often less than 30 feet, and it may amount to 100 feet. The initial tension in the rope is generally small, and is chiefly due to the weight of the rope, hence the reason for placing the pulleys so far apart.

Whenever possible the driving side of the rope is placed on the bottom, and the slack side on the top. This arrangement makes the arcs of contact between the rope and the pulleys greater, and therefore gives greater resistance to slipping.

We shall now illustrate and describe a rope pulley made by Mr. Benjamin Goodfellow, of Hyde, Manchester. This pulley acts as the fly-wheel of an engine. It is 32 feet in diameter, and has 36 grooves

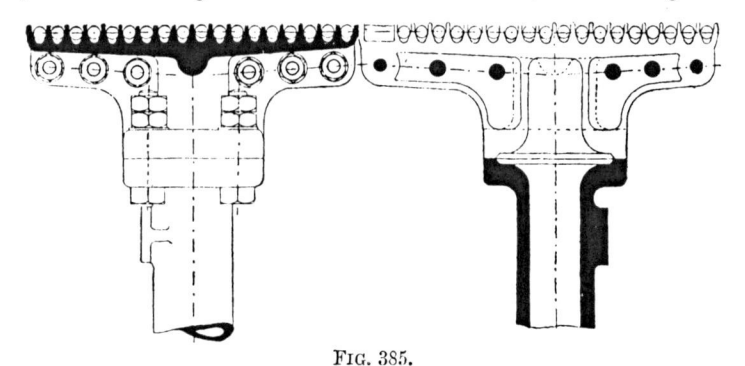

FIG. 385.

for ropes 1¾ inches in diameter. As will be seen from the section of the rim in Fig. 385, the wheel consists of two main parts, each with its set of arms. There are two separate bosses, which are bored and faced

158   MACHINE DRAWING AND DESIGN.

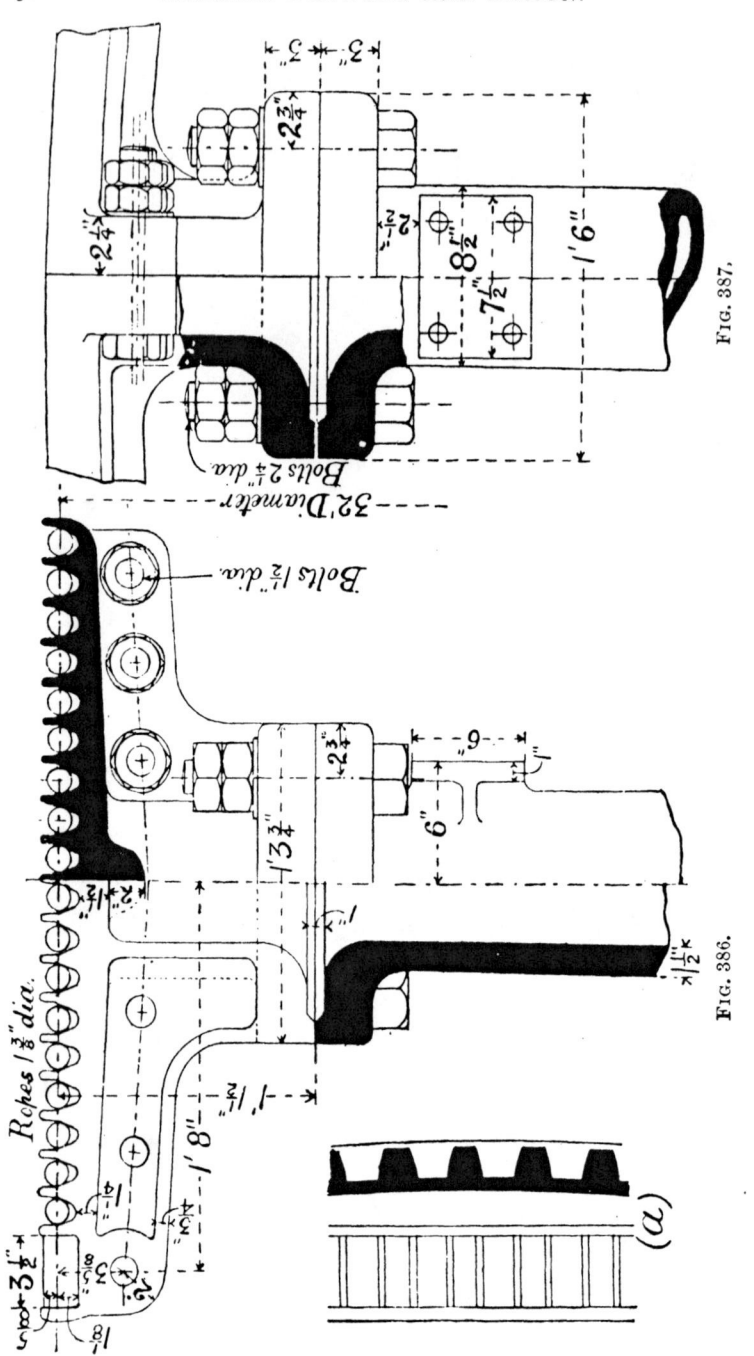

Fig. 387.

Fig. 386.

ROPE GEARING.

to receive the inner ends of the arms. There are in all 24 arms, 12 to each boss, which are attached to the bosses by steel cotters. Each arm is secured to a segment of the rim by four turned bolts, two of which are "fitting bolts." The rim is in 24 segments, one to each arm, which are faced to form close joints with one another, and with the outer ends of the arms. The segments of the rim are bolted together as shown. In the centre of the rim, and extending completely round it, is a toothed portion called a "barring rack," into which can be made to gear a pinion driven by a small "barring engine." The barring of an engine is resorted to when it is required to turn it very slowly, or to turn it with precision to some definite position. Figs. 386 and 387 give all the necessary dimensions of the rim and the outer ends of the arms. A plan and section of a portion of the barring rack is shown at (a).

191. Weight of Large Rope Pulleys.—The following table gives the weights and some other particulars of grooved fly-wheels of steam-engines as used in mills where the power is transmitted by ropes. These examples are taken from actual practice:—

Diameter of wheel in feet . .	12	20	28	30	30	30	30	32	34
Diameter of ropes in inches .	$1\frac{1}{2}$	$1\frac{5}{8}$	$1\frac{3}{4}$	$1\frac{3}{4}$	$1\frac{5}{8}$	$1\frac{3}{4}$	2	$1\frac{5}{8}$	$1\frac{3}{4}$
Number of grooves	7	36	32	25	27	35	27	35	32
Total weight of wheel in tons	14	45	74	70	$54\frac{1}{2}$	60	76	80	78

192. Friction of Rope on a Grooved Pulley.—The resistance to slipping of a rope in a wedge-shaped groove is greater than the resistance to slipping of the same rope on a cylindrical pulley, because in the latter case, Figs. 388 and 389, the pressure (r) between the rope and the pulley at any point is equal to the radial force (q) at the same point

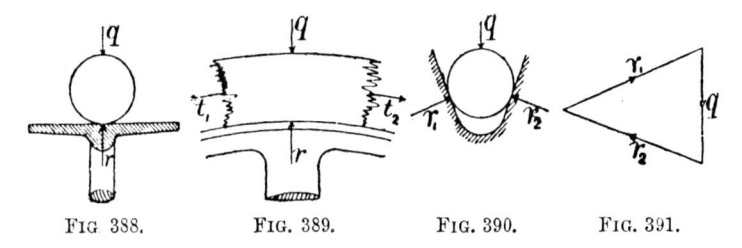

FIG. 388. FIG. 389. FIG. 390. FIG. 391.

produced by the tension in the rope, whereas in the former case, Fig. 390, the pressure (r_1) between the rope and the sides of the groove is much greater, as is shown by the triangle of forces, Fig. 391. If n is the number of degrees in the angle of the groove, then $r_1 = \dfrac{q}{2}$ co-secant $\dfrac{n}{2}$; and if f is the coefficient of friction between the rope and a flat metal surface, the resistance to slipping of the small portion of the rope which we are considering is $r_1 f$ on *each side of the groove*. But the rope, if it slipped, would slip on *both* sides of the groove. Therefore, the total resistance to slipping of this portion of rope is $2r_1 f$ or

qf co-secant $\frac{n}{2}$. The resistance to slipping of the same portion of rope on a cylindrical pulley would be qf. The ratio of the resistances to slipping in these two cases is therefore as co-secant $\frac{n}{2}:1$. The groove, therefore, has the same effect as increasing the coefficient of friction from f to f co-secant $\frac{n}{2}$ on a cylindrical pulley. If for f co-secant $\frac{n}{2}$ we put f_1, then questions on the friction of a rope on a grooved pulley may be treated in the same way, and by the same formulæ, as questions on the friction of a band on a cylindrical pulley if f_1 be used as the coefficient of friction instead of f. The following table gives values of f_1 corresponding to various values of f and n :—

Angle of Groove in Degrees $=n$.	Values of f_1 when $f=$				
	·15	·2	·25	·3	·35
30	·58	·77	·97	1·16	1·35
35	·50	·66	·83	1·00	1·16
40	·44	·58	·73	·88	1·02
45	·39	·52	·65	·78	·91

In practice, with hemp or cotton ropes on iron pulleys, the value of f varies from ·15 to ·35, and therefore for grooves with the usual angle 45 degrees f_1 varies from, say, ·4 to ·9. The value $f_1 = ·4$ is for greased pulleys. For ropes and pulleys in ordinary working condition f_1 may be taken at ·7, the angle of the grooves being 45 degrees.

The following table gives values of the ratio of the tensions on the tight and slack sides of a rope corresponding to various values of the coefficient of friction f_1 and the angle of contact n from the formula:—

$$\text{Log} \frac{T_1}{T_2} = ·00758 f_1 n,$$

which has been already given in connection with belt gearing. T_1 = tension on tight side, and T_2 = tension on slack side.

Values of the Ratio of the Tensions on the Tight and Slack Sides of a Rope when on the Point of Slipping.

Angle of Contact in Degrees $=n$.	Ratio of Tensions $=\frac{T_1}{T_2}$.					
	$f_1=·4$.	$f_1=·5$.	$f_1=·6$.	$f_1=·7$.	$f_1=·8$.	$f_1=·9$.
60	1·52	1·69	1·87	2·08	2·31	2·57
90	1·87	2·19	2·57	3·00	3·51	4·11
120	2·31	2·85	3·51	4·33	5·34	6·59
150	2·85	3·70	4·81	6·25	8·12	10·55
180	3·51	4·81	6·59	9·02	12·35	16·90
210	4·33	6·25	9·02	13·01	18·77	27·08
240	5·34	8·12	12·35	18·77	28·53	43·38

ROPE GEARING.

193. Power Transmitted by Ropes.—Let T_1 be the tension on the tight or driving side of a rope, and T_2 the tension on the slack side. The driving force is then $T_1 - T_2 = P$. Let H be the horse-power transmitted by one rope, and V the velocity of the rope in feet per minute; then

$$H = \frac{VP}{33,000}$$

and $P = \dfrac{33,000\,H}{V}$.

The following table has been calculated on the assumption that $\dfrac{T_1}{T_2} = 4$, and that the greatest stress on the rope is 140 lbs. per square inch of the gross section of the rope, so that the driving force $P = T_1 - T_2$, is 105 lbs. per square inch of the gross section of the rope. The formula for the horse-power for one rope then becomes $H = \dfrac{D^2 \times \cdot 7854 \times 105\,V}{33,000}$

$= \dfrac{D^2 V}{400}$ very nearly, where D is the diameter of the rope. If C, the circumference of the rope, be used instead of the diameter, then $H = \dfrac{C^2 V}{3949 \cdot 44}$ or say $\dfrac{C^2 V}{4000}$.

Diameter of Rope in Inches.	Circumference or Girth of Rope in Inches.	\multicolumn{7}{c}{Horse-Power Transmitted by one Rope when the Speed in Feet per Minute is:—}						
		3000	3500	4000	4500	5000	5500	6000
1	3·14	7·50	8·75	10·00	11·25	12·50	13·75	15·00
1⅛	3·53	9·48	11·07	12·66	14·24	15·82	17·40	18·98
1¼	3·93	11·72	13·67	15·62	17·58	19·53	21·48	23·44
1⅜	4·32	14·18	16·54	18·91	21·27	23·63	26·00	28·36
1½	4·71	16·87	19·69	22·50	25·31	28·12	30·94	33·75
1⅝	5·10	19·80	23·11	26·41	29·71	33·01	36·31	39·61
1¾	5·50	22·97	26·80	30·62	34·45	38·28	42·11	45·94
1⅞	5·89	26·37	30·76	35·15	39·55	43·95	48·34	52·73
2	6·28	30·00	35·00	40·00	45·00	50·00	55·00	60·00

Mr. Alfred Towler, of Leeds, gives the following formula for the power transmitted by ropes:—

$$H = \frac{C^2 V\,(N-1)}{5000}.$$

Where H = horse-power transmitted.
    C = circumference of rope in inches.
    V = velocity of rope in feet per minute.
    N = number of ropes.

L

CHAPTER XII.

WIRE-ROPE GEARING.

194. Telodynamic Transmission.—This is the name given to the method of transmitting power to long distances by means of pulleys and wire-ropes. This method was first introduced by M. Ch. F. Hirn in the year 1850, and since that time it has been used extensively, chiefly on the Continent of Europe. The ropes used in this system are made of iron or steel wires. A rope generally has six strands laid on a central hempen core, each strand consisting of six wires surrounding a hempen core, but the number of strands and the number of wires in each may be varied. The hempen cores give flexibility to the rope, but do not add to its strength. The cores are saturated with tar or boiled oil, free from acid. The wires vary in size from No. 25 to No. 14 Birmingham wire gauge, or from ·02 inch to ·083 inch in diameter. The diameter of the rope varies from $\frac{3}{8}$ in. to $\frac{3}{4}$ in. If D is the diameter of the rope, n the number of wires in it, and d their diameter, then, according to Unwin, $D = \frac{nd}{13} + 7d$.

M. Achard states that the ratio of the sectional area of the rope to its weight per foot of length differs little from a mean value of ·24. Hence, if w is the weight per foot of length, in lbs., $\frac{·7854 d^2 n}{w} = ·24$ or $w = 3·27 d^2 n$. The speed of the ropes varies from 3000 to 6000 feet per minute.

The pulleys are of large diameter, and each carries one rope, unless it is an intermediate pulley, when it carries two. The distance between

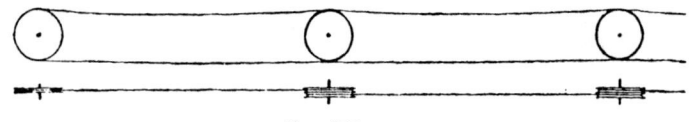

FIG. 392.

the pulleys varies from about 80 feet to 500 feet. When it is required to transmit power to a greater distance than 500 feet (the distance may amount to several miles), it is done either by relays, as shown in Fig. 392, or by introducing guide pulleys, as shown in Fig. 393, so that the rope is supported at intervals not exceeding 500 feet. The guide pulleys have a diameter not less than half the diameter of the other pulleys.

WIRE-ROPE GEARING. 163

A common practice is to make the guide pulleys under the driving side of the rope equal in diameter to the others, and to make those under the slack side half that diameter.

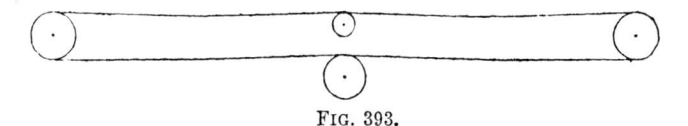

FIG. 393.

Each pulley is fixed on a shaft which is supported on bearings at each end. These bearings are carried by masonry, iron, or wood-framed supports, which are sufficiently high to prevent the rope touching the ground.

When it is required to change the direction of the rope, a horizona guide pulley (A) may be used near to two vertical pulleys (B and C), as shown in Fig. 394, or bevel wheels may be used, as shown in Fig. 395. The latter method is generally preferred.

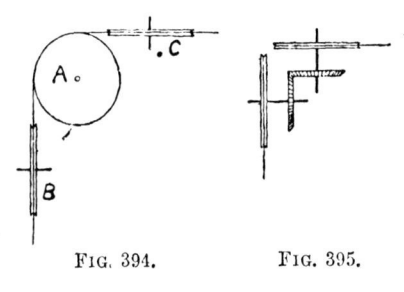

In practice, the friction between the rope and the pulley, and the arc of the contact, are such as to make the tension on the driving side of the rope double that on the slack side. Hence, if T_1 is the tension on the driving side, T_2 the tension on the slack side, and P the driving force, $T_1 = 2T_2$, and $P = T_1 - T_2 = T_2$.

FIG. 394. FIG. 395.

The horse-power transmitted when the rope has a velocity of V feet per minute is $H = \dfrac{PV}{33,000}$.

195. Tensions in a Suspended Flexible Rope.—The curve in which a flexible rope, ACB, Fig. 396, hangs when suspended from points A and B is known as a *catenary*, but for questions on wire-rope gearing it is

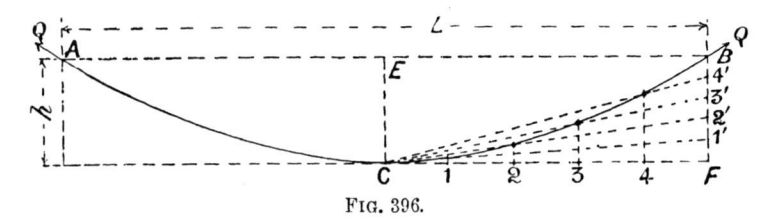

FIG. 396.

sufficiently accurate to substitute for this curve a *parabola*. If the points A and B are at the same level, the vertex C of the parabola is vertically under the middle point E of AB. The distances AB and EC being known, the curve is determined as follows :—Divide the horizontal line CF into any number of equal parts at the points 1, 2, etc., and divide the vertical BF into the same number of equal parts at the points 1', 2', etc. Lines drawn from C to the points 1', 2', etc., to meet the vertical lines from the points 1, 2, etc., determine points on the curve, and these being joined by a fair curve, the required parabola is obtained.

If the points A and B are not at the same level, Fig. 397, but are at

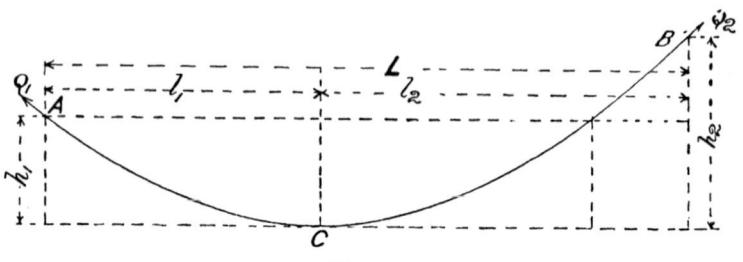

FIG. 397.

heights h_1 and h_2 respectively above the lowest point of the rope, the horizontal distances l_1 and l_2 of the vertex C from A and B are found by the following equations:—

$$l_1 = \frac{L\sqrt{h_1}}{\sqrt{h_1} + \sqrt{h_2}} \qquad l_2 = \frac{L\sqrt{h_2}}{\sqrt{h_1} + \sqrt{h_2}}$$

where L is the horizontal distance between A and B.

Having fixed the position of the point C, the two portions AC and BC of the parabola may be drawn as already explained.

The tensions in the rope are due to its weight, and vary from a minimum at the lowest point to a maximum at the highest point.

Let w be the weight of the rope per foot of length, and let all the distances be measured in feet, then when A and B are at the same level, Fig. 396, the tension at the point C is

$$T = \frac{wL^2}{8h}$$

When A and B are at different levels, Fig. 397, the tension at the point C is

$$T = \frac{wl_1^2}{2h_1} = \frac{wl_2^2}{2h_2}.$$

When a rope hangs in a catenary curve, the difference between the tensions at any two points in the curve is equal to the weight of a portion of the rope whose length is equal to the difference between the levels of the two points. Hence, neglecting the difference between the catenary and the parabola, if Q be the tension at A or B, Fig. 396, $Q - T = wh$. Therefore,

$Q = T + wh = \dfrac{wL^2}{8h} + wh$; and if Q_1 and Q_2 be the tensions at A and B respectively, Fig 397, $Q_1 - T = wh_1$. Therefore, $Q_1 = T + wh_1 = \dfrac{wl_1^2}{2h_1} + wh_1$, and $Q_2 = \dfrac{wl_2^2}{2h_2} + wh_2$. Also $Q_1 = Q_2 - w(h_2 - h_1)$.

If we solve the equation, $Q = \dfrac{wL^2}{8h} + wh$, for h we get

$$h = \dfrac{Q}{2w} \pm \sqrt{\dfrac{Q^2}{4w^2} - \dfrac{L^2}{8}}$$

which shows that for one value of Q there are two values of h, namely—

$$h = \dfrac{Q}{2w} + \sqrt{\dfrac{Q^2}{4w^2} - \dfrac{L^2}{8}}, \text{ and } h = \dfrac{Q}{2w} - \sqrt{\dfrac{Q^2}{4w^2} - \dfrac{L^2}{8}}$$

that is, there are two different deflections of the rope which will produce the same tension at the ends, except in the case where $\dfrac{Q^2}{4w^2} = \dfrac{L^2}{8}$; then $h = \dfrac{Q}{2w}$.

196. Stresses in a Wire-Rope when Transmitting Power.—The total stress on a wire rope when transmitting power is made up (1) of the stress due to the weight of the rope per foot of length, the distance between the pulleys and the depth of the curve in which the rope hangs; (2) of the stress due to the bending of the rope to the curve of the pulley rim; and (3) of the stress due to the centrifugal force.

The tension due to the weight of the rope has already been given, namely—

$$Q = \dfrac{wL^2}{8h} + wh \text{ for horizontal transmission, shown in Fig. 396,}$$

and $Q_2 = \dfrac{wl_2^2}{2h_2} + wh_2$ for inclined transmission, shown in Fig. 397.

The stress per square inch due to the above tension, when the rope contains n wires of diameter d, is

$$f_w = \dfrac{Q}{\cdot 7854 d^2 n} = \dfrac{wL^2 + 8wh^2}{6 \cdot 2832 d^2 nh} \text{ for horizontal transmission,}$$

and $f_w = \dfrac{Q_2}{\cdot 7854 d^2 n} = \dfrac{wl_2^2 + 2wh_2^2}{1 \cdot 5708 d^2 nh_2}$ for inclined transmission.

If for w we insert its value from the formula $w = 3 \cdot 27 d^2 n$, we get—

$$f_w = \dfrac{\cdot 52(L^2 + 8h^2)}{h} \text{ for horizontal transmission,}$$

and $f_w = \dfrac{2 \cdot 08(l_2^2 + 2h_2^2)}{h_2}$ for inclined transmission.

The stress per square inch due to the bending of the rope on the pulley rim is $f_b = \dfrac{Ed}{2R}$. Where R is the radius of the pulley in inches and E the modulus of elasticity = 29,000,000 for wrought-iron wire and 30,000,000 for steel wire.

The stress per square inch due to the centrifugal force is $f_c = \cdot 13v^2$, where v is the velocity of the rope in feet per second.

The sum of all these stresses, namely, $f_w + f_b + f_c$, should not exceed 25,600 lbs. per square inch.

EXAMPLE.—*Power is transmitted by a wire-rope containing* 36 *wires each* ·065 *in. in diameter. The velocity of the rope is* 4200 *ft. per minute. The pulleys are* 12 *ft. in diameter; they are at the same level and are* 400 *ft. apart. To find the horse-power transmitted and the deflections of the rope:—*

Velocity of rope in feet per second $= \dfrac{4200}{60} = 70$.

Stress due to centrifugal force $= f_c = \cdot 13 \ v^2 = \cdot 13 \times 70^2 = 637$.

Stress due to bending of rope on pulley $= f_b$.

$$f_b = \frac{Ed}{2R} = \frac{29{,}000{,}000 \times \cdot 065}{12 \times 12} = 13{,}090.$$

But $f_w + f_b + f_c = 25{,}600$;
therefore, $f_w = 25{,}600 - (f_b + f_c) = 25{,}600 - 13{,}727 = 11{,}873$.

The greatest tension in the rope due to its weight and the dimensions of the curve in which it hangs, is

$$T_1 = \cdot 7854 d^2 n f_w = \cdot 7854 \times \cdot 065^2 \times 36 \times 11{,}873 = 1418 \text{ lbs}.$$

The tension on the slack side of the rope due to its weight and the dimensions of the curve in which it hangs, is

$$T_2 = \tfrac{1}{2}T_1 = \frac{1418}{2} = 709 \text{ lbs}.$$

and the driving force $P = T_1 - T_2 = 1418 - 709 = 709$ lbs.

Therefore, horse-power $= \dfrac{709 \times 4200}{33{,}000} = 90 \cdot 2$.

The weight of the rope per foot of length in this example is
$w = 3 \cdot 27 d^2 n = 3 \cdot 27 \times \cdot 065^2 \times 36 = \cdot 5$ very nearly.

From the formula already given, namely—

$$h = \frac{Q}{2w} \pm \sqrt{\frac{Q^2}{4w^2} - \frac{L^2}{8}} \text{ we get}$$

deflection of rope on tight side $= \dfrac{1418}{2 \times \cdot 5} \pm \sqrt{\dfrac{1418^2}{4 \times \cdot 5^2} - \dfrac{400^2}{8}}$

$= 1418 \pm 1410 \cdot 9$
$= 2828 \cdot 9$ ft. or $7 \cdot 1$ ft.

The second answer, $7 \cdot 1$ ft. is obviously the one to take in this case.

Deflection of rope on slack side $= \dfrac{709}{2 \times \cdot 5} \pm \sqrt{\dfrac{709^2}{4 \times \cdot 5^2} - \dfrac{400^2}{8}}$

$= 709 \pm 694 \cdot 8$
$= 1403 \cdot 8$ ft. or $14 \cdot 2$ ft.

Here again the second answer, $14 \cdot 2$ ft., is obviously the one to take.

WIRE-ROPE GEARING.

197. Pulleys for Rope Gearing.—The pulleys used in the transmission of power by wire-ropes generally vary in diameter from 6 ft. to 18 ft. They may be made entirely of cast-iron, or they may have wrought-iron arms and cast-iron rims and naves. Sometimes the rims are also made of wrought-iron. Fig. 398 shows the form of the rim for a single rope, and Fig. 399 shows the form for an intermediate pulley which carries two ropes. At the bottom of the groove there is a dovetailed recess which goes right round the rim, and into this recess is packed some material which is softer than metal, and is generally leather, but wood, gutta-percha, tarred oakum and tarred jute-yarn are also used for this purpose. The use of the lining at the bottom of the groove is to increase the resistance to slipping between the rope and the pulley, and to diminish the wear of the rope. When leather is used for filling the recess, it is cut into pieces from the hide or from scrap by means of a die to the shape of the recess, and then placed in on edge. For a wheel 10 ft. in diameter, about 1500 of these pieces would be required if each

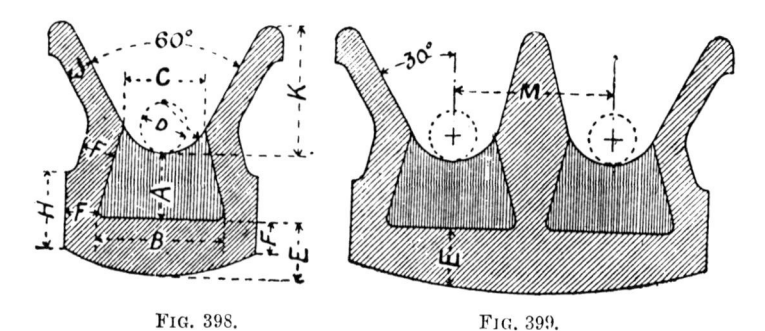

FIG. 398.   FIG. 399.

had a thickness of $\frac{1}{4}$ inch. After the recess has been filled, the pulley is again placed in the lathe, and the bottom of the groove turned to the correct form. According to M. Achard, this filling will last, on an average, three years.

It will be seen from Figs. 398 and 399 that the rope rests on the bottom of the groove, and is not wedged in like a hemp or cotton rope in ordinary rope gearing.

The following rules may be used in proportioning the rims of pulleys for wire ropes:—

D = diameter of rope.
A = ·7D + ·35.
B = 2·37D + ·2.
C = 1·73D.
E = ·5D + ·3.

F = ·3D + ·2.
H = ·8D + ·4.
J = ·2D + ·22.
K = 1·5D + ·7.
M = 2·7D + ·4.

198. Weight of Wire-Rope Pulleys.—According to M. Achard,[1] the

[1] Proceedings of the Institution of Mechanical Engineers, 1880.

weights of the most ordinary sizes of pulleys employed, including their shafts, are on an average as follows :—

Diameter.		Weight.
Ft.	In.	Single-Groove Pulley.	Double-Groove Pulley.
		Lbs.	Lbs.
18	0	6232	8267
14	9	5180	6988
12	4	2425	4078
7	0	798	1164

CHAPTER XIII.

FRICTION GEARING.

199. Ordinary Friction Gearing.—In friction gearing one wheel drives another with which it is in contact, by reason of the friction between their surfaces. When the wheels are cylindrical (spur) or conical (bevel), one of them should be faced with wood, leather, paper, or similar

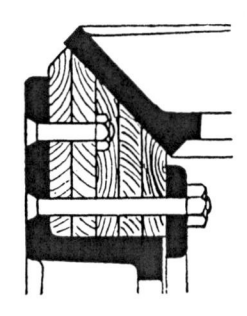

FIG. 400.   FIG. 401.

material, to increase the friction. If wood is used, it should be put round the rim in segments, as shown in Figs. 400 and 401, care being taken to make the grain of the wood lie in a tangential direction to the working surface. The different layers of wood are nailed together and further secured by glue or white lead. The wood best suited for this purpose is soft maple. Instead of wood, leather or paper are also used, put on in layers, the edges of the layers forming the working surface. The wheel which is faced with wood, leather, or paper should be the driver, the follower being made of cast-iron. If the follower was faced with wood, "flats" would soon be formed on it due to the slipping which generally takes place at starting.

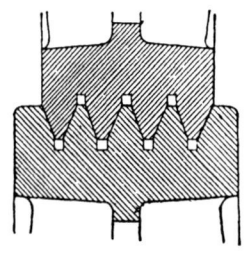

FIG. 402.

200. Robertson's Friction Gearing.—In Robertson's friction gearing both driver and follower are made of cast-iron, and their circumferences are grooved as shown in Fig. 402. The angle between the sides of the groove is usually 40 degrees.

170   MACHINE DRAWING AND DESIGN.

In the original description of this form of gearing, given by the inventor in the year 1856, he stated that the pitch of the grooves varied from $\frac{1}{8}$ in. to $\frac{3}{4}$ in., the ordinary pitch being $\frac{3}{8}$ in. A much larger pitch, however, has been adopted in some cases. In a paper read before the Institution of Mechanical Engineers in 1888, Mr. Griffith, of Dublin, related his experience with friction wheels having nine grooves of $1\frac{3}{4}$ in. pitch, transmitting a tangential force of 3960 lbs. at a circumferential speed of 500 ft. per minute. In this case, however, the gearing did not prove to be satisfactory, owing chiefly, no doubt, to the slow speed and necessarily high pressure between the surfaces in contact.

Friction gearing should only be used, like belt gearing, where the force to be transmitted is comparatively small and the speed high.

201. Force Required to Keep Friction Wheels in Gear.—Let P be the driving force at the circumference of the friction wheels, Q the

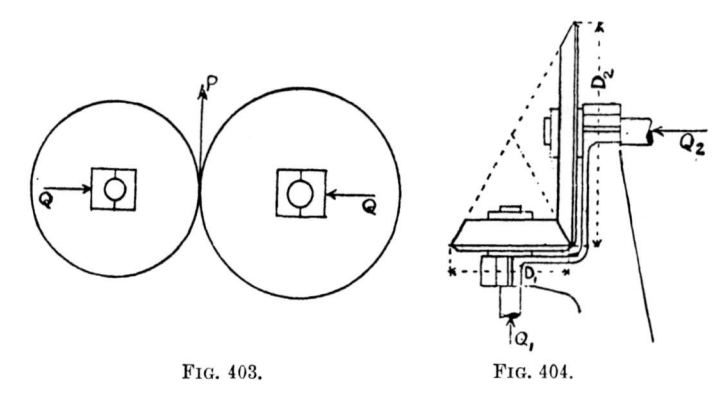

FIG. 403.   FIG. 404.

force pressing the wheels together, and f the coefficient of friction between the surfaces in contact. f may be taken at $\frac{1}{6}$ for metal on metal, and $\frac{1}{3}$ for wood on metal.

For spur wheels, Fig. 403, $P = Qf$ and $Q = \dfrac{P}{f}$, when one wheel is about to slide on the other. This shows that the force necessary on the bearings must be at least six times the tangential driving force when both wheels are of metal, and three times when one is faced with wood.

For Robertson's friction gearing, Q may be taken equal to 3 P when the angle of the grooves is 40 degrees.

For bevel wheels, Fig. 404, of diameters D_1 and D_2, the smallest forces, Q_1 and Q_2, acting parallel to the axes of the shafts, which are necessary to prevent slipping, are as follows :—

$$Q_1 = \frac{PD_1}{f\sqrt{D_1^2 + D_2^2}}, \quad Q_2 = \frac{PD_2}{f\sqrt{D_1^2 + D_2^2}}$$

P being the tangential driving force, and f the coefficient of friction as before.

202. Transmission of Power by Friction Gearing.—If V is the velocity of the surfaces in contact in feet per minute and if the horse-power transmitted, then $H = \dfrac{PV}{33,000}$ and $P = \dfrac{33,000 H}{V}$.

EXAMPLE.—*A friction spur wheel faced with wood drives another which is 2 feet in diameter at 400 revolutions per minute. The force pressing the wheels together is 600 lbs. To find the horse-power transmitted.*

Here $Q = 600$ lbs.; $f = \tfrac{1}{3}$; $P = Q f = \dfrac{600}{3} = 200$ lbs.;

$$V = 2 \times 3\cdot1416 \times 400.$$

Therefore, horse-power $= H = \dfrac{PV}{33,000} = \dfrac{200 \times 2 \times 3\cdot1416 \times 400}{33,000} = 15\cdot23.$

203. Professor Hele Shaw's Experiments.—Professor Hele Shaw, who has given great attention to the subject of roller friction, has obtained valuable and interesting results from experiments on cast-iron friction wheels. There were three series of experiments. In the first series the wheels or discs had faces 3-16ths inch wide. In the second series the wheels had faces 1-16th inch wide. In the third series the wheels were of chilled cast-iron, and the faces were ground to true spherical surfaces. The results are shown in the following table. P is the tangential force transmitted to the driven wheel, and Q is the pressure between the faces :—

	First Series.		Second Series.		Third Series.	
Q.	P.	$\dfrac{P}{Q}$	P.	$\dfrac{P}{Q}$	P.	$\dfrac{P}{Q}$
Lbs.	Lbs.		Lbs.		Lbs.	
315	45	·143	51	·162	57	·181
420	59	·140	70	·167	79	·188
525	81	·154	96	·183	105	·200
630	105	·167	124	·197	130	·206
735	125	·170	146	·199	158	·215
840	149	·177	161	·192	186	·221
945	172	·182	195	·206	248	·262
1,050	...	...	215	·205	268	·255
1,155	...	...	237	·205	...	...
1,260	...	...	258	·205	...	...
1,365	...	...	272	·199	...	...

The above table shows that the coefficient of friction $(P \div Q)$ increases with an increase in the force pressing the wheels together, and also that the coefficient of friction is greater the smaller the amount of the surfaces in contact. Another important result which Professor Shaw found was that with spherical surfaces and a large pressure between them, the presence of oil did not affect the coefficient of friction; in fact, with the highest pressures used in the third series of experiments, oil poured continuously upon the rolling surfaces had no appreciable effect whatever. With larger bearing surfaces and lower pressures, however, the presence of oil lowered the coefficient of friction, and therefore diminished the amount of power transmitted. It is obvious that small bearing surfaces can only be used with wheels made of very hard materials, such as chilled cast-iron or hardened steel, which offer great resistance to crushing.

CHAPTER XIV.

TOOTHED GEARING.

204. Pitch Surfaces and Pitch Lines of Toothed Wheels—Pitch of Teeth.—The *pitch surfaces* of two toothed wheels which gear with one another are the surfaces of two imaginary friction wheels which have the same axes, and which would have the same relative angular velocities as the wheels if one was the driver of the other.

A section of a pitch surface by a plane at right angle to its axes is called a *pitch line*, or a *pitch circle*, if that section should be a circle, which it is in most cases.

The *pitch* of the teeth is the distance between the centre of one tooth and the centre of the next, *measured along the pitch line*.

205. Relation between the Number and Pitch of the Teeth and the Diameter of the Pitch Circle.—Let D = diameter of pitch circle, n = number of teeth, and p = pitch of teeth.

The circumference of the pitch circle is equal to $3\cdot1416 D$, but it is also equal to np; therefore $3\cdot1416 D = np$ and

$$D = \frac{np}{3\cdot1416}; \text{ also } p = \frac{3\cdot1416 D}{n}.$$

206. Relation between the Speeds of Two Wheels in Gear.—Let N_1 and N_2 be the speeds of the driver and follower respectively in revolutions per minute, D_1 and D_2 their diameters, and let the driver contain n_1 teeth, and the follower n_2 teeth. Then, as in belt gearing—

$$\frac{\text{Speed of follower}}{\text{Speed of driver}} = \frac{N_2}{N_1} = \frac{D_1}{D_2} \text{ also } \frac{N_2}{N_1} = \frac{n_1}{n_2}.$$

207. Proportions of Teeth.—The rules given by various authorities for proportioning the teeth of wheels vary to some extent, but the following, for ordinary iron toothed wheels working together, may be taken as representing good average practice:—

Pitch of teeth . . . . . $= p\ $ = arc abc (Fig. 405.)
Thickness of tooth . . . . $= ab = \cdot 47p$.
Width of space . . . . $= bc = \cdot 53p$.
Total height of tooth . . . $= h\ = \cdot 7p$.
Height of tooth above pitch line . $= k\ = \cdot 3p$.
Depth of tooth below pitch line . $= l\ = \cdot 4p$.
Width of tooth . . . . $= 2p$ to $3p$.

The above proportions allow of a side clearance between the teeth when in gear amounting to $\cdot 06p$. With carefully constructed teeth, this side clearance may be diminished, and the teeth made correspondingly thicker; but, on the other hand, if they are roughly made, a larger side clearance would be necessary, and the teeth would be made correspondingly thinner.

The wood cogs of mortice wheels have a thickness = $\cdot 6p$, and the iron

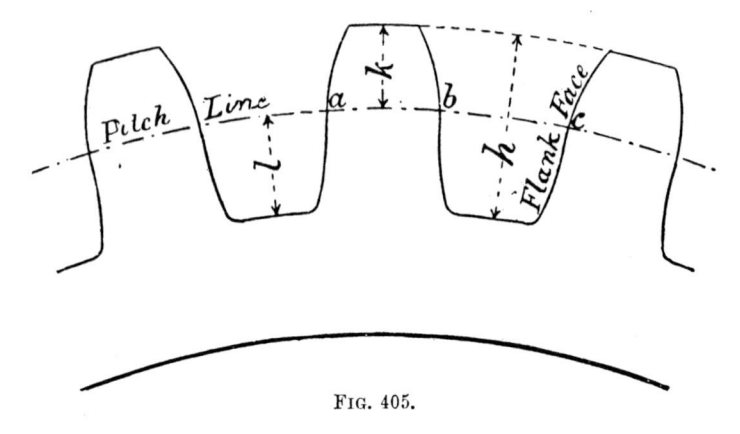

FIG. 405.

teeth gearing with them have a thickness = $\cdot 4p$, so that there is no side clearance when the teeth are new.

The ratio of the width of the teeth to the pitch is generally greater the greater the pitch. For very heavy mill gearing the width is sometimes as great as five times the pitch.

208. Condition to be Fulfilled by the Curves of the Teeth of Wheels in Order that they may Work Correctly.—Two toothed wheels, whose pitch lines are circles, are said to work correctly when the ratio of their speeds or angular velocities is exactly the same at *every instant*.

In Fig. 406, O_1 and O_2 are the centres of two wheels, the radii of whose pitch circles are R_1 and R_2. The shaded curves represent portions of two teeth, one on each wheel, which are in contact at the point ab, a being the point on the tooth of the first wheel which is in contact with the point b on a tooth of the other. It is a well-known fact in geometry that when two curves touch one another they have a common normal. Let M_1M_2 be the common normal of the curves of the teeth at a and b, and let O_1M_1 and O_2M_2 be perpendiculars from the centres of the wheels on to this common normal. The point a moves in a circle A whose centre is O_1, and the point b moves in a circle B whose centre is O_2. At the instant that the points a and b are in contact, the point a is moving in the direction of the tangent ca to the circle A, and the point b is moving in the direction of the tangent ad to the circle B. Let v_1 and v_2 be the linear velocities of the points a and b in the directions in which they are moving at the instant when they are in contact. Now, although the points a and b are moving in different lines with different velocities, the components of these velocities in the direction M_1M_2 must be the same, otherwise the points a and b would move relatively to each

TOOTHED GEARING.

other along the line M_1M_2. But for a small angular movement of either of the wheels, the only possible relative motion of a and b is in a direction at right angles to M_1M_2. If v is the component velocity of a, and also of b in the direction M_1M_2, then the ratio of the angular velocities of the two wheels must be $\dfrac{v}{O_1M_1} \div \dfrac{v}{O_2M_2} = \dfrac{O_2M_2}{O_1M_1} = \dfrac{O_2P}{O_1P}$, the point P being

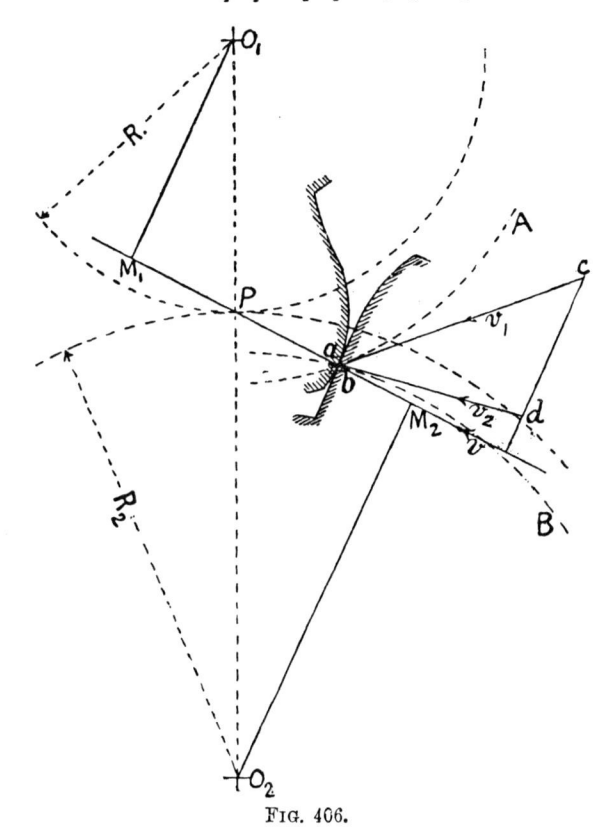

FIG. 406.

at the intersection of the lines O_1O_2 and M_1M_2. But the ratio of the angular velocities of the wheels is also equal to $\dfrac{R_2}{R_1}$, therefore $\dfrac{O_2P}{O_1P} = \dfrac{R_2}{R_1}$, or the point P is the point where the pitch circles touch one another. The condition to be fulfilled by the curves of the teeth is therefore that the common normal at their point of contact must pass through the pitch point—that is, the point where the pitch circles touch one another.

If the pitch circles are not circular, the above condition must still hold if the ratio of the speeds of the two wheels is to vary in exactly the same way as that of the friction wheels corresponding to the pitch lines.

209. Cycloidal Teeth.—Let APB and CPE, Fig. 407, be the pitch circles of two wheels. Let the outline of the flank of a tooth on APB be a portion of the hypocycloid aQb, described by the rolling of the circle PQM, on the inside of the pitch circle APB. Let the outline of the face of a tooth on CPE be a portion of the epicycloid cQe, described by the rolling of the circle PQM, on the outside of the pitch circle APE. Next let the face cQ be brought forward so as to touch the flank aQ, and let Q be the point of contact. The point Q must be on the rolling circle PQM when the latter touches both pitch circles, because at that instant the rolling circle is about to describe either the curve aQb or the curve cQe. Also QP is obviously the direction of the common normal to the curves aQb and cQe at their point of contact Q, therefore the wheels will work correctly if the faces of the teeth on one are epicycloids and the flanks of the teeth on the other are hypocycloids *described by the same rolling circle.*

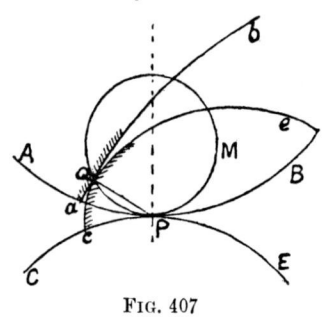

It is evidently not necessary that the flanks of the teeth on two wheels which gear together be described by the same rolling circle, but the rolling circle which describes the flanks on one wheel must be used to describe the faces on the other.

Since the hypocycloid becomes a straight line passing through the centre of the pitch circle when the diameter of the rolling circle is equal to the radius of the pitch circle, it follows that the flanks of wheel teeth may be made radial.

FIG. 407

If a number of wheels are to be interchangeable, that is, if any one of them is to be capable of gearing with any of the others, the faces and flanks of the teeth on each must be described by the same rolling circle.

210. The Path of Contact.—From an examination of Fig. 407 it is evident that the point of contact of two cycloidal teeth must be on one or other of the rolling circles when the latter are at the pitch point. It follows, therefore, that the path of contact of two teeth must be made up of arcs of these rolling circles.

If APB and CPE, Fig. 408, be the pitch circles of two wheels with cycloidal teeth in gear with one another, and if *ade* be the circle passing through the points of the teeth on the lower wheel, called the *addendum* circle, and *bfg* the corresponding circle for the other wheel, then, if the rolling circles be placed as shown, the points a and b where the circles *ade* and *bfg* cut the rolling circles are the extreme points of contact of the teeth, the upper wheel being the driver, and having its motion in the direction of the arrow. At the point a a point on the flank of a tooth on the driver will come in contact with a point on the face of a tooth on the follower. As the motion proceeds, the flank of the tooth on the driver will slide on the face of the tooth on the follower until the point of contact, which moves along the arc aP, reaches the point P. The face of the tooth on the driver will then slide on the flank of the tooth on the follower until the point of contact, which moves along the arc Pb, reaches the point b. The arc aP is called the *path of approach*, and the arc Pb *the path of recess*. If the driver move in the opposite direction the path of contact will be the line $a'Pb'$.

Since the common normal to the surfaces of the teeth in contact always passes through the pitch point P, and since the point of contact is always on the line aPb, it follows that the direction of the pressure between two teeth in contact varies. When contact begins, the pressure will act in the straight line joining a with P; and when the point of contact is at the pitch point, the pressure will act in the direction of the common tangent to the pitch circles.

FIG. 408.

211. Involute Teeth.—Let APB and CPE, Fig. 409, be the pitch circles of two wheels with involute teeth in gear with one another. Let aQb and cQd be the outlines of the surfaces of two teeth in contact at Q, these outlines being involutes of the base circles S_1 and S_2 respectively. Since a line drawn from any point on an involute to touch the base circle of that involute is a normal to the involute at that point, it follows that the common normal to the two involutes in contact at Q

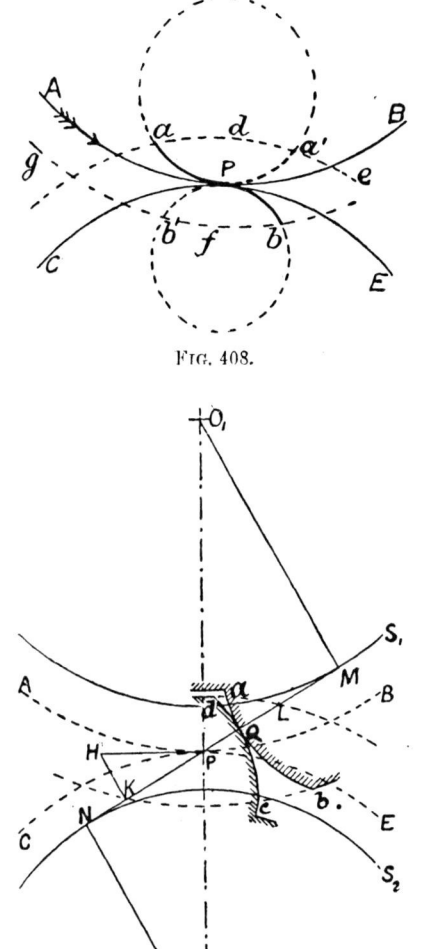

FIG. 409.

must be the common tangent MN of the two base circles. Hence the point of contact is always on the line MN, and a portion of that line is therefore the path of contact.

M

Comparing the similar triangles, O_1PM and O_2PN, it is clear that the ratio of the radii of the base circles is the same as the ratio of the radii of the pitch circles. If the centres of the wheels be pushed further apart or closer together, the wheels will still work correctly together, because this is equivalent to altering the radii of the pitch circles without altering their ratio. This is a special property of involute teeth, and is a valuable one in cases where the distance between the centres of two wheels cannot be maintained constant.

In designing involute teeth the direction of the path of contact is first fixed, so as to make an angle of 15 degrees with the common tangent to the pitch circles. The base circles are then drawn to touch the path of contact.

If on the tangent at P, PH be made equal to the pitch of the teeth, and HK be drawn perpendicular to MN, then PK is called the *normal pitch*. The normal pitch is equal to the pitch of the teeth measured on the base circles. If PL be made equal to PK, then KL is the minimum length of the path of contact, in order that at least two pairs of teeth may always be in contact. Circles drawn through the points H and L, as shown, determine the points and the inner ends of the flanks of the teeth. There should be a small clearance space between the root circle on one wheel and the addendum circle on the other.

The length of the path of contact is a maximum when the normal pitch, or the pitch measured on the base circle is equal to PN or PM, whichever is least. Let PN be less than PM, so that the lower wheel is the smaller. Let $O_2N = r$, $PN = p$, and let n be the number of teeth in the lower wheel, then

$$2 \times 3\cdot1416\, r = np$$
$$\text{but } p = r \tan 15 \text{ deg.}$$
$$\text{therefore } 2 \times 3\cdot1416\, r = nr \tan 15 \text{ deg.}$$
$$\text{therefore } n = \frac{2 \times 3\cdot1416}{\tan 15 \text{ degrees}} = \frac{2 \times 3\cdot1416}{\cdot2679} = 23\cdot45.$$

This shows that when the line of contact makes 15 degrees with the common tangent to the pitch circles the minimum number of teeth in a wheel is 24.

All wheels having involute teeth of the *same normal pitch* will work correctly together.

When the pitch circle becomes of infinite diameter, as in a rack, the base circle will also become of infinite diameter, and the involute will become a straight line. Hence, in a rack which gears with a wheel having involute teeth, the teeth are straight on face and flank, as shown in Fig. 411. The faces and flanks are at right angles to the path of contact, and therefore make an angle of 75 degrees with the pitch line.

212. Teeth of Bevel Wheels.—Bevel wheels are used to connect two shafts which intersect. The pitch surfaces of bevel wheels are portions of cones which have a common vertex, that vertex being at the point of intersection of the axes of the shafts, as shown at (*a*) Fig. 410.

In Fig. 410 is clearly shown the way to draw a bevel wheel in section, elevation, and plan. The pitch surface ABC is first drawn in elevation.

TOOTHED GEARING. 179

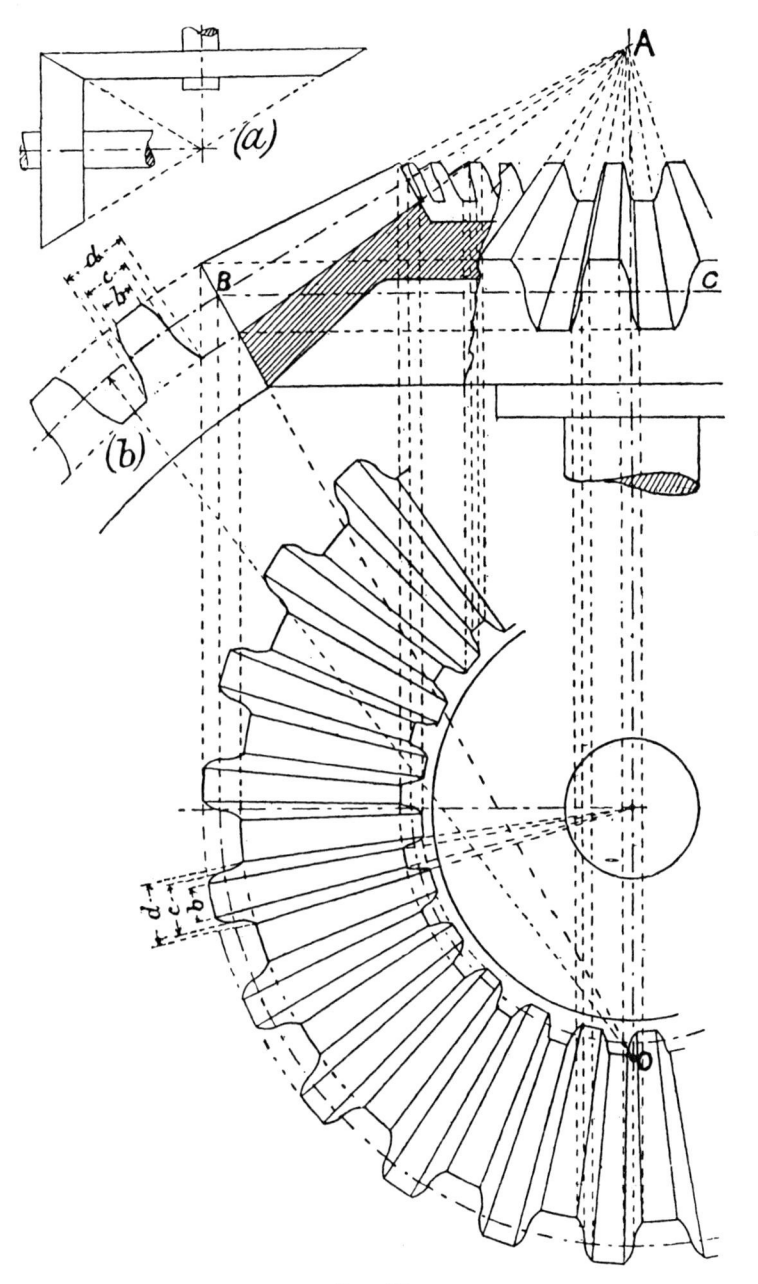

FIG. 410.

The vertex O of a cone, whose slant side is perpendicular to the slant side of the pitch surface, is then determined. On the development of the surface of this cone the teeth are drawn and proportioned as in ordinary spur wheels, as shown at (*b*). From this development the section, plan, and elevation of the teeth are drawn, as fully shown in the figure.

In speaking of the diameter of a bevel wheel, the largest diameter

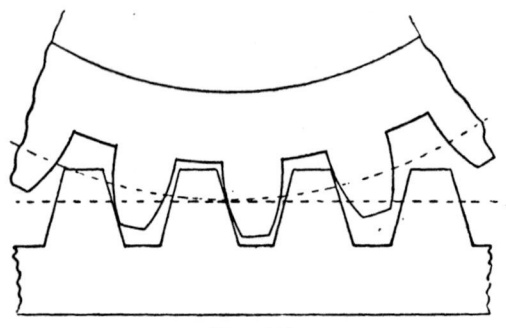

FIG. 411.

of the pitch cone is meant. In like manner the pitch of the teeth is taken as measured on the largest circle on the pitch cone.

213. Strength of Wheel Teeth.—In devising a formula for the strength of wheel teeth, there are two cases to consider: first, where all the pressure on a tooth comes on one corner, as in Fig. 412; secondly, where the pressure is uniformly distributed along the whole width of a tooth, as in Fig. 413.

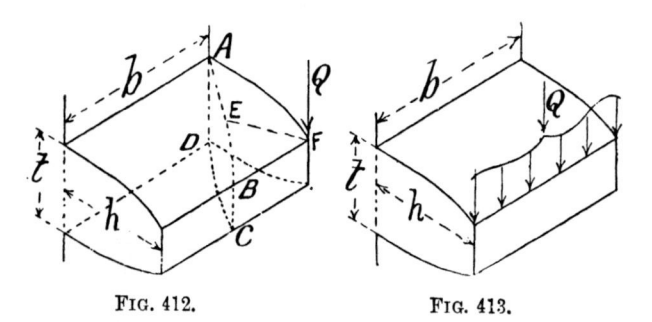

FIG. 412.   FIG. 413.

The pressure is liable to act at one corner of a tooth when the teeth are inaccurately constructed, or when the axes of the wheels are not properly set. In carefully constructed gearing on shafts accurately adjusted, the teeth come in contact along the whole width, with the result that the pressure between them is distributed along the width. It is obvious that the teeth in the second case will transmit a greater amount of power than the teeth in the first; but it is not always safe to assume that the pressure on the teeth is uniformly distributed along the width, even when the wheels are properly constructed and adjusted to begin with, because the wearing of the bearings of the shafts and the straining of the supports may cause imperfect contact between the teeth.

When the load Q on a tooth acts at one corner, as in Fig. 412, it may be shown that the tooth will give way at a section ABCD, which makes 45 degs. with its root, provided the width b is not less than twice the height h. Let EF be the perpendicular from F to AB. Then EF = AE = EB; let each of these be represented by s, then AB = $2s$. The bending moment of Q, with reference to the section ABCD = Qs, and assuming the tooth to be rectangular, the moment of resistance is $\frac{1}{6} \times 2st^2 f$, where f is the stress. Equating the bending moment to the moment of resistance, we get the relation

$$Qs = \tfrac{1}{6} \times 2st^2 f \text{ or } Q = \frac{t^2 f}{3} \text{ and } t^2 = \frac{3Q}{f}$$

which shows that the length or height of a tooth does not affect its strength if the load acts at one corner.

As the proportions of teeth are generally given in terms of the pitch, we will substitute for t in the above formula, its value in terms of p, the pitch. Allowing for wear, t may be taken at $\cdot 4p$ for iron teeth, then $\cdot 16\, p^2 = \frac{3Q}{f}$. The load Q on one tooth may be taken at $\tfrac{2}{3}$P, where P is the driving force at the pitch line, calculated from the horse-power and speed. The stress f may be taken at from 2500 to 4500 lbs. per square inch. Taking $f = 3500$, and Q = $\tfrac{2}{3}$P, we get from the formula above $p = \cdot 06 \sqrt{P}$ and P = $280p^2$.

Coming now to the case represented in Fig. 413, where the pressure is distributed uniformly along the width, we have the moment of resistance to bending equal to $\tfrac{1}{6}bt^2 f$, and a bending moment equal to Qh. Hence Q$h = \tfrac{1}{6}bt^2 f$. We may put $t = \cdot 4p$ and Q $= \tfrac{2}{3}$P as before, and $h = \cdot 7p$. Then P $= \cdot 0571\, bpf$ and $p = \dfrac{17 \cdot 5 P}{bf}$.

Putting $f = 3500$, we get P $= 200bp$, and $p = \dfrac{P}{200b}$. But b is generally expressed in terms of p; let $b = np$, then P $= 200np^2$, and $p = \sqrt{\dfrac{P}{200n}}$.

From these formulæ it is evident that when the teeth of different wheels are proportioned according to the same rules, the force which they are capable of transmitting is proportional to the square of the pitch. For example, if the pitch of the teeth on one wheel is double that on another, the latter will transmit four times the force at the pitch line, the ratio of breadth, length, and thickness to the pitch being the same for both wheels.

The rules given above are for spur wheels, but they will also apply to bevel wheels if the force P and pitch p be taken at the middle of the width of the pitch surface.

214. Shrouding of Wheel Teeth.—The teeth of a wheel are said to be shrouded when the rim is wider than the teeth and is carried outwards, as shown in Figs. 414, 415, and 416. The shrouding may extend right out to the points of the teeth, or it may extend to the pitch line only, and it may be on both sides or on one side only. If the shrouding extends to the points of the teeth and is on both sides of the wheel, it is obvious that only one of a pair of wheels in gear can be shrouded.

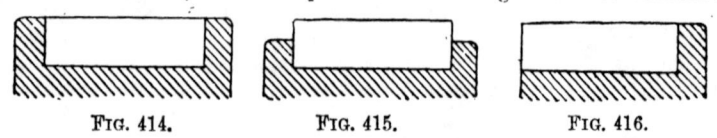

FIG. 414.   FIG. 415.   FIG. 416.

The effect of shrouding is generally to add to the strength of the teeth. The increase of strength due to shrouding varies with the form of the teeth and the width of the wheel. Teeth which are not thicker at the roots than at the pitch line may have their strength increased as much as 100 per cent. by shrouding out to the pitch line. On the other hand, teeth which are thicker at the roots than at the pitch line are less benefited by shrouding, and if the thickness of a tooth varies such that the square of the thickness is proportional to the distance from the point, there is no advantage at all in shrouding, unless the shrouding is carried right out to the point. The advantage of shrouding is greater the narrower the wheel.

If two wheels gearing together do not differ greatly in diameter, each may be shrouded to the pitch line on both sides; but when one is very much larger than the other, it is usual to shroud the smaller one only, because the teeth on the smaller are generally of a weaker form, and also because the wear of the teeth is greater on the smaller wheel.

215. Relative Pitches of Teeth of Equal Strength for Different Materials.—The following table gives the relative pitches of the teeth of wheels made of various materials, on the assumption that the teeth are proportioned according to the same rules in each case:—

Material.	Relative Pitches.
Cast-iron	1·0
Wrought-iron	0·6
Steel	0·5
Wood	1·3
Gun-metal	0·8

216. Transmission of Power by Toothed Wheels.—
Let P = driving force at pitch line in pounds.
  D = diameter of pitch circle in inches.
  V = velocity in feet per minute at pitch line.
  N = number of revolutions of wheel per minute.
  H = horse-power transmitted by wheel.

$$H = \frac{PV}{33,000} = \frac{3 \cdot 1416 DNP}{33,000 \times 12}.$$

$$P = \frac{33,000 H}{V} = \frac{33,000 \times 12 H}{3 \cdot 1416 DN}.$$

If in the above formulæ for the horse-power we insert the value of $P = 280p^2$, where p is the pitch of the teeth, we get—

$$H = \cdot 0085 p^2 V = \cdot 0022 p^2 DN$$

$$p = 10 \cdot 86 \sqrt{\frac{H}{V}} = 21 \cdot 21 \sqrt{\frac{H}{DN}}.$$

we use for the pitch the formula $p = \dfrac{P}{200b}$ and put $b = np$, then

$$H = \cdot 006 n p^2 V = \cdot 00159 n p^2 DN$$

$$p = 12 \cdot 845 \sqrt{\frac{H}{nV}} = 25 \cdot 1 \sqrt{\frac{H}{nDN}}.$$

217. Maximum Speed of Toothed Gearing.—According to Mr. Alfred Towler, of Leeds, the maximum safe speeds for toothed gearing, under favourable conditions, are as follows:—

Ordinary cast-iron wheels	1,800	feet per minute
Helical ,, ,,	2,400	,, ,,
Mortice ,, ,,	2,400	,, ,,
Ordinary cast-steel ,,	2,600	,, ,,
Helical ,, ,,	3,000	,, ,,
Special cast-iron machine-cut wheels	3,000	,, ,,

The above speeds are the linear velocities at the pitch circles. Higher speeds than these are, however, to be found in practice. In *Engineering*, vol. xlv. p. 223, is a notice of an example of iron wheels with double helical teeth running at a speed of 4319 feet per minute at the pitch line. These wheels are 6 feet 3 inches in diameter and 12 inches wide, and make 220 revolutions per minute with very little noise. A still higher speed than this is recorded in the "Transactions of the American Society of Mechanical Engineers" for 1885. In this case a spur wheel, 30 feet in diameter, makes fifty revolutions per minute, which is equivalent to a speed of 4712 feet per minute at the pitch line. The pitch of the teeth is 5·183 inches, and the width of face 30 inches.

From calculations made by Professor Unwin it would seem that the maximum safe speed for cast-iron spur wheels does not exceed 96 feet per second, or 5760 feet per minute at the pitch line.

218. Rims of Toothed Wheels.—For light spur wheels the rim section

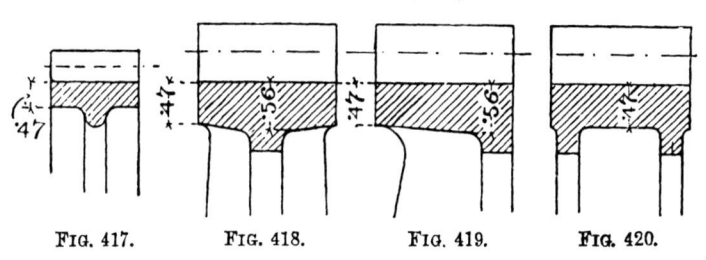

FIG. 417.   FIG. 418.   FIG. 419.   FIG. 420.

is generally as shown in Fig. 417. The section shown in Figs. 418, 419, and 420 are for the rims of heavy spur gearing. The proportions marked

184   MACHINE DRAWING AND DESIGN.

on these and the following figures are in terms of the pitch of the teeth. For a bevel wheel, Fig. 421, the thickest part of the rim may have a thickness $t_1 = \cdot 56p$.

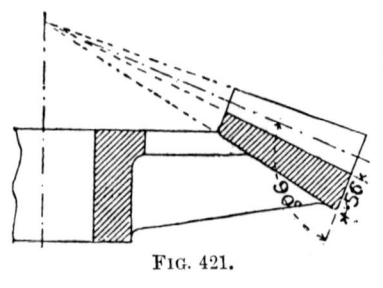

Figs. 422 and 423 show the form of the rim for a mortice wheel, and the most common method of securing the cogs. When the wheel is wide, double cogs are used, as shown in Fig. 424; t_2, the thickness of the rim $= p$. The thickness of the tenon varies from $\cdot 5p$ at the shoulder to $\cdot 45p$ at the point. Sometimes the side shoulder on the tooth is on one side only, as shown in Fig 425, and sometimes on both sides, as shown in Fig. 423.

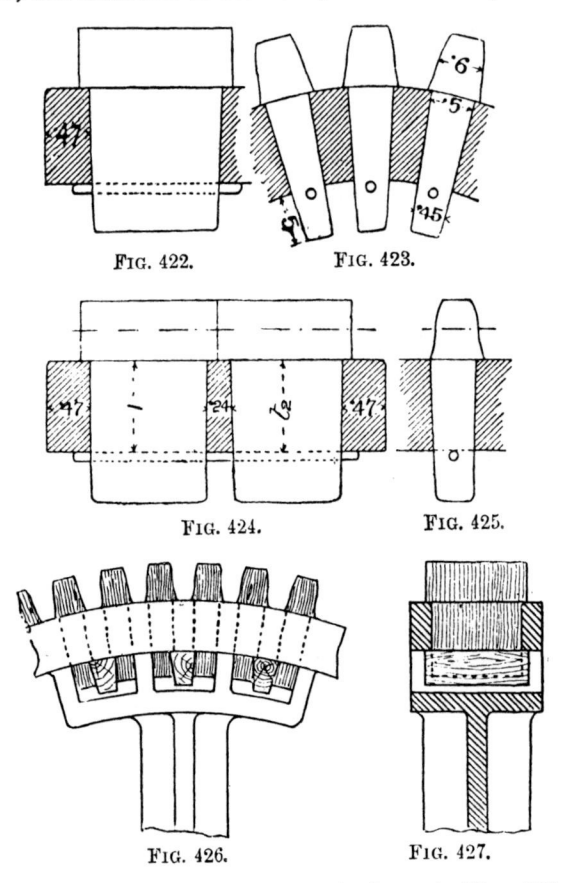

FIG. 421.

FIG. 422.   FIG. 423.

FIG. 424.   FIG. 425.

FIG. 426.   FIG. 427.

Another method of securing the cogs is shown in Figs. 426 and 427.

TOOTHED GEARING.

Here a dovetailed wooden key is driven in between the projecting ends of the tenons of the cogs. The same Figs. also show how the arms are connected to the rim so as to permit of these keys being used where the arms come in.

219. Arms of Wheels.—The force acting at the pitch line of a toothed wheel, due to the power transmitted, causes a bending action on the arms. The total bending moment on all the arms is equal to PR, where P is the tangential force at the pitch line and R the radius of the pitch line. If there are n arms, and we assume that the bending action is distributed equally to each arm, then the bending moment on one arm is equal to $\dfrac{PR}{n}$.

For light wheels the cross section of the arms is often of the form shown in Fig. 428. The section shown in Fig. 429 is a very common one for the arms of spur wheels. The section shown in Fig. 430 is often adopted for the arms of heavy spur wheels. For bevel wheels the section of the arms is generally that shown in Fig. 431. In each case the width B_1 or B is measured in the plane of the wheel—that is, in a plane

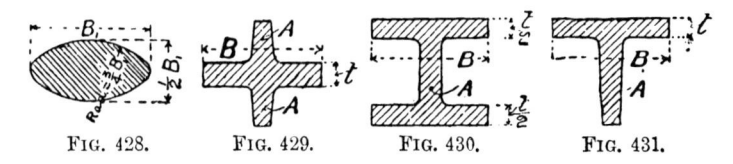

FIG. 428.   FIG. 429.   FIG. 430.   FIG. 431.

at right angles to the axis of the shaft. The feathers or ribs A are added to give lateral stiffness, but they add very little to the resistance of the arms to bending in the plane of the wheel, and their effect will be neglected in what follows.

As already stated under the head of "Arms of Pulleys," the moment of resistance of an arm to bending, whose cross section is that shown in Fig. 428, is approximately equal to $\cdot 05 B_1^3 f$, where f is the greatest stress in the arm due to the bending action.

For the sections shown in Figs. 429, 430, and 431, the moment of resistance to bending may be taken as equal to $\frac{1}{6} B^2 t f$.

It will be convenient to express the force P in terms of the pitch and breadth of the teeth from the formula already given, namely, $P = 200 bp$.

We now get the bending moment on one arm $= \dfrac{200 bpR}{n}$.

Then, for the arm with oval-shaped section,

$$\dfrac{200 bpR}{n} = \cdot 05 B_1^3 f \quad \text{and} \quad B_1 = \sqrt[3]{\dfrac{4000 bpR}{nf}}.$$

Taking $f = 2500$, which will allow for the initial straining actions due to unequal contraction in cooling in the mould, and also for the possible unequal distribution of the bending action on the arms, we get

$$B_1 = \sqrt[3]{\dfrac{1 \cdot 6 bpR}{n}}.$$

For the other forms of section given we have $\frac{200bpR}{n} = \frac{1}{6}B^2 tf$. The same value of f may be taken as before, namely, 2500, and the thickness t may be taken equal to the thickness of a tooth, so that $t = \cdot 48p$. We then get $\frac{200bpR}{n} = \frac{1}{6}B^2 \times \cdot 48p \times 2500$, and $B = \sqrt{\frac{bR}{n}}$.

The breadth given above is the breadth of the arm measured at the centre of the wheel, supposing the arm to be produced to that point. The breadth measured at the pitch line may be $B - \cdot 04R$. The thickness t is, however, the same throughout. For an oval-shaped arm, the taper on the width may be the same as that just given, and the thickness is also reduced towards the rim, so that at any point it is equal to half the width at that point.

The mean thickness of the ribs or web may be $\cdot 4p$; they are tapered slightly in thickness to facilitate the withdrawing of the pattern from the sand in moulding.

The number of arms is approximately $\frac{D}{36} + 4$, where D is the diameter of the pitch circle in inches. The nearest *even* number may be taken.

220. Naves of Wheels.—The thickness of the nave is sometimes proportioned according to the diameter of the shaft upon which it is fixed, but unless the wheel is capable of transmitting the full power of the shaft this is not a good method. In the following rule the thickness is made proportional to the diameter of a shaft, which would have the same strength as the wheel. The rule is, thickness of nave = $\frac{\sqrt[3]{bpR}}{3}$, where b is the width of the teeth, p the pitch of the teeth, and R the radius of the pitch circle, all in inches.

Another rule is, thickness of nave = $\cdot 8p + \cdot 02R$.

The length of the nave varies from b to $1 \cdot 4b$.

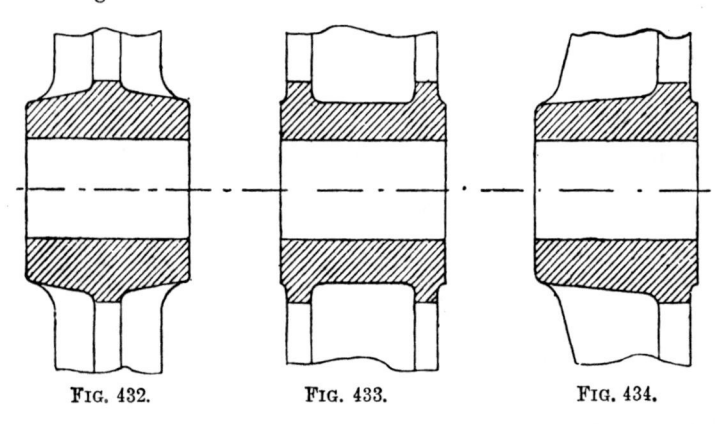

FIG. 432.   FIG. 433.   FIG. 434.

Examples of the forms of the naves of wheels corresponding to various forms of arms are shown in Figs. 432, 433, and 434. In the forms shown in Figs. 432 and 434, the thickness varies slightly to allow of the pattern being withdrawn freely from the sand in moulding.

To relieve large and heavy wheels from the initial strains due to unequal contraction in cooling, the nave is slotted across between the arms as shown in Figs. 435 and 436. The openings thus made are then

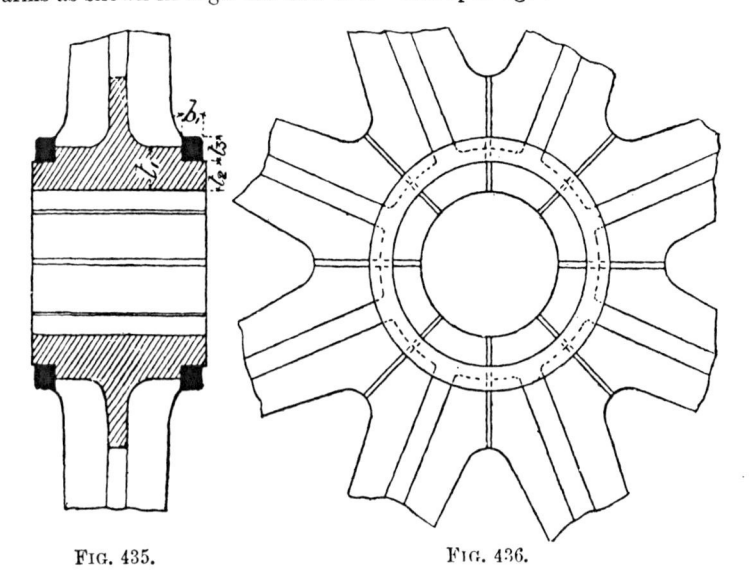

FIG. 435.   FIG. 436.

fitted with metal strips, and the segments of the nave are bound firmly together by wrought iron or steel rings, which are shrunk on in the positions shown.

The proportions of the rings and the thickness of the boss under it may be as follows : $t_3 = \frac{2}{3}t_1$; $t_2 = \frac{3}{4}t_1$; $b_1 = \frac{5}{8}t_3$.

221. Design for a Spur Wheel.—EXAMPLE.—As an example of the application of the rules which have been given for proportioning toothed wheels, we will work out a design for a cast-iron spur wheel, having given the following particulars. Diameter of pitch circle, 60 inches; number of revolutions per minute, 80 ; horse-power transmitted, 300.

From the formula $p = 25 \cdot 1 \sqrt{\dfrac{H}{nDN}}$, and taking $n = 2 \cdot 5$, we get the pitch of the teeth $= p = 25 \cdot 1 \sqrt{\dfrac{300}{2 \cdot 5 \times 60 \times 80}} = 3 \cdot 97$ inches.

With this pitch the number of teeth in the wheel would be equal to $\dfrac{60 \times 3 \cdot 1416}{3 \cdot 97} = 47 \cdot 48$. But as we cannot have a fraction of a tooth in the wheel, we must take the nearest whole number, namely, 47, as the number of teeth. This will necessitate increasing the pitch slightly, or diminishing the diameter of the pitch circle by a small amount. Keeping the diameter of the pitch circle as it is, we get the pitch corresponding to 47 teeth $= \dfrac{60 \times 3\cdot 1416}{47} = 4\cdot 01$ inches.

Having determined the pitch of the teeth, the other dimensions will be as follows:—

    Breadth of face $= b = 2\cdot 5p = 10\cdot 025$, say 10 inches.
    Thickness $= \cdot 48p = 1\cdot 92$ inches.
    Height above pitch line $= \cdot 3p = 1\cdot 2$,,
    Depth below pitch line $= \cdot 4p = 1\cdot 6$,,

The shaft for such a wheel as this would probably be subjected to bending as well as twisting, and its diameter would not be less than 6 inches. We will suppose that the diameter of the shaft is 7 inches, and that the wheel seat is $8\frac{1}{2}$ inches in diameter.

A wheel 60 inches in diameter would have six arms. From the equation $B = \sqrt{\dfrac{bR}{n}}$ we get the width of the arms measured at the centre of the wheel $= \sqrt{\dfrac{10 \times 30}{6}} = \sqrt{50} = 7\cdot 07$, say 7 inches. This is on the assumption that the thickness of an arm is equal to the thickness of a tooth—that is, $1\cdot 92$, say $1\frac{7}{8}$ inches. The breadth of the arms, measured at the pitch line, is obtained from the formula

    Breadth $= B - \cdot 04 R = 7 - \cdot 04 \times 30 = 5\cdot 8$, say $5\frac{3}{4}$ inches.
    Mean thickness of feathers or ribs $= \cdot 4p = 1\cdot 6$, say $1\frac{5}{8}$ inches.

Thickness of nave

$$= t_1 = \dfrac{\sqrt[3]{bp R}}{3} = \dfrac{\sqrt[3]{10 \times 4\cdot 01 \times 30}}{3} = \dfrac{\sqrt[3]{1203}}{3} = 3\cdot 54, \text{ say } 3\tfrac{1}{2} \text{ inches.}$$

Thickness of wrought-iron rings on nave
               $= t_3 = \tfrac{2}{3} t_1 = \tfrac{2}{3} \times 3\tfrac{1}{2} = 2\cdot 33$, say $2\tfrac{1}{4}$ inches.
Breadth of rings $= b_1 = \tfrac{7}{8} t_3 = \tfrac{7}{8} \times 2\tfrac{1}{4} = 1\tfrac{31}{32}$, say 2 inches.
Thickness of nave under rings $= t_2 = \tfrac{3}{4} t_1 = \tfrac{3}{4} \times 3\tfrac{1}{2} = 2\tfrac{5}{8}$ inches.
Length of nave, say $1\cdot 4 b = 1\cdot 4 \times 10 = 14$ inches.

The necessary working drawings of this wheel are shown in Figs. 437 and 438.

222. Built-up Wheels.—For convenience in manufacture, and for portability, large and heavy wheels are generally made in parts, which are secured together in various ways. Sometimes the boss or nave is cast by itself, so also is each arm, and the rim is made in segments. The arms, when made separate from the nave, may be secured to the latter by bolts, or they may be turned at their inner ends, and let into holes bored to receive them in the nave, and be secured by cotters. When the arms are cast separately, it is a common practice to cast with each a segment of the nave. The parts of the nave are then bound together by hoops shrunk on, as shown in Figs. 435 and 436, except that the segments come into direct contact without the intervening metal strips. Another practice is to cast the arms and nave together, the rim being

TOOTHED GEARING.

cast separately in segments, which are afterwards secured to one another and to the arms.

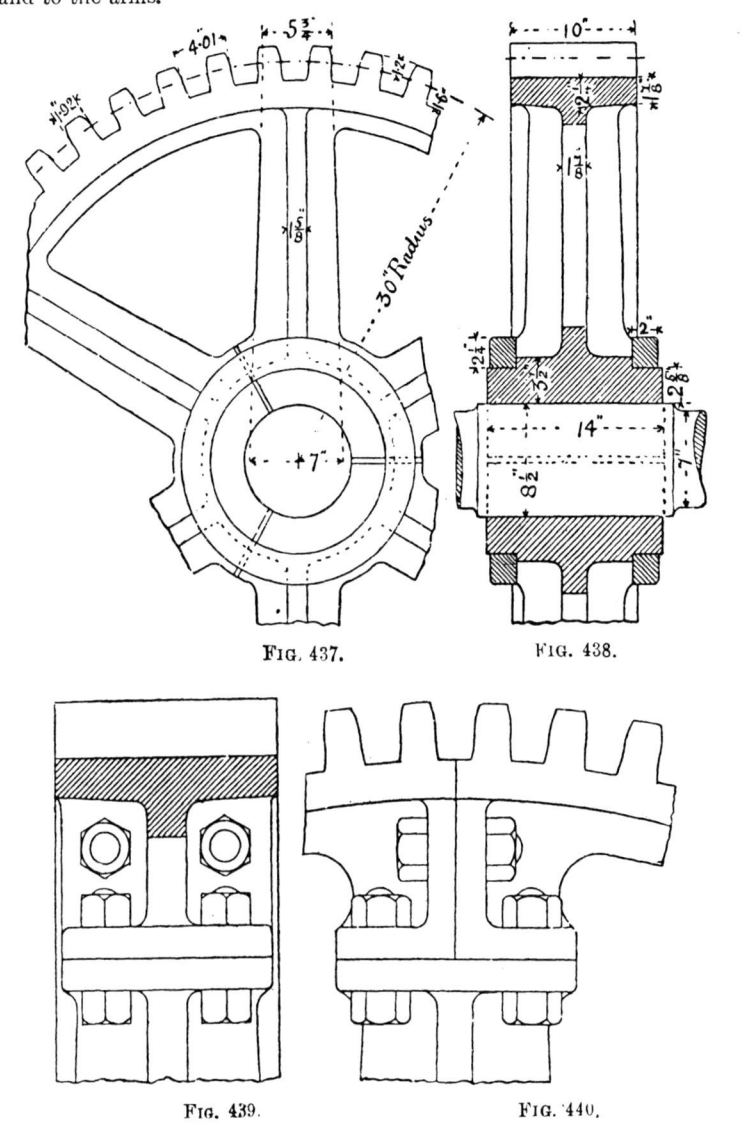

FIG. 437.

FIG. 438.

FIG. 439.

FIG. 440.

Frequently the wheel is cast in parts, each part consisting of an arm, a segment of the nave, and a segment of the rim.

Figs. 439 and 440 show one method of connecting the segments of the rim to one another and to the arms. Here the joints in the rim are at the arms.

Figs. 441, 442, 443, and 444 show a method frequently adopted when the joints in the rim come in between the arms.

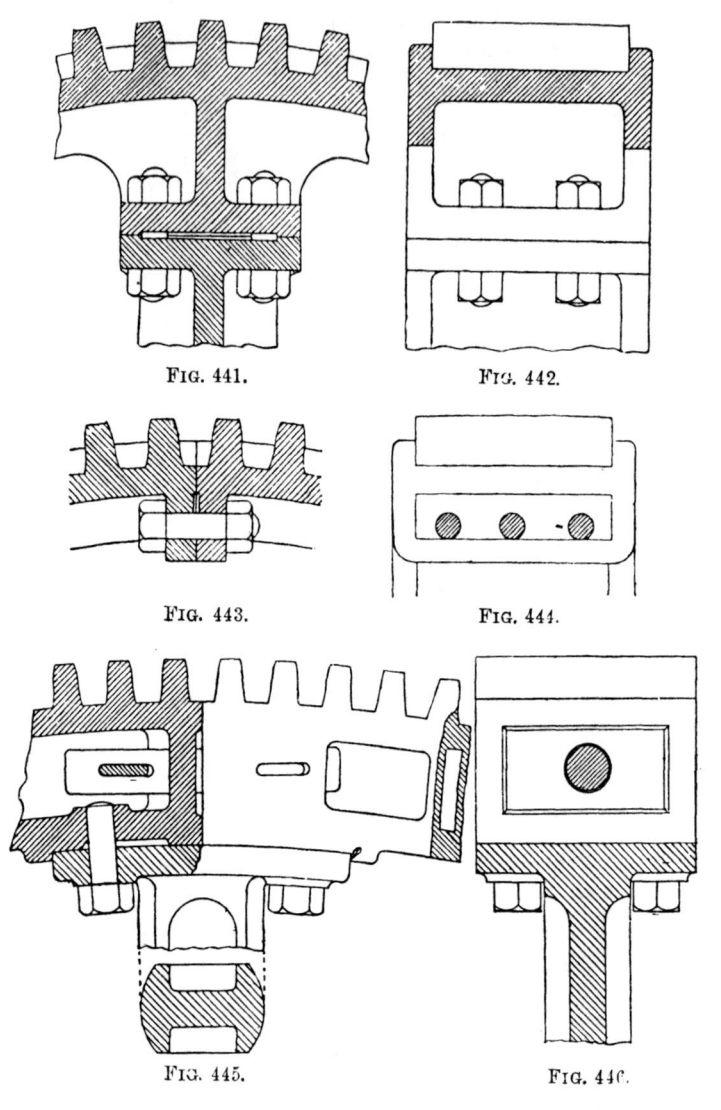

FIG. 441.

FIG. 442.

FIG. 443.

FIG. 444.

FIG. 445.

FIG. 446.

Another design is shown in Figs. 445 and 446. Here the segments of the rim are attached to the arms by bolts, and directly to one another

TOOTHED GEARING.

by dowels or cotter bolts. In the particular example here illustrated the pitch of the teeth is 4 in.; width of face, 16 in.; and diameter of pitch circle, 20 ft. 1·9 in.

223. Stepped Teeth.—The smaller the pitch of the teeth of two wheels in gear the smoother is the motion, but the teeth are weaker the smaller the pitch. To combine the smoothness of the motion due to fine pitched teeth with the strength due to coarse pitched teeth, Dr. Hooke invented stepped teeth. These teeth are shown in Figs. 447 and 448. Imagine a spur wheel having teeth of a pitch p to be divided into n discs of equal thickness by planes at right angles to the axis of the wheel, and let each disc be placed so that the teeth on it are $1 - n$th of the pitch p behind the teeth on the disc to the left of it. These discs would now form a spur wheel with stepped teeth, which would have the strength of teeth of pitch p, and which would work as smoothly as teeth of pitch $\dfrac{p}{n}$.

Fig. 447. 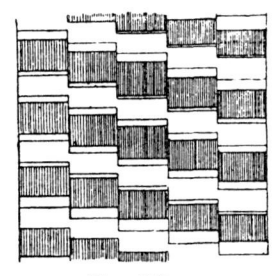   Fig. 448.

224. Helical Teeth.—If the number of steps on a stepped tooth be made infinite, the surface becomes a screw or helical surface, and teeth formed in this way would be called helical teeth. Fig. 449 shows the form of helical teeth. The outline of their section by a plane at right angles to the axis of the wheel is designed as for ordinary teeth, and their outline in the direction of the width of the wheel is drawn by the

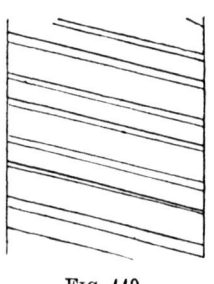

Fig. 449. 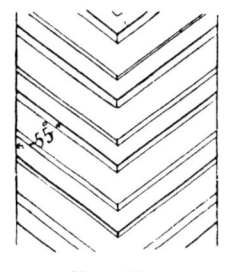   Fig. 450.

rule for drawing helices or screw-curves. It is obvious that two wheels gearing together and having helical teeth must have their teeth of "opposite hand," that is, one must be right-handed and the other left-handed. It is also evident that the inclinations of the helices must be the same.

The objection to the helical teeth, shown in Fig. 449, is that when at work there is a side pressure which tends to push the wheels out of gear. To overcome this difficulty, the double helical teeth, shown in Fig. 450, were introduced, and are now largely used. To ensure the proper bearing of the teeth on one another, the shaft of one of a pair of spur wheels having double helical teeth should have a slight amount of end play. There is a want of reliable data on the strength of helical teeth. They are at least 20 per cent. stronger than ordinary teeth of the same pitch and width.

225. Worm Gearing.—A screw which gears with the teeth of a wheel is called a worm, and the wheel is called a worm wheel. In worm gearing the axis of the worm is usually at right angles to the axis of the wheel, but this is not absolutely necessary.

If a section of the worm and wheel be taken by a plane containing the axis of the worm, the teeth of the wheel and the threads of the worm in this section should be the same as for an ordinary toothed wheel and a rack. The curves of the section of the teeth may be cycloidal, but since the teeth of a rack to gear with a wheel with involute teeth are straight from root to point, it is better to make the teeth of a worm wheel of the involute shape, because then the tool for cutting the worm is of a very simple form.

CHAPTER XV.

CRANKS, CRANKED SHAFTS, AND ECCENTRICS.

226. Strength and Proportions of Overhung Crank-Pins.—The dimensions of the overhung crank-pin are determined from two considerations. There is, first, the bending action on the pin, which is equal to $\frac{1}{2}$PL, where P is the load on the pin and L the length of its journal. Then, there is the consideration that the intensity of the pressure on the journal must not exceed a certain amount, so as to ensure durability. If the pressure per square inch on the journal is too great, the lubricant is forced out and the journal wears rapidly. The bearing surface of a journal is taken as its diameter multiplied by its length. This is also called the *projected area* of the journal, because it is the area of the bearing surface "projected" on to a plane containing its axis.

If p is the pressure on the journal in pounds per square inch of projected area, then $p\mathrm{DL} = \mathrm{P}$, and therefore $\mathrm{L} = \dfrac{\mathrm{P}}{p\mathrm{D}}$.

As stated above, the bending moment on the crank-pin journal is $\frac{1}{2}$PL, that is, $\dfrac{\mathrm{P}^2}{2p\mathrm{D}}$.

The moment of resistance of the journal to bending is $\dfrac{3\cdot 1416}{32}\mathrm{D}^3 f$, where f is the greatest stress due to the bending moment. Therefore, $\dfrac{\mathrm{P}^2}{2p\mathrm{D}} = \dfrac{3\cdot 1416}{32}\mathrm{D}^3 f$, which gives the result

$$\mathrm{D} = \sqrt[4]{\dfrac{16\mathrm{P}^2}{3\cdot 1416 pf}} = \dfrac{1\cdot 5\sqrt{\mathrm{P}}}{\sqrt[4]{pf}} = \mathrm{K}\sqrt{\mathrm{P}}.$$

From the formulæ $\mathrm{L} = \dfrac{\mathrm{P}}{p\mathrm{D}}$, and $\mathrm{D} = \dfrac{1\cdot 5\sqrt{\mathrm{P}}}{\sqrt[4]{pf}}$, we get $\dfrac{\mathrm{L}}{\mathrm{D}} = \dfrac{4}{9}\sqrt{\dfrac{f}{p}}$, and $\mathrm{L} = \dfrac{4}{9}\mathrm{D}\sqrt{\dfrac{f}{p}} = \mathrm{CD}$.

The following table gives values of K and C for various values of p and f:—

p.	f = 5000		f = 6500		f = 8000		f = 9500		f = 11,000		f = 12,500	
	K.	C.	K.	C.	K.	C.	K.	C.	K.	C.	K.	C.
200	·047	2·22	·044	2·53	·042	2·81	·040	3·06	·039	3·30	·038	3·51
300	·043	1·81	·040	2·07	·038	2·30	·037	2·50	·035	2·69	·034	2·87
400	·040	1·57	·037	1·79	·035	1·99	·034	2·17	·033	2·33	·032	2·48
500	·038	1·41	·035	1·60	·034	1·78	·032	1·94	·031	2·08	·030	2·22
600	·036	1·28	·034	1·46	·032	1·62	·031	1·77	·030	1·90	·029	2·03
700	·035	1·19	·032	1·35	·031	1·50	·030	1·64	·029	1·76	·028	1·88
800	·034	1·11	·031	1·27	·030	1·41	·029	1·53	·028	1·65	·027	1·76
900	·033	1·05	·030	1·19	·029	1·32	·028	1·44	·027	1·55	·026	1·66

The smallest value of p given in the foregoing table, namely, 200 lbs. per square inch, would be taken for small high-speed engines, and the largest value, 900 lbs. per square inch, for large low-speed engines.

If it is desired to make the length of the crank-pin less than that given by the foregoing rules, the diameter must be increased so as to satisfy the equation $pDL = P$. Let $L = nD$, then $npD^2 = P$, and $D = \sqrt{\dfrac{P}{np}}$. The crank-pin proportioned by this rule will have an excess of strength when n is less than $\dfrac{4}{9}\sqrt{\dfrac{f}{p}}$.

227. Proportions of Overhung Cranks.

The following rules are based on the assumption that the crank is made of wrought-iron or steel, and that the material of the shaft is the same as that of the crank. The letters refer to Figs. 451 and 452.

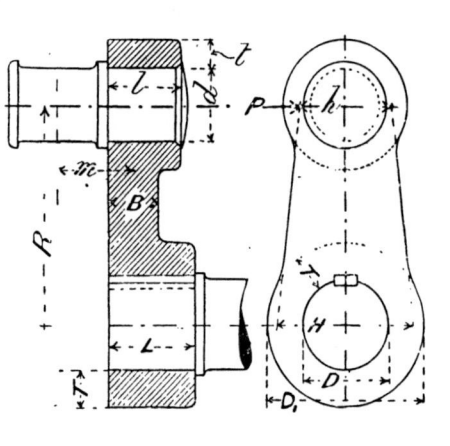

FIG. 451.   FIG. 452.

For the eye at the crank-shaft end, $L = CD$, where C varies from ·7 to 1·1. An average value of C is ·9. $T = KD$, where K depends on C. Values of K for various values of C are given in the following table:—

| C = | ·7 | ·8 | ·9 | 1·0 | 1·1 |
| K = | ·44 | ·40 | ·37 | ·34 | ·31 |

For the eye at the crank-pin end, $l = cd$, where c varies from ·9 to 1·4, but it is generally about 1·2. $t = kd$. Values of k for various values of c are given in the following table:—

| c = | ·9 | 1 | 1·1 | 1·2 | 1·3 | 1·4 |
| k = | ·62 | ·5 | ·41 | ·35 | ·30 | ·26 |

For the arm or web, the width H, measured at the centre of the shaft, may be taken equal to from $\cdot 7 D_1$ to D_1, where $D_1 = D + 2T$. The width at the crank-pin end may be $h = \cdot 7 d_1$ to d_1, where $d_1 = d + 2t$.

Having fixed upon the greatest width H of the arm, its thickness B is determined as follows :—First assume that the arm is subjected to bending only, the bending moment being equal to the twisting moment on the shaft. We may then take the resistance of the arm to bending equal to the resistance of the shaft to twisting, that is, $\frac{1}{6} BH^2 f_t = \frac{3 \cdot 1416}{16} D^3 f_s$.

If we assume $f_t = f_s$, we get $B = \frac{1 \cdot 18 D^3}{H^2}$.

Having determined approximately, as above, the thickness of the arm, the distance m between the centre of the crank-pin and the centre line of the crank-arm is also determined approximately.

Let P be the greatest force which acts on the crank-pin in a direction at right angles to the crank-arm. This force produces a bending moment on the arm $= M = PR$, and also a twisting moment $M_t = Pm$. Combining these we get the equivalent bending moment $M_e = \frac{1}{2} M_b + \frac{1}{2} \sqrt{M_b^2 + M_t^2}$.

Now recalculate the thickness of the crank-arm from the equation $\frac{1}{6} BH^2 f_t = M_e$, where f_t is the safe tensile stress. In fixing upon the value of f_t we use the factor of safety which was used in determining the diameter of the shaft.

The width of the key for securing the crank to the shaft does not generally exceed $\frac{1}{4}D$ for large cranks, and for small cranks it may be taken at $\frac{1}{4}D + \frac{1}{8}$. The thickness of the key varies from $\cdot 4$ to $\cdot 7$ of its width.

228. Connection of Crank-Pin to Crank-Arm.—The part of the crank-pin within the crank-arm may be parallel or slightly tapered. The most common method of securing the pin to the arm is by riveting, as shown in Fig. 453. The crank-arm is generally shrunk on to the pin, and if the part of the pin within the arm be parallel, as shown in Fig. 454, this is often quite sufficient to secure the pin. A good method is to make the

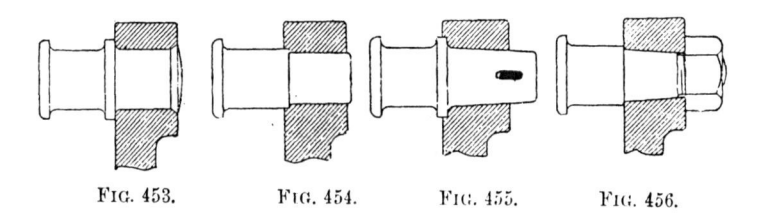

FIG. 453.   FIG. 454.   FIG. 455.   FIG. 456.

part of the pin within the crank parallel and slightly larger than the hole in the crank, and then force the pin into the crank by hydraulic pressure. With this method the difference between the diameters of the pin and the hole in the crank should be such that the pressure required to force the pin in is about 12 tons per inch of diameter. Sometimes the pin is kept from coming out by a cotter, as shown in Fig. 455, or by a nut screwed on to the tail end of the pin, as shown in Fig. 456. The last two methods have the advantage that it is easier to withdraw the pin for renewal or repairs.

If the end of the rod which works on the crank pin is solid, the outer collar on the pin must be made separate from the pin. Two methods of attaching this loose collar to the pin are shown in Figs. 457 and 458.

229. Forged Cranked Shafts.—The form of cranked shaft shown in Figs. 459 and 460 is largely used for marine and other kinds of engines. The following rules are based on a large number of examples chiefly taken from marine engines. The letters refer to Figs. 459 and 460.

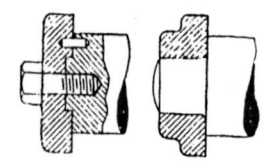

FIG. 457.   FIG. 458.

$T = aD$, where a varies from ·6 to ·8. Average value of $a = $ ·7.

$TW^2 = bD^3$, where b varies from ·8 to 1·06. Average value of $b = $ ·9.

If $T = $ ·7D and $TW^2 = $ ·9D^3, then $W = 1·134D$.

D_1 generally $= D$, but sometimes as large as 1·05D.

L generally $= D$.

Sometimes a hole is drilled through the crank-pin, as shown, the diameter of the hole being about one-third the diameter of the pin.

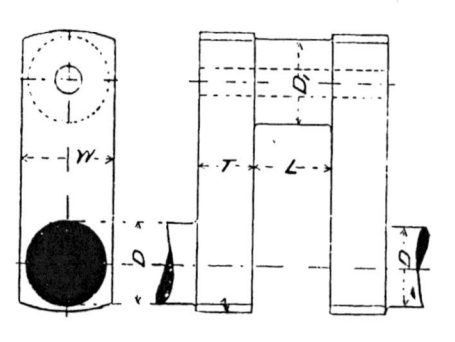

FIG. 459.   FIG. 460.

230. Locomotive Crank Axles.—The most common form of crank axle for locomotives with inside cylinders is represented in Figs. 461.

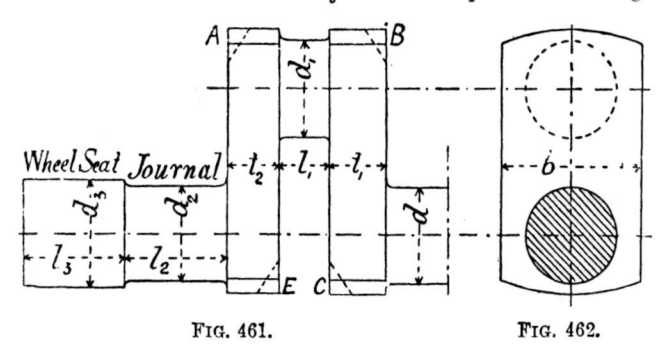

FIG. 461.   FIG. 462.

and 462, which show one half of the axle. The other half is exactly the same, except that the other crank is at right angles to the one

CRANKS AND ECCENTRICS.

shown. The ends of the webs are frequently bevelled or rounded more or less at A, B, C, and E, as shown by dotted lines. These axles are now generally made of steel. The following table gives the dimensions of locomotive crank-axles taken from actual practice. D is the diameter of the steam-cylinders, and L the stroke of the pistons. The other letters refer to Figs. 461 and 462. All the dimensions are in inches.

D	L	d	d_1	d_2	d_3	l_1	l_2	l_3	t_1	t_2	b
13	20	$5\frac{1}{4}$	$6\frac{1}{4}$	$5\frac{3}{4}$	$5\frac{3}{4}$	4	8	$6\frac{1}{2}$	$4\frac{1}{2}$	$4\frac{1}{8}$	8
16	22	7	$7\frac{1}{4}$	7	$8\frac{1}{2}$	$4\frac{1}{2}$	9	$6\frac{3}{4}$	$4\frac{3}{4}$	$4\frac{3}{8}$	10
$16\frac{1}{2}$	20	$6\frac{1}{2}$	$6\frac{1}{2}$	$6\frac{1}{2}$	$7\frac{1}{2}$	4	$8\frac{1}{2}$	$5\frac{1}{4}$	$4\frac{1}{4}$	4	10
17	24	$6\frac{3}{8}$	7	$6\frac{3}{4}$	8	4	9	7	4	4	11
$17\frac{1}{2}$	24	$6\frac{3}{4}$	7	$6\frac{3}{4}$	8	4	$7\frac{1}{2}$	$6\frac{3}{4}$	$4\frac{1}{2}$	4	12
$17\frac{1}{2}$	26	7	$7\frac{3}{4}$	$7\frac{1}{2}$	9	4	$7\frac{1}{2}$	$7\frac{1}{2}$	$4\frac{1}{2}$	$4\frac{1}{4}$	12
$17\frac{1}{2}$	26	7	$7\frac{1}{2}$	$7\frac{1}{2}$	$8\frac{1}{2}$	4	9	$7\frac{1}{32}$	$4\frac{1}{4}$	$4\frac{1}{4}$	13
18	24	7	8	$7\frac{1}{2}$	9	$4\frac{1}{2}$	9	$8\frac{1}{2}$	$4\frac{3}{4}$	$4\frac{1}{4}$	12
18	25	7	$7\frac{3}{4}$	$7\frac{1}{4}$	$8\frac{1}{2}$	4	9	7	$4\frac{1}{8}$	$4\frac{1}{8}$	14
18	26	$7\frac{1}{2}$	$7\frac{1}{2}$	$7\frac{1}{2}$	$8\frac{1}{2}$	4	$8\frac{1}{2}$	$7\frac{1}{8}$	$4\frac{1}{4}$	4	13
18	26	8	$8\frac{1}{2}$	8	$8\frac{3}{4}$	$4\frac{1}{4}$	$7\frac{1}{2}$	8	$4\frac{1}{2}$	$4\frac{3}{8}$	13
$18\frac{1}{4}$	26	7	$7\frac{1}{2}$	7	$8\frac{1}{2}$	4	7	$6\frac{7}{8}$	$4\frac{1}{4}$	$3\frac{3}{4}$	12
$18\frac{1}{4}$	26	$7\frac{3}{4}$	$8\frac{1}{4}$	8	$8\frac{1}{4}$	4	$8\frac{1}{8}$	$7\frac{1}{10}$	5	5	11
19	26	$7\frac{1}{2}$	$8\frac{1}{4}$	$8\frac{1}{4}$	$8\frac{1}{4}$	4	$8\frac{1}{2}$	$8\frac{3}{4}$	$5\frac{1}{4}$	$4\frac{7}{8}$	12
20	24	$7\frac{1}{2}$	8	8	9	5	10	$6\frac{3}{4}$	$4\frac{3}{4}$	$4\frac{3}{4}$	13

On some railways it is the practice to hoop the crank-webs. An example of this, from a locomotive on the Midland Railway, is shown in Figs. 463 and 464. These hoops, which are shrunk on, are not added so much with a view to strengthening the cranks, but as a safeguard

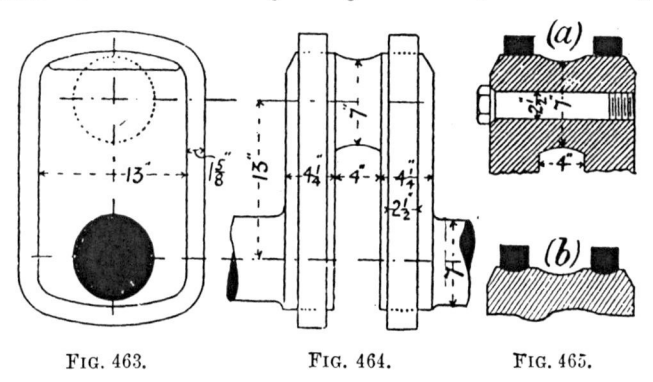

FIG. 463.   FIG. 464.   FIG. 465.

against accident should a crank break. For a similar reason a bolt is sometimes screwed into the crank-pin, as shown at (a), Fig. 465, which is also an example from a locomotive on the Midland Railway. Sometimes the bed for the hoop takes the form of a very shallow groove, as shown at (b), Fig. 465.

231. Worsdell's Cranked Axle.—The special feature of the form of cranked axle introduced on the Great Eastern and on the North-Eastern Railways by Mr. Worsdell is in the webs, which are made circular, as shown in Fig. 466. This enables the axle to be finished entirely in

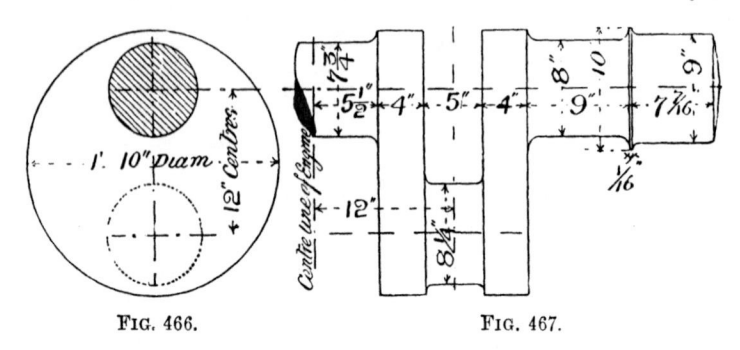

FIG. 466.   FIG. 467.

the lathe. The particular axle shown in Figs. 466 and 467 is for one of Mr. Worsdell's compound locomotives, having a high-pressure cylinder 20 inches in diameter, a low-pressure cylinder 28 inches in diameter, and a piston stroke of 2 feet. The axle is made of steel. The working pressure of steam in the boiler is 175 lbs. per square inch.

232. Built-up Cranked Shafts.—As already stated, the form of cranked shaft shown in Figs. 459 and 460 is largely used for marine engines, but for the very powerful engines now used in large steam-ships this design of shaft is very unreliable. For large marine engines the built-up cranked shaft shown in Figs. 468 and 469 is now generally used, although

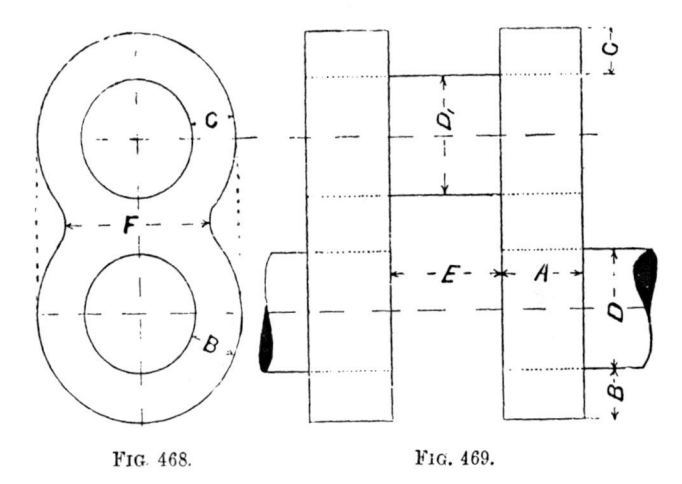

FIG. 468.   FIG. 469.

it is much heavier than the other form. It will be seen from the above Figs. that the crank-shaft, crank-arms, and crank-pin are made separately

The arms are shrunk on to the pin and shaft, and in addition are secured to the latter by sunk keys. Sometimes the arms are also keyed to the crank-pin, but most engineers consider this unnecessary. These built-up cranked shafts are generally made of steel.

The following rules give proportions agreeing well with average practice :—

$A = \cdot 75 D.$ $D_1 = D.$
$B = \cdot 43 D.$ $E = \cdot 95 D$ to $1 \cdot 1 D.$
$C = \cdot 4 D.$ $F = 1 \cdot 3 D.$

Sometimes the arms are straight on the sides, as shown by the dotted lines in Fig. 468, and the parts of the shaft and pin within the arms are frequently enlarged, the diameter of the enlarged part being about 1·05 times the diameter of the shaft or pin.

233. Foster's Cranked Shaft.—In this form of cranked shaft the crank-pin and crank-webs are cast in one piece of mild steel, as shown in Figs. 470 and 471. The webs are keyed to the two parts of the shaft, the keys being driven from the outside. To permit of the keys being driven from the outside, the parts of the shaft where they enter the webs are

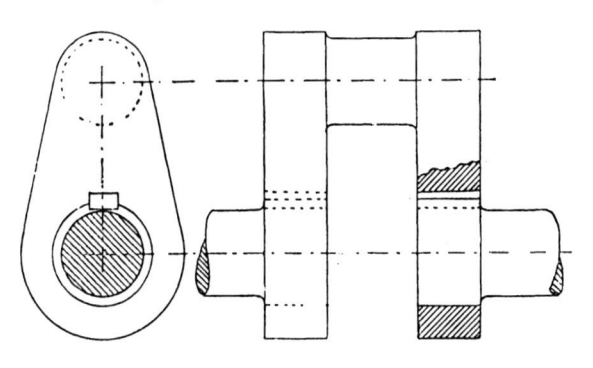

FIG. 470.   FIG. 471.

enlarged in diameter by an amount equal to the thickness of the keys. This form of crank has been applied successfully in repairing broken forged cranked shafts. Forged cranked shafts generally give way at the junction of the crank-pin and one of the webs. Should this happen, the webs are cut off and the ends of the two parts of the shaft at the webs are turned to receive the new webs.

The crank-webs in the above design of cranked shaft may be proportioned by the rules already given for built-up cranked shafts.

234. Dickinson's Cranked Shaft.—The cranked shaft designed by Mr. J. Dickinson, shown in Figs. 472 and 473, resembles Foster's cranked shaft in having the crank-pin and crank-webs in one piece, separate from the shaft; the webs are, however, not keyed to the shaft, but are bolted to flanges formed on the two parts of the shaft, these flanges being sunk into the webs as shown.

200   MACHINE DRAWING AND DESIGN.

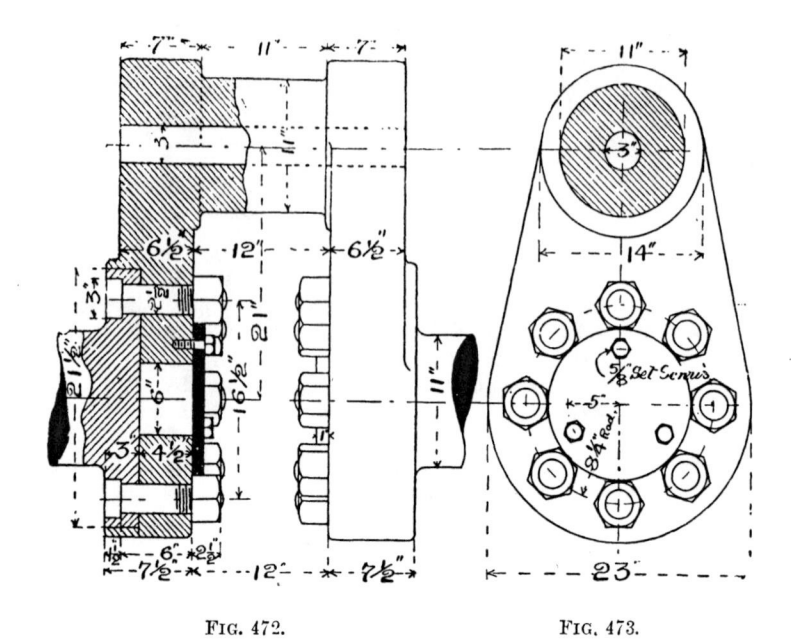

FIG. 472.   FIG. 473.

235. Crank-Discs.—In small engines, and in engines which run at a high speed, a cast-iron disc frequently takes the place of the ordinary

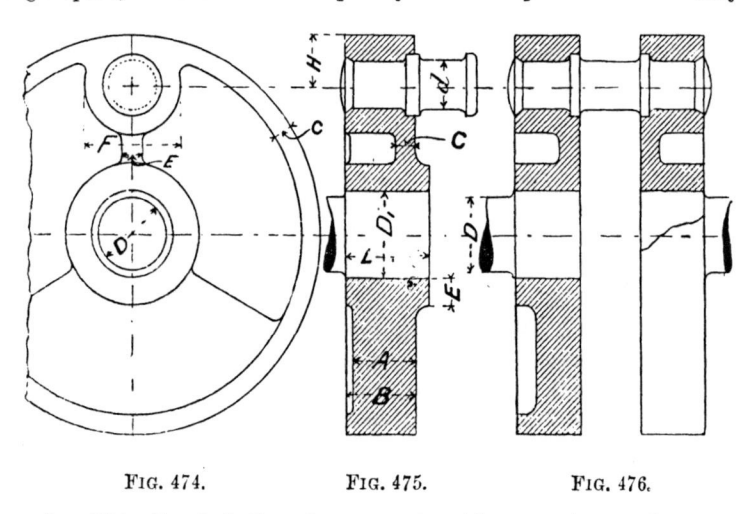

FIG. 474.   FIG. 475.   FIG. 476.

crank. This disc is hollowed out on the side next the crank-pin, as shown in Figs. 474 and 475, and the extra weight on the opposite side balances the crank-pin, and the part of the connecting-rod which revolves with it. Two crank-discs, Fig. 476, may be used to take the place of a double-armed crank.

The following rules may be used in proportioning crank-discs :—

D_1 = D to 1·2D.
B = ·7D to D.
A = ·9B.
C = ·25D.

E = ·4D.
F = 2d.
H = d to 1·5d.
L = B to 1·25B.

236. Board of Trade Rules for Shafts of Steamships.—Main, tunnel, propeller, and paddle shafts should not be passed if less in diameter than that found by the following formulæ, without previously submitting the whole case to the Board of Trade for their consideration. It will be found that first-class makers generally put in larger shafts than those obtained by the formulæ.

For compound condensing engines with two or more cylinders, when the cranks are not overhung :—

$$S = \sqrt[3]{\frac{C \times P \times D^2}{f\left(2 + \frac{D^2}{d^2}\right)}}$$

$$P = \frac{f \times S^3}{C \times D^2}\left(2 + \frac{D^2}{d^2}\right).$$

Where S = diameter of shaft in inches.

d^2 = square of diameter of high-pressure cylinder in inches or sum of squares of diameters when there are two or more high-pressure cylinders.

D^2 = square of diameter of low-pressure cylinder in inches or sum of squares of diameters when there are two or more low-pressure cylinders.

P = *absolute* pressure in pounds per square inch, that is, boiler pressure plus 15 lbs.

C = length of crank in inches.

f = constant from following table (p 202).

Note.—Intermediate-pressure cylinders do not appear in the formulæ.

For ordinary condensing engines, with one, two, or more cylinders, when the cranks are not overhung :—

$$S = \sqrt[3]{\frac{C \times P \times D^2}{3 \times f}}$$

$$P = \frac{3 \times f \times S^3}{C \times D^2}.$$

Where S = diameter of shaft in inches.

D^2 = square of diameter of cylinder in inches or sum of squares of diameters when there are two or more cylinders.

P = *absolute* pressure in pounds per square inch, that is, boiler pressure plus 15 lbs.

C = length of crank in inches.

f = constant from following table :—

For Two Cranks. Angle between Cranks.	For Crank, Thrust, and Propeller Shafts.[1] $f=$		For Tunnel Shaft $f=$
90°		1,047	1,221
100°	For paddle	966	1,128
110°	engines of	904	1,055
120°	ordinary type	855	997
130°	multiply con-	817	953
140°	stant in this	788	919
150°	column suitable	766	894
160°	for angle of	751	877
170°	cranks by 1·4	743	867
180°		740	864
For three cranks. 120°		1,110	1,295

Note.—When there is only one crank the constants applicable are those in the table opposite 180°.

237. Eccentrics.—The eccentric is a modification of the crank, the difference being that in the eccentric the crank-pin is so large that it embraces the crank-shaft. This difference is shown in Figs. 477 and 478, where A is the centre of the crank-shaft and B the centre of the crank-pin, or what is equivalent to the crank-pin.

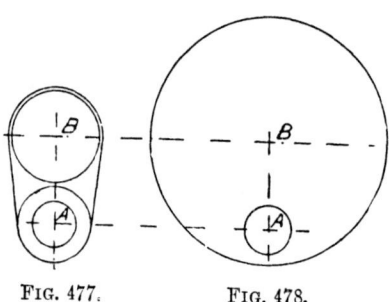

The eccentric is used for converting a circular into a reciprocating motion, and it has an advantage over the crank for this purpose, in that no break or gap is needed in the shaft, such as is made when a shaft is cranked at an intermediate point of its length.

Fig. 477.   Fig. 478.

The eccentric proper is generally called the eccentric sheave or eccentric pulley, and the end of the rod which encircles the eccentric, and which corresponds to the connecting-rod end on a crank-pin, is called the eccentric strap.

The eccentric sheave is generally made of cast-iron, and if it can be put on over the end of the shaft it is made in one piece. But if, as is most frequently the case, the eccentric has to be placed on the shaft between two cranks, or if there are enlargements on the shaft which would prevent a solid sheave being placed in position, the sheave is made in two parts, which are bolted together after being placed in position on the shaft. The smaller of the two parts of the sheave is made solid, and is frequently made of wrought-iron or steel. The larger part in all except very small eccentrics is made with a rim and boss, connected by one or

[1] The constants in this column should be reduced by 15 per cent. when dealing with propeller shafts of new vessels.

more arms, as shown in Figs. 479, 480, and 481. The two parts of the sheave are connected by bolts or studs and cotters or nuts, as shown in the same Figs.

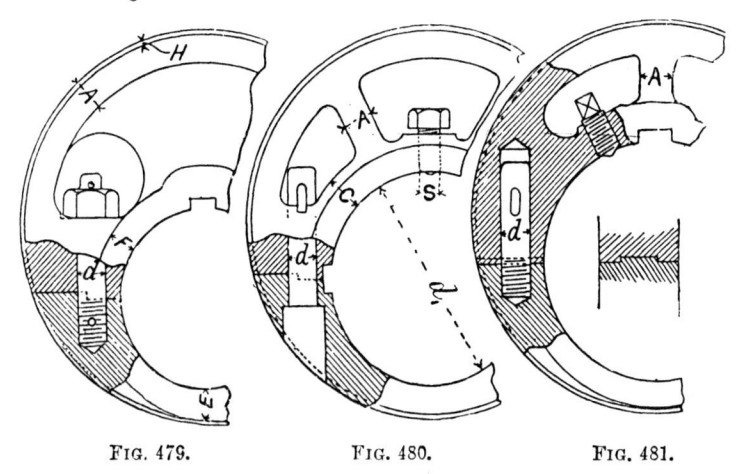

FIG. 479.   FIG. 480.   FIG. 481.

The sheave is secured to the shaft by a key, and very often by one or more set-screws in addition, as shown.

The most common form of the section of the rim of the sheave is that shown at a, Fig. 482. The next most common form is that shown at b. Other forms less frequently used are shown at c, d, e, and f.

238. Proportions of Eccentric Sheaves.—The breadth, B, of the sheave is first determined from the force required to work the valve which the eccentric has to drive. Let l = length and b = breadth of the slide valve, and p = pressure of steam in pounds per square inch on the back of the valve, then

$B = \cdot 017 \sqrt{lbp}$ to $\cdot 02 \sqrt{lbp}$ for locomotive engines.

FIG. 482.

Taking the mean of a large number of examples from practice, we find $B = \cdot 0185 \sqrt{lbp}$ for locomotive engines.

For marine engines $B = \cdot 01 \sqrt{lbp}$.

Since the area of the slide valve is generally proportional to the area of the piston in the same class of engines, we may express the breadth B of the eccentric in terms of the diameter of the piston and the steam pressure.

Let D = diameter of piston, then from a large number of examples we find—

$B = \cdot 0132 D \sqrt{p}$ for locomotive engines.

$B = \cdot 009 D \sqrt{p}$ to $\cdot 015 D \sqrt{p}$ for marine engines.

For stationary engines the breadth of the eccentric should not be less than that given for locomotive or marine engines; but since space is not so limited in stationary as in locomotive and marine engines, the width of the eccentric for the former may be $B = \cdot 013D \sqrt{p}$ to $\cdot 03D \sqrt{p}$.

Having decided on the breadth of the eccentric, the details of the sheave may be proportioned by the following rules:—

$A = \cdot 55B$.
$C = \cdot 65B$.
$d = \cdot 4B$.
$S = \cdot 32B$ for two set-screws.
 $= \cdot 4B$ for one set-screw.
$E = \cdot 4B$ (minimum for wrought-iron or steel).
 $= \cdot 5B$ (minimum for cast-iron).
$F = \cdot 4B$.
$H = \cdot 13B$.

Length or width at boss $= L = B$ to $1 \cdot 6B$.
Thickness of key $= \cdot 2B$ to $\cdot 3B$.
Breadth of key $= 1$ to 2 times its thickness, generally $1\frac{1}{2}$ times.
Breadth of cotters in bolts $= d$ to $1\frac{1}{4}d$.
Thickness of cotters in bolts $= \dfrac{d}{4}$.

Diameter of sheave $= d_1 + 2r + 2E$, where r is the radius or eccentricity, or half-throw of the eccentric.

The letters in the above rules refer to Figs. 479 to 481.

The proportions for the eccentric strap will be found in Art. 248, p. 215.

CHAPTER XVI.

CONNECTING-RODS.

239. Thrust on Connecting-Rod.—
Let D = diameter of piston in inches.
    l = length of stroke in inches.
    L = length of connecting-rod in inches.
    P = effective pressure of steam on piston in lbs. per square inch.

Then, total load on piston = $\cdot 7854 D^2 P = Q$. At the beginning of the stroke, when the piston-rod and connecting-rod are in a straight line, this will also be the thrust on the connecting-rod; but as the piston moves the connecting-rod becomes inclined to the centre line of the piston-rod, and the thrust on the connecting-rod will increase, and its magnitude for any position of the crank is found as follows:—

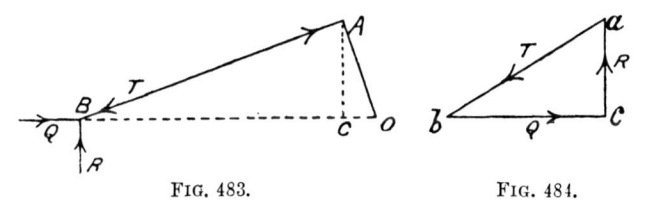

FIG. 483.   FIG. 484.

Let AO, Fig. 483, be the position of the crank, and AB the corresponding position of the connecting-rod. Draw AC perpendicular to BO. The forces acting at B are—Q, the load on the piston; R, the reaction of the guide on the crosshead; and T, the thrust along the connecting-rod. Drawing the triangle of forces *abc*, Fig. 484, we get

$$\frac{T}{Q} = \frac{ba}{bc} = \frac{BA}{BC} \text{ and } T = Q\frac{BA}{BC}.$$

If the steam is not cut off before AO is at right angles to BO, T will be a maximum when the angle AOB is 90 degrees.

Then when AO is at right angles to BO,

$$T = Q\frac{BA}{BO} = Q\frac{BA}{\sqrt{BA^2 - AO^2}} = Q\frac{L}{\sqrt{L^2 - \frac{l^2}{4}}} = Q\frac{2L}{\sqrt{4L^2 - l^2}}.$$

Let $L = nl$, then maximum value of

$$T = Q\frac{2nl}{\sqrt{4n^2l^2 - l^2}} = Q\frac{2n}{\sqrt{4n^2 - 1}} = cQ.$$

The following table gives values of c corresponding to various values of n :—

$n =$	1·75	2	2·5	2·75	3
$c =$	1·044	1·033	1·021	1·017	1·014

240. Length of Connecting-Rod.—The length of a connecting-rod is measured from the centre of the crosshead pin or gudgeon to the centre of the crank-pin. For stationary engines the most usual length of the connecting-rod is from 2·5 to 3 times the length of the stroke of the piston, but the ratio of length of rod to length of stroke is sometimes as small as 2, and sometimes as large as 3·75 for this class of engine. For ordinary marine engines the ratio of length of rod to stroke is generally about 2, but it is sometimes as large as 2·5. For the engines of war-ships, which require to occupy as small a space as possible, the connecting-rod is sometimes only 1·75 times the length of the stroke of the piston. In locomotive engines the length of the connecting-rod is generally about three times the stroke of the piston, but sometimes this ratio is as small as 2·7, and in some exceptional cases it is as large as 4·8. In portable engines 3·5 is a common ratio for the length of the connecting-rod to the length of stroke.

241. Strength of Connecting-Rods.—A connecting-rod is in the condition of a column or strut jointed or free at both ends, and in determining its strength, the ratio of its length to its diameter should be taken into account.

For a rod of circular section the relation between the safe thrust T and the diameter d and length l of the rod will be of the form—

$$T = \frac{k_1 d^2}{k_2 + \frac{l^2}{d^2}} = \frac{k_1 d^2}{k_2 + r^2} \text{ and } d = \sqrt{T} \frac{\sqrt{k_2 + r^2}}{\sqrt{k_1}}$$

where k_1 and k_2 are constants to be determined from experiment, or from a comparison of successful examples in practice. From an examination of the dimensions of connecting-rods made by different leading engineers, we find that the following formula gives good results :—

$$d = \frac{\sqrt{T} \sqrt{300 + r^2}}{1000} = c_1 \sqrt{T}.$$

If for T we put its value $cQ = c \times \cdot 7854 D^2 P$ and take $c = 1\cdot025$ we get—

$$d = \frac{\sqrt{300 + r^2}}{1115} D \sqrt{P} = c_2 D \sqrt{P}.$$

The following table gives values of c_1 and c_2 corresponding to various values of r :—

$r = \frac{l}{d} =$	8	10	12	14	16	18	20
$c_1 =$	·019	·020	·021	·022	·024	·025	·026
$c_2 =$	·017	·018	·019	·020	·021	·022	·023

If, as is generally the case, the crank-pin end of the connecting-rod is heavier than the crosshead end, the rod is made with a straight taper; the diameter at the crosshead end being not less than ·9d, where d is the diameter at the middle, calculated by the rules which have just been given. Sometimes a rod with a straight taper has a diameter at the crosshead end equal to ·95d, but generally it is about ·92d.

When the two ends of the rod are equally heavy and the rod is comparatively short it is usually made parallel. Long rods with equally heavy ends are generally made barrel-shaped, the diameter of each end is then from $\frac{3}{4}d$ to $\frac{7}{8}d$.

Sometimes the rod is tapered from the crosshead end to the middle, and is parallel for the remainder of its length.

For rods of rectangular section the formula for the strength will be of the form—

$$T = \frac{k_1 lt}{k_2 + \frac{l^2}{t^2}} = \frac{k_1 bt}{k_2 + r^2}$$

where b is the breadth and t the thickness of the rod (t being less than b), and k_1 and k_2 are constants to be determined in the same way as for round rods.

Connecting-rods of rectangular section are not much used except on locomotives. From a comparison of a number of locomotive connecting-rods we find that the following formula gives good results:—

$$T = \frac{6500^2 bt}{5800 + r^2}$$

If for b we put nt, then

$$T = \frac{6500^2 nt^2}{5800 + r^2} \quad \text{and} \quad t = \frac{\sqrt{5800 + r^2}}{6500} \frac{\sqrt{T}}{\sqrt{n}} = \frac{c_3 \sqrt{T}}{\sqrt{n}}.$$

If for T we put its value $cQ = c \times ·7854 D^2 P$ and take $c = 1·014$, which corresponds to a connecting-rod having a length three times the length of the stroke, a usual ratio for locomotives, then—

$$t = \frac{\sqrt{5800 + r^2}}{7280 \sqrt{n}} D \sqrt{P} = \frac{c_4 D \sqrt{P}}{\sqrt{n}}.$$

The following table gives values of c_3 and c_4 corresponding to various values of r:—

$r = \dfrac{l}{t} =$	35	40	45	50	55
$c_3 =$	·0129	·0132	·0136	·0140	·0145
$c_4 =$	·0115	·0118	·0121	·0125	·0129

In practice n varies from 1·5 to 2·4, an average value being 2.

The rod is either uniform in width or it has a straight taper. In the latter case the width at the crosshead end is from $\frac{3}{4}b$ to $\frac{7}{8}b$, where b is the width at the middle.

The thickness is nearly always uniform throughout the length.

242. Common Strap-End.—This form of connecting-rod end, which is not now so common as it used to be, is shown in Fig. 485. All the parts are held together by means of a gib and cotter. Sometimes two gibs are used, one on each acting face of the cotter.

Let T be the maximum thrust or pull on the connecting-rod, and f the safe stress on the material in pounds per square inch, then the working strength of the thinnest part of the strap is given by the equation $2bt_1 f = T$ and $t_1 = \dfrac{T}{2bf}$. The width b is generally equal to, or a little greater than the diameter of the adjacent end of the round part of the rod. For wrought-iron f may be taken at 5000, and for steel at 7000. t_2, the thickness of the strap at the crown, varies from $1·17t_1$ to $1·5t_1$.

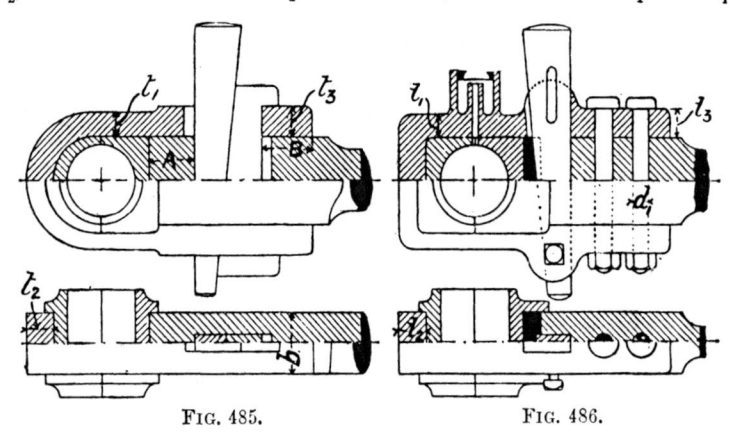

FIG. 485.   FIG. 486.

t_3, the thickness of the strap at the cotter, varies from $1·2t_1$ to $1·5t_1$ and it should be such that the area of the cross section at the cotter hole is not less than the area of the cross section at the thinnest part. The total area of the cross section of the gibs and cotter should not be less than bt_1. The thickness of the cotter is generally about ·25b, then t_3 should not be less than $1·33t_1$. The distance A from the butt end of the rod to the cotter hole may be about $2t_1$, and the distance B from the end of the strap to the gib may be $2·5t_1$.

243. Fixed-Strap End.—In the fixed strap form of connecting-rod end, the strap does not slide on the connecting-rod when the brasses are brought together, but it is secured to the rod by bolts as in Fig. 486,

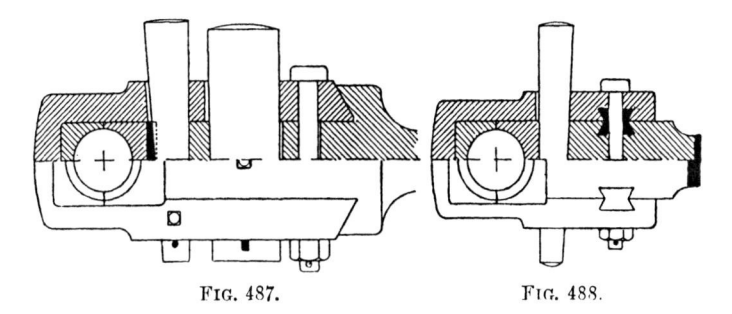

FIG. 487.   FIG. 488.

or by a bolt and cotter as in Fig. 487, or by a bolt and keys as in Fig. 488. The arrangement shown in Fig. 486 is a very common one. The cotter for tightening up the brasses is driven in between the inner brass and the butt end of the rod, and it is in compression only. In order to distribute the pressure of the cotter on the brass, a steel bearing-piece is generally introduced, as shown in Figs. 486 and 487.

The bolts shown in Fig. 486 are in "double shear," and their diameter d_1 is calculated from the formula

$$4 \times ·7854 d_1^2 f = \text{T}, \text{ or } d_1 = ·564 \sqrt{\frac{\text{T}}{f}}$$

where f is the safe shearing stress on the bolts, and T the greatest pull on the connecting-rod; f may be taken at 4000 for wrought-iron and 5500 for steel.

The thickness of the cotter for tightening the brasses may be $·25b$, and its mean width two to three times its thickness. The dimensions t_1, t_2, and t_3, are determined as for the common strap shown in Fig. 485. The thickness of the steel bearing-piece may be equal to the thickness of the cotter.

The keys shown in Fig. 488 may have a breadth equal to $1·9t_1$ and a thickness $1·3t_1$, and the diameter of the bolt in this design may be $·7t_1$.

244. Solid or Box-End.—In the solid or box-end, shown in Figs. 489 and 490, the part corresponding to the strap in the preceding examples is made in one piece with the connecting-rod. It is obvious that since the brasses in this design must enter from the side, they must be free from flanges on one side except at one end, as shown in Fig. 489.

At the cross-head end of a connecting-rod, the space is often very limited, and there is not sufficient room for an ordinary cotter. When this is the case, the brasses may be secured by a wrought-iron or steel wedge-shaped block shown in Fig. 490. This block is adjusted by a bolt screwed into it. The same arrangement may be used on a strap end, as

O

shown in Fig. 491. In this Fig. two bolts are shown, screwed in from opposite sides, one serving as a locking arrangement to the other. The side of the block next the brass has a slope of 1 in 4 to 1 in 7, generally

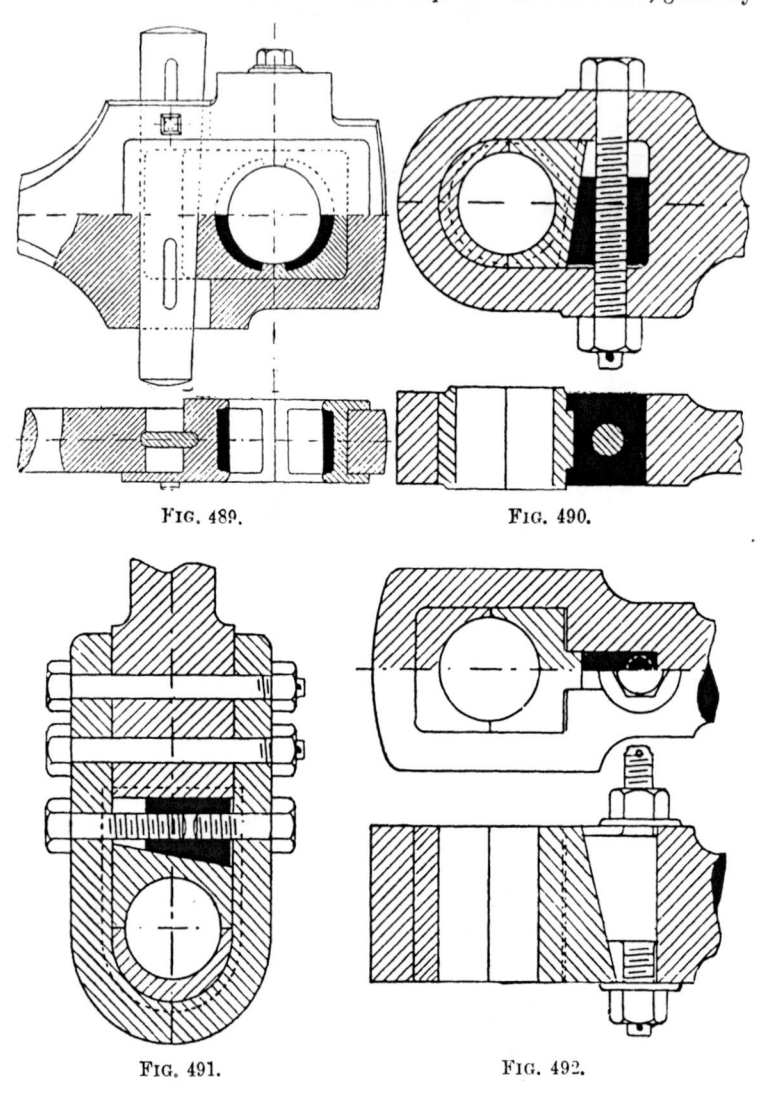

FIG. 489.   FIG. 490.

FIG. 491.   FIG. 492.

1 in 6. The diameter of the bolt passing through the block varies from ·2*d* to ·3*d*, where *d* is the diameter of the bearing.

Fig. 492 shows a solid end in which the axis of the cotter is parallel to the axis of the bearing. The cotter in this example has considerable taper, and it is screwed at both ends, and provided with nuts and washers.

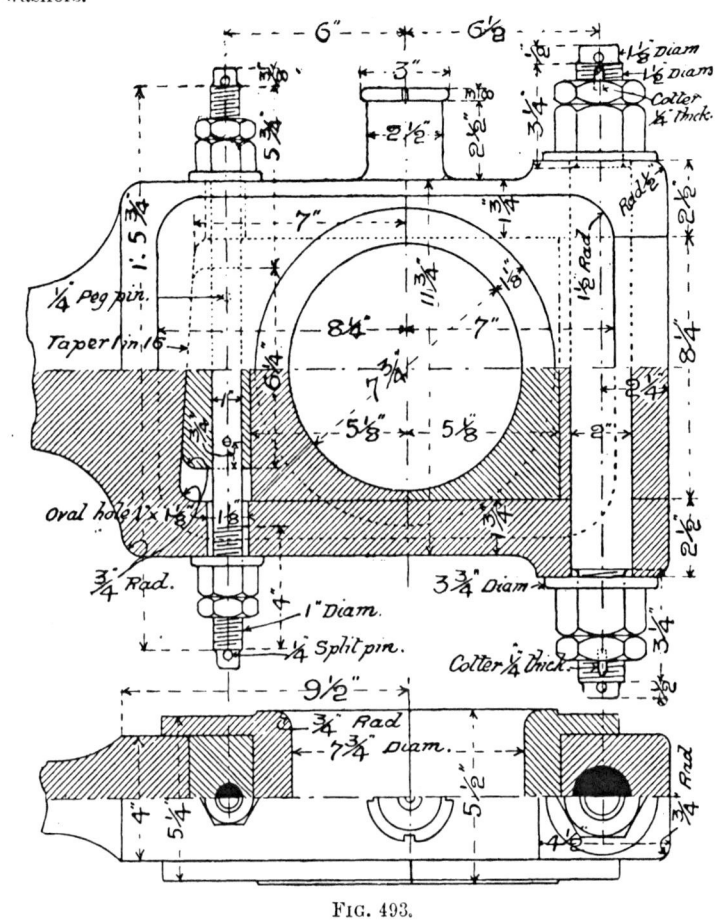

FIG. 493.

A modified form of box end is shown in Fig. 493. This is the design adopted on the London and North-Western Railway for the large or crank-pin ends of the connecting-rods. The end of the rod is forked, and the open end is closed by a block which is held in position by a bolt. This bolt has a slight taper, and is screwed at both ends and provided with nuts, so that it may be easily tightened up, and also easily withdrawn. The brasses, which have flanges on each side, are tightened up by a wedge-shaped block attached by two pins to a bolt screwed at both ends.

Since the angular motion of a long connecting-rod is small, the wear of the brasses at the cross-head end is slight. For this reason the bearing

212   MACHINE DRAWING AND DESIGN.

at the cross-head end of locomotive connecting-rods is sometimes made solid. Fig. 494 shows the form of connecting-rod end with a solid bearing adopted on the London and North-Western Railway. The end of the rod is lined with a steel bush $\frac{1}{4}$ in. thick, which works on a phosphor bronze sleeve $\frac{11}{16}$ in. thick, which is keyed to the cross-head pin. The steel bush is prevented from rotating by the end of the shank of the lubricator, which is screwed into the end of the rod over the bearing. The dimensions given in Figs. 493 and 494 apply to a connecting-rod 5 ft. 8 in. long for a goods engine having cylinders 18 in. in diameter with a piston stroke of 24 in.

245. Marine Connecting-Rod End.—The forms of connecting-rod ends shown in Figs. 495, 496, 497, and 498 are largely used in marine engines, and on that account they are generally termed marine connecting-rod ends; but these forms are also now extensively used in stationary engines, and sometimes on locomotives.

FIG. 494.

In the form shown in Figs. 495 and 496, the end is forged in one solid piece with the rod. The rod and all the faces of the end except the sides are turned in the lathe. The sides of the end are then planed, and the holes for the bolts drilled or bored. If the brasses are to be of uniform thickness, the hole to receive them is then bored to the size required, and the cap is slotted off. If the brasses are to be thicker at the top and bottom than at the sides, the hole to receive them is not finished until the cap has been slotted off and bolted on again without the distance pieces. The distance pieces have a thickness equal to the thickness of the slot formed in cutting off the cap.

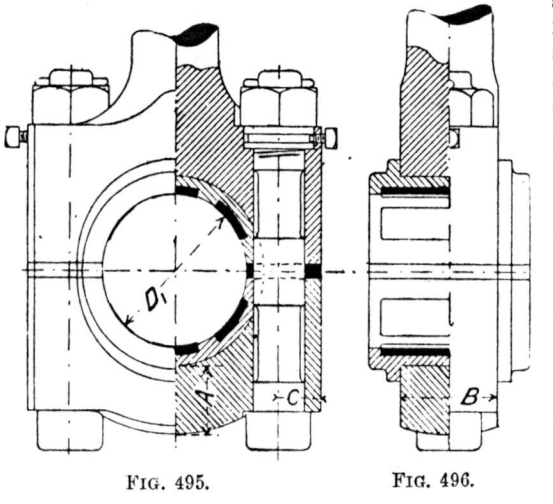

FIG. 495.   FIG. 496.

CONNECTING-RODS. 213

In the form shown in Figs. 497 and 498, the rod and cap are made separately. A great weight of brass is required for the brasses in this

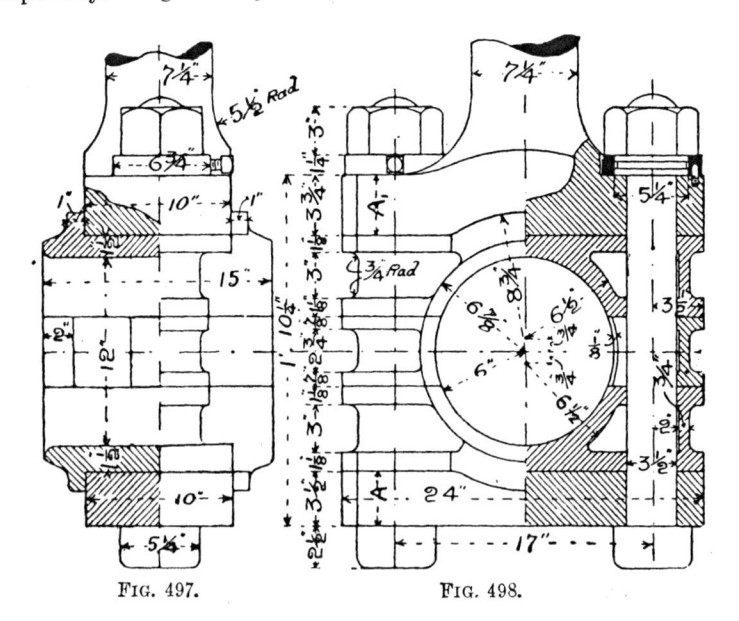

Fig. 497.   Fig. 498.

design, which with the large amount of work on them makes them expensive to renew when worn.

The diameter of the bolts is given by the equation $2 \times \cdot 7854 d_1^2 f = T$, or $d_1 = \cdot 798 \sqrt{\dfrac{T}{f}}$, where d_1 is the diameter of the screwed part of the bolt at the bottom of the thread, f the safe stress, and T the maximum pull on the connecting-rod. For bolts 2 in. in diameter and upwards, f may be taken at 5000 for wrought-iron and 7000 for steel. But since smaller bolts are more liable to be over-strained in screwing up, the safe stress should be diminished as the diameter of the bolt diminishes. For bolts under 1 in. in diameter, the safe stress in the above formula should not exceed 3000 for wrought-iron and 4200 for steel.

If for T we put its value $cQ = c \times \cdot 7854 D^2 P$, and we take $c = 1 \cdot 03$ we get $d_1 = \cdot 72 D \sqrt{\dfrac{P}{f}}$ where D is the diameter of the piston and P the effective pressure of the steam per square inch.

Portions of the unscrewed part of each bolt are generally turned down to the same diameter as the screwed part at the bottom of the thread, as shown in Fig. 495. This not only reduces the weight of the bolt without weakening it, but makes it more elastic and less liable to break under a suddenly applied load.

The breadth B of the rod end is generally a little greater than the diameter of the adjacent part of the rod. B should not be less than $1\cdot 73d + \cdot 15$, where d is the diameter of the bolt over the screw thread.

If we assume the cap to be in the condition of a beam of length D_1 fixed at the ends and carrying a load at the middle equal to twice the load on one bolt, we get its thickness $A = 1\cdot 086 d_1 \sqrt{\dfrac{D_1}{B}}$, a rule which agrees very well with a number of examples from actual practice.

The inner side of the bolt is generally at the middle of the thickness of the brass, as shown in Fig. 495, and the distance C varies from $\cdot 9d$ to d, and should not be less than half the minimum value of B given above.

A_1, the thickness of the palm on the end of the rod, Figs. 497 and 498, varies from A, the thickness of the cap, to $1\cdot 2A$.

246. Connecting-Rod Brasses.—The "brasses" or *steps* for connecting-rod ends are frequently made of phosphor bronze, and they are often lined with white metal. The following rules for the thickness of the step at the bottom, where the wear is greatest, agree very well with actual practice.

$t = \tfrac{1}{8}d + \tfrac{1}{4}$ for bearings up to 8 inches in diameter,
$t = \tfrac{1}{16}d + \tfrac{3}{4}$ for bearings above 8 inches in diameter,

where t is the thickness of the step and d is the diameter of the bearing.

The thickness of the steps at the sides is generally from $\cdot 5t$ to t, but is sometimes as small as $\cdot 2t$.

In steps with light flanges for bearings of large diameter there is a danger of the steps springing away from their seats at the sides and binding on the journal, causing undue friction. To prevent this the steps are sometimes secured at the sides to their seats by screws, as shown in Fig. 499.

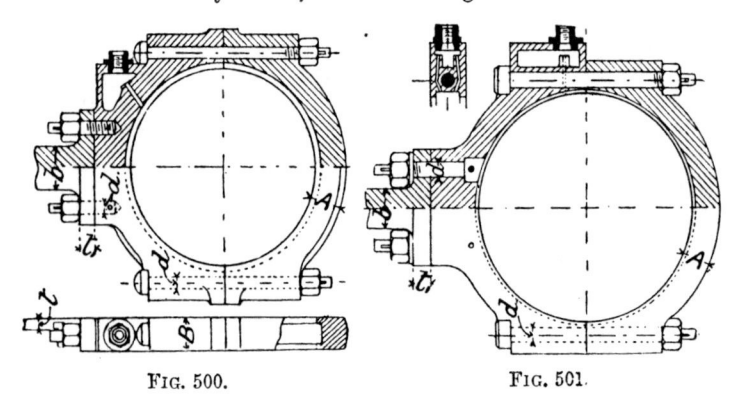

FIG. 499.

FIG. 500.

FIG. 501.

247. Eccentric Straps.—Just as an eccentric is a particular form of crank, so an eccentric strap is a particular form of connecting-rod end.

CONNECTING-RODS.

The best material for the strap is cast-iron. With a cast-iron strap working on a cast-iron sheave, no lining of any kind is required. For small eccentrics the strap is often made of brass or bronze. Malleable cast-iron, cast-steel, and wrought-iron are also used as materials for eccentric straps. Figs. 500 and 501 show good forms of eccentric straps made of cast-iron. The strap is in two pieces, which are bolted together

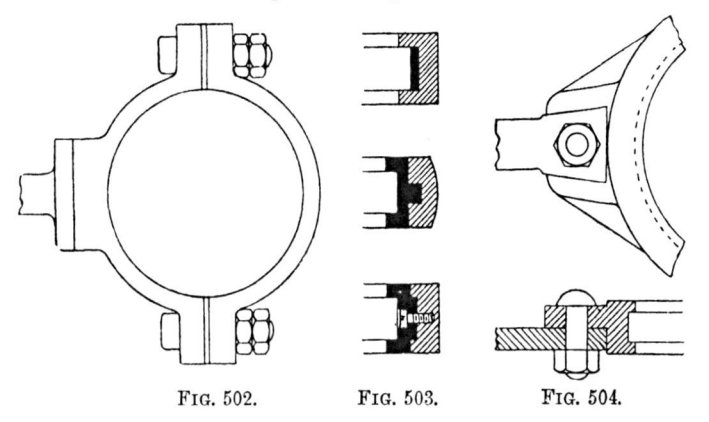

FIG. 502.   FIG. 503.   FIG. 504.

in the same way as a marine connecting-rod end. Fig. 502 shows the usual form of a wrought-iron strap. Fig. 503 shows three forms of brass liners for eccentric straps.

The most common way of connecting the eccentric rod to the eccentric strap is that shown in Figs. 500 and 501. Four other methods of connecting the eccentric rod to the strap are shown in Figs. 504, 505, 506, and 507.

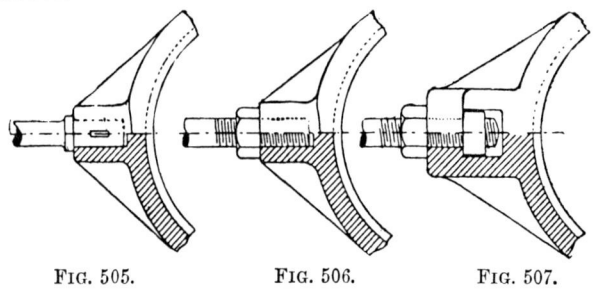

FIG. 505.   FIG. 506.   FIG. 507.

248. Proportions of Eccentric Straps.—The breadth B of the strap is first determined by one of the rules given under the head of Eccentrics. The following rules may then be used in proportioning the other parts.

Thickness of strap if of cast-iron $= A = \cdot 7B$ to $\cdot 9B$.
Thickness of strap if of wrought-iron or steel $= A = \cdot 5B$ to $\cdot 6B$.
Thickness of strap if of brass or malleable cast-iron $= A = \cdot 6B$ to $\cdot 8B$.
Diameter of bolts or studs, Figs. 500 and 501, $= d = \cdot 4B$.

Thickness of palm on end of eccentric rod $= t_1 = \cdot 45 B$.
Breadth of eccentric rod (if rectangular) at strap end $= b = 1\cdot 3B$ to $1\cdot 5B$.
Thickness of eccentric rod $= t = \cdot 4B$.
Diameter of eccentric rod (if round) at strap end $= \cdot 9B$ to B. For the design shown in Fig. 504 the bolts for fastening the eccentric rod to the strap may have a diameter $= \cdot 45B$ for two bolts, $\cdot 37B$ for three bolts, and $\cdot 32B$ for four bolts.

249. Forked Ends.—Certain forms of cross-head require that the cross-head end of the connecting-rod be forked. When this is the case, the jaws of the forked end may be solid, as shown in Fig. 508, or they may be provided with brasses like any of the ends which have been already described. Fig. 509 shows a forked end with brasses, caps, and bolts, as in the marine connecting-rod end.

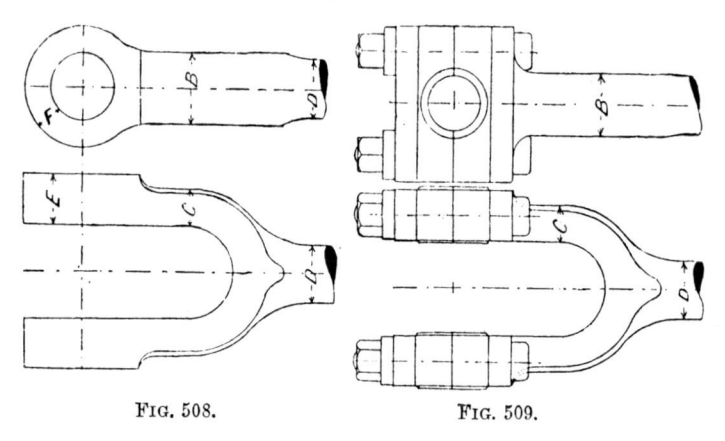

FIG. 508.   FIG. 509.

The total area of the section of the jaws is usually from 1·25 to 1·8 times the area of the section of the adjacent part of the rod, and in the case of the solid forked end, the total area of the section through the eyes is usually from 1·4 to 1·9 times the area of the section of the rod. If D is the diameter of the rod near the fork, the area of the section at that part is $\cdot 7854 D^2$. The following proportions make the total section through the jaws about $1\cdot 4 \times \cdot 7854 D^2$, and the total section through the eyes about $1\cdot 5 \times \cdot 7854 D^2$.

B = D to 1·2D, average value = 1·1D.
C = ·5D when B = 1·1 D.
E = C to 1·35C, average value = 1·2C = ·6D, when C = ·5D.
F = ·5 D, when E = ·6 D.

250. Coupling-Rods.—A rod used to transmit the motion of one crank to another is called a *coupling-rod*. A familiar example of the use of coupling-rods is found in locomotives, where one or more pairs of wheels are coupled to the wheels on the crank axle in order to increase the adhesion of the engine on the rails. Coupling-rods are made of wrought-iron or steel, and are generally either of rectangular or **I**-section. The ends may be like any of the connecting-rod ends which have been described, but they are now generally made solid and lined with solid brass bushes, *without any adjustment for wear*. This form of coupling-rod end is found to answer very well in locomotive practice where the workmanship and arrangements for lubrication are excellent. When the brass bush becomes worn it is replaced by a new one. When coupling-rod ends are not made solid with solid bushes, they are generally similar to the connecting-rod ends shown in Figs. 489 and 490. With these arrangements, however, it is necessary that at one end the cotter or wedge should be on the outside, and at the other end on the inside, so that the rod may remain of the same length when the bushes are readjusted after having become worn.

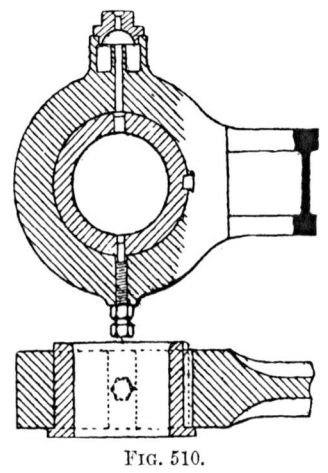

Fig. 510 shows a common form of solid coupling-rod end. The brass bush is secured by a key, and sometimes also in addition by a pin screwed into the rod from below as shown.

When more than two pairs of wheels are coupled together, two coupling-rods meet. In this case the two rods may be joined together on the same pin, as shown in Fig. 511, or another pin may be used, as shown in Fig. 512. The arrangement shown in Fig. 511 is that adopted on the London and North-Western Railway. The other design is that in use on the London, Brighton, and South Coast Railway. It will be observed

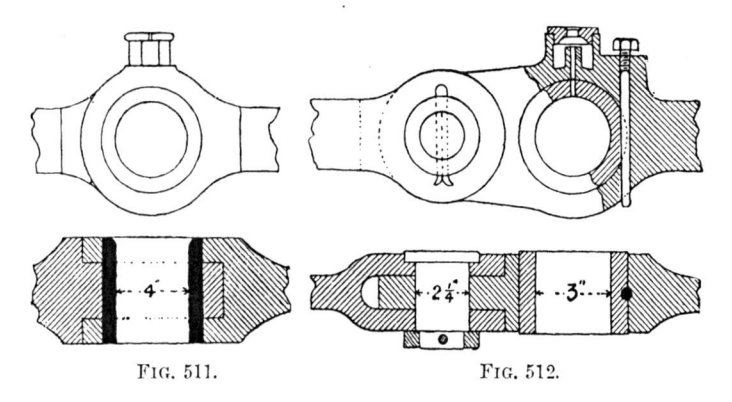

FIG. 510.

FIG. 511.   FIG. 512.

that the bush in the latter design is secured by a pin which has a slight taper (1 in 48), this pin being in turn secured by a set-screw which presses on the larger end of the pin.

CHAPTER XVII.

CROSS-HEADS AND GUIDES.

251. Pressure on Rubbing Surface of Slide Block.—The forces acting on a cross-head are, the force Q transmitted through the piston-rod from the piston, the force T acting along the connecting-rod, and the reaction R

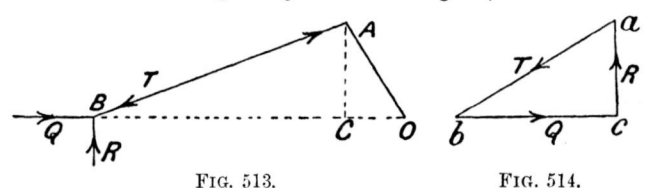

FIG. 513.   FIG. 514.

of the guide-bar on the slide-block. Drawing the triangle of forces, we get $\frac{R}{Q} = \frac{ac}{bc} = \frac{AC}{BC}$. Assuming Q to be constant throughout the stroke, $\frac{R}{Q}$ will be a maximum when the angle AOB is a right angle. When AO is perpendicular to BO, $\frac{AC}{BC} = \frac{AO}{BO} = \frac{AO}{\sqrt{AB^2 - AO^2}}$. Let l = length of stroke, and let $AB = nl$. Then, when $\frac{R}{Q}$ is a maximum, $\frac{R}{Q} = \frac{AO}{\sqrt{AB^2 - AO^2}} = \frac{1}{\sqrt{4n^2 - 1}}$ and $R = \frac{Q}{\sqrt{4n^2 - 1}} = \frac{\cdot 7854 D^2 P}{\sqrt{4n^2 - 1}}$, where D is the diameter of the piston and P the effective pressure of the steam per square inch.

If s is the number of square inches of rubbing surface upon which the force R acts, and p is the pressure on the rubbing surface in pounds per square inch, then $R = ps$ and $s = \frac{R}{p}$.

In practice p varies from 25 to 100.

The area s should, however, depend on the speed of the piston as well as on the force Q, being greater the greater the speed. Since the indicated horse-power is proportional to Q and to the speed of the piston in the same type of engine working with the same initial steam pressure and the same cut-off, the following formula is more satisfactory, namely, $s = \frac{CH}{n}$, where H is the indicated horse-power developed in the cylinder or in the two cylinders of a tandem engine, and C is a multiplier to be determined from good examples of the type of engine which is being designed.

252. Diameter of Cross-Head Pin or Gudgeon.—The journal on the cross-head pin is either supported at both ends, as in Fig. 515, or there are two separate end journals, as in Fig. 516. In the former case the load on the journal is equal to T, and in the latter case the load on each end journal is equal to $\frac{T}{2}$, where T is the

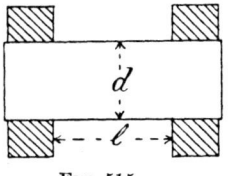

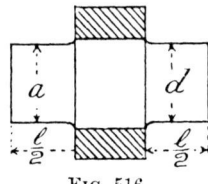

FIG. 515.   FIG. 516.

thrust or pull on the connecting-rod. The load may be taken as equally distributed over the journal, and in either case the bending moment is equal to $\frac{Tl}{8}$. The moment of resistance to bending is $\frac{3\cdot1416}{32}d^3 f$. Hence $\frac{3\cdot1416}{32}d^3 f = \frac{Tl}{8}$, and $d = \sqrt[3]{\frac{1\cdot273 Tl}{f}}$.

If p is the pressure per square inch on the journal, then $pdl = T$, and $l = \frac{T}{pd}$, then $d = \sqrt[4]{\frac{1\cdot273}{pf}}\sqrt{T}$.

If D is the diameter of the cylinder, and P the maximum pressure of the steam in pounds per square inch, it will be near enough for our present purpose to take $T = \cdot 8 D^2 P$, then $d = \frac{\cdot 9 D \sqrt{P}}{\sqrt[4]{pf}}$.

p may be taken at 1200, and f at 5000 for wrought-iron and 7000 for steel. Then
$$d = \cdot 0182 D \sqrt{P} \text{ for wrought-iron,}$$
$$d = \cdot 0167 D \sqrt{P} \text{ for steel.}$$

Hence $l = 2d$ for wrought iron, and $l = 2\cdot 4 d$ for steel. But these values of l in terms of d do not agree with practice. In practice $l = d$ to $1\cdot 3d$. Hence, if l is less than $2d$, the pin will have an excess of strength if its diameter is determined from the formula $pdl = T = \cdot 8 D^2 P$, and this is the formula which should be used.

Let $l = nd$, then $npd^2 = \cdot 8 D^2 P$, and $d = \frac{\cdot 9 D \sqrt{P}}{\sqrt{np}} = k D \sqrt{P}$.

The following table gives values of k for various values of n and p:—

p	n						
	·9	1	1·1	1·2	1·3	1·4	1·5
800	·0335	·0318	·0303	·0290	·0279	·0269	·0260
1000	·0300	·0284	·0271	·0260	·0249	·0241	·0232
1200	·0274	·0260	·0248	·0237	·0228	·0220	·0212
1400	·0254	·0241	·0229	·0220	·0211	·0203	·0196

253. Forms of Cross-Heads.—Cross-heads vary greatly in form, but they may be roughly classified according to the number of slide or guide bars upon which they work. With one class four guide-bars are used, with a second class there are two guide-bars; with a third class there is only one guide-bar, and with a fourth class the cross-head is attached to a piece called a slipper or slipper-slide, which is guided by an arrangement equivalent to three guide-bars. Differences in the design of cross-heads also arise from arrangements for taking up the wear of the slide-blocks. The form of a cross-head also depends on the form of the cross-head pin and the manner of fixing it, either to the cross-head or to the connecting-rod.

254. Cross-Heads for Four Guide-Bars.—The cross-head shown in Fig. 517 is forged in one piece with the piston-rod. This cross-head is either shrunk on to the pin or it is forced on by hydraulic pressure. The parts A and B of the cross-head pin enter into the slide-blocks which work between the guide-bars. The proportions of the different parts are marked on the Fig., the unit being the diameter of the piston-rod.

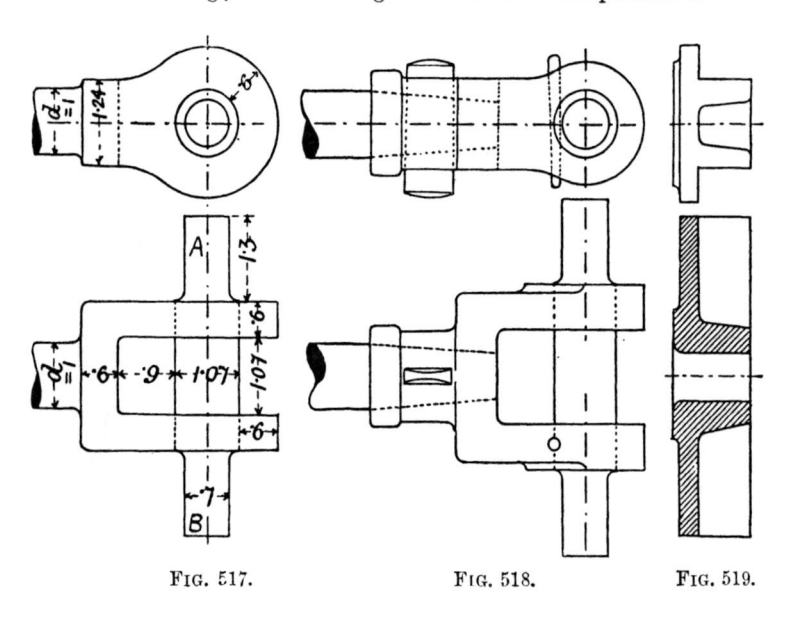

Fig. 517. Fig. 518. Fig. 519.

The design shown in Fig. 518 is somewhat similar to that just described, but it is made separate from the piston-rod, to which it is attached by a cotter. When this design of cross-head is used for a small piston-rod, the rod and cross-head are often connected by screwing them together Fig. 519 shows the usual form of the slide-blocks. The slide blocks may be of cast-iron or brass, or they may be of cast-iron faced on their rubbing surfaces with brass or white metal. Hard cast-iron is considered the best material for slide-blocks.

CROSS-HEADS AND GUIDES.

Figs. 520 and 521 show a form of cross-head used on the London and North-Western Railway. In this form the cross-head and slide-blocks are made in one piece of cast-steel. The cross-head pin is tapered, and carries a phosphor bronze bush or sleeve at its centre, upon which works

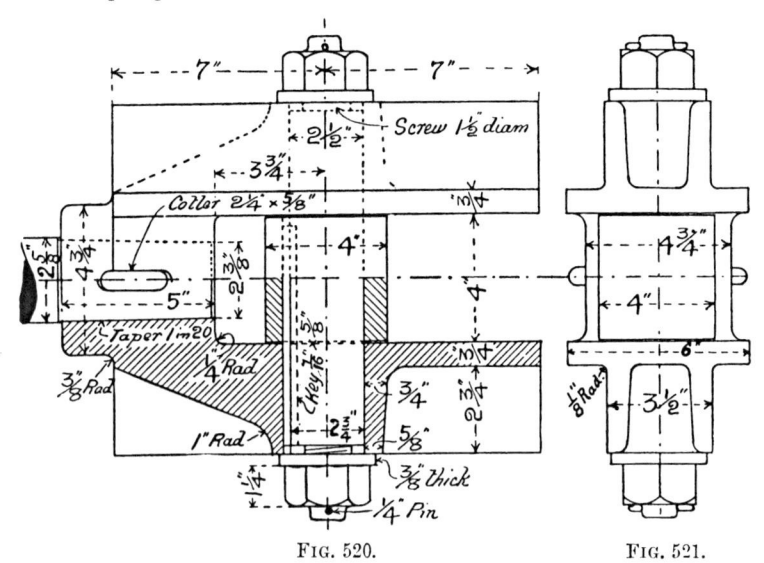

FIG. 520.　　　　　　　　FIG. 521.

the connecting-rod end, the latter being solid and fitted with a hardened steel bush. The phosphor bronze bush is prevented from rotating on the cross-head pin by a sunk key, which also prevents the pin from rotating.

A form of cross-head frequently used in American practice is shown

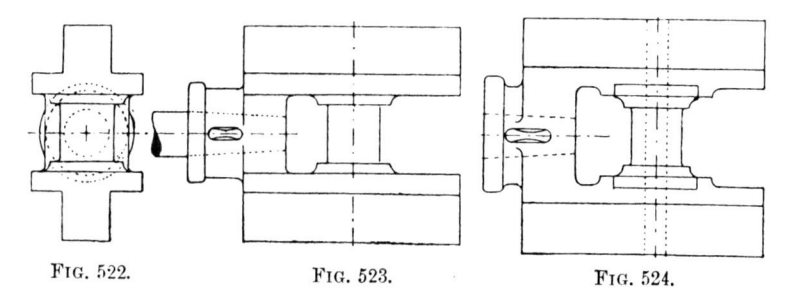

FIG. 522.　　　FIG. 523.　　　　　FIG. 524.

in Figs. 522 and 523. Like the one just described, the slide-blocks are made in one piece with the cross-head, but the pin is not made separate from the cross-head, but is cast in one piece with it. A special contrivance has to be adopted for turning the pin in this case. This form of cross-head is generally made of hard cast-iron, but it may be made of cast-steel. The objection to this design is that the pin when worn cannot be renewed; it may, however, be returned if it is made sufficiently large at first to allow of this. A modification of this design is shown in Fig. 524. The pin in this example is inserted and secured by a central steel pin as shown.

255. Cross-Heads for Two Guide-Bars.—Figs. 525 and 526 show a simple form of cross-head for two guide-bars. This design of cross-head is that used on the locomotives on the Highland Railway. The cross-head is forged on the end of the piston-rod. Projecting pieces forged on the top and bottom of the cross-head carry the slide-blocks which work on the guide-bars. The cross-head pin, shown in Fig. 526, is secured to the connecting-rod, which is forked at the cross-head end.

The bush in the cross-head which forms the bearing for the pin is of wrought-iron case-hardened, and is prevented from rotating by the key shown. The piston-rod and cross-head are of wrought-iron, and the slide-

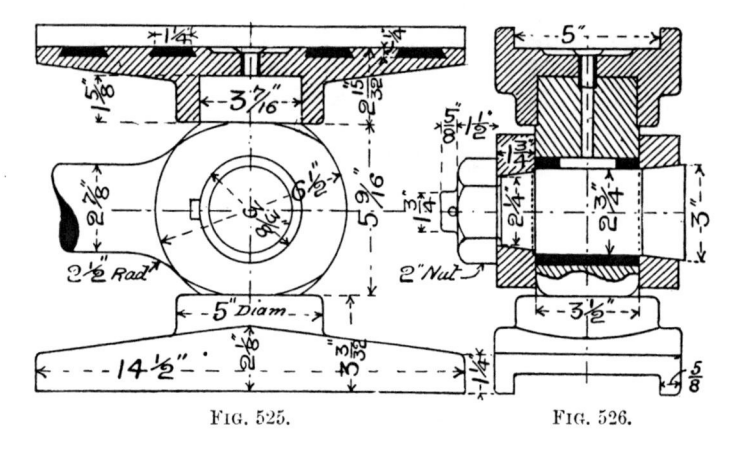

FIG. 525.   FIG. 526.

blocks are of cast-iron, and provided with white metal strips as shown. A short brass tube leads oil from the upper slide-block into a hole in the cross-head, which carries it to a slot in the bush, which distributes it over the cross-head pin.

The cross-head shown in Fig. 527 is made of cast-steel, but it may be made of cast-iron or wrought-iron. When the slide-blocks are made of cast-steel or wrought-iron, they should be lined with a softer metal. When the liners are made of brass or bronze, they are comparatively thick ($\frac{3}{8}$ in. to $\frac{1}{2}$ in), and they add to the weight of the cross-head without adding to its strength. In American practice it is common to tin the rubbing surfaces of cast-steel and wrought-iron slide-blocks to a depth of $\frac{1}{16}$ inch, as shown in Fig. 527.

Another example of a two-bar cross-head is shown in Figs. 528 and 529. This is taken from a single-cylinder condensing engine by Messrs. Hick, Hargreaves, & Co., Bolton. The cylinder is 40 inches in diameter, and the stroke of the piston is 10 feet. The length of the connecting-rod is 28 feet. The great width between the guides, 55½ inches, is necessitated by the great length of stroke. The details are so fully shown that we need only add that the piston-rod and cross-head are made of steel and the slide-blocks of cast-iron.

Many first-class makers of stationary engines cast the guides with the framing, and face the working surfaces with a boring bar, so that the cross-sections of the guiding surfaces are arcs of a circle. For guides of this construction the slide-blocks are turned after they are attached to the cross-head. As an example of a cross-head with turned slide-blocks, we take the design shown in Figs. 530 and 531, which represents the

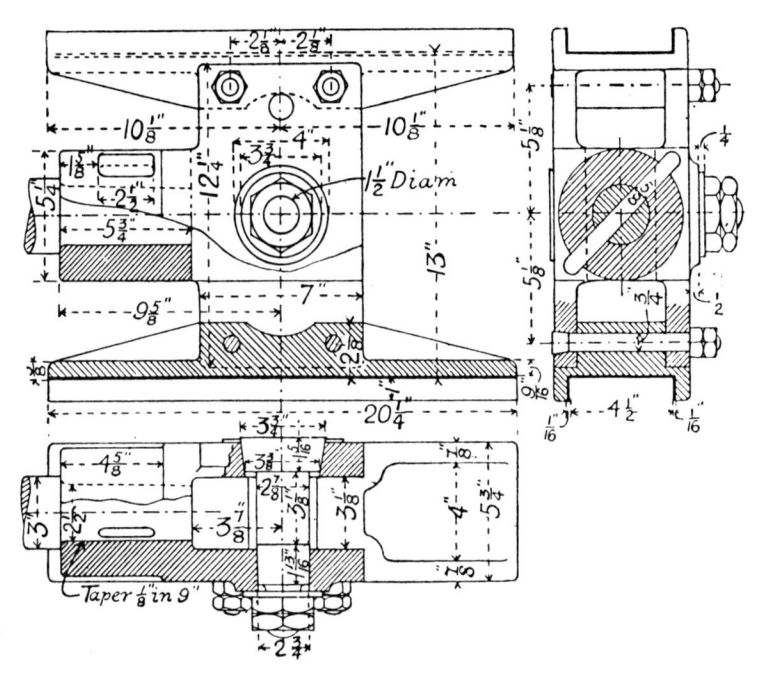

FIG. 527.

practice of Messrs. Marshall, Sons, & Co., Gainsborough. The description and illustrations are taken from Mr. D. K. Clark's Treatise on the Steam Engine.

"The cross-head, Figs. 530 and 531, is of malleable cast-iron, forked to take the connecting-rod. The gudgeon is of steel, 2¾ inches in diameter, and 3¾ inches long between the sides of the cross-head, which are each 1½ inches thick to give bearing for the pin. The pin is let in with a taper head, 3 inches and 3/32 inches in diameter, tapering 3/32 inch, or at the rate of 1 in 15. The gudgeon is fastened with a washer and a nut ⅝ inch thick, and the head is prevented from turning by a

224   MACHINE DRAWING AND DESIGN.

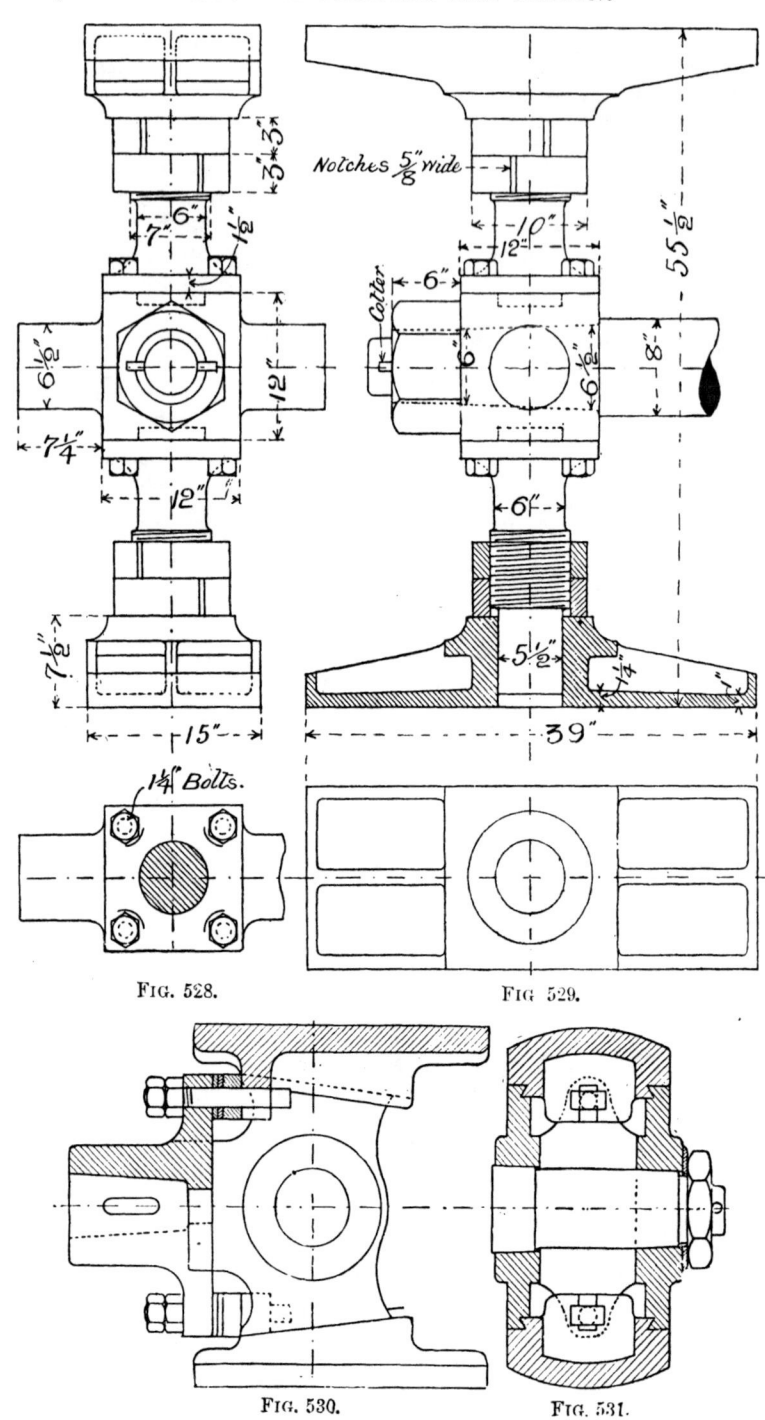

FIG. 528.

FIG. 529.

FIG. 530.

FIG. 531.

short key, $\frac{3}{8}$ inch square, driven in from the outer side. The cross-head is let on the end of a piston-rod for a length of $4\frac{1}{2}$ inches, tapering from $2\frac{3}{8}$ inches to $1\frac{7}{8}$ inches, or $\frac{1}{2}$ inch in $4\frac{1}{2}$ inches, at the rate of 1 in 9. It is fastened with a cotter 2 inches by $\frac{1}{2}$ inch thick, and a nut $\frac{3}{4}$ inch thick against the end of the socket. The slides are of wrought-iron, with large wearing surfaces, 6 inches wide, 11 inches long, and turned to a radius of $6\frac{1}{4}$ inches to fit the guides. They fit on, and are notched into the upper and lower faces of the cross-head, which are wedge-form, so that by longitudinal adjustment the slides may be adapted to the exact radius of the guides, and so compensate for wear. The adjustment is effected longitudinally by bolts and double nuts, which engage in eyes cast on the cross-head and the slides."

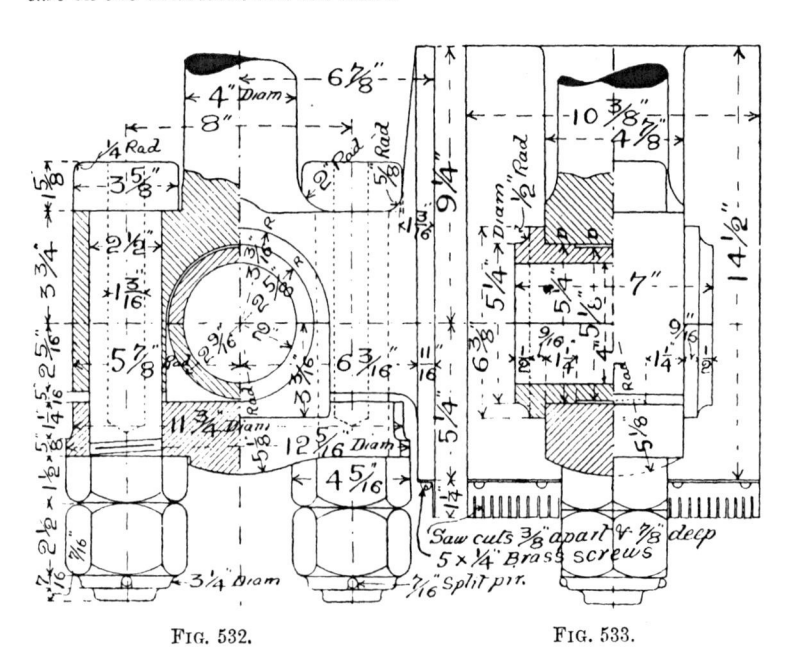

FIG. 532.   FIG. 533.

256. Cross-Heads with Slipper Slides.—One form of cross-head with a slipper slide is shown in Figs. 532 and 533. In this example, which is from a vertical marine engine, the slipper is forged with the cross-head and piston-rod. The slitted strip of sheet brass shown attached to the lower end of the slipper dips into an oil-box at the end of each downward stroke, and lifts a portion of oil which is distributed over the surface of the guide during the upward stroke.

The cross-head shown in Figs. 534 and 535 differs from that just described in having the slipper a separate piece from the cross-head to which it is attached by four screws having cylindrical slotted heads.

An excellent form of cross-head for a slipper slide is shown in Figs.

P

226   MACHINE DRAWING AND DESIGN.

536 and 537. This was designed by Mr. Lindley of Messrs. Browett, Lindley, & Co., Manchester. A rectangular block is forged on the end

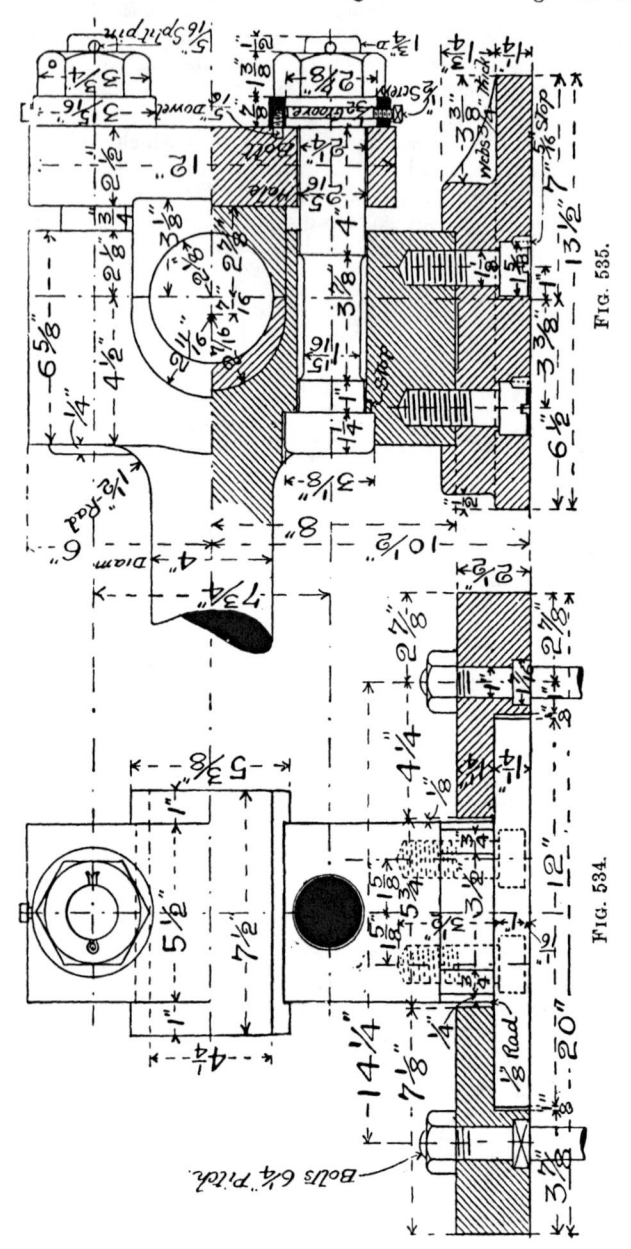

CROSS-HEADS AND GUIDES. 227

of the steel piston-rod, and is slotted out to receive the brass bushes. A forged steel cap is secured by two bolts, which also serve to secure the slipper, which is of cast-iron. It will be observed that the cap is lipped over at the ends so as to hold the jaws of the block together. The consequence of this arrangement is, that the bolts are relieved from the driving force acting through the piston-rod. Also the cap and block

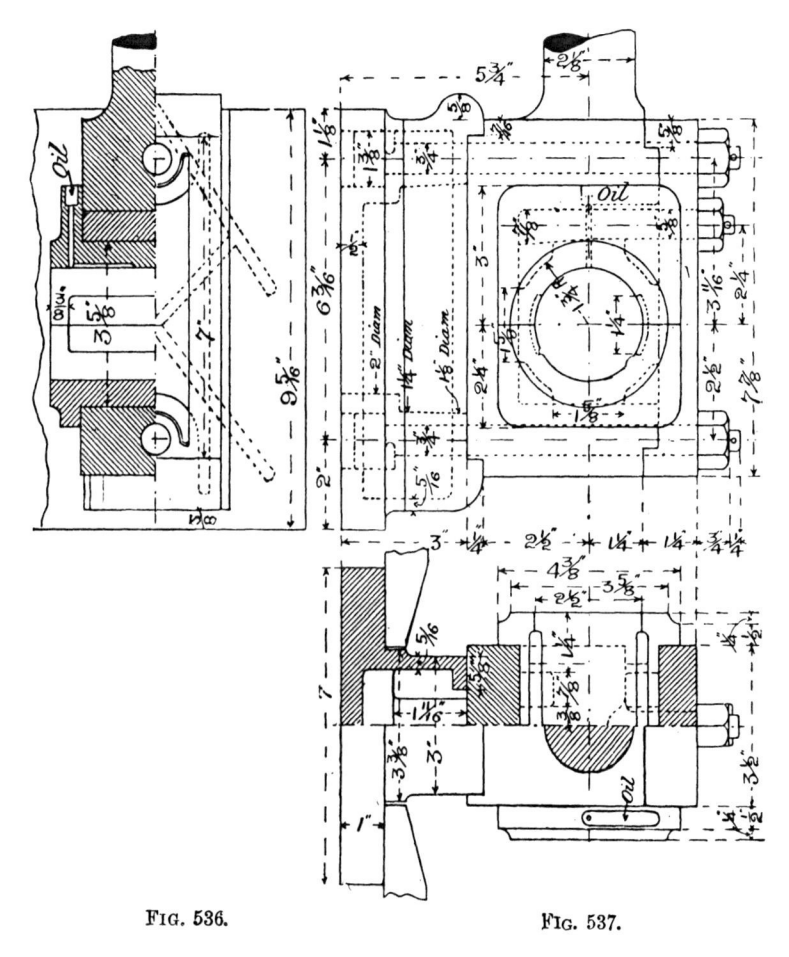

FIG. 536.   FIG. 537.

being bolted up "metal to metal," there is less danger of the nuts working loose. The bushes are tightened up by a broad steel wedge, which is provided with a screw and nut as shown. The example illustrated is for a vertical engine, but by rearranging the oil-box it can be used equally well on a horizontal engine.

257. Guides for Cross-Heads.—The guides or slide bars for cross-heads may be made of cast-iron, wrought-iron, or steel. When the guides are not cast with the framing of the engine, they are generally of rectangular or T-section. Fig. 538 shows a form of cast-iron slide bar of T-section, and Fig. 539 shows a form of wrought-iron slide bar of rectangular section.

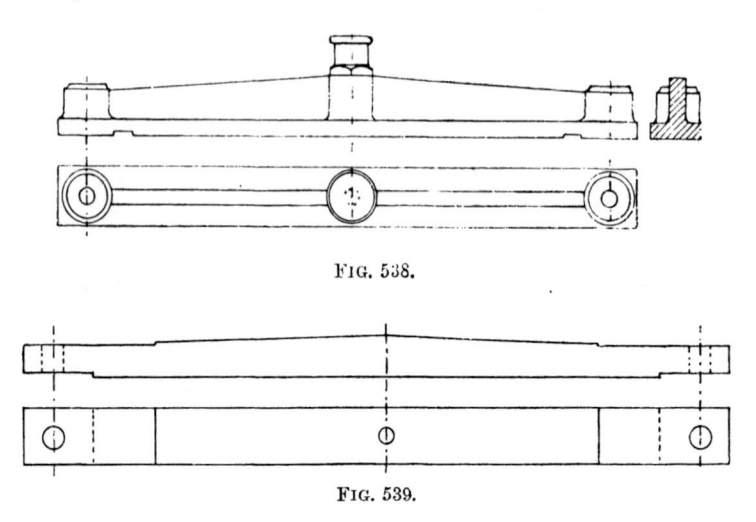

FIG. 538.

FIG. 539.

As shown in Art. 251, p. 218, the maximum load on the guide bar is $R = \dfrac{Q}{\sqrt{4n^2-1}}$, where Q is the load on the piston and n the ratio of the length of the connecting-rod to the stroke of the piston. If we assume that the maximum load on the guide bar acts at its centre, then the maximum bending moment on the bar is $\dfrac{RS}{4} = \dfrac{QS}{4\sqrt{4n^2-1}}$, where S is the distance between the supports of the bar. Equating the bending moment to the moment of resistance of the section of the bar, as explained in Art. 63, p. 32, the dimensions of the section of the bar may be determined. The stress f may be taken at 3000 for cast-iron and 6000 for wrought-iron or steel.

CHAPTER XVIII.

PISTONS AND PISTON-RODS.

258. Pistons, Buckets, and Plungers.—A *piston*, Fig. 540, is generally a cylindrical piece which slides backwards and forwards inside a hollow cylinder. The piston may be moved by the action of fluid pressure upon it as in a steam-engine, or it may be used to give motion to a fluid as in a pump.

The piston is usually attached to a rod called the *piston-rod*, which

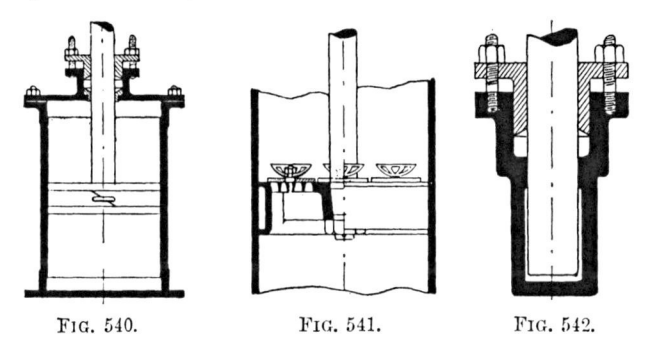

FIG. 540.   FIG. 541.   FIG. 542.

passes through the end of the cylinder inside which the piston works, and which serves to transmit the motion of the piston to some piece outside the cylinder, or *vice versâ*.

A piston which is provided with valves which allow the fluid to pass from one side of the piston to the other is called a *bucket* (Fig. 541).

If a piston-rod be enlarged until it becomes of the same diameter as its piston, we get a plunger (Fig. 542).

259. Pistons without Packing.—The simplest form of piston is a true cylindrical piece fitting accurately into another which is bored truly parallel. The piston used in the steam-engine indicator is of this description. This form of piston is also often used in pumps. The leakage is diminished and the piston is lubricated to a certain extent by having a number of circumferential grooves on it. The great objection to this simple form of piston is that there is no adjustment for wear, but on the other hand, as there is no radial pressure of the piston on the cylinder such as occurs with pistons which are packed to prevent leakage, the wear is small, and is diminished by having the piston of considerable depth, so that the wearing surface is large. The friction being small, the power lost through this cause is slight.

The depth of this form of piston should not be less than one-third of its diameter, and it is often equal to and sometimes greater than the diameter.

260. Piston Packings.—In the early days of the steam-engine, when the pressure of steam used was low, the piston was made steam-tight by filling a wide groove on its circumference with hempen rope soaked in

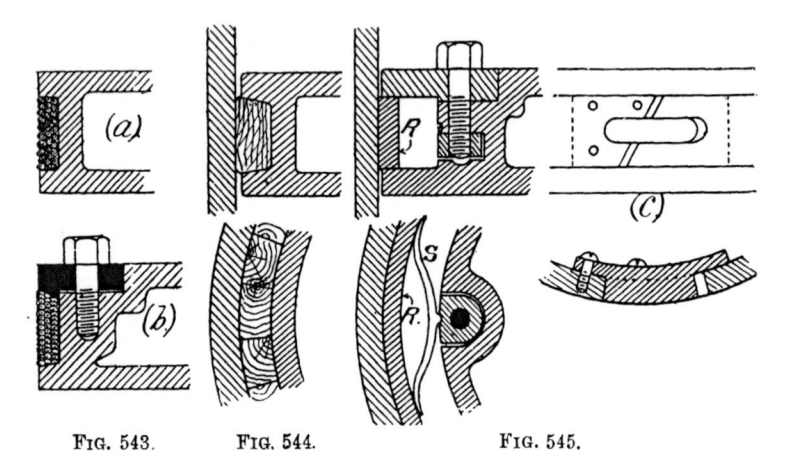

FIG. 543.   FIG. 544.   FIG. 545.

tallow, as shown at (*a*) Fig. 543. To permit of this packing, or *junk*, as it was called, being tightened up from time to time as it became worn, and also to allow of its being removed without withdrawing the piston from the cylinder, a *junk ring*, shown at (*b*) Fig. 543, was added. This junk ring is still used on many pistons in which metallic packing has taken the place of junk. Hempen packing is not at all suitable for the high pressure, and consequently high temperature, steam now used in steam-engines, as it rapidly becomes charred and worn. This form of packing is, however, still used for pump pistons and buckets.

Wood packing put in in short staves, as shown in Fig. 544, is sometimes used for pump pistons and buckets. The staves are jammed in tight, and are held in position by friction only. Care must be taken to make the diameter over the wood packing a little less than the diameter of the barrel of the pump, to allow for the expansion which takes place when the wood becomes saturated with the water.

With the introduction of higher steam pressure, and more perfect workmanship in the boring of the cylinder, came metallic packing for the piston. A form of metallic packing which was at one time very largely used is shown in Fig. 545. The wide groove on the piston contained a cast-iron packing ring, R, split at one part as shown at (*c*), to permit of its expansion against the cylinder by the action of a series of coach-springs, S, placed behind it. Leakage at the split in the packing ring was prevented by a brass tongue-piece formed on a plate screwed or riveted to one end of the ring as shown.

The great objection to the metallic packing just described is that there is no automatic arrangement for taking up the wear, and preventing leakage between the edges of the packing ring and the junk ring and the flange of the piston. Another objection is that the force exerted by the springs cannot be adjusted without a great deal of trouble. Still another objection, which also applies to several modern forms of pistons, is that there is a reaction between the springs and the body of the piston which does not allow of the axis of the piston-rod and piston being out of line with the axis of the cylinder, without causing unequal pressure at different parts of the ring. In practice, the axis of the piston and piston-rod may not coincide with the axis of the cylinder through inferior workmanship, or through the unequal wearing of the cylinder.

There are a great many designs of metallic piston packing now in the market, but the majority of them consist essentially of two cast-iron rings which are pressed outwards against the cylinder, and also away from one another, one against the flange of the piston, and the other against the junk ring. In some pistons these two motions are produced simultaneously by the same springs, while in others they are produced by separate and independent springs.

In *Buckley's* packing, Fig. 546, a flattened spiral spring presses on the inside sloping faces of the flanges of two packing rings, so that they

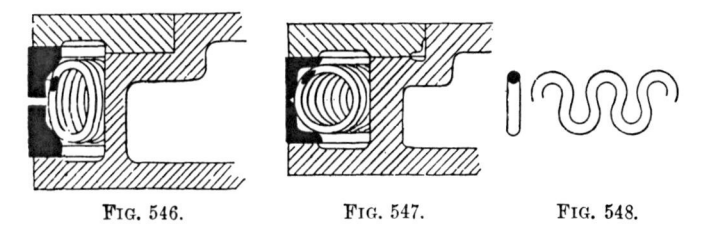

FIG. 546.   FIG. 547.   FIG. 548.

are pushed apart and outwards against the cylinder at the same time. In *Lancastar's* packing, Fig. 547, the spring is an ordinary spiral spring bent round the piston, and placed inside the packing rings, which are of ⌊-section. These rings are kept apart because of the resistance of the coils to being flattened, and they are pressed outwards against the cylinder by the force which the spring exerts in trying to straighten itself. In *Prior's* packing the action is the same as in Buckley's, but the spring used is of the form shown in Fig. 548.

Mr. Thomas Mudd, of West Hartlepool, has introduced a simple design of metallic piston packing, of which the following is his description.[1] "The piston rings, Fig. 549, are of square cross-section, and their size is 2 inches square for all pistons from 18 inches up to 80 inches diameter. The cross-section 2 inches square gives a large bearing surface on the junk ring and the piston flange. The springs are mere

[1] Proceedings of the Institution of Mechanical Engineers, 1891, p. 281.

short helical springs of one universal size—that is to say, the springs for all positions of any one piston, for all pistons in any one engine, and for all engines from small to large powers are identical; and this universality of spring is of the greatest benefit in many ways. The spring pressures which regulate the force applied between the junk ring and piston flange, and the force applied tangentially pressing the rings against the cylinder walls, are each applied independently of the other, and can therefore be separately regulated in amount. The position of the vertical springs being in the centre of the breadth of the rings, the force applied by them

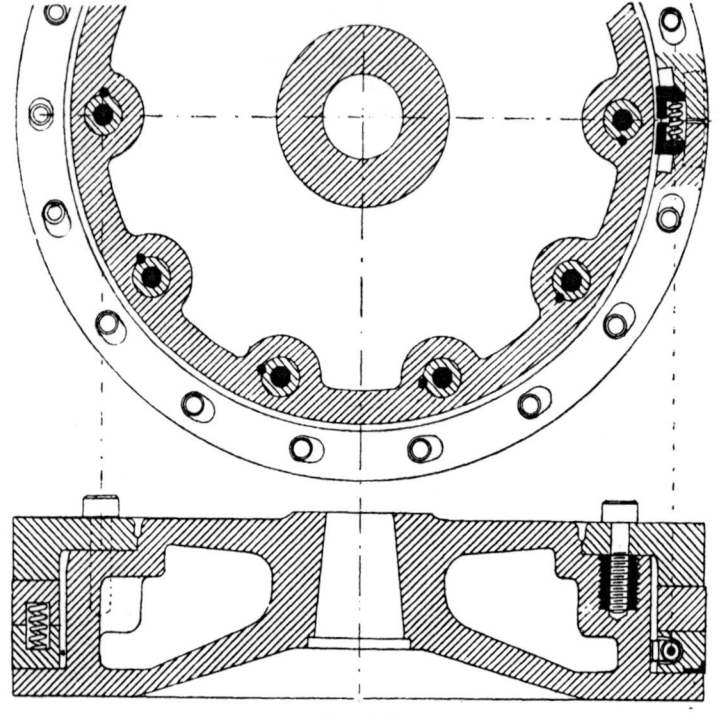

FIG. 549.

is fair and central on the bearing surfaces, thus not encouraging any rocking or twisting of the rings. The rings being so stiff, require no springs behind them against the piston body, and hence they partake of the quality known as floating; that is to say, in running up and down the cylinder they can follow the line of the cylinder, irrespective of where the piston may be; and the oscillation of the piston from side to side of the cylinder has no effect in putting more pressure on the packing rings at one side of the cylinder than at the other."

261. Construction and Proportions of Pistons.—The material chiefly

PISTONS AND PISTON-RODS.

used for pistons is cast-iron, but wrought-iron, brass, bronze, and cast-steel are also used. When made of wrought-iron or steel, it is desirable that the construction of the piston should be as simple as possible. It is not usual for wrought-iron pistons to be provided with ribs, but in some recent examples of wrought-iron pistons in American locomotive practice, the designs are more like what would be adopted for cast-iron pistons. Figs. 550 and 551, which have been prepared from illustrations

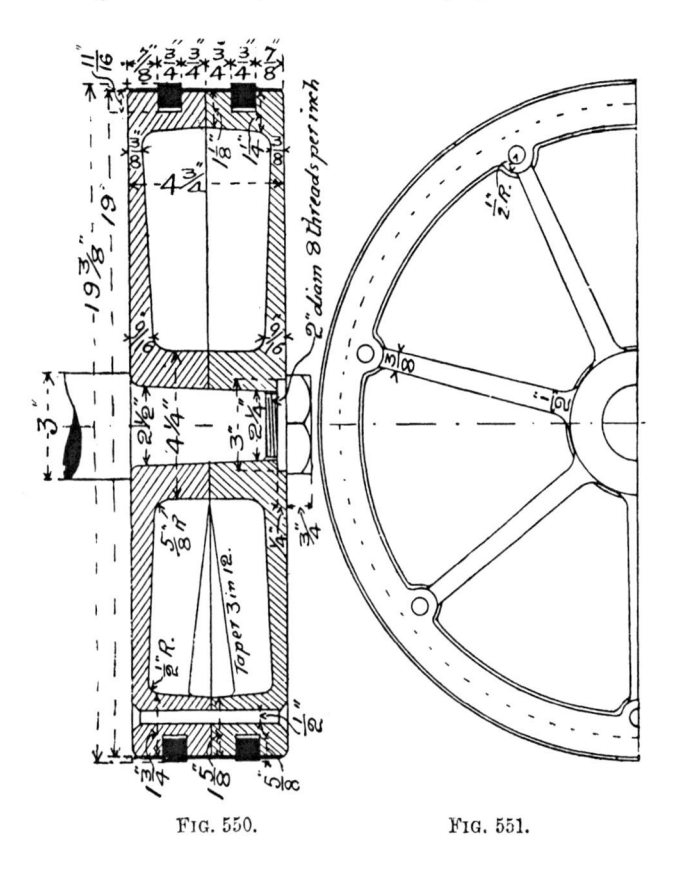

FIG. 550.   FIG. 551.

given in the *Railroad Gazette*, show one of these wrought-iron pistons. The surface of the piston in contact with the cylinder is tinned to a depth of $\frac{1}{16}$ inch. The weight of this piston without the rod is given as 171 lbs.

Small and Medium-Sized Pistons.—Figs. 552, 553, and 554 show examples of cast-iron pistons taken from locomotive practice. Fig. 555 shows a form of brass piston used on the London, Brighton, and South Coast Railway.

234   MACHINE DRAWING AND DESIGN.

The proportions for the pistons shown in Figs. 552 to 555 may be as

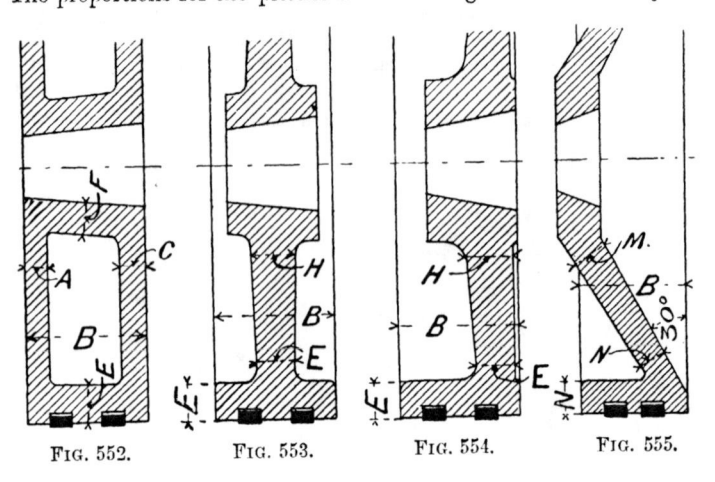

FIG. 552.   FIG. 553.   FIG. 554.   FIG. 555.

follows, the unit being $\dfrac{D \sqrt{P}}{100}$ where D is the diameter of the cylinder, and P the steam pressure :—

$A = \cdot 4.$
$B = 1 \cdot 5 \text{ to } 2 \cdot 5.$
$C = \cdot 45.$
$E = \cdot 65.$

$F = \cdot 55.$
$H = \cdot 75.$
$M = \cdot 75.$
$N = \cdot 6.$

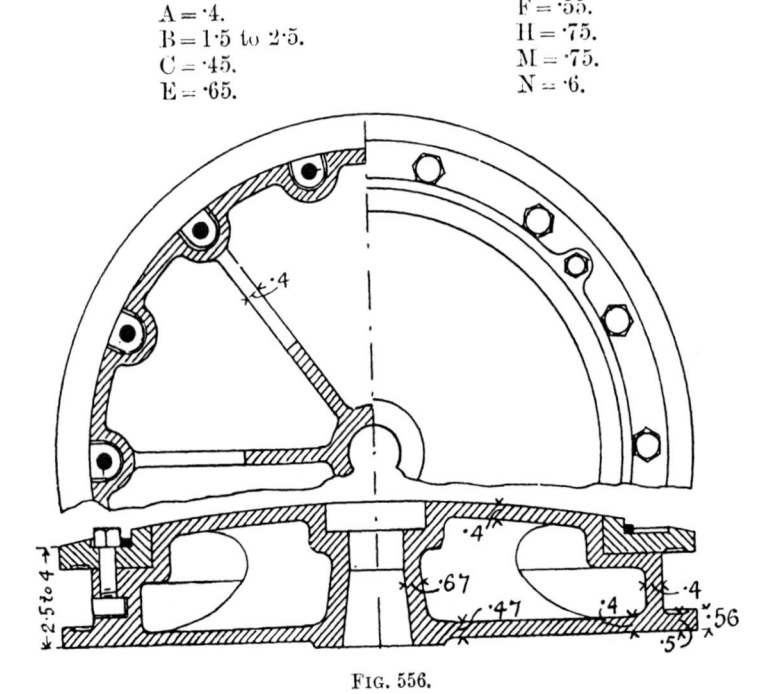

FIG. 556.

PISTONS AND PISTON-RODS.

Large Cast-Iron Pistons.—These are made hollow and are strengthened by internal radial ribs. Fig. 556 shows a design for a large cast-iron piston. The dimensions marked on the Fig. are in terms of the unit $\dfrac{D\sqrt{P}}{100}$. The junk ring is secured by wrought-iron or steel bolts and brass nuts. The diameter of the junk ring bolts may be $\dfrac{\cdot 28 D \sqrt{P}}{100} + \tfrac{1}{4}$ inch, and they may be placed at a pitch of seven to ten times their diameter. The number of ribs or webs may be about $\dfrac{D}{10} + 2$, and their thickness $\dfrac{\cdot 4 D \sqrt{P}}{100}$. The size of the space for the packing will depend on the design of packing adopted.

Cast-Steel Pistons.—These are now extensively used in large marine engines. They are of conical form, to secure the rigidity which in large cast-iron pistons is obtained by adopting the hollow or cellular form, a form which cannot be adopted for cast-steel pistons on account of the difficulty of ensuring a sound casting with a shape so complicated. The brass nuts for the junk ring bolts are screwed into the body of the piston, and each nut is locked by a small screw as shown in the upper portion of Fig. 558.

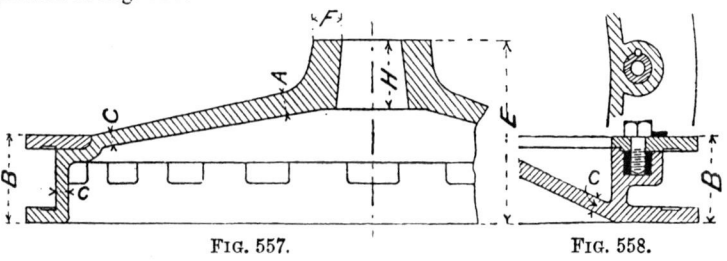

FIG. 557.   FIG. 558.

The following proportions for cast-steel pistons are based on a number of examples in triple expansion marine engines. The unit is $\dfrac{D\sqrt{P}}{100}$, and in compound and triple expansion engines it is assumed that this unit is the same for each cylinder, D being the diameter of the cylinder in inches, and P the initial pressure in pounds per square inch.

$A = \cdot 48$ for high-pressure cylinder piston.
$A = \cdot 54$ for intermediate-pressure cylinder piston.
$A = \cdot 64$ for low-pressure cylinder piston.
$C = \cdot 33$ for high-pressure cylinder piston.
$C = \cdot 34$ for intermediate-pressure cylinder piston.
$C = \cdot 38$ for low-pressure cylinder piston.
$B = 1 \cdot 8$ to $3 \cdot 1$, average $2 \cdot 2$.
$E = 3 \cdot 8$ to $5 \cdot 4$, average $4 \cdot 6$.
$F = \cdot 74$.
$H = 1 \cdot 5$ to $2 \cdot 7$, average $1 \cdot 7$.

In a compound or triple expansion engine the dimensions B, E, F, and H are generally the same for each piston.

262. Hydraulic Pistons.—Pistons for hydraulic machines, where the water is under great pressure, are generally packed with cup-leathers, as shown in Fig. 559. The pressure of the water acting on the inside of the cup forces the leather against the inside of the cylinder in which the piston works. The advantage of this form of packing is that the greater the pressure of the water the more firmly is the leather pressed against the cylinder. It will be observed that there are two cup-leathers—one acts during the forward stroke and the other during the return stroke. U-leathers, described on page 245, may also be used for the packing of pistons.

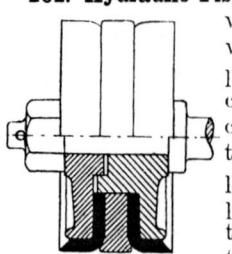

FIG. 559.

263. Diameter of Piston-Rod.—A piston-rod is in the condition of a column fixed at one end and guided at the other. The breaking load of such a column is given approximately by the formula—

$$w = \frac{fS}{1 + \frac{16}{9} a \frac{l^2}{d^2}}$$

Where w = crushing load in lbs.

f = crushing stress = 36,000 lbs. per square inch for wrought-iron and mild steel.

S = sectional area of rod in square inches = $\cdot 7854 d^2$.

$a = \frac{1}{2250}$ for wrought-iron and mild steel.

l = length of rod in inches.

d = diameter of rod in inches.

Since the load on a piston-rod is suddenly applied and continually changing, the safe stress should not exceed 4000 lbs. per square inch; and if we assume that the above formula gives the safe load on the rod, when for f we insert 4000, we get

$$\text{Safe load in lbs.} = \frac{4000 \times \cdot 7854 d^2}{1 + \frac{16}{9 \times 2250} \times \frac{l^2}{d^2}} = \frac{3976087 \cdot 5 d^2}{1265 \cdot 6 + \frac{l^2}{d^2}}$$

But the safe load on the rod is the load on the piston, which is $\cdot 7854 D^2 p$, where D is the diameter of the piston and p the pressure of the steam per square inch; therefore

$$\cdot 7854 D^2 p = \frac{4000 \times \cdot 7854 d^2}{1 + \frac{16}{9 \times 2250} \times \frac{l^2}{d^2}}$$

If for $\frac{l}{d}$ we put r, then

$$d = \frac{\sqrt{20250 + 16 r^2}}{9000} D \sqrt{p} = c D \sqrt{p}.$$

The following table gives values of c corresponding to various values of r :—

$r =$	10	15	20	25	30
$c =$	·0164	·0171	·0181	·0193	·0207

The screwed end of the rod is subjected to tension only, and the diameter of the screw at the bottom of the thread may be ·014D $\sqrt{p}$ for wrought-iron and ·012D $\sqrt{p}$ for steel.

For the piston-rods of oscillating cylinders d is usually ·022D $\sqrt{p}$.

264. Connection of Piston-Rod to Piston.—For small pistons the rod may be screwed in, as shown in Fig. 560, and slightly riveted over. The rod is stronger and more likely to be true to the piston if the un-

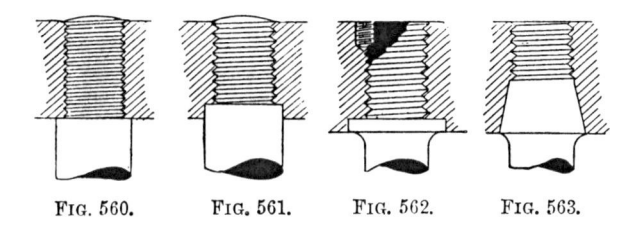

FIG. 560.   FIG. 561.   FIG. 562.   FIG. 563.

screwed part enters and fits tightly into a recess in the piston, as shown in Fig. 561. In the rods shown in Figs. 560 and 561 the shoulder at the end of the screw is obtained by making the screw smaller in diameter than the rod. In Fig. 562 the shoulder on the rod is obtained by forging a collar on it, and in Fig. 563 an equivalent to the shoulder is obtained by making part of the rod within the piston conical.

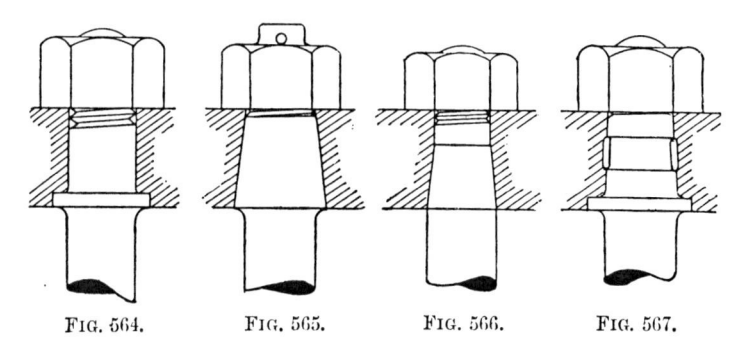

FIG. 564.   FIG. 565.   FIG. 566.   FIG. 567.

The most approved method of securing the piston-rod to the piston is by a nut, as shown in Figs. 564 to 569. The part of the rod within the piston may be parallel or tapered, or partly tapered and partly parallel.

The amount of the taper varies considerably in practice. When the tapered part is long, the taper or variation in the diameter may be 1 in 24. When the tapered part is short, the taper may be as much as 1 in 4.

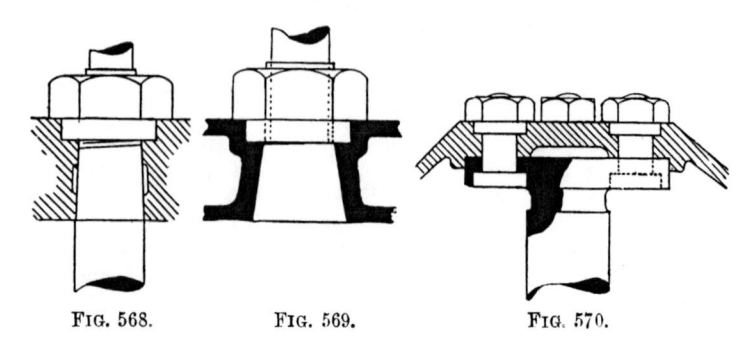

FIG. 568.   FIG. 569.   FIG. 570.

The smaller the taper the more firmly can the rod be secured to the piston, but it is the more difficult to remove. Fig. 570 shows Kirk's method of securing a piston-rod to a steel piston.

CHAPTER XIX.

STUFFING-BOXES.

265. A Stuffing-Box is used where a sliding or rotating piece passes through the end or side of a vessel containing a fluid under pressure. The

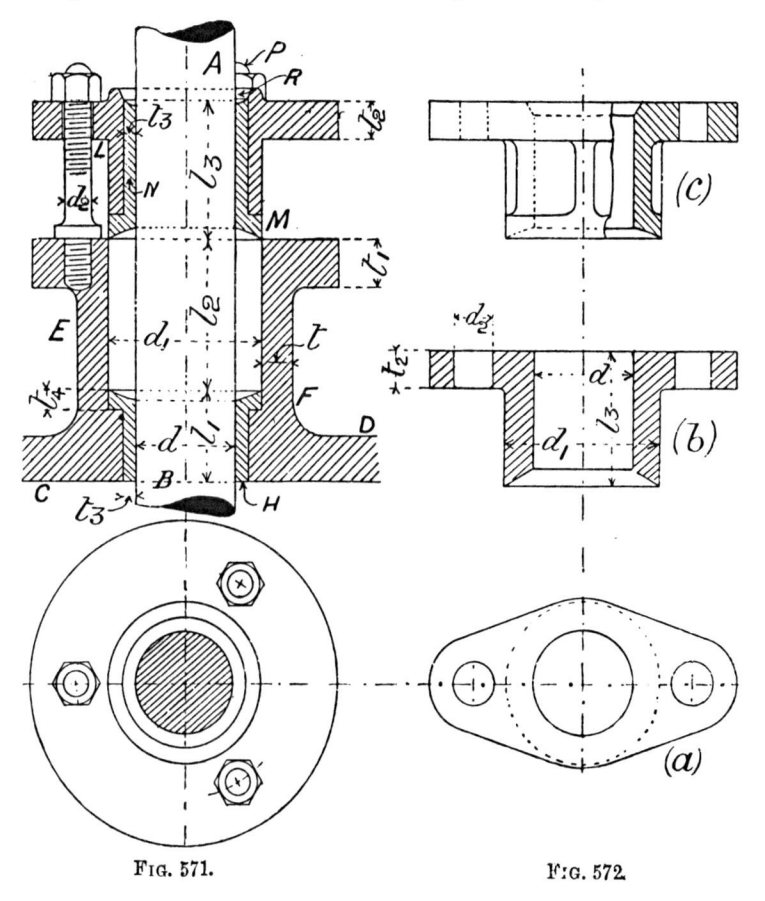

FIG. 571.

FIG. 572.

stuffing-box allows the sliding or rotating piece to move freely, without allowing any leakage of the fluid. Applications of the stuffing-box are found in the steam-engine where the piston-rod passes through the cylinder cover, and where the valve spindle passes through the valve casing; also at the trunnions of the cylinder in an oscillating engine, and where the shaft of a centrifugal pump passes through the pump casing. Stuffing-boxes are also used to permit of the expansion and contraction of steam pipes.

Fig. 571 shows an ordinary form of stuffing-box for the piston-rod of a vertical engine. AB is the piston-rod, CD a portion of the cylinder cover, and EF the stuffing-box. Fitting into the bottom of the stuffing-box is a brass bush, H. The space around the rod AB is filled with *packing*, of which there are a great many kinds, the simplest being greased hempen rope. The packing is compressed by screwing down the gland LM by means of the bolts P. When more than two bolts are used, the gland flange is generally circular, as shown in the lower part of Fig. 571, but for three bolts it is sometimes triangular, and for four bolts it is sometimes square, the bolt holes being near the corners of the triangle or square. For two bolts the gland flange is generally of the form shown at (*a*) Fig. 572.

When the gland is made of cast-iron, it is lined with a brass bush as shown in Fig. 571. When the gland is made of brass, no liner is necessary, and it then has the form shown at (*b*) Fig. 572. When heavy glands are made of brass, they may be lightened in the manner shown at (*c*) Fig. 572.

266. Proportions of Ordinary Stuffing-Boxes and Glands.—From the examination of a large number of stuffing-boxes and glands, we find that their proportions vary very much, even in cases where the conditions under which they work are the same. We believe, however, that the following rules will be useful, as they are based on the averages of a great many examples from actual practice. The letters refer to Figs. 571, 572.

d = diameter of rod.
$d_1 = 1\cdot 22d + \cdot 6$.
$l_1 = \cdot 4d + 1$.
$l_2 = d + 1$.
$l_3 = \cdot 75 l_2 = \cdot 75d + \cdot 75$.
d_2 = diameter of bolts = $\cdot 12d + \cdot 5$ when two are used.

$t = \cdot 1d + \cdot 6$.
$t_1 = 1\cdot 4t = \cdot 14d + \cdot 84$.
$t_2 = t$.
$t_3 = \cdot 04d + \cdot 2$, but not to exceed $\frac{1}{2}$ in.
$t_4 = \cdot 1d + \cdot 13$, but not to exceed 1 in.

For n bolts $d_2 = \dfrac{1\cdot 6}{\sqrt{n}}(\cdot 12d + \cdot 5)$, where n is greater than two.

267. Example of a Large Inverted Stuffing-Box.—Fig. 573 shows an example of a stuffing-box used on a large marine engine with inverted cylinders of the type to be found in large ocean steamers. The stuffing-box is cast separate from the cylinder cover, to which it is afterwards bolted. The lubricant is first introduced to the oil-boxes, from which it passes to the recess underneath, where it comes in contact with the piston-rod. To prevent the lubricant from being wasted by running down the rod, the main gland is provided with a shallow gland and stuffing-box, which is filled with soft cotton packing, which soaks up the lubricant.

The main gland is screwed up by means of six bolts, and to prevent the gland from locking itself in the stuffing-box, it is necessary that the nuts should be turned together. This is done in a simple and ingenious

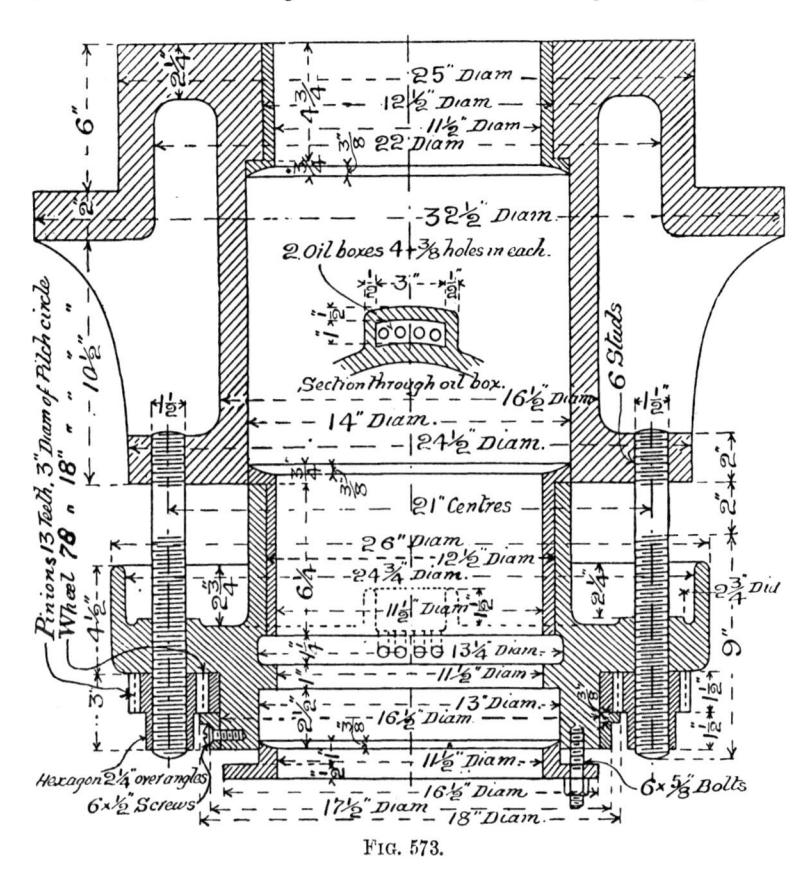

FIG. 573.

manner. One-half of each nut is provided with teeth, and these gear with a toothed wheel which has a rim only; this rim is held up by a ring secured to the main gland as shown. When one nut is turned, all the rest follow in the same direction.

268. Stuffing-Box for Oscillating Cylinder.—In ordinary cases the lateral pressure of the piston-rod on the gland and stuffing-box is comparatively small, but in oscillating engines the swinging of the piston-rod is communicated to the cylinder through the gland and stuffing-box, and when the cylinder is large and consequently heavy, there is considerable lateral pressure on the gland and stuffing-box. This necessitates a greater depth of stuffing-box and gland to give sufficient bearing surface.

An example of a stuffing-box for a large oscillating engine made by the

242   MACHINE DRAWING AND DESIGN.

Fairfield Shipbuilding and Engineering Co., Glasgow, is shown in Fig. 574. It will be observed that lateral support is given to the outer end

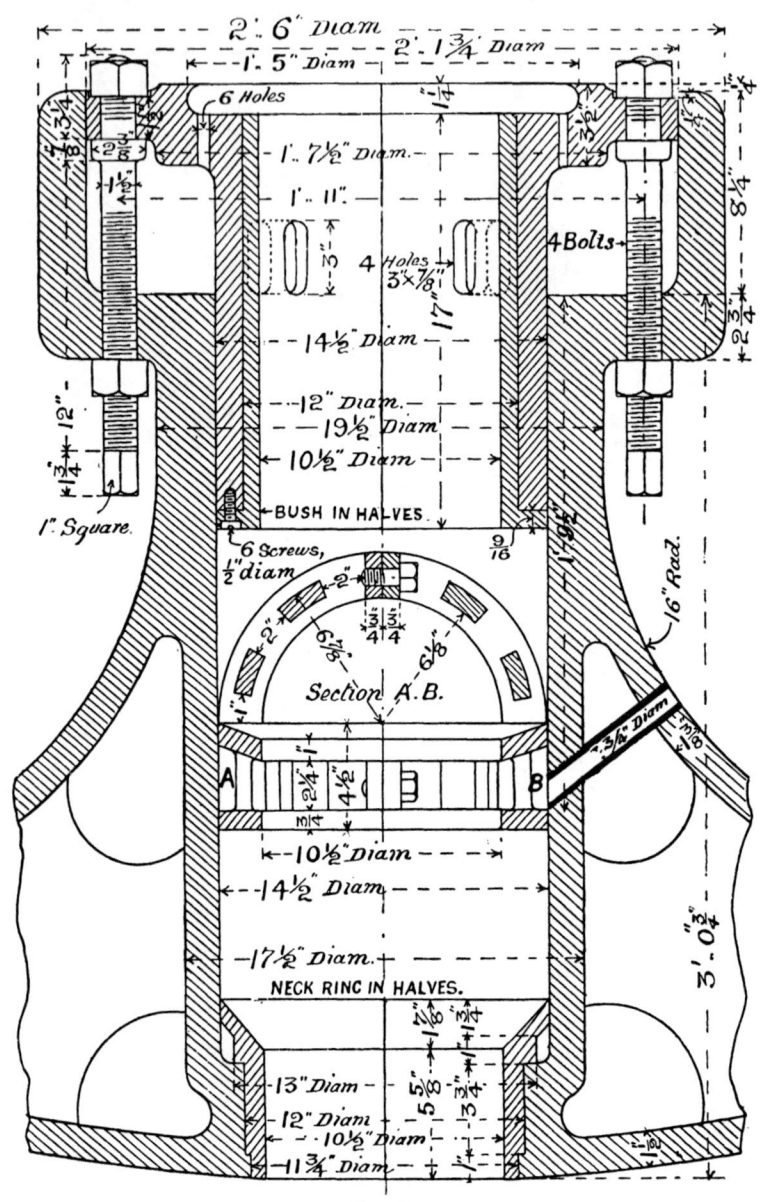

FIG. 574.

of the gland by an enlarged prolongation of the stuffing-box, which fits round the flange of the gland. It will also be noticed that in the centre of the packing space there is a *lantern brass*, which is inserted after the lower part of the box has been filled with packing. Lubricant is introduced into the lantern brass through the pipe shown, and by this means the piston-rod and the packing are effectively lubricated. Oil is introduced at the top of the gland and through four holes in the side of the gland, as shown.

269. Screwed Stuffing-Boxes.—For small rods the stuffing-box is sometimes made of brass, and screwed externally to receive a brass nut by means of which the gland, also of brass, is screwed down, as shown in Fig. 575. This is a very neat and compact arrangement. The proportions may be as follows :—

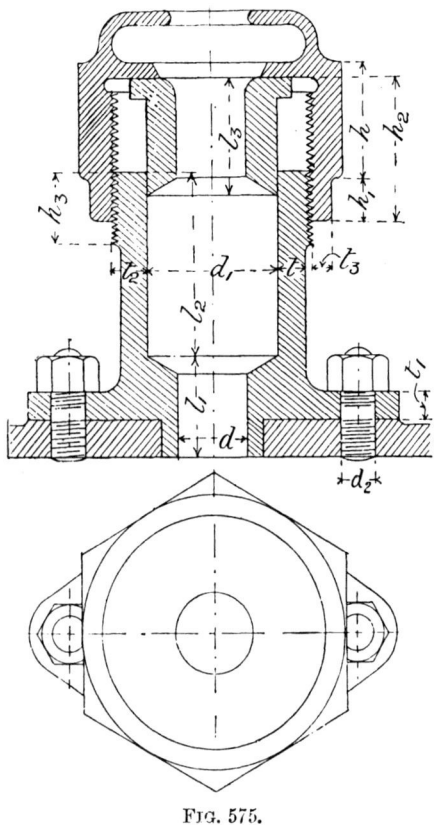

d = diameter of rod.
$d_1 = 1 \cdot 3d + \cdot 6$.
$d_2 = \cdot 15d + \cdot 5$.
$l_1 = \cdot 4d + 1$.
$l_2 = d + 1 \cdot 5$.
$l_3 = \cdot 6d + 1$.

$t = \cdot 1d + \cdot 3$.
$t_1 = \cdot 15d + \cdot 5$.
$t_2 = \cdot 13d + \cdot 4$.
$h = \cdot 6d + 1$.
$h_1 = \cdot 14d + \cdot 4$.
$h_2 = \cdot 67d + 1 \cdot 2$.
$h_3 = \cdot 3d + \cdot 7$.

FIG. 575.

270. Metallic Packing.—Under the head of pistons it was pointed out that in the early days of the steam-engine, the piston was made to work steam-tight in the cylinder by means of soft hempen packing. For pistons the soft packing was soon superseded by metallic packing, which has been greatly improved from time to time. For stuffing-boxes the original form of packing was the soft hempen packing used for pistons, and it is remarkable that while the objections to the use of soft packings in pistons also apply to their use in stuffing-boxes, the majority of stuffing-boxes are still packed with some form of soft packing. Many designs for metallic packing for stuffing-boxes have been brought forward, but very few of them have met with much favour. The design shown in Figs. 576 and 577 is one of the forms of packing known as the United States metallic packing, made in England by the United States Metallic Packing Co., Bradford. The illustrations show the packing as used for locomotive piston-rods. The packing consists of three split Babbitt metal rings, A, B, C, which are contained by a cup D. In section, the first ring, A, is wedge-shaped, the second, B, is nearly rectangular, but is partly wedge-shaped, and the third, C, is rectangular. The packing rings are forced into the cup D by the spiral spring shown, and by the steam pressure acting on the follower E. The wedge-shaped packing rings on being forced into the correspondingly shaped cup, are of course pressed against the piston-rod. The cup D fits against a ring F, which has a spherical bearing on the head of the casing. This spherical bearing permits of the piston-rod changing its direction, and the joint between

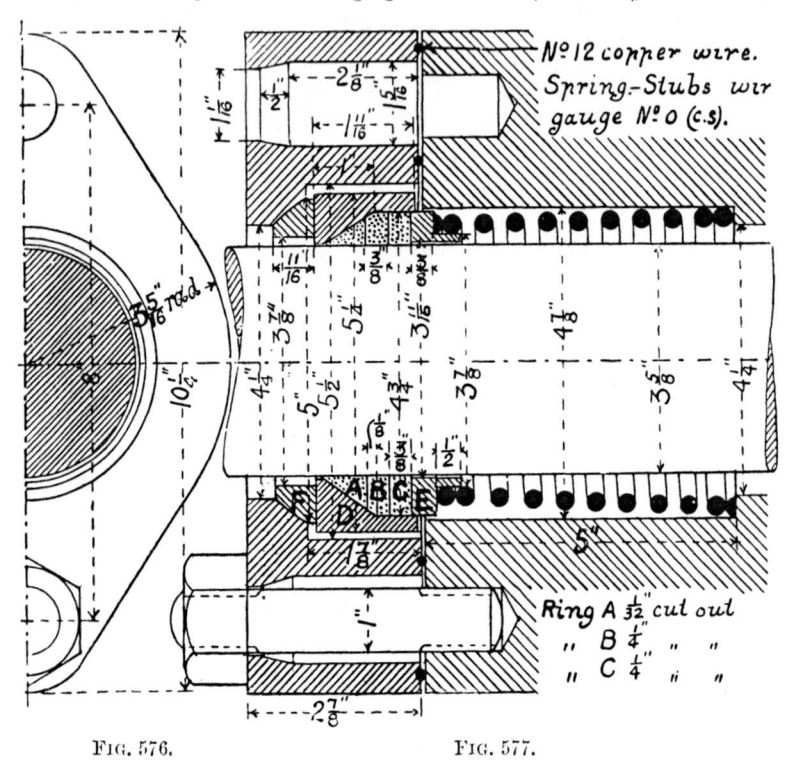

FIG. 576.   FIG. 577.

the cup D and the ring F allows the rod to move sideways without straining either the rod or the packing. It will thus be seen that the rod is not constrained in any way by the packing, and this is as it should be. The packing rings A, B, C, are the only parts which come in contact with the rod. This packing has been extensively used in America, and it is now having a very fair trial in England; it has many advantages, and is highly spoken of by many engineers who have used it. For large stationary and marine engines, the design of the packing which we have described above is modified. In the modified design the packing is pressed against the rod by short spiral springs.

271. Hydraulic Leather Packings.—For pump-rods and plungers, and the rams of hydraulic machines, where the pressure of the water is great, the ordinary form of stuffing-box with hempen, asbestos, or other soft packing, is objectionable on account of the excessive friction. The soft packing must be made sufficiently tight to prevent leakage at the highest pressure under which the machine may be worked, and the friction between the packing and the rod, plunger, or ram, will be the same at the lower pressures at which the machine may be worked as at the highest. Then, again, it is never certain that for the highest pressure the packing is not too tight. These objections are overcome by using particular forms of leather packing.

The U-leather packing, shown in Fig. 578, is a simple and ingenious device due to Bramah. As will be seen, this packing is in the form of a hollow ring, the cross section of the ring being of U-form. This packing

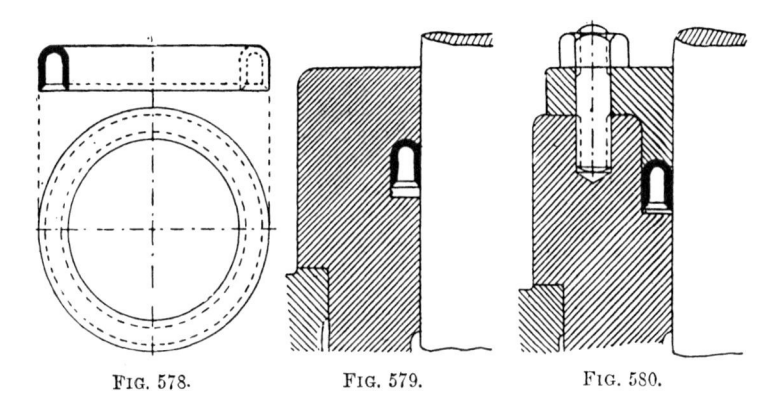

FIG. 578.   FIG. 579.   FIG. 580.

is made from a solid disc of leather by pressing it in a moist state into a mould. After the leather is taken from the mould the superfluous portions are cut away and the edges bevelled. The U-leather packing is placed in a groove, Fig. 579, or in a stuffing-box, Fig. 580. The external diameter of the packing should be slightly larger than that of the groove or stuffing-box, and the internal diameter should be slightly smaller than that of the ram, so as to ensure that the packing will act promptly as soon as the water is admitted. When the water enters the cylinder, it leaks past the ram as far as the packing. Entering inside the packing-ring, the water presses one side against the ram and the other against the opposite side of the recess. The greater the pressure of the water, and therefore the greater the tendency to leakage, the greater is the pressure exerted by the packing. It will be seen that this form of packing is automatic in its action, and the friction will be proportional to the load.

The width of the cross section of the U-leather is usually from $\frac{1}{2}$ inch to $\frac{3}{4}$ inch, and the height about 1·6 times the width.

The diameter of the bolts for the gland, Fig. 580, is determined as follows:—

D_1 = diameter of ram, in inches.
D_2 = internal diameter of stuffing-box, in inches.
d_1 = diameter of bolts at bottom of screw-thread, in inches.
n = number of bolts.
p = maximum working pressure of water, in pounds per square inch.
f = safe stress on bolts, in pounds per square inch.

$$\cdot 7854(D_2^2 - D_1^2)p = \cdot 7854 d_1^2 nf$$

$$d_1 = \sqrt{\frac{(D_2^2 - D_1^2)p}{nf}}$$

Another form of leather packing, known as a *hat-leather*, is shown in Fig. 581. This works on the same principle as the U leather, but it is generally only used for small rams or plungers.

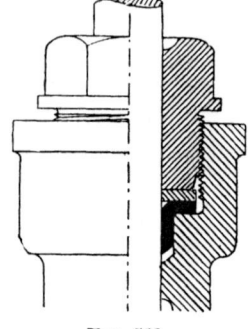

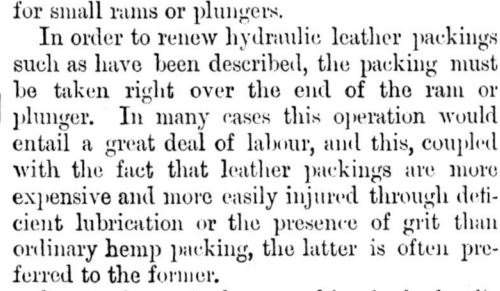

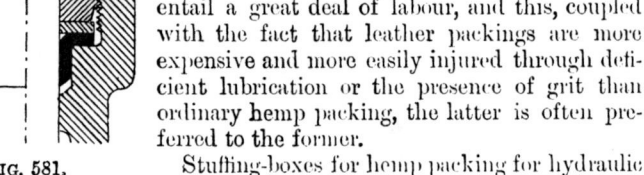

In order to renew hydraulic leather packings such as have been described, the packing must be taken right over the end of the ram or plunger. In many cases this operation would entail a great deal of labour, and this, coupled with the fact that leather packings are more expensive and more easily injured through deficient lubrication or the presence of grit than ordinary hemp packing, the latter is often preferred to the former.

FIG. 581.

Stuffing-boxes for hemp packing for hydraulic work may have a depth equal to $2\sqrt{D}$, and a diameter equal to $D + 1\cdot 62\sqrt{D}$, where D is the diameter of the ram or plunger.

CHAPTER XX.

VALVES.

272. Classification of Valves.—A valve is a piece of mechanism for controlling the magnitude or direction of the motion of a fluid through a passage. The *seat* of a valve is the surface against which it presses when closed, and the *face* of the valve is the portion of its surface which comes in contact with its seat. Valves may be classified according to the means by which they are moved. In one class the valves are moved by the pressure of the fluid. Nearly all pump valves belong to this class. The valves belonging to this class are automatic in their action, and they permit the fluid to pass in one direction only. Flap valves, flexible disc valves, lift valves (including rigid disc valves with flat seats, disc valves with bevelled edges fitting on conical seats, and ball valves), belong to the class of valves moved by the fluid. In another class the valves are moved by hand or by external mechanism, so that the motion of the valve is independent of the motion of the fluid.

Since many of the valves belonging to the first class mentioned above may be converted into valves of the second class without altering the construction of the valves themselves, a better classification is as follows: (1) flap valves, which bend or turn upon a hinge; (2) lift valves, which rise perpendicularly to the seat; (3) sliding valves, which move parallel to the seat.

273. Flap or Clack Valves.—At one time flap valves were very common, but they are now used to a very limited extent. These valves may be entirely of brass or other metal, but they are generally made of leather stiffened by metal plates. Fig. 582 shows a simple form of flap valve. The valve forms the central part of a disc of leather which is held between the flanges of the pipes or valve-box. The disc has a piece cut out, leaving a portion which forms the hinge for the valve. The valve is stiffened by metal plates above and below the leather, to which they are secured by rivets as shown.

Large flap valves work very much smoother if made as shown in Fig. 583. An opening is made in the main valve, and this is covered by a supplementary valve which is hinged on the side opposite to the hinge of the main valve as shown. The area of the opening of the supplementary valve may be one-third of the area of the opening of the main valve. A valve or clack of this description is known as a relief or break clack. Mr. Henry Teague of Lincoln states[1] that break clacks 15 inches in diameter

---
[1] Proceedings of the Institution of Mechanical Engineers, 1887.

have worked incessantly for five years without changing a leather, and without showing the least sign of leakage, under 350 feet head of water, and without the slightest concussion. For a velocity of 160 feet per

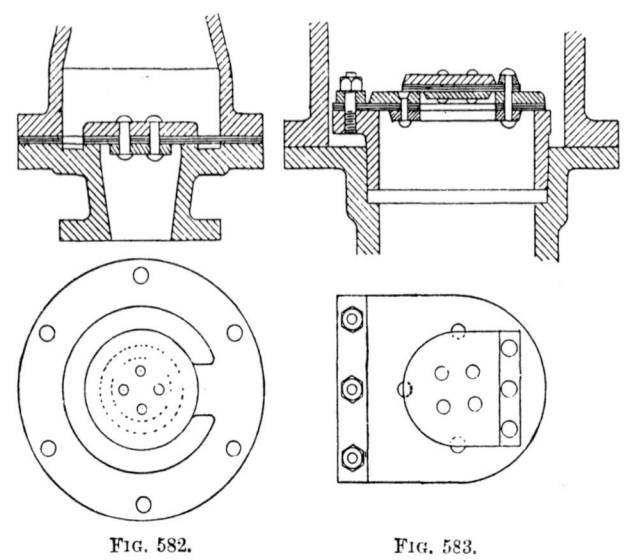

FIG. 582.   FIG. 583.

minute of the pump piston, Mr. Teague found that the weight of the flap should be about 2 lbs. per square inch.

The width of the seat for a flap valve may be from one-eighth to one-twelfth of the diameter of the valve, and the flap should open about 35°.

274. India-rubber Disc Valves.—Examples of india-rubber disc valves are shown in Figs. 743 to 746. The thickness of the india-rubber is generally $\frac{3}{8}$ inch to $\frac{1}{2}$ inch for small valves, and $\frac{5}{8}$ inch to $\frac{7}{8}$ inch for large valves. The seat for the india-rubber is in the form of a grating, and the area of the seat in contact with the india-rubber should be such that the intensity of the pressure between the two does not exceed 40 lbs. per square inch. The guard which limits the lift of the valve is perforated, and may either be conical or spherical. If conical, the slant side may slope to the valve seat at an angle of 30°, and if spherical it may have a radius equal to the diameter of the india-rubber disc. The distance between the seat and the guard at the centre should be about $\frac{1}{8}$ inch greater than the thickness of the india-rubber, so that the latter may have room to bend freely. The hole in the india-rubber should be $\frac{1}{16}$ inch to $\frac{1}{8}$ inch larger than the central bolt or stud for the same reason, and also that the india-rubber may be free to rotate, and so change the bearing surface of the valve on its seat. To facilitate this rotary motion of the valve, the openings in the grating are sometimes made at an angle, so that the water impinges on the valve obliquely. With the same object in view, the edge of the disc is sometimes serrated obliquely.

275. Single-Beat Direct-Lift Valves.

—The face of a single-beat direct-lift valve is generally conical, and forms the edge of a disc, as shown in Figs. 584, 585, and 586 ; but the face may be flat, as shown in

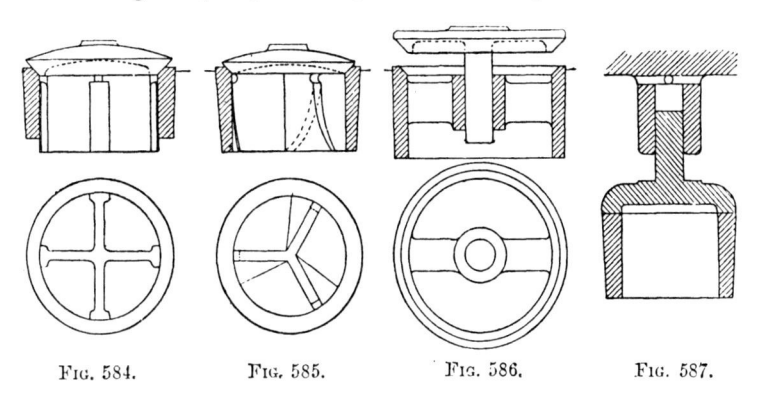

Fig. 584.   Fig. 585.   Fig. 586.   Fig. 587.

Fig. 587, or spherical, as shown in Fig. 588. When the valve face is conical, its inclination to the axis of the valve is generally 45°.

The disc valve is guided in rising and falling either by feathers on its under side, which fit into the cylindrical part of the seat, as shown in Figs. 584 and 585, or by a central spindle, attached to the valve, which slides freely in a hole formed either in a bridge stretching across the seat, as in Fig. 586, or in the valve casing, as shown in Fig. 587.

The ball valve, Fig. 588, does not require to be guided so precisely as the disc valve; but to prevent it moving too far from its seat, it is surrounded by a guard, which is in the form of a cage.

In Fig. 585 the guiding feathers are shown of a screw form. This enables the fluid to give the valve a rotary motion as it rises, and thus prevents the parts of the valve face from always beating on the same parts of the valve seat, and thus prevents unequal wear.

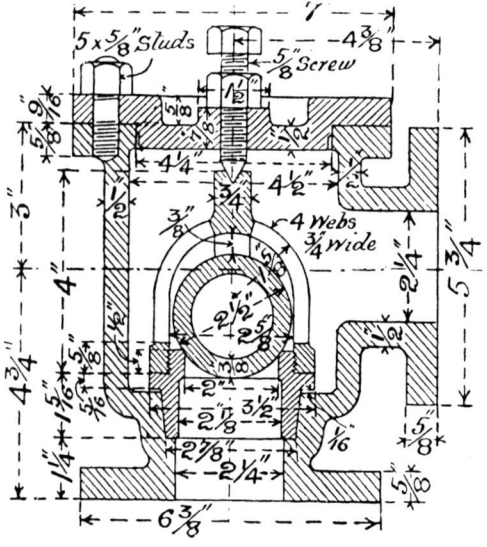

Fig. 588.

The lift of a single-beat valve should not exceed one-fourth of the diameter, because with this lift the area of the opening is equal to the area of the passage below the valve. This is proved as follows:—Let D = diameter of passage below valve, called the diameter of the valve; H = lift of valve. Area of valve = $D^2 \times \cdot 7854$. Area of opening = H × D × 3·1416. If these areas are equal, then $H \times D \times 3 \cdot 1416 = D^2 \times \cdot 7854$, and $H = \dfrac{D}{4}$.

When a single-beat valve is controlled mechanically, and not automatically by the action of the fluid, the lift may be as much as one-fourth of the diameter of the valve. But a lift so large as one-fourth of the diameter of the valve is generally objectionable when the valve is opened and closed by the pressure of the fluid upon it, because of the shock produced by the valve striking on its seat; and when the pressure on the valve is great, the lift should not exceed $\frac{1}{4}$ inch.

The force required to open a single-beat valve is equal to its area multiplied by the effective pressure per unit of area.

The width of the valve seat may be as small as $\frac{1}{32}$ inch, and is sometimes as much as $\frac{1}{2}$ inch. The narrower the seat, the easier is it to make the valve tight, but the area of the seat must be sufficient to prevent the material of the valve and of the seat from being crushed.

276. Double-Beat Partially Balanced Valves.—When a single-beat valve is large, the force required to open it is great, and it must have a comparatively large lift to get the full opening of the valve. These difficulties are overcome in double-beat valves, of which we give two examples. Fig. 589[1] shows a form of double-beat valve used as a steam valve on an engine at the Birmingham Waterworks. It will be observed that the valve is guided by feathers, which fit inside the upper part of the seating, and by a central spindle cast on the bottom of the seating.

Another form of double-beat valve is shown in Fig. 590. This is a design by Mr. A. F. Nagle, described by him in a paper read at a meeting of the American Society of Mechanical Engineers in 1889. This valve is one of a series for a pump pumping water under an effective pressure of about 52 lbs. per square inch. The valve is worked automatically by the water, and the lift is limited by the collar and nut at the top of the central spindle. The seats were designed to be $\frac{3}{8}$ inch wide, but they were actually chamfered so that only $\frac{1}{8}$ inch bearing surface remained, as shown at (*a*). This valve and its seat are made of bronze.

Let D_1 and D_2 be the diameters of the larger and smaller seats of a double-beat valve, H the maximum lift, and P the effective pressure of the fluid in lbs. per square inch. Then, $H = \dfrac{D_1^2}{4(D_1 + D_2)}$, and the force required to open the valve = $\cdot 7854 (D_1^2 - D_2^2) P$, neglecting the width of the seats and the weight of the valve. From this it is evident that

---

[1] This illustration has been prepared from particulars given in *Engineering*, vol. xl.

VALVES

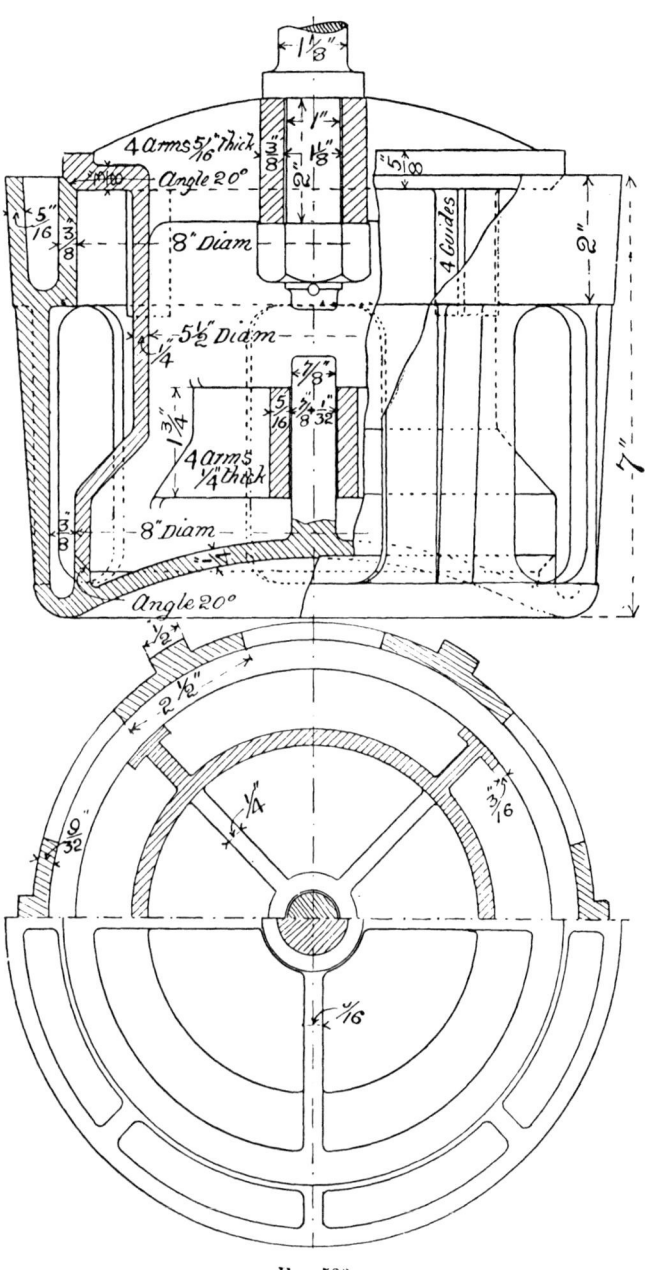

FIG. 589.

by making the difference between D_1 and D_2 small, the force required to open the valve will also be small. Also if D_2 is nearly equal to D_1, the lift of the double-beat valve to give the maximum opening will be about one-half the lift of a single-beat valve of the same diameter.

These values and their seatings are made of brass or bronze.

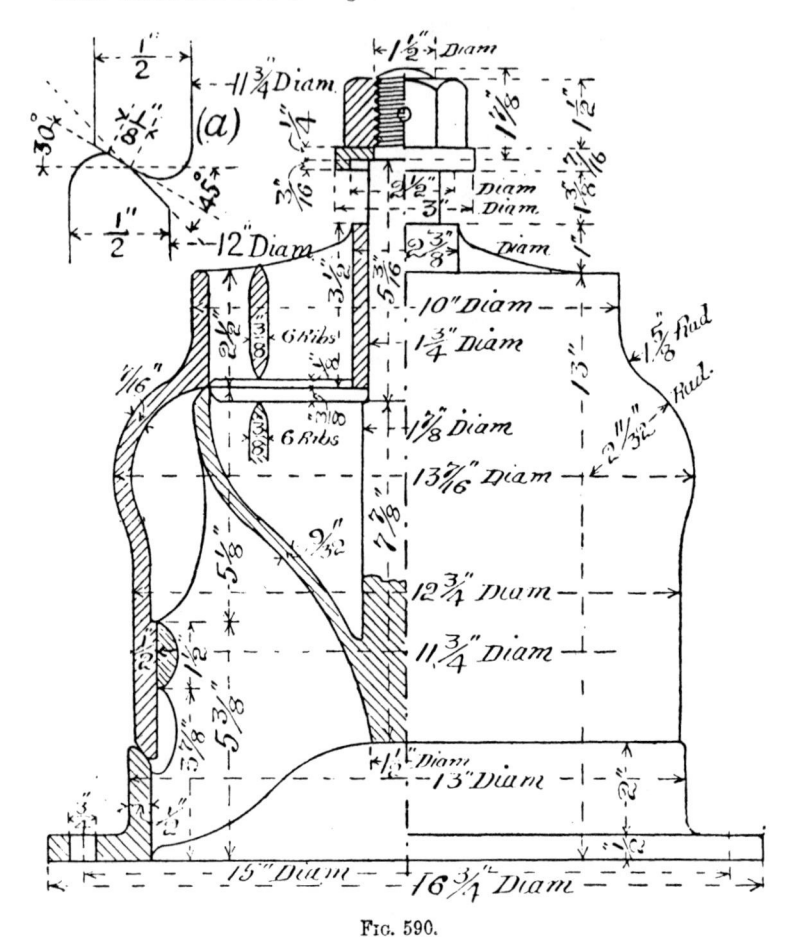

FIG. 590.

277. Simple Slide Valve.—The valve which has been most extensively used for the distribution of the steam in steam-engines is the simple three-ported slide valve.

The form and action of the common slide valve will be understood by reference to the example shown in Figs. 591 and 592. It will be seen that the valve is in the form of a box open on the under side, and when in its middle position two of its edges cover two rectangular steam ports

S_1, S_2, while the hollow part is opposite to the exhaust port E. The valve is usually driven by an eccentric, and it is easy to show that if the valve no more than covers the steam ports when in its middle position, the eccentric must be placed 90° in advance of the crank of the engine. As shown in Fig. 592, the piston is at the left-hand end of the cylinder,

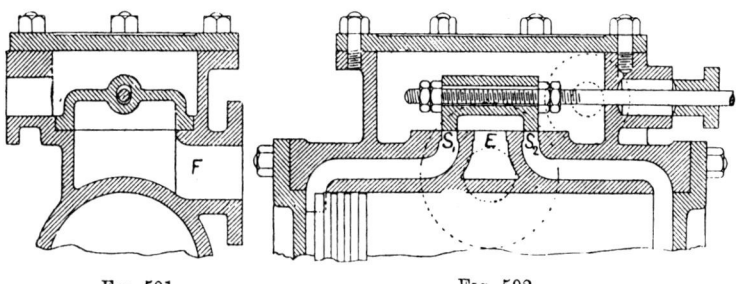

Fig. 591.   Fig. 592.

the valve is in its middle position, and is moving towards the right, and is therefore about to open the steam port S_1 and allow steam to pass into the left-hand end of the cylinder, which will force the piston towards the right. Meanwhile the hollow part of the valve will come over the steam port S_2, and the steam in the cylinder to the right of the piston which caused the piston to make the preceding stroke will escape into the exhaust port E, and thence through the exhaust pipe F shown in Fig. 591. When the piston is near the middle of its stroke, the valve will have moved to the extreme right, and will then commence to move to the left, and by the time that the piston has reached the right-hand end of the cylinder, the valve will again be in its middle position. During the return stroke of the piston the valve will move from its middle position to the extreme left, and back again to its middle position. It is evident that in the above example one end of the cylinder will be open to steam from the valve-chest, and the other end will be open to exhaust during the whole of the stroke of the piston, so that there is no expansive working of the steam.

In practice, the valve more than covers the steam ports when in its middle position. The amount which the valve projects over the steam port on the outside, the valve being in its middle position, is called the *outside lap* of the valve, and the amount which it projects on the inside is called the *inside lap*. When the term *lap* is used without any qualification, outside lap is to be understood. Fig. 603, page 265, shows a slide valve with both outside and inside lap.

The addition of lap to a slide valve necessitates a change in the position of the eccentric in relation to the crank because the valve must be on the point of opening the steam port when the piston is at the beginning of its stroke, and therefore the valve must be away from its middle position by an amount equal to the lap when the piston is at the beginning of its stroke. Hence the eccentric must be more than 90° in advance of the crank. The correct position of the eccentric in relation to the crank is found by the construction shown in Fig. 593. AO represents the crank, and the circle CDE the path of the centre of the eccentric sheave. On AO produced mark off OB equal to the lap of the valve. Draw BD at right angles to OB, meeting the circle CDE at D. OD is the position of the eccentric if the motion is in the direction of the arrow. If the motion is in the opposite direction, then OE will be the position of the eccentric. If OC be drawn at right angles to OA, the angle COD is called the *angle of advance* of the eccentric.

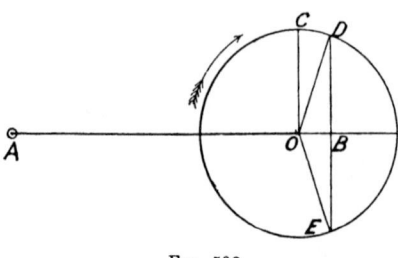

Fig. 593.

If, when the piston is at the beginning of its stroke, the steam port is partly open on the steam side, the amount of this opening is called the *lead* of the valve. To give a valve lead, the angle of advance of the eccentric must be increased, so as to make OB, Fig. 593, equal to the lap, plus the lead.

Slide-valves are generally made of cast-iron, brass, or bronze.

278. Double-Ported Slide-Valve.—If a large steam cylinder was fitted with a simple slide-valve, the valve would require a long travel in order to give a sufficient port opening for the steam, and consequently a large amount of work would be expended in moving the valve. By making the slide-valve and the steam passages double ported, the same amount of opening will be obtained with half the travel which a simple slide-valve would require, and consequently much less work will be required to drive it. Fig. 594 is a longitudinal section, and Fig. 595 is a transverse section, of a double-ported slide-valve for the low-pressure cylinder of a large marine engine. In Fig. 594 the valve is shown in the middle of its travel; and it will be observed that in this position the centre of the valve is not opposite the centre of the exhaust port, but is 7-16ths of an inch higher. It will also be noticed that if the valve was placed with its centre opposite the centre of the exhaust port, it would not cover the steam ports completely, but would leave them all open to the exhaust 9-16ths of an inch. The object of these points in the design is to admit the steam earlier and release it later on the under side of the piston than on the upper side, so as to counteract the effect of the great weight of the piston, piston-rod, cross-head, and connecting-rod, the engine being of the vertical inverted cylinder type, common in marine engine practice.

The weight of the valve is counterbalanced by the pressure of the steam on the under side of a piston attached to the upper end of the valve spindle. This piston works in a cylinder 11 inches in diameter (not shown in the illustrations), open on the under side to the valve casing.

279. Regulator Valve for Locomotive Engine.—The valve shown in

FIG. 594.

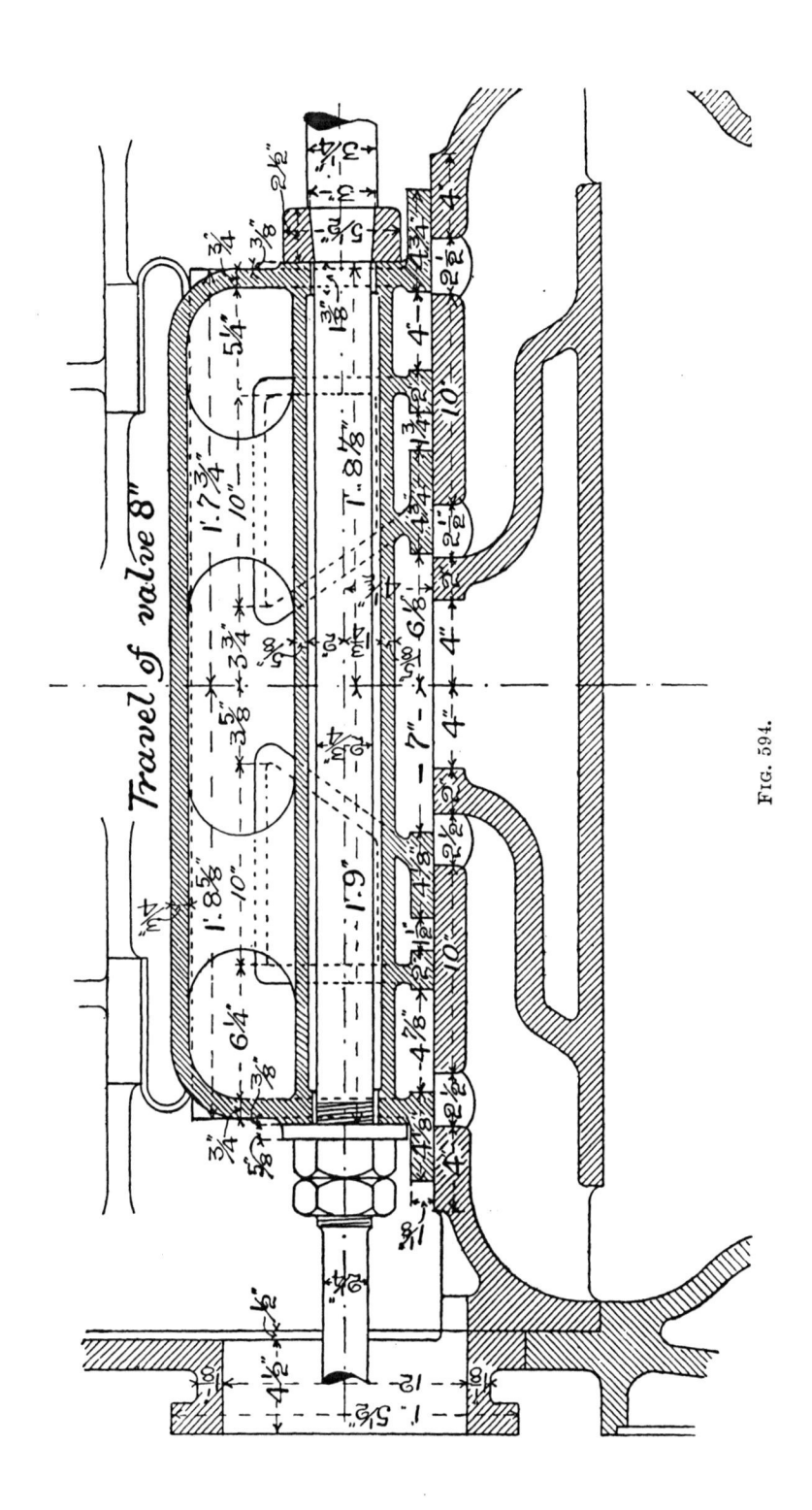

FIG. 595.

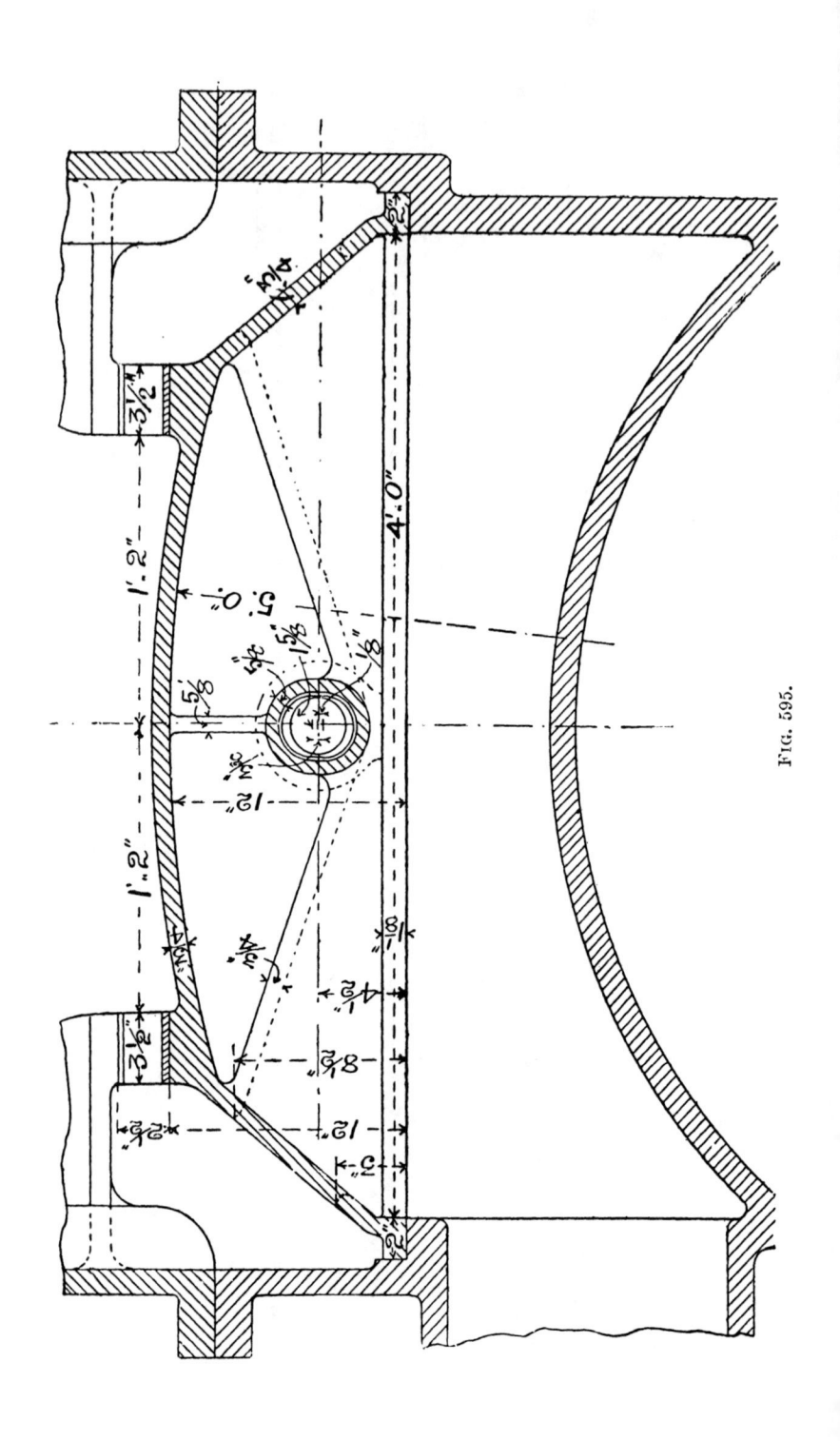

VALVES. 257

Figs. 596 and 597 is a three-ported slide valve used for regulating the supply of steam to a locomotive engine. A is the valve, made of brass, which works on the side of a vertical cast-iron pipe, BC, placed in the dome of the boiler. On the back of the valve A there is a supple-

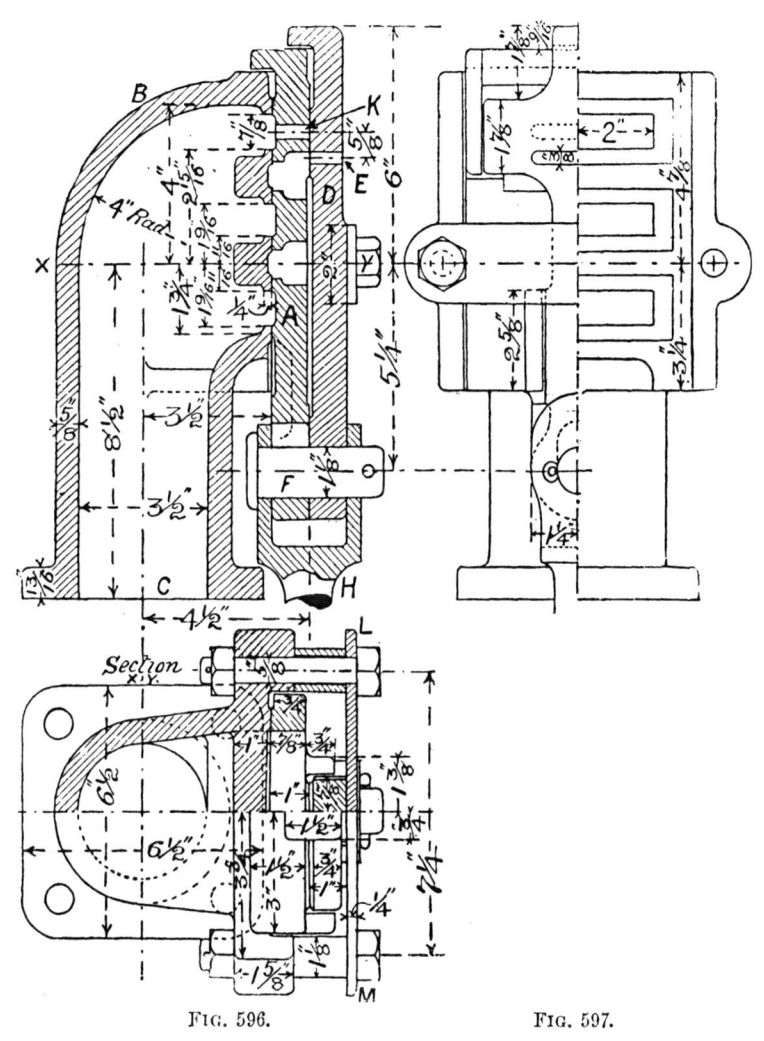

FIG. 596. FIG. 597.

mentary slide valve, D, made of cast-iron, having one small port, E. A pin, F, in the end of the link H passes through a slotted hole in the main valve A, and through a fitting hole in the supplementary valve D. When the link H is raised, the valve D moves, but the valve

R

A remains at rest until the port E in the supplementary valve is opposite to the port K in the main valve, when the two valves move together. The area of the supplementary valve being small, this valve is easily opened, and once open, the steam reaches both sides of the main valve, which is then nearly balanced, and a comparatively small force will then be sufficient to open it. The valves are kept on their seats by a flat steel spring plate, LM. The link H is moved by means of a handle in the cab, and intermediate mechanism not shown.

280. Friction of Slide Valves.—Let l equal the length and b the breadth of a slide valve, both in inches, and let P be the pressure of the steam on the back of the valve in pounds per square inch. Then the force pressing the valve on its seat is lbP, diminished by the force which tends to lift the valve off its seat due to the steam pressures in the steam and exhaust ports. The latter force will vary throughout the travel of the valve, but its average value may be taken equal to the area of one steam port multiplied by the pressure in the valve chest, plus the area of the other steam port multiplied by the pressure of the exhaust steam, plus the area of the exhaust port multiplied by the pressure of the exhaust steam. Using symbols—

Let R = effective force pressing the valve on its seat, in pounds.
A = area of valve, in square inches = length l × breadth b.
S = area of one steam port in a single-ported valve, in square inches.
  = area of two steam ports in a double-ported valve, in square inches.
E = area of exhaust port, in square inches.
P = steam pressure in valve chest, in pounds per square inch.
p = pressure of exhaust steam, in pounds per square inch.
F = force required to move valve, in pounds.
f = co-efficient of friction between the valve and its seat.

Then, approximately—

$$R = AP - (SP + Sp + Ep) = P(A - S) - p(S + E).$$
And, $F = Rf$.

The coefficient of friction varies considerably in different cases, depending on the condition of the sliding surfaces and on the lubrication. It may be taken as varying from ·05 to ·1.

281. Piston Valves.—To overcome the friction of a slide valve of large area, working under a high steam pressure, requires the expenditure of a considerable amount of power. This waste of power may be reduced by fitting the back of the valve with a relief frame containing packing rings which press against the back of the valve casing, and so cut off a portion of the area of the valve from the pressure of the steam; but this and other arrangements for partly balancing the valve have not proved quite satisfactory.

The great waste attending the use of large slide valves under high pressure steam has led engineers to adopt piston valves, which, although usually much more expensive to construct than ordinary slide valves, require very much less power to work them.

A piston valve is really a slide valve in which the face and seat are complete cylindrical surfaces. A piston valve is perfectly balanced so far as the steam pressure is concerned, and the only friction to be overcome is that due to the small side pressure necessary to keep the valve steam tight. Piston valves are made steam-tight in much the same way as ordinary pistons, but since the amount of the motion of the valve is small compared with that of the engine piston, the wear is small, and the springs therefore may be stiff with very little range.

Figs. 598 and 599 illustrate a form of piston valve designed and made by that eminent firm of marine engineers, Messrs. Robert Napier & Sons, Glasgow. In this valve there are two pistons, one for each steam port, connected by a pipe. Only the lower piston is shown here. The piston consists of a stiff cast-iron ring, flanged internally as shown for bolting to the connecting pipe already mentioned. This ring is slotted across in a sloping direction, the opening made being of wedge-shape both longitudinally and transversely. Into this opening is carefully fitted a gun-metal wedge-piece, A, which is held firmly in place by the studs and nuts B and bridge-piece C. The whole is then turned to a working fit, after which the bridge-pieces D, shown in Fig. 598 by dotted lines, are cut away. It will be observed that in this design the adjustment for wear is not automatic, but has to be made when required by tightening up the wedge-piece, which causes the piston ring to expand.

The particular valve which we have just described was made for the high-pressure cylinder of a large marine engine. The intermediate and low-pressure cylinders were also fitted with piston valves. The pistons for the low-pressure valve had a depth of $8\frac{1}{4}$ inches, and the diameters of the upper and lower pistons were 3 ft. 9 in. and 3 ft. 7 in. respectively. The piston rings for all the valves were constructed in the same way, and except as regards depth and diameter, they were of the same dimensions.

Another design for a piston valve by Messrs. Robert Napier & Sons is shown in Figs. 600 and 601. In this design stiff cast-iron spring rings are used for making the pistons steam tight. To prevent the packing rings from springing into the ports, the latter have a series of sloping bars across them. These bars are $1\frac{1}{2}$ inches broad, and they slope to the right and left alternately at 74° to the horizontal; they are placed at $6\frac{3}{16}$ inches pitch. The total effective areas of the upper and lower ports are 884 square inches and 822 square inches respectively. The greatest openings to steam in the upper and lower ports are 591 square inches and 659 square inches respectively.

282. Cocks.—A cock is a rotary slide valve, and consists of a plug which is a frustum of a cone, and which fits into a casing of similar shape cast on a pipe. Through the plug there is a hole, which may be made to form a continuation of the hole in the pipe by turning the plug into one position. When in this position the cock is open, and the fluid may pass through the pipe. By turning the plug into another position, its solid part comes across the hole in the pipe, and the cock is closed, and the passage of the fluid through the pipe is stopped.

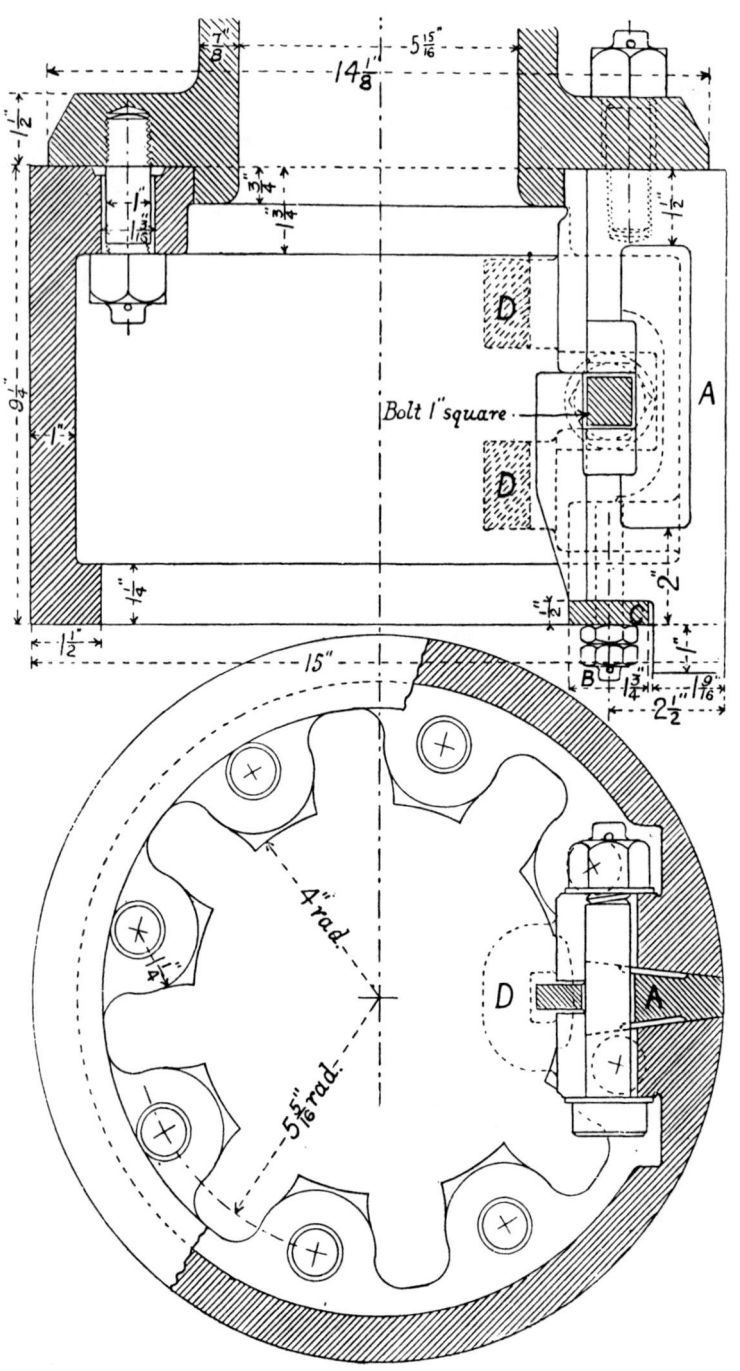

FIG. 598.

VALVES.

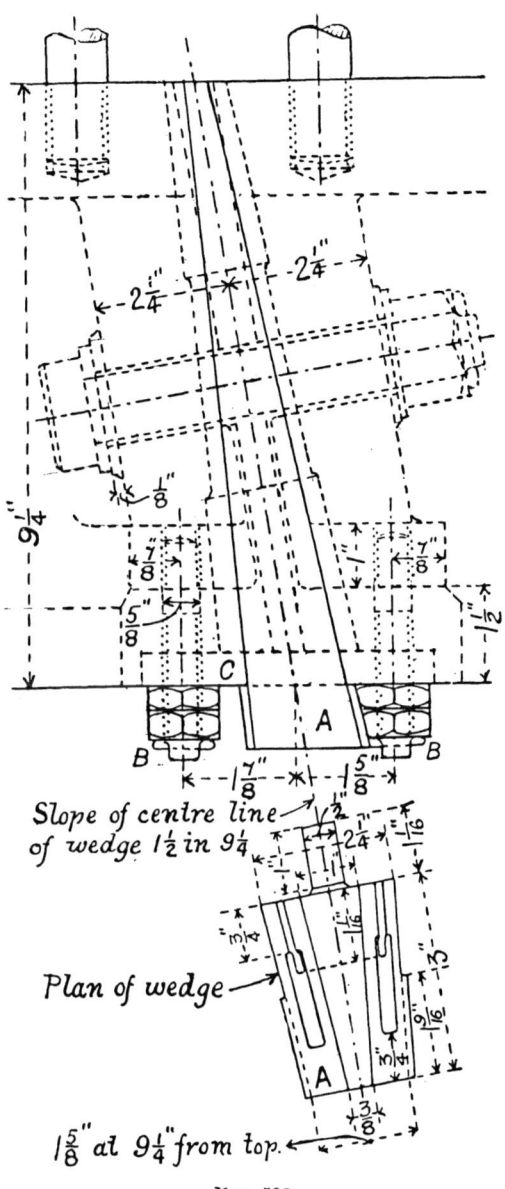

Fig. 599.

FIG. 600.

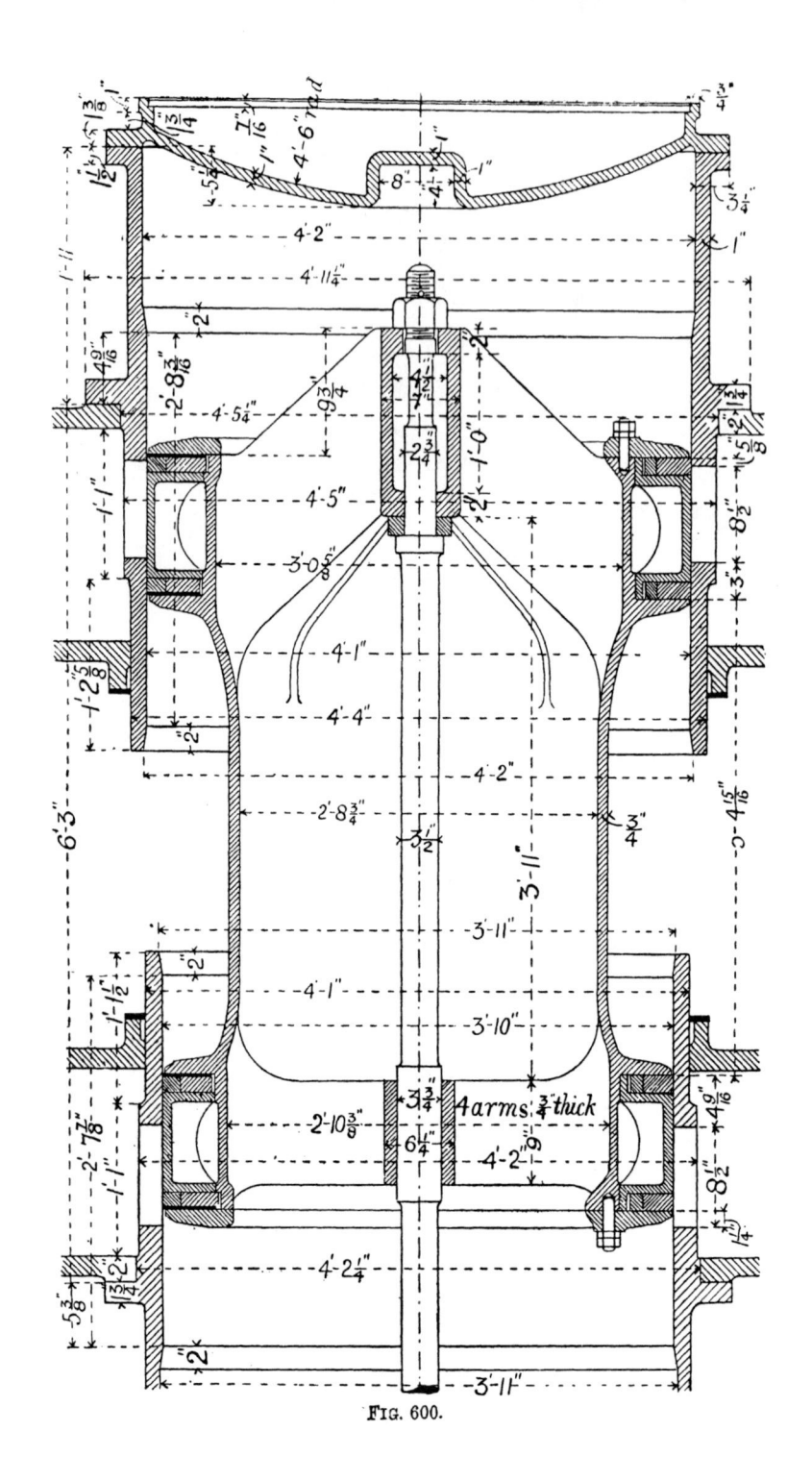

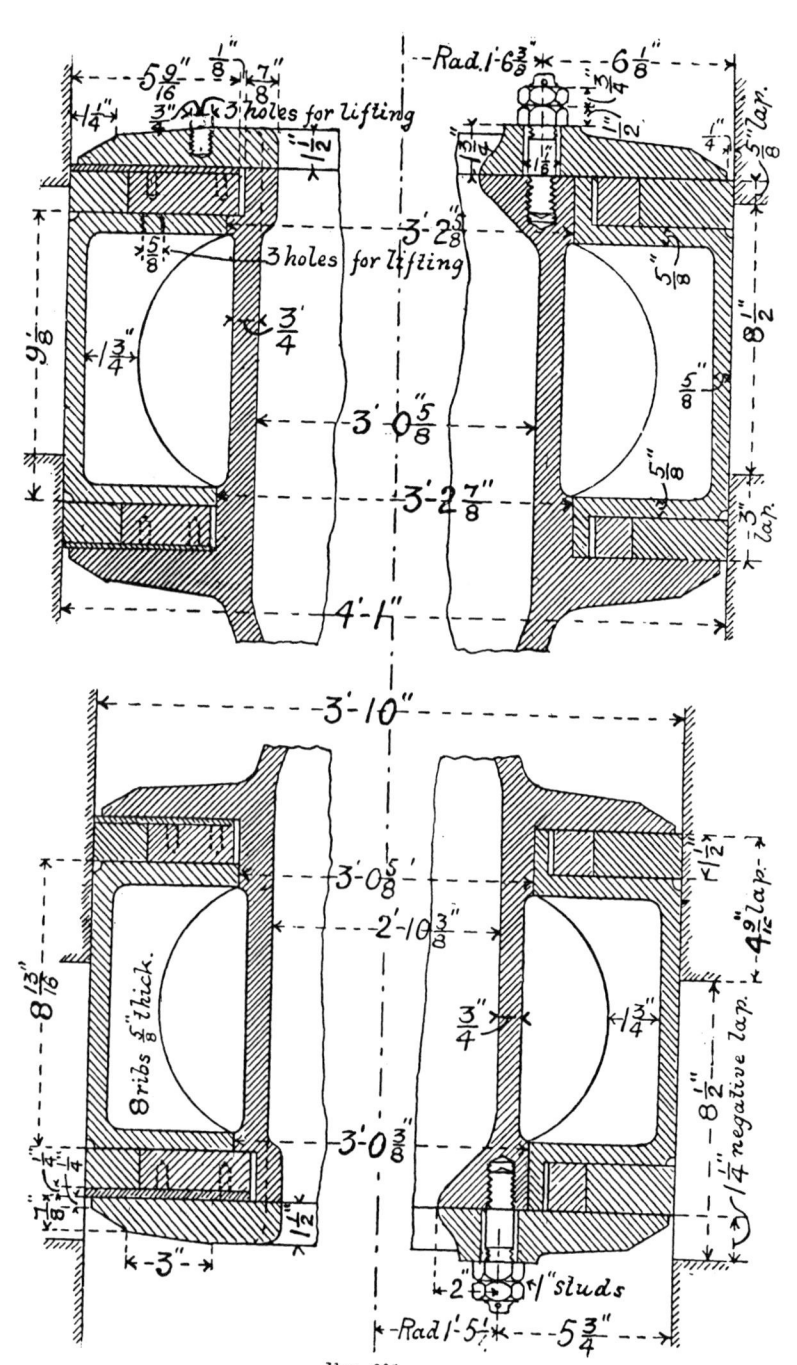

FIG. 601.

Fig. 602 shows a good example of a cock, suitable for steam or water, and adjacent to it is a table of dimensions, taken from actual practice, for four different sizes. The two parts of the pipe meeting at the plug may be in the same straight line, or one may proceed from the bottom of the casing, as shown by the dotted lines.

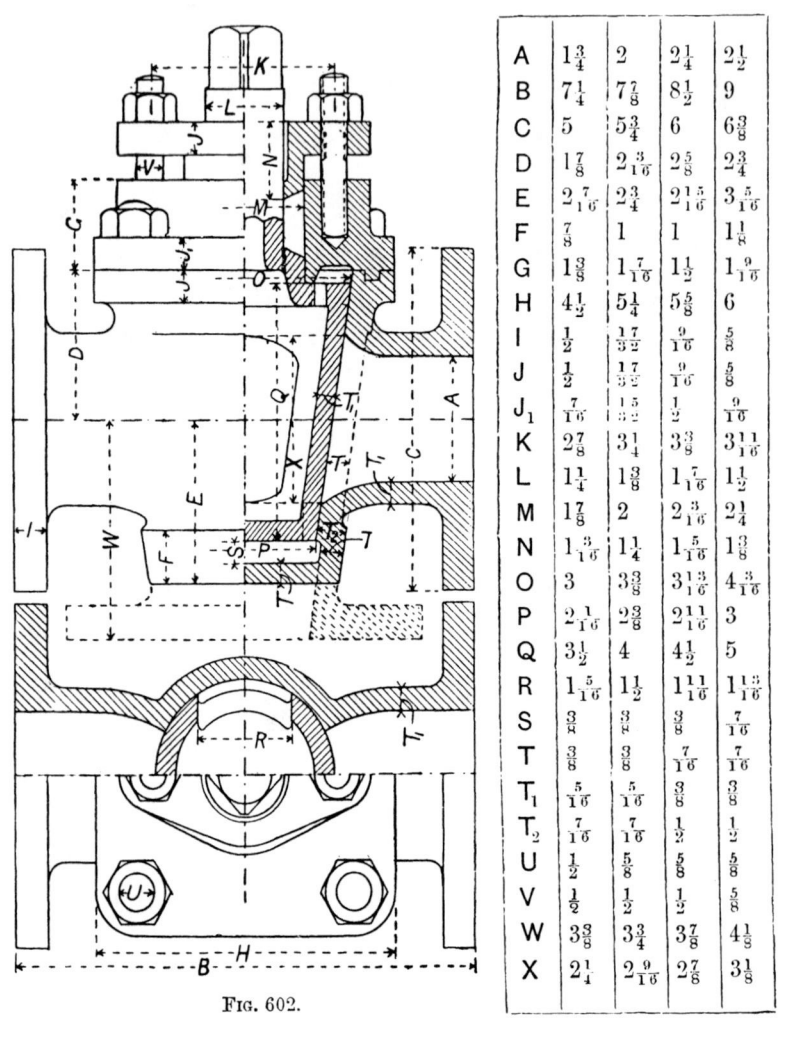

A	$1\frac{3}{4}$	2	$2\frac{1}{4}$	$2\frac{1}{2}$
B	$7\frac{1}{4}$	$7\frac{7}{8}$	$8\frac{1}{2}$	9
C	5	$5\frac{3}{4}$	6	$6\frac{3}{8}$
D	$1\frac{7}{8}$	$2\frac{3}{16}$	$2\frac{5}{8}$	$2\frac{3}{4}$
E	$2\frac{7}{16}$	$2\frac{3}{4}$	$2\frac{15}{16}$	$3\frac{5}{16}$
F	$\frac{7}{8}$	1	1	$1\frac{1}{8}$
G	$1\frac{3}{8}$	$1\frac{7}{16}$	$1\frac{1}{2}$	$1\frac{9}{16}$
H	$4\frac{1}{2}$	$5\frac{1}{4}$	$5\frac{5}{8}$	6
I	$\frac{1}{2}$	$\frac{17}{32}$	$\frac{9}{16}$	$\frac{5}{8}$
J	$\frac{1}{2}$	$\frac{17}{32}$	$\frac{9}{16}$	$\frac{5}{8}$
J_1	$\frac{7}{16}$	$\frac{15}{32}$	$\frac{1}{2}$	$\frac{9}{16}$
K	$2\frac{7}{8}$	$3\frac{1}{4}$	$3\frac{3}{8}$	$3\frac{11}{16}$
L	$1\frac{1}{4}$	$1\frac{3}{8}$	$1\frac{7}{16}$	$1\frac{1}{2}$
M	$1\frac{7}{8}$	2	$2\frac{3}{16}$	$2\frac{1}{4}$
N	$1\frac{3}{16}$	$1\frac{1}{4}$	$1\frac{5}{16}$	$1\frac{3}{8}$
O	3	$3\frac{3}{8}$	$3\frac{13}{16}$	$4\frac{3}{16}$
P	$2\frac{1}{16}$	$2\frac{3}{8}$	$2\frac{11}{16}$	3
Q	$3\frac{1}{2}$	4	$4\frac{1}{2}$	5
R	$1\frac{5}{16}$	$1\frac{1}{2}$	$1\frac{11}{16}$	$1\frac{13}{16}$
S	$\frac{3}{4}$	$\frac{3}{4}$	$\frac{3}{8}$	$\frac{7}{16}$
T	$\frac{3}{8}$	$\frac{3}{8}$	$\frac{7}{16}$	$\frac{7}{16}$
T_1	$\frac{5}{16}$	$\frac{5}{16}$	$\frac{3}{8}$	$\frac{3}{8}$
T_2	$\frac{7}{16}$	$\frac{7}{16}$	$\frac{1}{2}$	$\frac{1}{2}$
U	$\frac{1}{2}$	$\frac{5}{8}$	$\frac{5}{8}$	$\frac{5}{8}$
V	$\frac{1}{2}$	$\frac{1}{2}$	$\frac{1}{2}$	$\frac{5}{8}$
W	$3\frac{3}{8}$	$3\frac{3}{4}$	$3\frac{7}{8}$	$4\frac{1}{8}$
X	$2\frac{1}{4}$	$2\frac{9}{16}$	$2\frac{7}{8}$	$3\frac{1}{8}$

FIG. 602.

283. Zeuner's Slide Valve Diagram.—Fig. 603 shows a simple slide valve, the valve being in its middle position. The circle ACBD, Fig. 604, represents the path of the centre of the sheave of the eccentric for driving this valve. When the valve is in its central position and moving towards the left, the eccentric will be in the position OC, and will be moving in the direction of the arrow. Suppose the eccentric to move into the position OP. Draw PM perpendicular to the horizontal line AOB. Then, if the effect of the obliquity of the eccentric rod be neglected, OM will be the distance which the valve has moved from its central position. Make OQ equal to OM. If this construction be per-

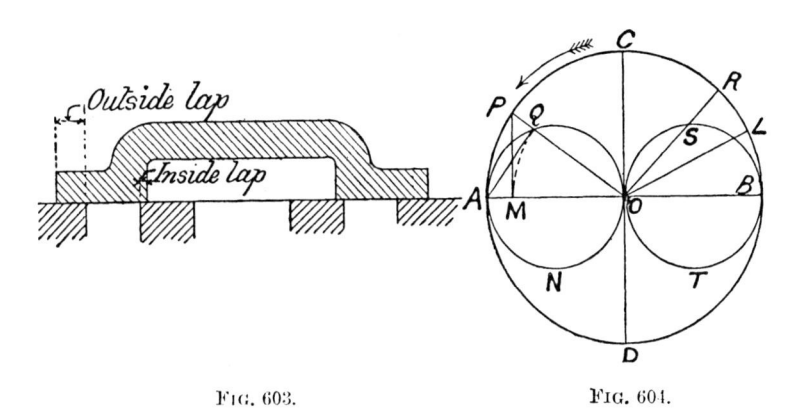

Fig. 603.　　　　Fig. 604.

formed for a number of positions of the eccentric as it makes a complete revolution, and if the points like Q obtained be joined by a fair curve, it will be found that this curve is the two circles described on OA and OB as diameters. By means of these valve circles the position of the valve can be found at once for any given position of the eccentric. For instance, if OR is the position of the eccentric, then the valve will be at a distance OS from its central position.

It is easy to prove that the locus of Q is the circles OQAN, and OSBT, for, comparing the triangles AQO and PMO, it will be seen that they are equiangular, the angle AQO being equal to the angle PMO. But the angle PMO is a right angle, therefore the angle AQO is a right angle, and therefore the locus of Q, while P moves through the arc CAD is the circle OQAN. In like manner the locus of Q while P moves through the arc DBC is the circle OSBT.

To find the position of the crank of the engine for any given position OP of the eccentric, make the angle POL equal to the angle between the eccentric and the crank (equal to 90° plus the angle of advance), then OL is the position of the crank when the eccentric is in the position OP.

If the length OQ, equal to OM, be marked off on OL instead of on OP, and the operation be repeated for a number of positions of the crank as it makes one revolution, it will be found that the locus of Q will be two circles described on OE and OF (Fig. 605) as diameters, EOF being a straight line making the angle COE equal to the angle of advance of the eccentric.

From this diagram, Fig. 605, the position of the valve for any given position of the crank can be found at once. For instance, if the crank is in the position OL, then the valve is at a distance OQ from its central position. It follows that if we can find from this diagram the position of the valve for any position of the crank, we can also find the position of the crank for any position of the valve. For example, when the

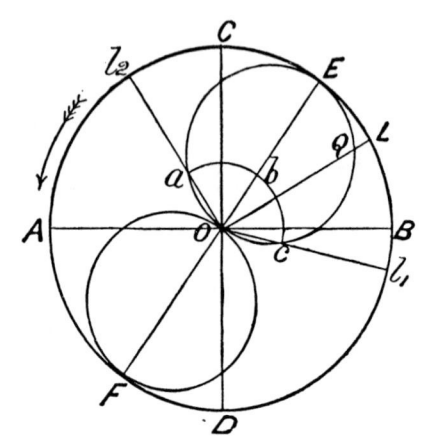

FIG. 605.

valve is just on the point of opening or just on the point of closing, it will be at a distance from its middle position equal to the lap. Hence if the arc abc be described with centre O, and radius equal to the lap, Ocl_1 will be the position of the crank when steam is admitted, and Oal_2 will be its position when the steam is cut off. Knowing the positions of the crank when steam is admitted and cut off, the corresponding positions of the piston can easily be determined.

Fig. 606 shows the complete valve diagram for a simple slide valve and the probable form of the indicator diagram; the length of the connecting-rod is supposed to be infinite.

OA = half travel of valve = radius of eccentric; Ob = outside lap; bd = lead; angle COE = angle of advance of eccentric; Og = inside lap; $km = gh$ = width of steam port.

Ocl_1 position of crank when steam is admitted.
Oal_2 ,, ,, ,, cut off.
Ofl_3 ,, ,, ,, released.
Oel_4 ,, ,, ,, compressed.

Port is full open for steam while the crank moves through the angle qOr.

Port is full open for exhaust while the crank moves through the angle tOs.

VALVES.

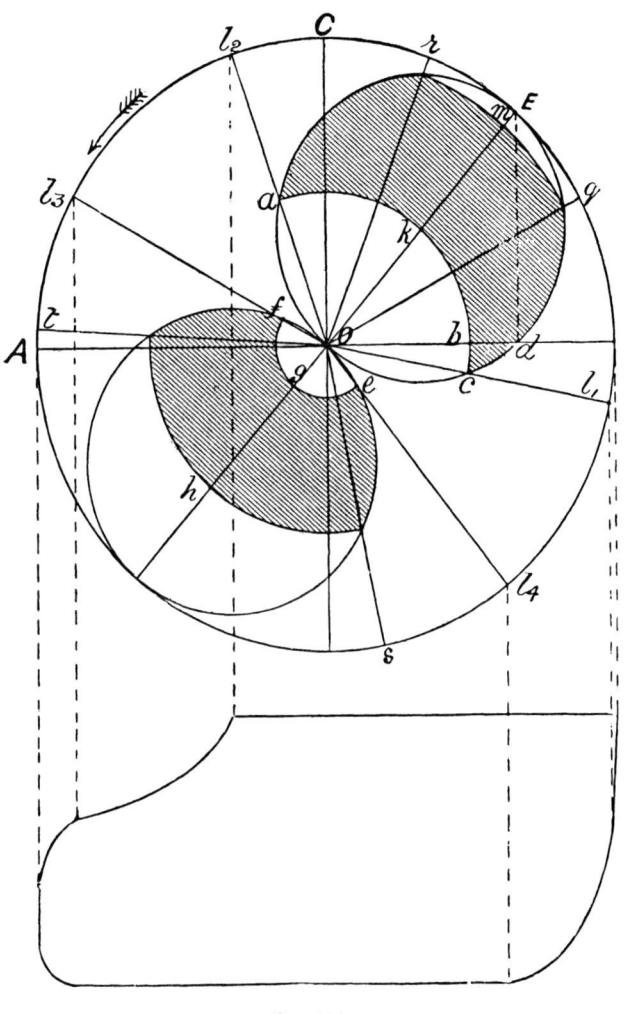

FIG. 606.

CHAPTER XXI.

RIVETED JOINTS.

284. Forms and Proportions of Rivet Heads.—The most common form of rivet head is the cup or spherical head shown in Fig. 607. The upper part of Fig. 608 shows the conical head. This is the form produced by hand-hammering. When it is desired to form a cup head by hand-hammering, it must be finished with a die or snap and a heavy hammer

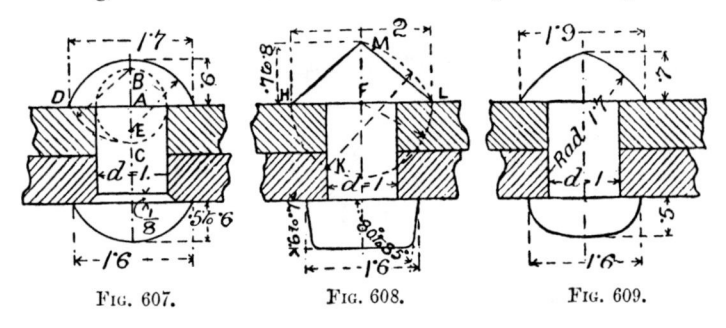

FIG. 607.   FIG. 608.   FIG. 609.

The lower part of Fig. 608 shows a pan head. Fig. 609 shows modified cup heads, the upper one being conoidal and the lower one ellipsoidal.

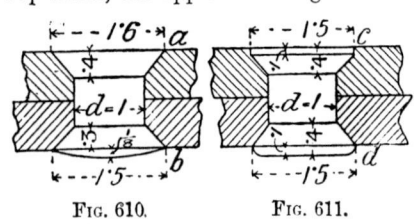

FIG. 610.   FIG. 611.

Countersunk heads are shown in Figs. 610 and 611. There is a slight objection to the sharp edge of the countersunk head shown at a, as it is liable to spring away from the plate. This objection is overcome in the designs shown at b, c, and d. The usual proportions for rivet heads are marked on the above Figs., the diameter of the rivet being taken as the unit in each case. If the rivet hole be slightly countersunk, as shown in the lower part of Fig. 607, the strength of the rivet head is increased, and it may therefore be reduced slightly in size.

The upper parts of Figs. 607 and 608 show geometrical constructions for spherical and conical heads, which give nearly the same proportions as those marked on the same figures.

For the spherical head the construction is, with centre A and radius equal to half diameter of rivet, describe a circle, cutting the centre line of the rivet at B and C. With centre B and radius BC, describe the arc

CD. Make BE equal to AD. E is the centre, and ED the radius for the curved outline of the head.

For the conical head the construction is, with centre F, and radius equal to diameter of rivet, describe the semicircle HKL. With centre K and radius KL, describe the arc LM to cut the centre line of the rivet at M, join HM and LM.

The length of rivet required to form a rivet head is about $1\frac{1}{4}d$ for conical and snap heads, where d is the diameter of the rivet. For countersunk heads the allowance is from $\frac{3}{4}d$ to d.

As the rivet expands when heated, its diameter when cold should be less than the diameter of the hole. In practice, the diameter of the rivet is generally one-sixteenth of an inch less than the diameter of the hole for rivets $\frac{3}{4}$ inch in diameter and upwards.

In calculating the strength of a riveted joint, the diameter of the rivet is assumed to be the same as that of the hole, which it ought to be if the riveting is well done.

285. Rivet Holes.—Formerly the practice was to punch all the rivet holes in plates, but now, for boiler-work, they are generally made by drilling. Punching has an injurious effect upon plates, especially when they are made of an inferior quality of wrought-iron, or when they are made of steel, unless the steel is very mild. It is found that the injury due to punching, provided the plates have not been cracked by the process, is removed by afterwards annealing them. It is also found that if a punched hole be enlarged by rymering or drilling to the extent of $\frac{1}{16}$ inch on the diameter, the injury due to punching is removed. It is generally cheaper to punch holes than to drill them; but in a well equipped workshop, having multiple drilling machines, the difference between the cost of drilling and punching is not great.

It is more difficult to ensure the correct spacing of the holes when they are made by punching. A punched hole is always tapered, because the hole in the bolster or die is necessarily a little larger than the punch. The widest end of the punched hole is that next to the bolster. This taper in a punched hole has a certain advantage, because, when two plates with punched holes are put together, with the narrowest ends of the holes adjacent, as shown in Fig. 612, the force binding the plates together which the rivet exerts in contracting does not all act through the heads. In fact, if the heads get broken or worn off, the rivet will still hold the plates together.

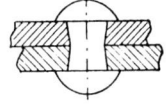

FIG. 612.

In the best boiler-work the rivet holes are drilled after the plates have been bent or flanged and put together in their proper places. This ensures that the corresponding holes in the different plates shall be exactly opposite to one another. After drilling, the plates are taken asunder, and any burr which has been formed at the edges of the holes is removed. It has been found that the shearing strength of the rivets is increased if the sharp edges of the holes between the plates are taken off by countersinking *very slightly*. If the holes are countersunk too much on the inside between the plates, the rivets when riveted up will spread out into the recess formed, and tend to separate the plates.

286. Riveting.—Riveting may be performed either by hand-hammering or by a machine. Hydraulic riveting machines are the best, because the steady pressure produced on the end of the rivet staves it up better, and causes it to fill the hole more completely than is likely to be the case with steam or hand riveting, where the action is percussive.

287. Caulking and Fullering.—However well a joint may be riveted up, it will seldom be perfectly steam-tight. To make the joint steam-tight, the edges of the plates and the heads of the rivets are burred down with a caulking tool, as shown in Fig. 613. The caulking tool resembles

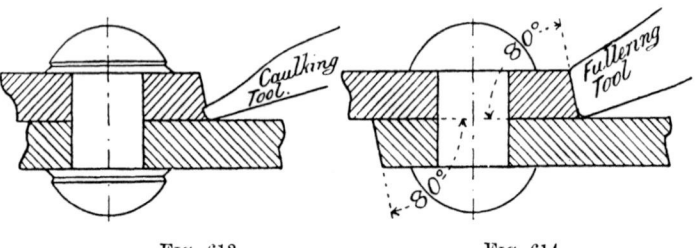

FIG. 613.   FIG. 614.

a chisel except that the point or caulking end is flat. The caulking tool is forced into the plate by hand-hammering. Unless this caulking is done with care, the joint may be seriously injured, because the tool may indent the plates unnecessarily, and may also open the joint instead of closing it. A better method of making the joint steam-tight is to use a tool having a thickness at the point equal to the thickness of the plates, as shown in Fig. 614. This is known as fullering. To facilitate the caulking or fullering of a riveted joint, the edges of the plates are generally planed to a slight bevel, as shown in Fig. 614.

288. Strength of a Riveted Joint.—A riveted joint may give way

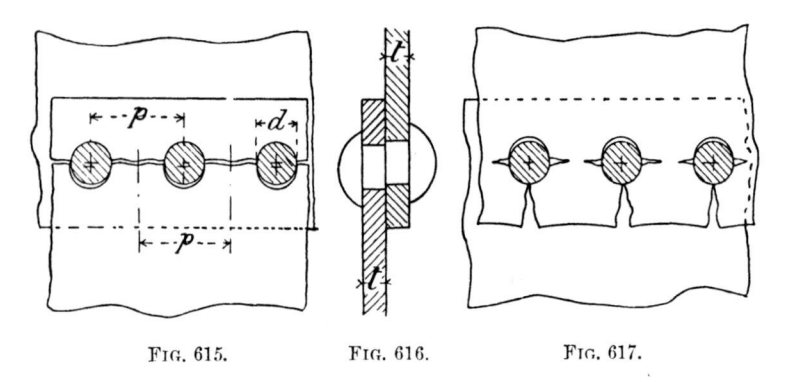

FIG. 615.   FIG. 616.   FIG. 617.

(1) by the tearing of the plates between the rivets, as shown in Fig. 615;
(2) by the shearing of the rivets, as shown in Fig. 616; (3) by the

breaking of the plate between its edge and the rivet holes, as shown in Fig. 617; and (4) by the crushing of the rivet or the plate.

Let p = pitch of rivets.
d = diameter of rivets.
t = thickness of plates.
f_t = tensile stress, or resistance of plates to tension.
f_s = shearing stress, or resistance of rivets to shearing.
f_c = crushing stress, or resistance of rivets or plates to crushing.

Consider a width of the joint equal to the pitch of the rivets.

The resistance of this portion to tearing is = $(p - d) t f_t$.
,, ,, shearing is = $\cdot 7854 d^2 f_s$.
,, ,, crushing is = $d t f_c$.

If the resistance to tearing be equal to the resistance to shearing, then $(p - d) t f_t = \cdot 7854 d^2 f_s$, and $p = \dfrac{\cdot 7854 d^2 f_s}{t f_t} + d$. If in a width of the joint equal to p there are n rivets in single shear, then $(p - d) t f_t = \cdot 7854 d^2 n f_s$, and $p = \dfrac{\cdot 7854 d^2 n f_s}{t f_t} + d$. If the rivets are in double shear, then $p = \dfrac{\cdot 7854 d^2 \times 2 n f_s}{t f_t} + d$.

If the resistance to shearing be equal to the resistance to crushing, then, $\cdot 7854 d^2 f_s = d t f_c$, and $d = \dfrac{t f_c}{\cdot 7854 f_s}$. Since f_c is about double f_s, this gives $d = 2 \cdot 54 t$. If the rivet is in double shear, then for equal shearing and crushing resistances, $d = 1 \cdot 27 t$. It follows that the crushing resistance need not be considered when, for lap joints, d is less than $2 \cdot 54 t$, and, for butt joints with double cover straps, when d is less than $1 \cdot 27 t$.

To resist the cross-breaking of the plates in front of the rivets (Fig. 617), it has been proved by experiment that if l, the distance of the centre of the rivet from the edge of the plate, is equal to $1\frac{1}{2}d$, this is sufficient, and this is the rule which is followed in practice.

289. Tensile Strength of Boiler Plates.—The following table gives the tenacity of plates used in the construction of steam boilers. In the case of wrought-iron the tenacity given is for plates in the direction of the fibre. The tenacity of wrought-iron plates across the fibre, that is, at right angles to the direction of rolling, is from 85 to 95 per cent. of the tenacity in the direction of the fibre.

Material.	Tenacity in Pounds per Square Inch.
Wrought-iron plates	43,000 to 58,000
Steel plates.	58,000 to 68,000
Copper plates	30,000

The Board of Trade allows for wrought-iron plates a tenacity of 47,000 lbs. per square inch along the fibre, and 40,000 lbs. per square inch across the fibre. For steel plates the Board of Trade allows a tenacity of 27 tons to 32 tons per square inch.

290. Strength of Rivets.—The shearing resistance of wrought-iron rivets is about equal to the tenacity of wrought-iron plates. The shearing resistance of steel rivets is about $\frac{8}{10}$ths of the tenacity of steel plates. If f_s = shearing resistance of rivet, and f_t = tenacity of plates, then—

$$\frac{f_s}{f_t} = 1 \text{ for iron plates and iron rivets,}$$

$$= \cdot 8 \text{ for steel plates and steel rivets.}$$

The Board of Trade rules are—

$$\frac{f_s}{f_t} = 1 \text{ for iron plates and iron rivets,}$$

$$\frac{f_s}{f_t} = \frac{23}{28} \text{ for steel plates and steel rivets.}$$

The experiments of Professor Kennedy have proved that the shearing resistance, in pounds per square inch of section sheared, is the same whether the rivet be in single or double shear. Hence the force required to shear a rivet in double shear, as in a butt joint with double cover straps, is twice that required to shear the same rivet in single shear, as in a lap joint. The Board of Trade, however, only allow a load on a rivet in double shear equal to 1·75 times the load allowed on a rivet in single shear.

In the rules and tables which are given in this chapter it is assumed that $\frac{f_s}{f_t} = 1$ for iron rivets and iron plates, and $\frac{f_s}{f_t} = \cdot 8$ for steel plates and steel rivets. It is also assumed that the shearing resistance per square inch of total section to be sheared is the same for rivets in double shear as for rivets in single shear.

291. Efficiency of a Riveted Joint.—Let T equal the resistance which a riveted joint of given width offers to tearing, S the resistance which the rivets of the same joint offer to shearing, and let R denote the resistance to tearing of the solid plate of the same width as the joint. Then $\frac{T}{R}$ or $\frac{S}{R}$, whichever is least, is the ratio of the strength of the joint to that of the solid plate. This ratio is called the efficiency of the joint. It is usual to express the efficiency of the joint as a percentage.

$$\text{Then, efficiency per cent.} = \frac{100T}{R} = R_t$$

$$\text{or, efficiency per cent.} = \frac{100S}{R} = R_s.$$

It is assumed that the resistance to crushing is not less than T or S.

RIVETED JOINTS. 273

292. Single Riveted Lap Joints.—Fig. 618 shows the connection of two plates by means of a single riveted lap joint. When the plates are arranged as shown in the lower section, the tension in the plates causes a bending action on them at the lap. To avoid this the plates have sometimes a set at the lap as shown in the upper section.

For this form of joint $R_t = \dfrac{100(p-d)}{p}$,

and $R_s = \dfrac{100 \times \cdot 7854 d^2 f_s}{p t f_t}$.

If $R_t = R_s$, then $p = \dfrac{\cdot 7854 d^2 f_s}{t f_t} + d$.

For joints which have to be made steam-tight the pitch of the rivets is generally less than that given by the above formula.

The following table gives proportions for single riveted lap joints suitable for boiler work.

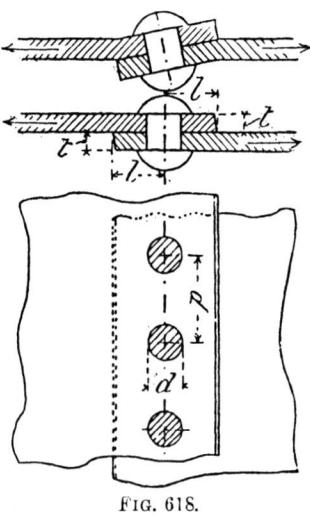

FIG. 618.

Dimensions of Single Riveted Lap Joints.

t.	Iron Plates and Iron Rivets.				Steel Plates and Steel Rivets.			
	d.	p.	R_t.	R_s.	d.	p.	R_t.	R_s.
$\tfrac{5}{16}$	$\tfrac{5}{8}$	$1\tfrac{1}{2}$	58·3	65·5	$\tfrac{5}{8}$	$1\tfrac{7}{16}$	56·5	54·6
	$\tfrac{3}{8}$	$\tfrac{3}{4}$	$1\tfrac{3}{4}$	57·1	67·3	$\tfrac{3}{4}$	$1\tfrac{11}{16}$	55 6	55·9
$\tfrac{7}{16}$	$\tfrac{13}{16}$	$1\tfrac{7}{8}$	56·7	63·2	$\tfrac{7}{8}$	2	56·2	55·0
	$\tfrac{1}{2}$	$\tfrac{7}{8}$	2	56·2	60·1		$1\tfrac{15}{16}$	$2\tfrac{1}{16}$	54·5	53·6
$\tfrac{9}{16}$	$1\tfrac{5}{16}$	$2\tfrac{1}{8}$	55·9	57·7	1		$2\tfrac{1}{8}$	52·9	52·6
	$\tfrac{5}{8}$	1		$2\tfrac{1}{4}$	55·6	55·8	$1\tfrac{1}{16}$	$2\tfrac{1}{4}$	52·8	50·4
$\tfrac{11}{16}$	$1\tfrac{1}{16}$	$2\tfrac{3}{8}$	55·3	54·3	$1\tfrac{1}{8}$	$2\tfrac{3}{8}$	52·6	48·7

In each case $l = 1\tfrac{1}{2}d$.

293. Double Riveted Lap Joints.—Figs. 619 and 620 show double riveted lap joints.

The arrangement of the rivets in Fig. 619 is known as zigzag riveting, while that shown in Fig. 620 is known as chain riveting.

For the above forms of joints, $R_t = \dfrac{100(p-d)}{p}$

and $R_s = \dfrac{100 \times 2 \times \cdot 7854 d^2 f_s}{p t f_t}$.

If $R_t = R_s$, then $p = \dfrac{2 \times \cdot 7854 d^2 f_s}{t f_t} + d$.

s

In all riveted joints the distance between adjacent rivets, measured from centre to centre, whether in the same or different rows, should not

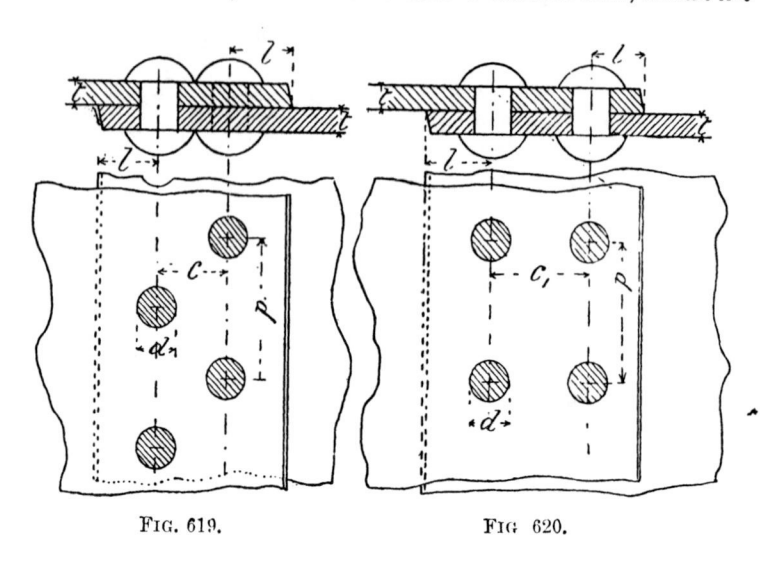

FIG. 619.   FIG. 620.

be less than $2d$. Hence, for zigzag riveting the distance c between the rows of rivets should not be less than $\dfrac{\sqrt{16d^2 - p^2}}{2}$, and for chain riveting the distance c_1 between the rows of rivets should not be less than $2d$.

The Board of Trade rules for c and c_1 are as follows:—

$$c = \dfrac{\sqrt{(11p + 4d)(p + 4d)}}{10}. \qquad c_1 = 2d \text{ to } 2d + \tfrac{1}{2}.$$

On the results of experiments on riveted joints Professor Kennedy has stated that the net section of metal in the plate, measured zigzag, should be from 30 to 35 per cent. in excess of that measured straight across. This gives a diagonal pitch of $\dfrac{2p + d}{3}$. The value of c corresponding to this diagonal pitch is $\dfrac{\sqrt{(7p + 2d)(p + 2d)}}{6}$.

The following table gives the proportions of double riveted lap joints for plates from $\tfrac{3}{8}$ inch to 1 inch thick. The diameters of the rivets are such as will be found in average practice, and the pitch in each case has been calculated so as to make R_t about equal to R_s.

Dimensions of Double Riveted Lap Joints.

t.		Iron Plates and Iron Rivets.					Steel Plates and Steel Rivets.				
		d.	p.	c.	c_1.	R_t.	R_s.	d.	p.	c.	c_1.	R_t.	R_s.
$\tfrac{3}{8}$		$1\tfrac{1}{16}$	$2\tfrac{5}{8}$	$1\tfrac{5}{16}$	$1\tfrac{7}{8}$	73·8	75·4	$\tfrac{3}{4}$	$2\tfrac{5}{8}$	$1\tfrac{3}{8}$	2	71·4	71·8
	$\tfrac{7}{16}$		$2\tfrac{3}{4}$	$1\tfrac{3}{8}$	2	72·7	73·4	$1\tfrac{3}{16}$	$2\tfrac{3}{4}$	$1\tfrac{7}{16}$	$2\tfrac{1}{8}$	70·5	69·0
$\tfrac{1}{2}$		$1\tfrac{3}{16}$	$2\tfrac{7}{8}$	$1\tfrac{1}{2}$	$2\tfrac{1}{8}$	71·7	72·1	$\tfrac{7}{8}$	$2\tfrac{13}{16}$	$1\tfrac{1}{2}$	$2\tfrac{1}{4}$	68·9	68·4
	$\tfrac{9}{16}$	$\tfrac{7}{8}$	3	$1\tfrac{9}{16}$	$2\tfrac{1}{4}$	70·8	71·3	$\tfrac{15}{16}$	$2\tfrac{7}{8}$	$1\tfrac{9}{16}$	$2\tfrac{3}{8}$	67·4	68·3
$\tfrac{5}{8}$		$1\tfrac{5}{16}$	$3\tfrac{1}{8}$	$1\tfrac{5}{8}$	$2\tfrac{3}{8}$	70·0	70·7	1	3	$1\tfrac{3}{8}$	$2\tfrac{1}{2}$	66·7	67·0
	$\tfrac{11}{16}$	1	$3\tfrac{1}{4}$	$1\tfrac{3}{4}$	$2\tfrac{1}{2}$	69·2	70·3	$1\tfrac{1}{16}$	$3\tfrac{1}{4}$	$1\tfrac{11}{16}$	$2\tfrac{5}{8}$	66·0	66·0
$\tfrac{3}{4}$		$1\tfrac{1}{16}$	$3\tfrac{7}{16}$	$1\tfrac{13}{16}$	$2\tfrac{5}{8}$	69·1	68·8	$1\tfrac{1}{8}$	$3\tfrac{1}{4}$	$1\tfrac{3}{4}$	$2\tfrac{3}{4}$	65·4	65·2
	$\tfrac{13}{16}$	$1\tfrac{1}{8}$	$3\tfrac{9}{16}$	$1\tfrac{7}{8}$	$2\tfrac{3}{4}$	68·4	68·7	$1\tfrac{3}{16}$	$3\tfrac{3}{8}$	$1\tfrac{13}{16}$	$2\tfrac{7}{8}$	64·8	64·6
$\tfrac{7}{8}$		$1\tfrac{3}{16}$	$3\tfrac{11}{16}$	2	$2\tfrac{7}{8}$	68·3	67·5	$1\tfrac{1}{4}$	$3\tfrac{1}{2}$	$1\tfrac{15}{16}$	3	64·3	64·1
	$\tfrac{15}{16}$	$1\tfrac{1}{4}$	$3\tfrac{7}{8}$	$2\tfrac{1}{16}$	3	67·7	67·6	$1\tfrac{5}{16}$	$3\tfrac{5}{8}$	2	$3\tfrac{1}{8}$	63·8	63·7
1		$1\tfrac{5}{16}$	4	$2\tfrac{1}{16}$	$3\tfrac{1}{8}$	67·2	67·6	$1\tfrac{3}{8}$	$3\tfrac{3}{4}$	$2\tfrac{1}{8}$	$3\tfrac{1}{4}$	63·3	63·4

In each case $l = 1\tfrac{1}{2}d$.

294. Treble Riveted Lap Joints.—Figs. 621 and 622 show treble riveted lap joints. In Fig. 621 the riveting is zigzag, and in Fig. 622 chain.

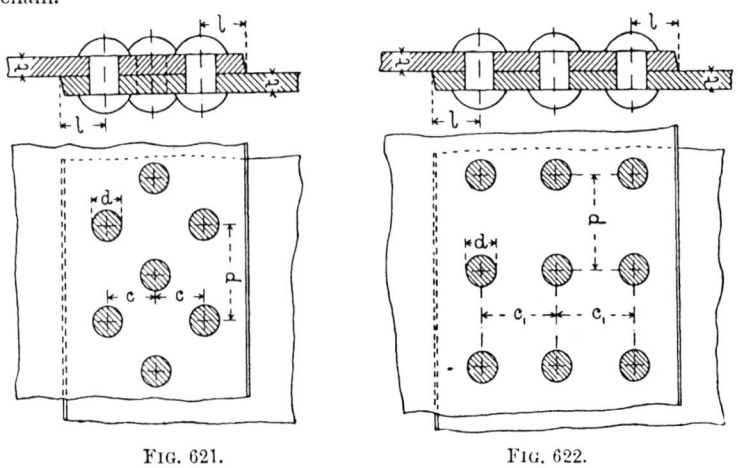

FIG. 621. FIG. 622.

For the above forms of joints, $R_t = \dfrac{100(p-d)}{p}$

and $R_s = \dfrac{100 \times 3 \times \cdot7854 d^2 f_s}{p t f_t}$

If $R_t = R_s$, then $p = \dfrac{3 \times \cdot7854 d^2 f_s}{t f_t} + d$.

The rules for the distances c and c_1 between the rows of rivets are the same as for double riveted lap joints.

Dimensions of Treble Riveted Lap Joints.

		Iron Plates and Iron Rivets.					Steel Plates and Steel Rivets.				
	d.	p.	c.	c_1.	R_t.	R_s.	d.	p.	c.	c_1.	R_t.	R_s.
$\tfrac{5}{8}$	$\tfrac{13}{16}$	$3\tfrac{1}{4}$	$1\tfrac{5}{8}$	$2\tfrac{1}{8}$	75·0	76·6	$\tfrac{7}{8}$	$3\tfrac{1}{4}$	$1\tfrac{5}{8}$	$2\tfrac{1}{4}$	72·0	73·9
$\tfrac{11}{16}$	$\tfrac{7}{8}$	$3\tfrac{1}{2}$	$1\tfrac{3}{4}$	$2\tfrac{1}{4}$	75·0	75·0	$\tfrac{15}{16}$	$3\tfrac{3}{8}$	$1\tfrac{3}{4}$	$2\tfrac{3}{8}$	72·2	71·4
$\tfrac{3}{4}$	$\tfrac{15}{16}$	$3\tfrac{11}{16}$	$1\tfrac{7}{8}$	$2\tfrac{3}{8}$	74·6	74·9	1	$3\tfrac{1}{2}$	$1\tfrac{13}{16}$	$2\tfrac{1}{2}$	71·4	71·8
$\tfrac{13}{16}$	1	$3\tfrac{7}{8}$	$1\tfrac{15}{16}$	$2\tfrac{1}{2}$	74·2	74·8	$1\tfrac{1}{16}$	$3\tfrac{11}{16}$	$1\tfrac{15}{16}$	$2\tfrac{5}{8}$	71·2	71·0
$\tfrac{7}{8}$	$1\tfrac{1}{16}$	$4\tfrac{1}{8}$	$2\tfrac{1}{16}$	$2\tfrac{5}{8}$	74·2	73·7	$1\tfrac{1}{8}$	$3\tfrac{7}{8}$	2	$2\tfrac{3}{4}$	71·0	70·4
$\tfrac{15}{16}$	$1\tfrac{1}{8}$	$4\tfrac{5}{16}$	$2\tfrac{3}{16}$	$2\tfrac{3}{4}$	73·9	73·8	$1\tfrac{3}{16}$	4	$2\tfrac{1}{16}$	$2\tfrac{7}{8}$	70·3	70·9
1	$1\tfrac{3}{16}$	$4\tfrac{1}{2}$	$2\tfrac{1}{4}$	$2\tfrac{7}{8}$	73·6	73·8	$1\tfrac{1}{4}$	$4\tfrac{3}{16}$	$2\tfrac{3}{16}$	3	70·1	70·3
$1\tfrac{1}{16}$	$1\tfrac{1}{4}$	$4\tfrac{11}{16}$	$2\tfrac{3}{8}$	3	73·3	73·9	$1\tfrac{5}{16}$	$4\tfrac{3}{8}$	$2\tfrac{1}{4}$	$3\tfrac{1}{8}$	70·0	69·9
$1\tfrac{1}{8}$	$1\tfrac{5}{16}$	$4\tfrac{7}{8}$	$2\tfrac{1}{2}$	$3\tfrac{1}{8}$	73·1	74·0	$1\tfrac{3}{8}$	$4\tfrac{1}{2}$	$2\tfrac{3}{8}$	$3\tfrac{1}{4}$	69·4	70·4

In each case $l = 1\tfrac{1}{2}d$.

The strength of a treble riveted lap joint may be increased by making the pitch of the outer rows of rivets double the pitch of the inner row as shown in Figs. 623 and 624.

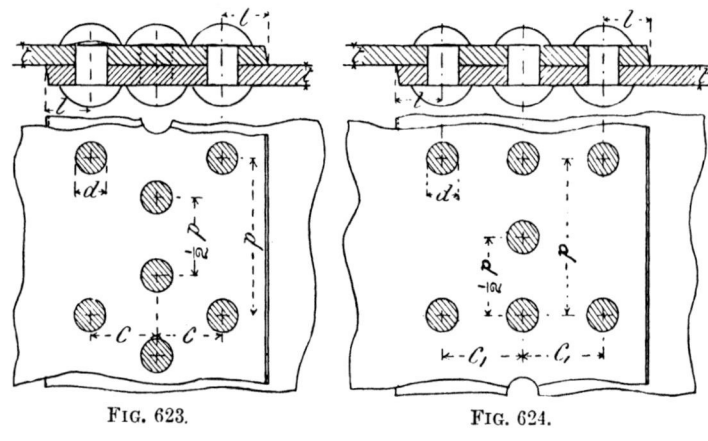

FIG. 623.   FIG. 624.

With this arrangement of the rivets, $R_t = \dfrac{100(p-d)}{p}$

and $R_s = \dfrac{100 \times 4 \times \cdot 7854 d^2 f_s}{p t f_t}$.

If $R_t = R_s$, then $p = \dfrac{4 \times \cdot 7854 d^2 f_s}{t f_t} + d$.

This form of joint might give way by the shearing of the rivets in the outer rows and the tearing of the plates between the rivets of the inner row. Let the strength of the joint to resist this combined shearing and tearing, expressed as a percentage of the strength of the solid plate, be denoted by R. Then,

$$R = \dfrac{100\{\cdot 7854 d^2 f_s + (p - 2d) t f_t\}}{p t f_t}.$$

If $R = R_t = R_s$, then $p = \dfrac{4 \times \cdot 7854 d^2 f_s}{t f_t} + d$, as before;

and $d = \dfrac{t f_t}{\cdot 7854 f_s} = 1\cdot 27 t$ for iron plates and iron rivets,

$\qquad\qquad = 1\cdot 59 t$ for steel plates and steel rivets.

If the diameter of the rivets is greater than that just given, the strength of the joint must be taken as R_t or R_s; but if the diameter is less than that just given, then the strength of the joint must be taken as R.

The Board of Trade rules for c and c_1 are as follows:—
$$c = \sqrt{(\tfrac{11}{20}p + d)(\tfrac{1}{20}p + d)}. \quad c_1 = 2d \text{ to } 2d + \tfrac{1}{2}.$$

295. Quadruple Riveted Lap Joints.—Using the same notation as before, we have for a quadruple riveted lap joint,
$$R_t = \dfrac{100(p-d)}{p}, \text{ and } R_s = \dfrac{100 \times 4 \times \cdot 7854 d^2 f_s}{p t f_t}.$$

If $R_t = R_s$, then $p = \dfrac{4 \times \cdot 7854 d^2 f_s}{t f_t} + d$.

296. Butt Joints with Single Cover Straps.—When a butt joint has a cover strap on one side only, as shown in Fig. 625, it is really composed of two lap joints, and it may be proportioned by the rules already given for lap joints. It is evident that with this form of joint the tension on the plates will tend to bend the cover strap. For that reason the cover strap is made thicker than the plates.

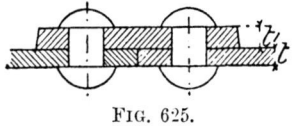

FIG. 625.

If $t_1 =$ thickness of cover strap, and $t =$ thickness of plates, then $t_1 = 1\tfrac{1}{8} t$.

297. Single Riveted Butt Joints with Double Cover Straps.—A butt joint, single riveted, with double butt straps, is shown in Fig. 626. The usual rule for the thickness of each butt strap is $t_1 = \tfrac{5}{8} t$.

$$R_t = \dfrac{100(p-d)}{p}$$ and, the rivets being in double shear,

$$R_s = \dfrac{100 \times 2 \times \cdot 7854 d^2 f_s}{p t f_t}.$$

If $R_t = R_s$, then $p = \dfrac{2 \times \cdot 7854 d^2 f_s}{t f_t} + d$.

The diameter of the rivets for different thicknesses of plates may be as follows:—

$d = t + \tfrac{1}{4}$ for iron plates and iron rivets.
$d = t + \tfrac{5}{16}$ for steel plates and steel rivets.

298. Double Riveted Butt Joints with Double Cover Straps.—Fig. 627 shows a double riveted butt joint with double cover straps, the pitch of the rivets being the same in each row. If alternate rivets in the outer rows be removed, we get the arrangement shown in Fig. 628, which is stronger.

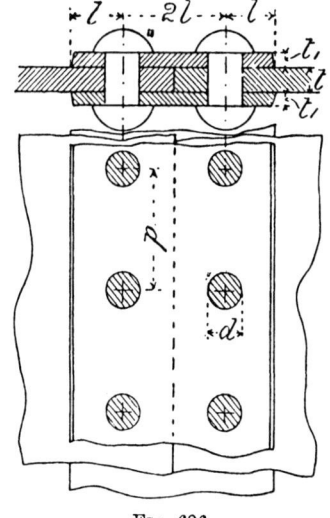

FIG. 626.

For Fig. 627, $R_t = \dfrac{100(p-d)}{p}$, and $R_s = \dfrac{100 \times 4 \times \cdot 7854 d^2 f_s}{p t f_t}$.

If $R_t = R_s$, then $p = \dfrac{4 \times \cdot 7854 d^2 f_s}{t f_t} + d$.

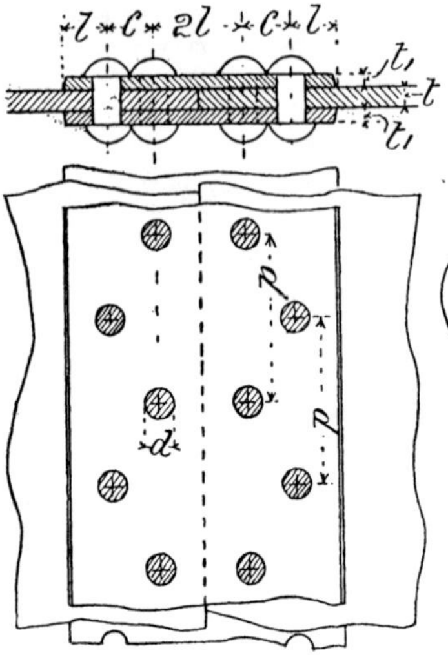

FIG. 627. 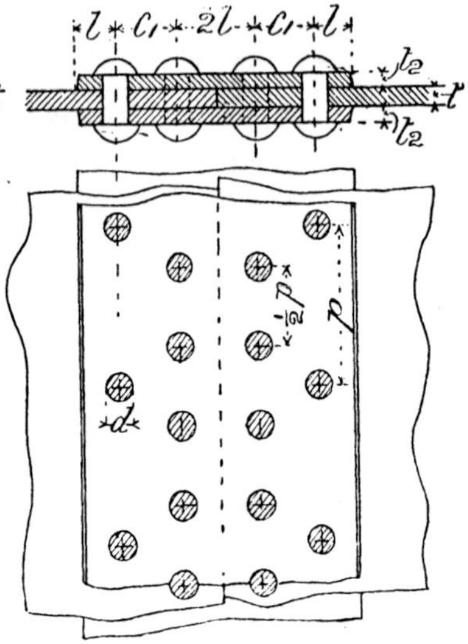   FIG. 628.

The diameter of the rivets (Fig. 627) for different thicknesses of plates may be as follows :—

$d = t + \tfrac{3}{16}$ for iron plates and iron rivets.
$d = t + \tfrac{1}{4}$ for steel plates and steel rivets.

$c = \dfrac{\sqrt{(11p + 4d)(p + 4d)}}{10}$, and $t_1 = \tfrac{5}{8} t$ (Board of Trade rules).

For Fig. 628, $R_t = \dfrac{100(p-d)}{p}$, and $R_s = \dfrac{100 \times 6 \times \cdot 7854 d^2 f_s}{p t f_t}$.

If $R_t = R_s$, then $p = \dfrac{6 \times \cdot 7854 d^2 f_s}{t f_t} + d$.

The diameter of the rivets (Fig. 628) for different thicknesses of plates may be as follows :—

$d = t + \tfrac{1}{8}$ for iron plates and iron rivets.
$d = t + \tfrac{3}{16}$ for steel plates and steel rivets.

For Fig. 627 the ratio of the section of the cover straps to the section of the plates to resist tearing is $\dfrac{2 t_1 (p - d)}{t (p - d)} = \dfrac{2 t_1}{t}$. For Fig. 628 the corresponding ratio is $\dfrac{2 t_2 (p - 2d)}{t (p - d)}$. Hence, in order that the ratio of the strength of the straps to the strength of the plates may be the same for both arrangements, we must have $\dfrac{2t_2(p-2d)}{t(p-d)} = \dfrac{2t_1}{t} = \dfrac{2 \times 5t}{8t} = \dfrac{5}{4}$, and, therefore, $t_2 = \dfrac{5t(p-d)}{8(p-2d)}$.

$$c_1 = \sqrt{(1\tfrac{1}{20}p+d)(\tfrac{1}{20}p+d)}.$$ (Board of Trade rule.)

The form of joint shown in Fig. 628 might give way by the shearing of the rivets in the outer rows, and the tearing of the plates between the rivets of the inner rows. Let the strength of the joint to resist this combined shearing and tearing, expressed as a percentage of the strength of the solid plate, be denoted by R'.

Then, $R' = \dfrac{100\{2 \times \cdot 7854 d^2 f_s + (p-2d)\, tf_t\}}{ptf_t}$

If $R' = R_t = R_s$, then $p = \dfrac{6 \times \cdot 7854 d^2 f_s}{tf_t} + d$, as before,

and $d = \dfrac{tf_t}{2 \times \cdot 7854 f_s} = \cdot 64t$ for iron plates and iron rivets.

$= \cdot 8t$ for steel plates and steel rivets.

As the diameter of the rivets is never less than the thickness of the plates, it follows that R' will always be greater than R_t or R_s, therefore the strength R' need not be considered in this form of joint.

299. Treble Riveted Butt Joints with Double Cover Straps.—Fig. 629 shows a treble riveted butt joint with double cover straps, the pitch of the rivets being the same in each row. For this form of joint,

$R_t = \dfrac{100(p-d)}{p}$, and,

$R_s = \dfrac{100 \times 6 \times \cdot 7854 d^2 f_s}{ptf_t}$.

If $R_t = R_s$, then

$p = \dfrac{6 \times \cdot 7854 d^2 f_s}{tf_t} + d.$

The diameter of the rivets for different thickness of plates may be as follows:—

$d = t + \tfrac{1}{16}$ for iron plates and iron rivets.

$d = t + \tfrac{1}{8}$ for steel plates and steel rivets.

As for double riveted butt joints with double cover straps, $c = \dfrac{\sqrt{(11p+4d)(p+4d)}}{10}$, and $t_1 = \tfrac{5}{8}t$; also, $l = 1\tfrac{1}{2}d$.

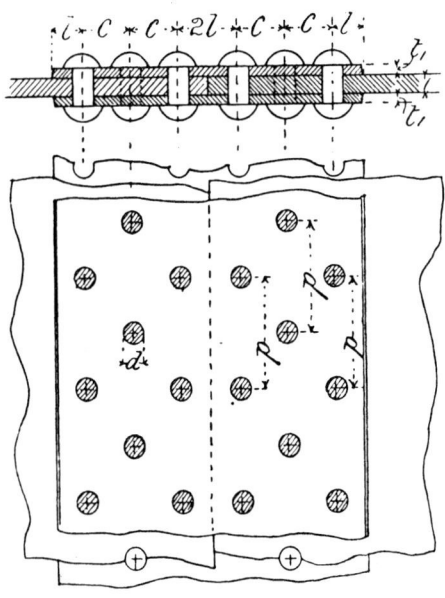

FIG. 629.

EXAMPLE.—*To find the dimensions and strength of a treble riveted butt joint with double cover straps for plates $\frac{3}{4}$ inch thick. Plates and rivets to be of steel.*

$d = t + \frac{1}{8} = \frac{3}{4} + \frac{1}{8} = \frac{7}{8}$ inch.

$$p = \frac{6 \times \cdot 7854 d^2}{t} \times \frac{f_s}{f_t} + d = \frac{6 \times \cdot 7854 \times (\frac{7}{8})^2}{\frac{3}{4}} \times \cdot 8 + \frac{7}{8}$$

$= 3 \cdot 848 + \cdot 875 = 4 \cdot 723$ inches, say $4\frac{11}{16}$ inches.

$$c = \frac{\sqrt{(11p + 4d)(p + 4d)}}{10} = \frac{\sqrt{(11 \times 4\frac{11}{16} + 4 \times \frac{7}{8})(4\frac{11}{16} + 4 \times \frac{7}{8})}}{10} = \frac{\sqrt{450 \cdot 82}}{10}$$

$= 2 \cdot 12$, say $2\frac{1}{8}$ inches.

$t_1 = \frac{5}{8}t = \frac{5}{8} \times \frac{3}{4} = \frac{15}{32}$ inch. $l = 1\frac{1}{2}d = 1\frac{1}{2} \times \frac{7}{8} = 1\frac{5}{16}$ inches.

$$R_t = \frac{100(p - d)}{p} = \frac{100(4\frac{11}{16} - \frac{7}{8})}{4\frac{11}{16}} = 81 \cdot 3 \text{ per cent.}$$

$$R_s = \frac{100 \times 6 \times \cdot 7854 d^2}{pt} \times \frac{f_s}{f_t} = \frac{100 \times 6 \times \cdot 7854 \times (\frac{7}{8})^2}{4\frac{11}{16} \times \frac{3}{4}} \times \cdot 8 = 82 \cdot 1 \text{ per cent.}$$

If, instead of assuming that the resistance of a rivet in double shear to shearing is twice its resistance when in single shear, we take the Board of Trade multiplier 1·75,

then $p = \dfrac{3 \times 1 \cdot 75 \times \cdot 7854 d^2}{t} \times \dfrac{f_s}{f_t} + d = 4\frac{1}{4}$ inches (very nearly),

$R_t = 79 \cdot 4$ per cent., and $R_s = 79 \cdot 2$ per cent.

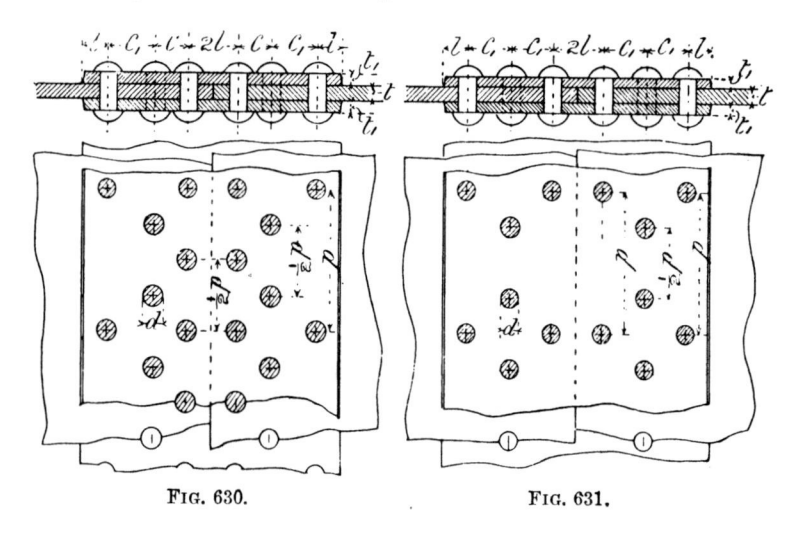

FIG. 630.   FIG. 631.

Fig. 630 shows a treble riveted butt joint with double cover straps, the pitch of the rivets in the outer rows being twice the pitch of the rivets in the other rows.

For this form of joint, $R_t = \dfrac{100\,(p-d)}{p}$, and $R_s = \dfrac{100 \times 10 \times \cdot 7854 d^2 f_s}{p t f_t}$.

If $R_t = R_s$, then $p = \dfrac{10 \times \cdot 7854 d^2 f_s}{t f_t} + d$.

If R' denote the resistance of this joint to shearing of the outer row of rivets and tearing of the plate between the rivets of the next row, then $R' = \dfrac{100\{2 \times \cdot 7854 d^2 f_s + (p - 2d) t f_t\}}{p t f_t}$.

As for the joint shown in Fig. 628, R' may be neglected when d is greater than $\cdot 64 t$ for iron plates and iron rivets or greater than $\cdot 8 t$ for steel plates and steel rivets.

The thickness of the cover straps for the joint in Fig. 630 is given by the same formula as for the cover straps for the joint shown in Fig. 628, namely, $t_1 = \dfrac{5t(p-d)}{8(p-2d)}$.

$c = \dfrac{\sqrt{(11p + 8d)(p + 8d)}}{20}$, and $c_1 = \sqrt{(\tfrac{11}{20}p + d)(\tfrac{1}{20}p + d)}$.

Fig. 631 shows a treble riveted butt joint with double cover straps, the pitch of the rivets in the outer and inner rows being twice the pitch of the rivets in the intermediate rows.

For this form of joint, $R_t = \dfrac{100\,(p-d)}{p}$, and $R_s = \dfrac{100 \times 8 \times \cdot 7854 d^2 f_s}{p t f_t}$.

If $R_t = R_s$, then $p = \dfrac{8 \times \cdot 7854 d^2 f_s}{t f_t} + d$.

$t_1 = \tfrac{5}{8} t$. $c_1 = \sqrt{(\tfrac{11}{20}p + d)(\tfrac{1}{20}p + d)}$.

300. Intersecting Riveted Joints.—In the shells of steam boilers there is a complication where the longitudinal seams meet the circumferential seams. Examples of the intersection of longitudinal and circumferential riveted joints are shown in the following figures. Figs. 632 and 633 show the intersection of two single riveted butt joints with single cover straps. In both cases the end of the strap A is tucked under the strap B. In the first arrangement, Fig. 632, the end of the longitudinal strap A is thinned down by forging, and the part of the circumferential strap B, under which the thinned end of A fits, is set up to make room for it. In the second case, Fig. 633, the end of the longitudinal strap is thinned down by planing, and a piece is planed off the under side of the circumferential strap. The second method is the one generally preferred for steel plates, especially when the plates are thick, because of the injury to which steel plates are liable when heated locally.

Fig. 634 shows the junction of three plates, the joints being single riveted lap joints. Fig. 635 shows the junction of three plates with lap joints, one being double riveted, the other single riveted.

The junction of four plates is shown in Fig. 636, the joints being single riveted lap joints.

The intersection of the longitudinal seams of a boiler shell with the circumferential seam is shown in Fig. 637. The longitudinal seams are double riveted butt joints with double cover straps, while the circumferential seam is a single riveted lap joint.

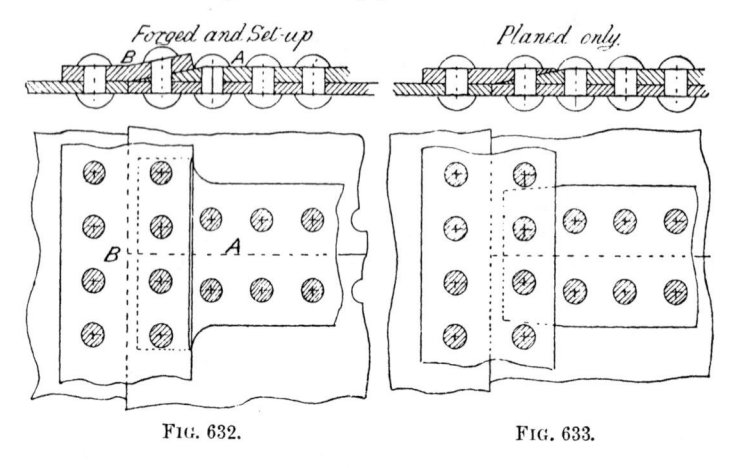

Forged and Set-up *Planed only.*

FIG. 632.   FIG. 633.

The connection of the end plate to the shell in a large marine boiler, at the point where the longitudinal seam comes in, is shown in Fig. 639. The longitudinal seam is a treble riveted butt joint with double cover

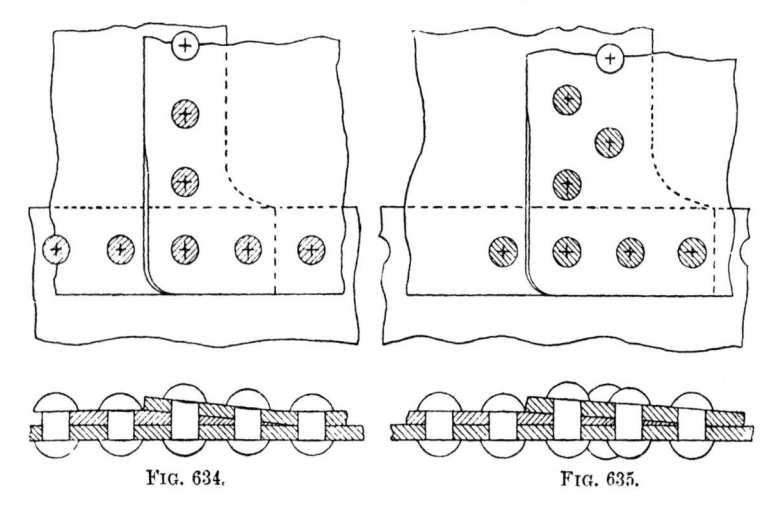

FIG. 634.   FIG. 635.

straps. The end plate is flanged and connected to the shell by a double riveted lap joint. The different rings of the shell are connected by treble riveted lap joints as shown in Fig. 638. All the plates and rivets in this example are of steel.

RIVETED JOINTS. 283

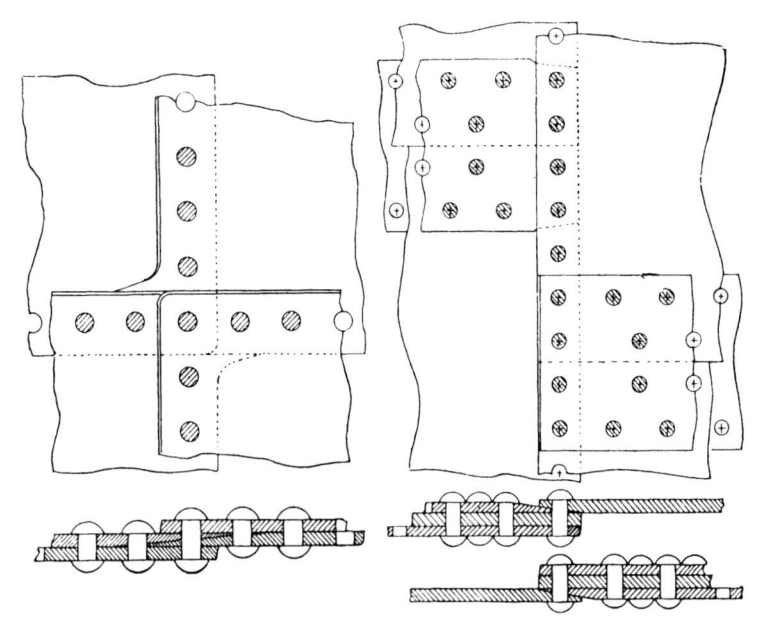

FIG. 636.

FIG. 637.

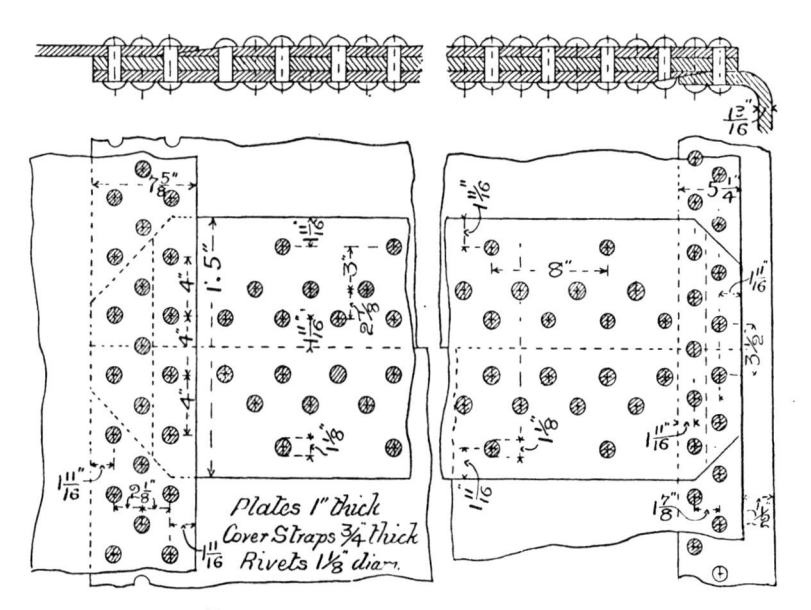

FIG. 638.

Plates 1" thick
Cover Straps ¾ thick
Rivets 1⅛ diam.

FIG. 639.

301. Plates with Scolloped Edges.—It has already been pointed out that a riveted joint is stronger when the rivets of the outer rows have a wider pitch. But with this wide pitch the joint is more difficult to keep steam-tight. If, however, the edges of the plates are scolloped as shown in Fig. 640, the joint may be more securely caulked.

FIG. 640.

302. Connection of Plates at Right Angles. —A very common way of connecting two plates which are at right angles is by means of an angle iron or angle bar as shown in Figs. 641 and 642. Angle bars can be obtained in many sizes with equal or unequal sides. Angle bars usually increase in thickness from the edge to the corner, the taper being at the rate of about 1 in 24. The mean

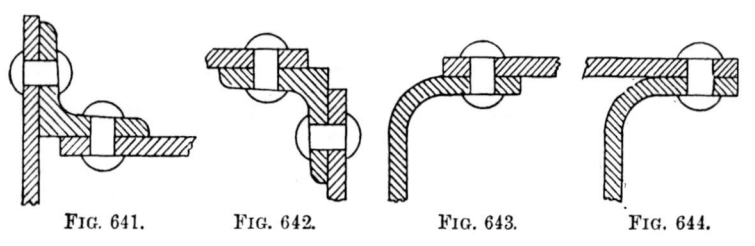

FIG. 641.   FIG. 642.   FIG. 643.   FIG. 644.

thickness is generally a little greater, about $\frac{1}{16}$ inch, than the thickness of the plates to which they are riveted.

Connections by flanging one of the plates are shown in Figs. 643 and 644. Only plates of good quality will stand flanging, and the radius of the inside curve should not be less than twice the thickness of the plate.

303. Connection of Parallel Plates.—Figs. 645 to 650 show the various ways of connecting the lower part of the internal fire-box to the external shell in vertical, locomotive, and other boilers. The joints shown in Figs. 647 to 650 are also used round the openings for the fire-

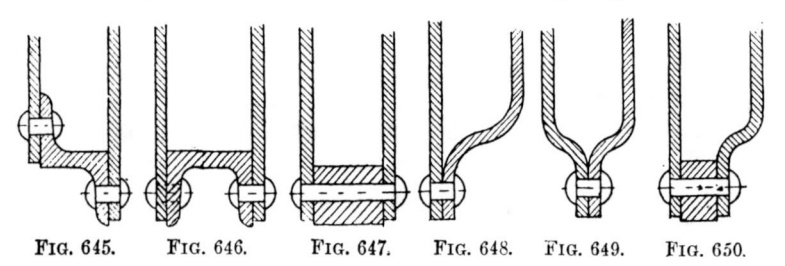

FIG. 645.   FIG. 646.   FIG. 647.   FIG. 648.   FIG. 649.   FIG. 650.

door. The Z-iron shown in Fig. 645 is not generally used when the steam pressure exceeds 80 lbs. per square inch. The design shown in Fig. 647, in which a solid rectangular bar is used, is adopted more extensively than any other in locomotive boilers.

CHAPTER XXII.

STEAM BOILERS.

304. Weight of Steam required by an Engine.—In designing a steam boiler for a given engine, the probable indicator diagram is first drawn, and from this the volume of steam used, at a particular pressure, in one stroke is determined. From a table of the properties of steam which is usually given in treatises on the steam-engine, the weight of a cubic foot of this steam may be found, and from this the weight of steam used in one stroke is calculated. Multiplying the weight of steam used in one stroke by the number of strokes per hour, the weight of steam used per hour is determined.

In actual practice, the amount of water evaporated in the boiler is from 1·2 to 1·8 times the amount shown by the indicator diagram of the engine.

Properties of Saturated Steam.

Absolute Pressure in lbs. per sq. inch.	Temperature in degrees Fahrenheit.	Total Units of Heat in 1 lb. weight of Steam.	Weight of 1 cubic foot of Steam in lbs.	Absolute Pressure in lbs. per sq. inch.	Temperature in degrees Fahrenheit.	Total Units of Heat in 1 lb. weight of Steam.	Weight of 1 cubic foot of Steam in lbs.
14·7	212·0	1178·1	·0380	85	316·1	1209·9	·1980
25	240·1	1186·6	·0625	95	324·1	1212·3	·2198
35	259·3	1192·5	·0858	110	334·6	1215·5	·2521
45	274·4	1197·1	·1089	125	344·2	1218·4	·2867
55	287·1	1201·0	·1314	145	355·6	1221·9	·3294
65	298·0	1204·3	·1538	165	366·0	1224·9	·3714
75	307·5	1207·2	·1759	185	375·3	1227·8	·4142

EXAMPLE.—*To find the weight of water which must be evaporated per hour by a boiler to supply steam to an engine having a cylinder* 10 *inches in diameter, with a piston stroke of* 18 *inches. Steam cut off at one-third of the stroke. Revolutions per minute* 100. *Pressure of steam* 50 *lbs. per square inch by gauge, or* 65 *lbs. per square inch absolute.*

$$\text{Volume of steam used in one stroke} = \frac{10^2 \times \cdot 7854 \times 18}{3} \text{ cubic inches.}$$

$$\text{,, \qquad ,, \qquad ,, \qquad ,,} = \frac{10^2 \times \cdot 7854 \times 18}{3 \times 1728} \text{ cubic feet.}$$

Weight of one cubic foot of steam at an absolute pressure of 65 lbs. per square inch, from Table = ·1538 lb.

Weight of steam used in one stroke,
$$= \frac{10^2 \times \cdot 7854 \times 18 \times \cdot 1538}{3 \times 1728} \text{ lbs.}$$
Weight of steam used in one minute,
$$= \frac{10^2 \times \cdot 7854 \times 18 \times \cdot 1538 \times 2 \times 100}{3 \times 1728} \text{ lbs.}$$
Weight of steam used in one hour,
$$= \frac{10^2 \times \cdot 7854 \times 18 \times \cdot 1538 \times 2 \times 100 \times 60}{3 \times 1728} \text{ lbs}$$
$$= 503 \cdot 3 \text{ lbs.}$$

In addition to the above quantity of steam, there will be that required to fill the clearance spaces in the cylinder, also a quantity to make up for the loss due to condensation in the cylinder and steam pipe. Allowing a margin of 40 per cent., the boiler will require to evaporate $503 \cdot 3 \times 1 \cdot 4 = 704 \cdot 6$ lbs., say 705 lbs. of water per hour under the given pressure.

305. Equivalent Evaporation from and at 212°.—For the purpose of comparison it is usual to state the weight of steam produced in a boiler on the assumption that the feed water is supplied at a temperature of 212° F., and that it is evaporated at this temperature.

Let W = weight of steam produced at any given temperature or pressure.
H = total heat in 1 lb. weight of this steam.
t = temperature of feed water.
W_1 = equivalent weight of steam at 212° temperature produced from feed water at 212°.

Then $W_1(1178 \cdot 1 - 212) = W(H - t)$

and $W_1 = \dfrac{W(H - t)}{966 \cdot 1}$

EXAMPLE.—*To determine the equivalent evaporation from and at 212° in the boiler for the engine in the example worked out in the preceding article, supposing that the feed water is supplied at a temperature of 60°.*

$$W_1 = \frac{705 (1204 \cdot 3 - 60)}{966 \cdot 1} = \frac{705 \times 1144 \cdot 3}{966 \cdot 1} = 835 \text{ lbs.}$$

306. Evaporative Performances of Steam Boilers.—Theoretically 1 lb. of coal should evaporate from 12 to 16 lbs. of water from and at 212°, the amount depending on the quality of the coal. The following table gives the weight of water evaporated from and at 212° by 1 lb. of average coal in various types of boilers.

Type of Boiler.	Water evaporated from and at 212° per pound of Coal Lbs.
Vertical cross-tube boilers	5 to 7
Vertical multitubular boilers	6 ,, 9
Cornish boilers	7 ,, 9
Lancashire boilers	8 ,, 10½
Sectional or water-tube boilers	8 ,, 10
Marine boilers	8 ,, 10
Locomotive boilers	8½ ,, 11
Torpedo-boat boilers	7 ,, 8½

STEAM BOILERS.

307. Rate of Combustion.—The rate of combustion of fuel in the furnaces of steam boilers is usually expressed in pounds per square foot of grate surface per hour. With natural or chimney draught the usual rate of combustion in factory boilers is from 15 to 20 lbs. of coal per square foot of grate. In ordinary marine boilers, with natural draught, the rate of combustion is from 11 to 20 lbs., but with forced draught the rate may be as high as 40 lbs. per square foot of grate. In locomotive boilers and in the boilers of torpedo-boats, where forced draught is used, the rate of combustion is from 40 to 100 lbs. per square foot of grate.

308. Vertical Cross Tube Boilers.—Vertical boilers possess the advantage of taking up a comparatively small amount of floor area. In its simplest form the vertical boiler consists of a cylindrical shell surrounding a nearly cylindrical fire-box, in the bottom of which is the grate. A tube, called an uptake, passes from the crown of the fire-box to the crown of the shell, where it is connected to a chimney. To increase the amount of heating surface and improve the circulation of the water, and also to increase the strength of the fire-box, the latter is fitted with one or more cross-tubes as shown in Fig. 651.

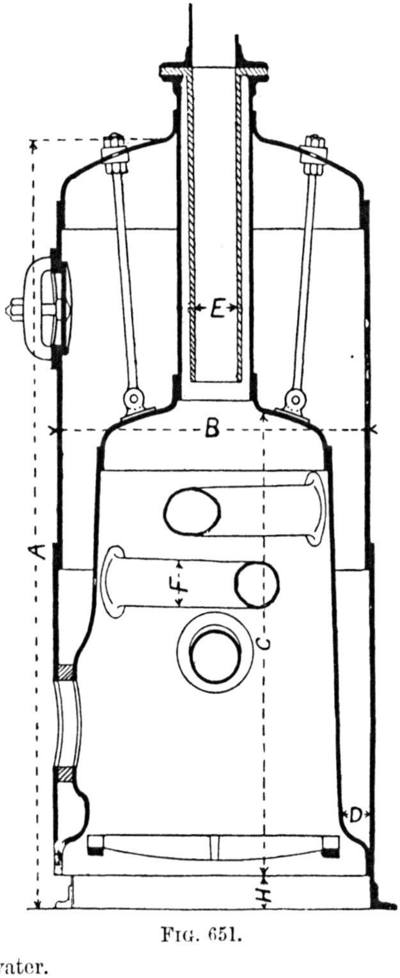

The vertical seam of the fire-box shell may be a single riveted joint, but in the best work the joint is lap welded. The fire-box shell slopes to the vertical at the rate of $\frac{1}{2}$ inch to $\frac{3}{4}$ inch per foot of length.

The cross-tubes are either flanged and riveted to the fire-box or they are welded to it. With welded joints in the fire-box double thicknesses of plates are avoided and the chances of overheating at the joints reduced. When the cross tubes are welded in, however, it is more difficult to renew them if they should wear out before the fire-box.

The cross tubes are placed slightly inclined to ensure a more efficient circulation of the water.

Fig. 651.

The uptake is connected to the crown of the fire-box and to the crown of the outside shell by flanged joints as shown in Fig. 651, or by welded angle iron rings which are riveted to the crowns and to the uptake.

The part of the uptake which passes through the steam space is very liable to become overheated, and to prevent this it is enlarged in diameter from the top to the low-water level and lined with fire-clay. A cast-iron liner is sometimes introduced for the same purpose. The cast-iron liner may fit close into the uptake, or it may be separated from it all round by an air space not exceeding $\frac{3}{4}$ inch wide as shown in Fig. 651.

Opposite the lower end of each cross tube there is a hand hole in the shell, in addition to one or more just above where the shell joins the fire-box. There is also a man-hole in the shell above the water-level.

The crowns of the fire-box and shell are strengthened by being dished. The radius of curvature may be about equal to the diameter of the crown. In the larger sizes the crowns are further strengthened by introducing bar stays between them.

The following table gives the principal dimensions of vertical cross-tube boilers. This table has been constructed after comparing the dimensions of a large number of examples by different makers, and may, therefore, be taken as representing average practice.

Dimensions of Vertical Cross Tube Boilers.

Height of Boiler.	Diameter of Boiler.	Height of Fire-box.	Width of Water-space round Fire-box.	Diameter of Uptake.	Number of Cross Tubes.	Diameter of Cross Tubes.	Height of bottom of Fire-box above Ground.	Heating Surface (approximate).	Grate Area.
A.	B.	C.	D.	E.		F.	H.		
ft. in.	ft. in.	ft. in.	inches.	ft. in.		inches.	inches.	sq. ft.	sq. ft.
5 0	2 3	3 0	2	0 8	1	6	7	19	2·7
5 8	2 6	3 4	2	0 8½	1	7	7½	24	3·5
6 4	2 9	3 10	2	0 9½	2	7	7½	35	4·4
7 0	3 0	4 3	2¼	0 10	2	8	8	43	5·2
7 8	3 3	4 9	2¼	0 11	3	8	8	57	6·2
8 4	3 6	5 1	2¼	0 11½	3	8	8½	65	7·4
9 0	3 9	5 8	2½	1 0	3	9	8½	80	8·4
9 8	4 0	6 3	2½	1 0½	4	9	9	100	9·7
10 4	4 3	6 7	2¾	1 1	4	9	9	110	10·8
11 0	4 6	6 11	2¾	1 1½	4	9	9½	120	12·1
11 8	4 9	7 6	3	1 2	4	10	9½	138	13·7
12 4	5 0	7 10	3	1 3	4	10	10	152	15·4

309. Cochran's Vertical Multitubular Boiler.—Messrs. Cochran & Co., Birkenhead, have manufactured for many years a design of vertical multitubular boiler, which has given excellent economical results in practice, and which is of a very strong form, there being no flat surfaces, except the tube plates, requiring to be supported by stays. Figs. 652, 653, and 654 are fully dimensioned illustrations of a Cochran boiler 6 feet 6 inches in diameter, and 14 feet 6 inches high, for a working steam pressure of 80 lbs. per square inch. It will be observed that the crowns of the fire-box and external shell are of hemispherical shape,

STEAM BOILERS.

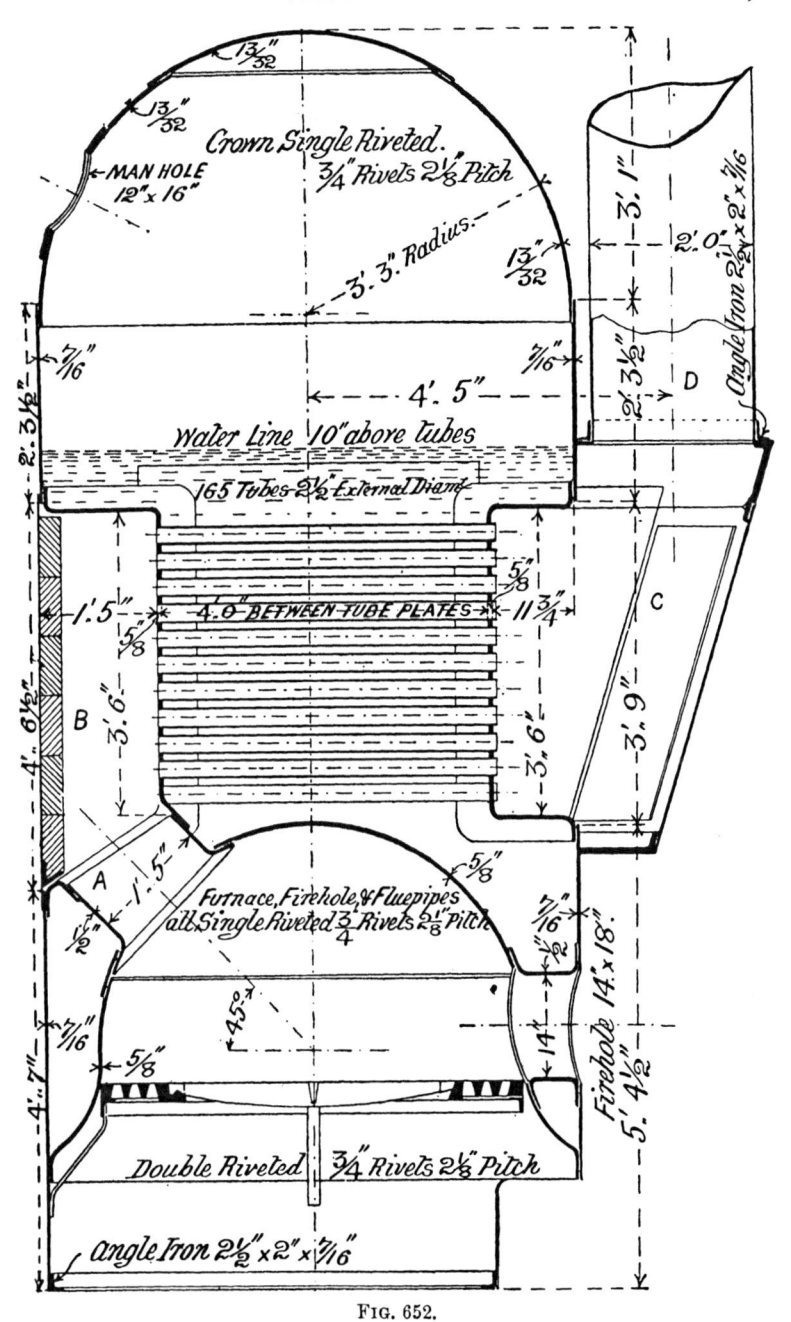

FIG. 652.

a shape which, for a given weight of material in the form of plates, gives the maximum volume, and the maximum strength. The heated gases from the fuel pass from the fire-box through the oval-shaped flue-pipe A into the combustion-chamber B, and from thence through numerous horizontal tubes to the smoke-box C, from which they pass into the chimney D. The combustion-chamber, on the side next the shell, is lined with fire-brick.

There are 165 tubes, $2\frac{1}{2}$ inches diameter, externally. Of these 165 tubes 22 are stay tubes, $\frac{5}{16}$th inch thick. The remaining tubes have a thickness No. 11 B.W.G., or ·12 inch.

The grate, which is made of cast-iron, consists of four segmental

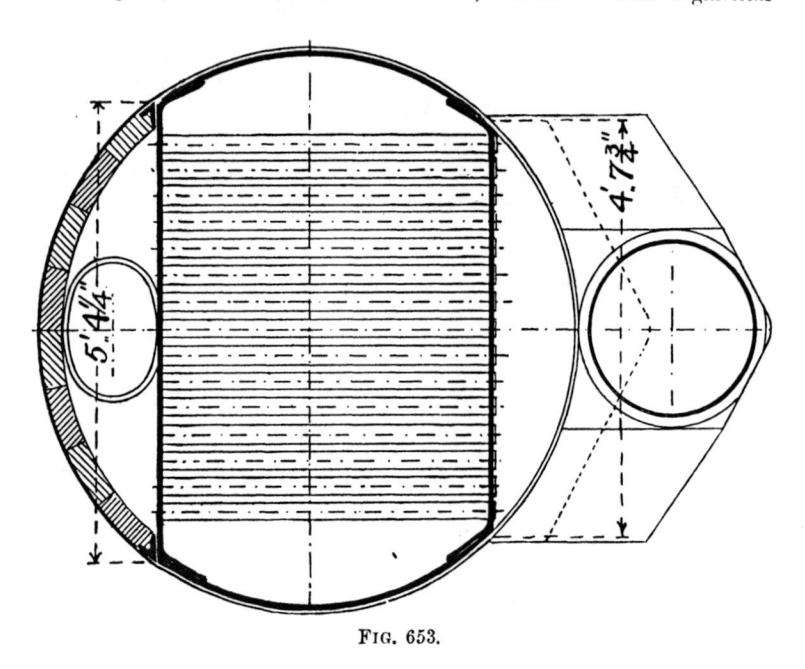

FIG. 653.

gratings surrounding a rectangular space containing ordinary fire-bars, 3 inches deep at the middle, and $1\frac{1}{4}$ inch thick at the top, tapering to $\frac{1}{2}$ inch at the bottom. The segmental gratings rest on an angle-iron ring, which is supported on brackets as shown.

With the exception of the plate at the back of the fire-brick lining in the combustion-chamber and the angle irons, which are made of wrought-iron, the boiler is constructed of steel.

The circumferential seams in the shell are single riveted and the longitudinal or vertical seams are double riveted. The rivets throughout are $\frac{3}{4}$ inch in diameter, placed at $2\frac{1}{8}$ inches pitch, except in the double riveted seams, where the pitch is $2\frac{1}{2}$ inches. The rivet-holes are $\frac{13}{16}$ inch in diameter.

Fig. 654.

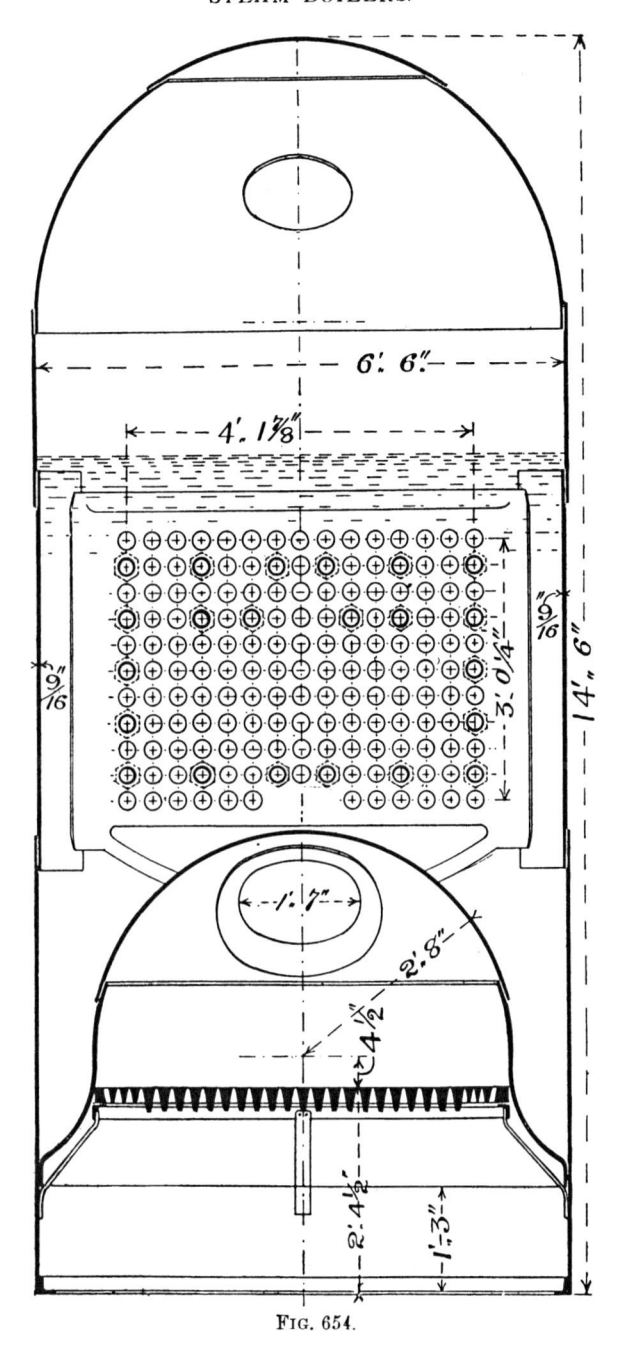

The area of the grate is 21 square feet, and the heating surface amounts to 500 square feet. Of the 500 square feet of heating surface, the tubes present 428, and the plates 72.

As already stated this boiler is designed for a working pressure of 80 lbs. per square inch. It is tested under a water-pressure of 160 lbs. per square inch.

The following table gives the principal dimensions of boilers, made by Messrs. Cochran, of the type which we have described.

Dimensions of Cochran's Vertical Multitubular Boilers.

Height, including Ashpit.	Internal Diameter.	Tubes. No.	Tubes. Diameter.	Heating Surface.	Grate Area.	Approximate Weight.
ft. in.	ft. in.		inches.	sq. ft.	sq. ft.	cwt.
5 3	2 3	32	$1\frac{1}{2}$	30	$2\frac{3}{4}$	10
5 6	2 6	40	$1\frac{1}{2}$	40	$3\frac{1}{4}$	12
6 9	3 0	54	$1\frac{3}{4}$	60	5	18
7 6	3 3	58	$1\frac{3}{4}$	80	$5\frac{3}{4}$	25
8 6	3 9	54	2	100	$8\frac{1}{4}$	32
9 0	4 0	60	2	120	9	38
9 6	4 3	60	$2\frac{1}{2}$	140	$9\frac{3}{4}$	40
10 0	4 6	69	$2\frac{1}{2}$	160	10	45
10 3	4 9	78	$2\frac{1}{2}$	200	12	50
10 9	5 0	86	$2\frac{1}{2}$	220	$12\frac{1}{2}$	60
11 9	5 3	96	$2\frac{1}{2}$	250	$13\frac{1}{2}$	70
12 3	5 6	108	$2\frac{1}{2}$	300	15	80
13 0	5 9	127	$2\frac{1}{2}$	350	18	85
13 0	6 0	114	$2\frac{1}{2}$	350	$18\frac{1}{2}$	95
14 0	6 0	140	$2\frac{1}{2}$	400	$18\frac{1}{2}$	100
14 6	6 6	165	$2\frac{1}{2}$	500	21	125
14 6	7 0	210	$2\frac{1}{2}$	640	$23\frac{3}{4}$	145
15 0	7 6	238	$2\frac{1}{2}$	720	$36\frac{3}{4}$	165
16 0	8 0	238	$2\frac{1}{2}$	800	28	190
18 0	8 0	270	$2\frac{1}{2}$	900	28	210

310. Marine Boilers.—The ordinary type of marine boiler now in use has a cylindrical shell with flat ends. This shell contains one or more cylindrical or corrugated furnaces which lead into combustion-chambers which are more or less rectangular in shape. From the combustion-chambers a large number of tubes lead the products of combustion back to the front of the boiler into the smoke-box. This type of boiler is sometimes as large as 16 feet in diameter, and may be either single or double ended, that is, it may have furnaces in one or in both ends. Figs. 655 to 660 show the general form and arrangement of the furnaces and combustion-chambers in single and double ended marine boilers.

Number and Diameter of Furnaces.—Boilers up to 9 feet in diameter may have one furnace in an end. Boilers from 8 feet to 13 feet in diameter may have two furnaces in an end. Boilers from 11 feet 6 inches to 15 feet 6 inches in diameter may have three furnaces in an

FIG. 655.

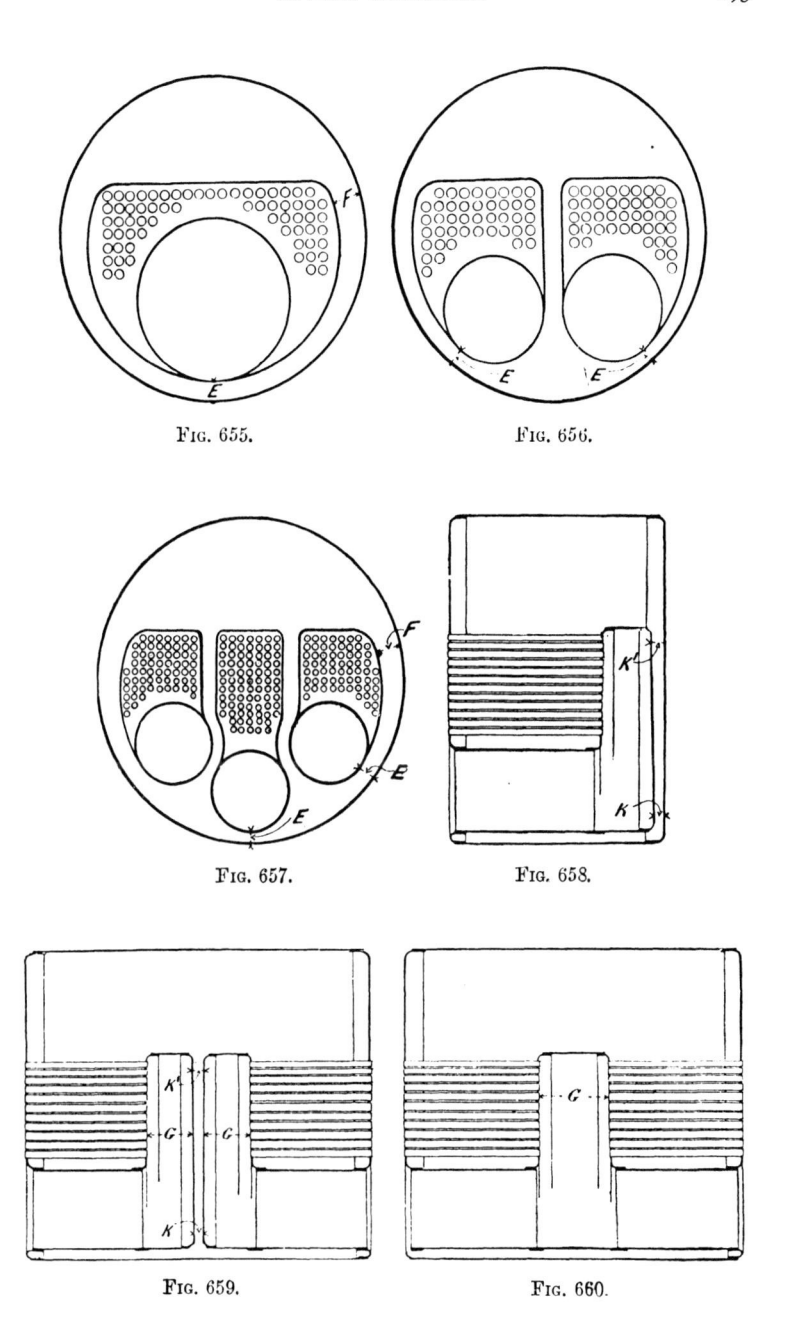

FIG. 656.

FIG. 657.

FIG. 658.

FIG. 659.

FIG. 660.

end, and boilers above 14 feet 6 inches in diameter may have four furnaces in an end.

The following empirical rules, based on a large number of examples, give the average relation between the diameter of the shell and the diameter of the furnace.

A = diameter of shell in inches,
B = diameter of furnace in inches.
$B = \frac{1}{2}A$ when there is one furnace in an end.
$B = \frac{1}{4}A + 5$ when there are two furnaces in an end.
$B = \frac{1}{4}A - 1$ when there are three furnaces in an end.

Combustion-Chambers.—The height C of the crown of the combustion-chamber above the centre of the boiler is usually about $\frac{1}{6}$th of the diameter of the shell. The length G of the combustion-chamber should not exceed $\frac{7}{8}$ B; it is generally about $\frac{3}{4}$ B, but in boilers having not more than two furnaces in an end it is sometimes as small as $\frac{1}{2}$ B. For double-ended boilers, where two furnaces from opposite ends lead into one combustion-chamber (Fig. 660), its length G' varies from $\frac{5}{6}$ B to $1\frac{1}{6}$ B.

Water-Spaces.—The spaces E between the shell and the furnaces should not be less than 5 inches (sometimes they are as small as $4\frac{1}{2}$ inches), and they need not exceed 6 inches. The spaces F between the shell and the combustion-chambers vary from 5 inches to 7 inches. The spaces F' between the sides of the combustion-chambers also vary from 5 inches to 7 inches. The width K between the back of the combustion-chamber and the end of the boiler at the bottom in single-ended boilers, or between the backs of the combustion-chambers at the bottom in double-ended boilers, varies from 5 inches to 7 inches. The width K' at the top may be the same as at the bottom, but it is usually wider, and is sometimes as much as 12 inches.

Tubes.—The tubes used in boilers of merchant steamers are made of wrought-iron or steel, and are usually either $3\frac{1}{4}$ inches or $3\frac{1}{2}$ inches in external diameter. They are, however, sometimes as small as $2\frac{1}{2}$ inches, and occasionally as large as 4 inches in diameter. In the navy the tubes are generally made of brass, and are then from $2\frac{1}{2}$ inches to 3 inches in diameter.

To facilitate the inserting or withdrawing of the tubes the front or smoke-box ends are swelled $\frac{1}{16}$ inch larger in diameter.

The thickness of the iron tubes, of 3 inches to $3\frac{1}{2}$ inches diameter, is usually No. 9 or No. 8 Birmingham wire gauge for steam-pressures of 80 lbs. per square inch and upwards.

The length of the tubes between the tube plates varies in practice from 5 ft. 9 in. to 7 ft. 6 in. The average length is about 6 ft. 9 in.

The tubes are arranged in horizontal and vertical rows, as shown in Fig. 661. The pitch is generally $1\frac{3}{8}$ times the external diameter of the tubes ($p = 1\frac{3}{8}d$).

The clear space H between the nests of tubes is usually about 11 inches. Of a considerable number of examples which we tabulated we found the minimum value of H to be $10\tfrac{1}{2}$ inches and its maximum value $12\tfrac{1}{2}$ inches.

The ordinary tubes, that is, tubes other than stay tubes, are secured to the tube plates by expanding them at their ends by means of a tool called a tube expander.

Stay Tubes. — To give support to the flat tube plates, a certain number of the tubes are made to act as stays, and are called stay tubes. These tubes are of the same external diameter as the others, but they are from $\tfrac{1}{4}$ inch to $\tfrac{3}{8}$ inch in thickness. They are screwed at both ends, and are either screwed into both plates or screwed into the back plate, and secured to the front plate by two nuts, one on each side of the plate. Frequently in addition to screwing the stay tubes into the tube plates nuts are put on, sometimes one at each end and sometimes one on the combustion-chamber end only. Usually from $\tfrac{1}{5}$th to $\tfrac{1}{3}$rd of all the tubes are stay tubes.

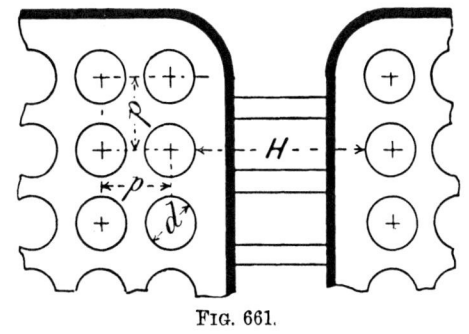

FIG. 661.

The latest practice in the construction of modern high pressure marine boilers is to dispense with the nuts on the stay tubes altogether, the tubes being simply screwed into the plates. About $\tfrac{3}{8}$ths of all the tubes are then stay tubes.

311. Example of a Marine Boiler.—The marine boiler which we shall now illustrate and describe is an example from the recent practice of Messrs. James Howden & Co., marine engineers, Glasgow. As will be seen from Figs. 662 and 663, the boiler is of the cylindrical single-ended type, having three furnaces, and three combustion-chambers or fire-boxes. The shell has a mean diameter of 14 ft. 3 in., and a length over all of 11 ft. 6 in.

The furnace tubes have an internal diameter of 3 ft. 6 in., each being made up of five flanged rings, four of which are each 1 ft. $7\tfrac{11}{32}$ in. long. These rings have no longitudinal seams; they are $\tfrac{19}{32}$ inch thick, and their flanges are riveted together with rivets $1\tfrac{3}{16}$ inch diameter, placed at $2\tfrac{1}{4}$ inches pitch. The caulking rings between the flanges are $\tfrac{5}{16}$ inch thick.

There are in all 350 tubes $2\tfrac{1}{2}$ inches in diameter externally. These tubes are arranged in three nests, one nest to each fire-box. The central nest contains 112 tubes arranged in fourteen horizontal rows, each row containing eight tubes. The side nests each contain 119 tubes, arranged in ten rows, each row containing twelve tubes, except the top row, which contains eleven. The tubes in the central nest are placed horizontally at a pitch of $3\tfrac{7}{8}$ inches, and vertically at a pitch of $3\tfrac{3}{4}$ inches. The tubes in the side nests have a uniform vertical pitch of $3\tfrac{3}{4}$ inches, but horizontally the pitch varies uniformly from $3\tfrac{13}{16}$ inches in the top row to $3\tfrac{7}{8}$ inches in the bottom row.

296   *MACHINE DRAWING AND DESIGN.*

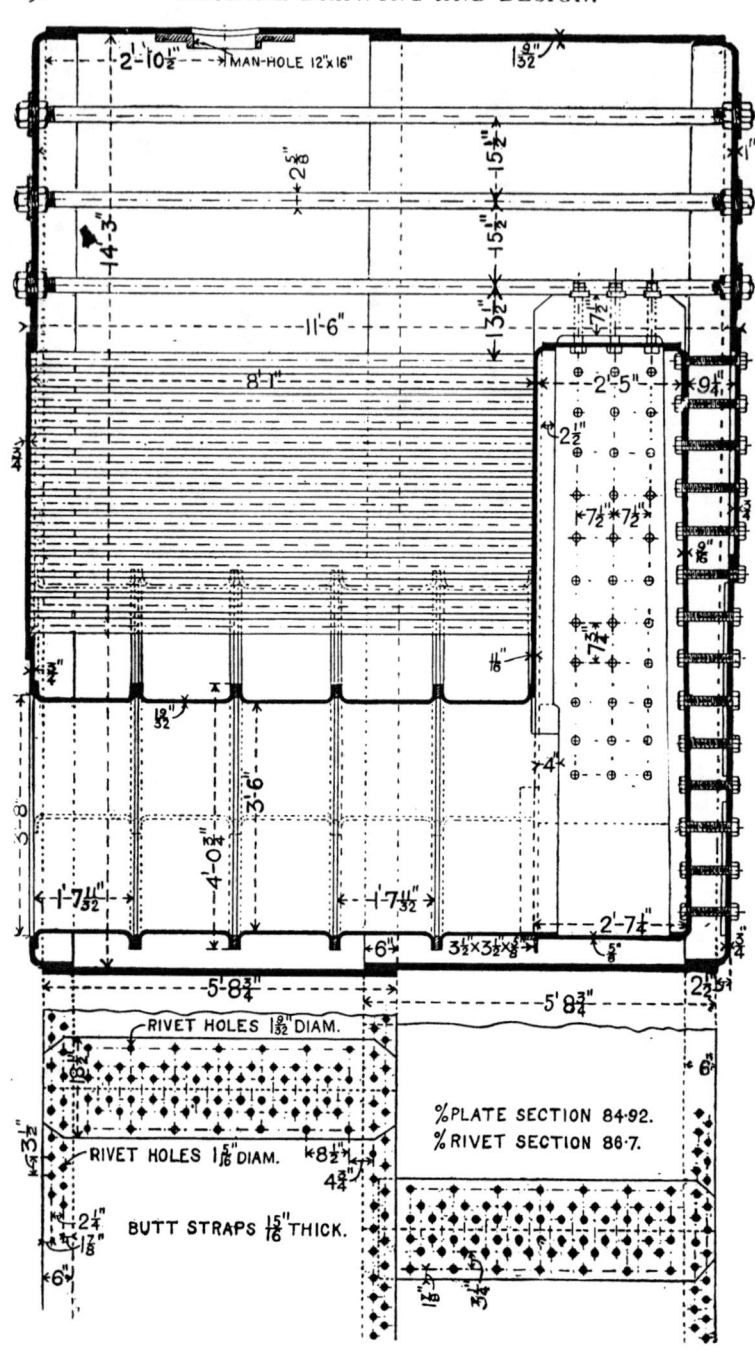

FIG. 662.

STEAM BOILERS. 297

Of the 350 tubes, 243 are common tubes, No. 9 I.S.W.G. (Imperial standard wire gauge) thick (·144 inch). These tubes have an external diameter of $2\frac{1}{4}$ inches, and at one end they are swelled to $2\frac{5}{8}$ inches diameter. The remaining 107 tubes are stay tubes. These are shown

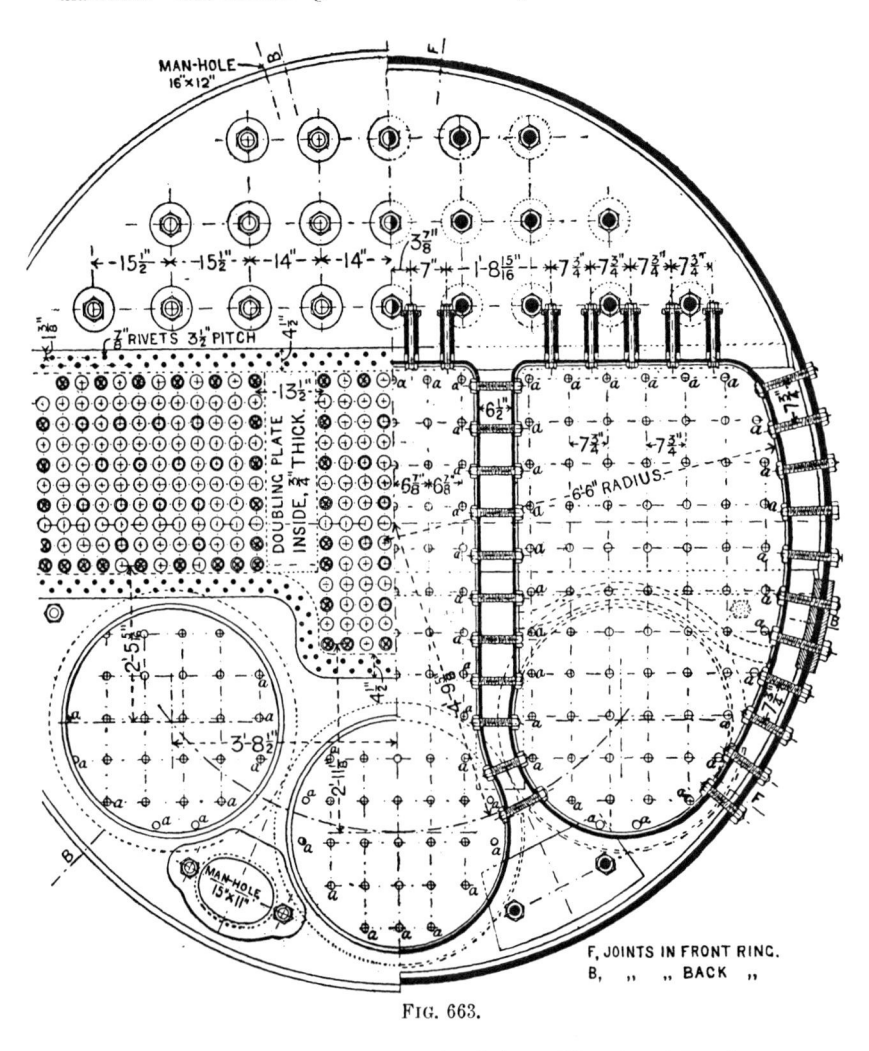

FIG. 663.

in Fig. 663, each by two concentric circles. The stay tubes have an external diameter of $2\frac{1}{4}$ inches, and they are screwed into both tube plates. One screwed end of each stay tube has a diameter of $2\frac{1}{2}$ inches, and the other a diameter of $2\frac{3}{4}$ inches. The screwed ends have ten threads per inch. The inner stay tubes are $\frac{1}{4}$ inch thick, and the outer stay tubes, marked thus (X), are $\frac{5}{16}$ inch thick. All the tubes are made of wrought-iron.

The cylindrical shell is made up of two belts or rings, each belt consisting of three plates 5 ft. $8\frac{3}{4}$ in. wide and $1\frac{9}{32}$ in. thick. The riveted joints for these plates are fully shown in the lower part of Fig. 662. The thicknesses of the other plates are as follows:—End plates in steam space 1 inch, other end plates $\frac{3}{4}$ inch, front tube plate $\frac{3}{4}$ inch, back tube plates $\frac{11}{16}$ inch, back plates of combustion-chambers $\frac{9}{16}$ inch, doubling plates $\frac{3}{4}$ inch. There are two doubling plates $9\frac{1}{2}$ inches wide, riveted on the inside of the front tube plate between the nests of tubes. Opposite to these there are two doubling plates of the same width riveted on the inside of the back end plate. Immediately under the latter there are two similar doubling plates riveted on the inside of the lower back end plate.

The bar stays in the steam-space are each $2\frac{5}{8}$ inches diameter screwed at the ends, the screws having six threads per inch. Area of stay at bottom of screw thread, 4·101 square inches. These bar stays are screwed into the front end plate, and at each end they are provided with nuts outside and inside the boiler. Under the outside nuts there are loose washers 8 inches diameter and $\frac{11}{16}$ inch thick.

The screwed stays for the combustion-chambers are $1\frac{3}{8}$ inches diameter, except those marked a in Fig. 663, which are $1\frac{5}{8}$ inches diameter. Each of the screwed stays, in addition to being screwed into the plates, has a nut at each end.

The crown stays for the combustion-chambers are 2 ft. $6\frac{1}{2}$ in. long, and are made from plates $\frac{3}{4}$ inch thick. They are $7\frac{1}{2}$ inches deep, and, except at the ends where they rest on the crown, they stand $1\frac{1}{4}$ inches above the crown. The bolts are $1\frac{3}{8}$ inches diameter.

With the exception of the tubes, which are of wrought-iron, this boiler is made entirely of steel, and is designed for a working pressure of 160 lbs. per square inch.

When worked with Mr. Howden's well-known and successful forced draught system, this boiler will, at sea, supply steam to a triple expansion engine developing over 1100 indicated horse-power.

312. Locomotive Boilers.—The ordinary locomotive boiler used on English railways consists of a cylindrical shell or barrel from 3 ft. 6 in. to 4 ft. 6 in. in diameter, united to a rectangular fire-box at one end, and to a cylindrical smoke-box at the other. The fire-box is in two parts, the inner part being separated from the outer at the sides, back, and front by water-spaces from $2\frac{1}{2}$ inches to 4 inches wide. The smoke-box is from 3 inches to 10 inches larger in diameter than the barrel, and from 2 ft. 3 in. to 3 ft. long. The inside fire-box is about 3 ft. 4 in. wide and from 4 ft. 3 in. to 6 ft. long.

A large number of tubes pass through the barrel and are secured to a tube plate at the smoke-box end, and to the inner fire-box at the other. These tubes are from $1\frac{1}{2}$ inches to 2 inches in diameter outside, and from 8 ft. 3 in. to 11 ft. 9 in. long, and they are arranged zigzag with water-spaces between them about $\frac{3}{4}$ inch wide.

The tubes are usually No. 11 B.W.G. (·125 inch) thick at the fire-box end, and No. 13 B.W.G (·095 inch) thick at the smoke-box end. At the smoke-box end the tubes are generally a little larger, to allow of them being put in position or withdrawn easily. The tubes are secured to the smoke-box tube plate by means of a tube expander, and they are allowed to project into the smoke-box from $\frac{1}{8}$ inch to $\frac{1}{2}$ inch. The tubes are also expanded into the fire-box tube plate, but in addition they sometimes have driven into them steel ferrules about $\frac{1}{8}$ inch thick, which serve not only to secure the tubes but to protect their ends from the action of the flame in the fire-box. Fig. 664 shows the fire-box end of a tube, beaded over and fitted with a steel ferrule.

The inner fire-box is made of copper. The tubes are usually made of brass or copper, but steel tubes are now sometimes used. The outer fire-box, the barrel, and smoke-box are made of wrought-iron or steel.

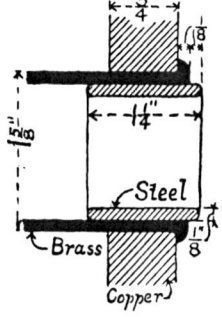

FIG. 664.

When a steam dome is added it is from 1 ft. 3 in. to 2 ft. in diameter, and from 1 ft. 9 in. to 2 ft. 6 in. high.

The usual working pressure is 140 lbs. per square inch for simple engines, and 175 lbs. per square inch for compound engines.

The table on p. 300 gives the principal dimensions of eight locomotive boilers representing recent practice on eight different English railways.

In American locomotive practice the general design of the boiler is much the same as that adopted for English locomotive boilers, but in the details there are several important points of difference. The principal distinguishing features in American locomotive boilers are as follows :— The tubes are made of wrought-iron or steel, and so also are the plates of the inside fire-box. The top of the outside fire-box stands from 9 to 12 inches higher than the top of the barrel, to which it is connected by a sloping plate, this design being known as the "wagon-top" pattern. Water-tube grates are used where anthracite coal is burned, and for bituminous coal the grate has rocking bars which may be agitated by moving a lever in the cab, for the purpose of breaking up the clinker. The smoke-box is of great length, and the grate is usually much longer than obtains in English locomotive practice.

The following are the principal dimensions of the boiler for a compound locomotive recently constructed for the Michigan Central Railroad :— Diam. of barrel, 4 ft. 10 in. Diam. of tubes, outside, 2 in. Length of tubes, 12 ft. Number of tubes, 247. Thickness of plates of inside fire-box—crown, $\frac{3}{8}$ in. ; tube plate, $\frac{1}{2}$ in. ; sides and back, $\frac{5}{16}$ in. Size of inside fire-box, inside, 8 ft. $0\frac{3}{16}$ in. long, by 3 ft. $6\frac{7}{8}$ in. wide. Inside diam. of chimney, 1 ft. 6 in. Heating surface—tubes, 1540·3 sq. ft. ; fire-box, 137·1 sq. ft. ; total, 1677·4 sq. ft. Grate surface, 28·5 sq. ft. All the boiler plates are made of steel, and the tubes are made of "semi-steel." The boiler is of the wagon-top form. Working pressure, 180 lbs. per sq. in. Diam. of cylinders—H. P., 20 in. ; L. P., 29 in. Stroke of pistons, 2 ft. Diam. of driving wheels, 5 ft. 8 in.

Principal Dimensions of Locomotive Boilers from Actual Practice.

	Express Passenger.				Goods.		Compound.	
	ft. in.	ft. in.	ft. in.	ft. in.	ft. in.	ft. in.	ft. in.	ft. in.
Diameter of cylinders	1 6	1 $6\frac{1}{4}$	1 7	1 6	1 5	1 6	{1 6 {2 2	{1 2 {1 2 {2 6
Length of stroke	2 0	2 2	2 2	2 2	2 0	2 2	2 0	2 0
Diameter of driving wheels	7 0	6 6	7 0	7 4	4 4	5 $0\frac{7}{8}$	7 $1\frac{1}{4}$	7 1
Boiler pressure in lbs. per square inch	140	140	150	160	160	160	175	175
Diameter of barrel	4 $1\frac{1}{8}$	4 5	4 4	4 2	4 1	4 2	4 2	4 3
Diameter of tubes (outside)	0 $1\frac{3}{4}$	0 $1\frac{1}{2}$	0 $1\frac{5}{8}$	0 $1\frac{5}{8}$	0 $1\frac{3}{4}$	0 $1\frac{3}{4}$	0 $1\frac{3}{4}$	0 $1\frac{7}{8}$
Length of tubes between tube plates	11 $9\frac{1}{4}$	10 6	10 $0\frac{1}{16}$	10 $8\frac{9}{16}$	9 $11\frac{1}{2}$	10 $9\frac{3}{8}$	10 $11\frac{1}{8}$	11 3
Number of tubes	201	331	202	244	192	220	203	225
Thickness of smoke-box tube plate	0 $\frac{3}{4}$	0 1	0 $\frac{7}{8}$	0 $\frac{3}{4}$	0 $\frac{3}{4}$	0 $\frac{7}{8}$	0 $\frac{3}{4}$	0 $\frac{3}{4}$
Thickness of fire-box tube plate	0 1	0 1	0 $\frac{3}{4}$	0 $\frac{13}{16}$	0 $\frac{3}{4}$	0 1	0 1	0 $\frac{7}{8}$
Length of inside fire-box at bottom	5 4	5 $11\frac{3}{4}$	5 $0\frac{5}{8}$	5 $9\frac{15}{16}$	4 7	5 $4\frac{5}{8}$	5 $3\frac{5}{8}$	6 3
Width of inside fire-box at bottom	3 3	3 $4\frac{3}{4}$	3 $3\frac{7}{8}$	3 $4\frac{1}{2}$	3 $5\frac{3}{4}$	3 6	3 3	3 5
From grate to centre line of barrel, front	5 $1\frac{1}{2}$	5 0	5 0	4 $8\frac{1}{2}$	3 $10\frac{1}{2}$	4 $6\frac{1}{2}$	4 7	4 9
From grate to centre line of barrel, back	5 $1\frac{1}{2}$	3 9	4 6	4 2	3 0	4 $6\frac{1}{2}$	4 7	4 9
From centre line of barrel to crown of fire-box	0 $8\frac{1}{2}$	0 $9\frac{3}{4}$	0 $6\frac{1}{2}$	0 $9\frac{1}{2}$	0 9	0 $8\frac{1}{4}$	0 $9\frac{1}{2}$	0 11
Diameter of smoke-box	4 10	4 $8\frac{1}{4}$	5 $1\frac{1}{4}$	4 1	4 10	5 $0\frac{1}{2}$	5 0	4 9
Length of smoke-box	2 $7\frac{1}{2}$	2 7	2 $8\frac{3}{8}$	2 $10\frac{7}{16}$	2 $7\frac{1}{4}$	2 8	2 $7\frac{1}{8}$	2 $8\frac{1}{4}$
Internal diameter of chimney (minimum)	1 3	1 5	1 4	1 $4\frac{1}{8}$	1 $3\frac{1}{2}$	1 $3\frac{3}{4}$	1 2	1 4
Heating surface, tubes, square feet	1082·5	1373	917	1123·6	875·8	1102·3	1026·1	1242·4
Heating surface, fire-box, square feet	117·5	112·5	103·5	117	81	107·7	110	159·1
Heating surface, total, square feet	1200	1485·5	1020·5	1240·6	956·8	1210	1136·1	1401·5
Grate area, square feet	17·3	20·6	16·8	19·7	16·3	18·75	17·2	20·5
Railway	G.E.R.	L.B. & S.C.R.	S.E.R.	M.R.	N.L.R.	L. & Y.R.	N.E.R.	L. & N.W.R.

STEAM BOILERS.

313. Semi-Portable Boiler—Locomotive Type.—Boilers of the locomotive type are now largely used for supplying steam to engines for many purposes on land, and they are also used in torpedo boats. Boilers of this class, when used in a fixed position on land, are sometimes called "semi-portable," and sometimes "semi-fixed." Figs. 665, 666, and 667 illustrate a semi-portable boiler of the locomotive type manufactured by Messrs. Marshall, Sons & Co., Gainsborough. Fig. 665 is a half end elevation and half transverse section through the fire-box. Figs. 666 and 667 show a longitudinal section of the boiler. The shell plates are of steel, and the internal fire-box of Bowling, Low Moor, or Farnley iron or steel of special quality. The plates are flanged by hydraulic machinery, and angle-irons are dispensed with, except at the front end of the smoke-box, and even there a flanged ring is often used.

The edges of the plates are planed, and the outer face of the flange of the smoke-box tube plate is turned. All the rivet-holes are drilled after the plates have been bent in position, and the riveting is done by machinery.

The circumferential seams of the barrel are double riveted, zigzag, with rivets $\frac{3}{4}$ inch diameter at $3\frac{1}{4}$ inches pitch. The width of the lap is $3\frac{1}{8}$ inches, and the distance between the centre lines of the two rows of rivets is $1\frac{5}{8}$ inches "bare."

The longitudinal seams of the barrel are double riveted, zig-zag, butt joints, strapped outside and inside. The rivets are $\frac{3}{4}$ inch diameter at $3\frac{3}{16}$ inches pitch. The distance between the centre lines of the two inner of the four rows of rivets is 3 inches. The distance between the centre line of each outer row and the centre line of the adjacent inner row is $1\frac{7}{8}$ inches. The total width of the butt straps is $9\frac{3}{8}$ inches.

The tubes have an external diameter of $2\frac{1}{2}$ inches, and they are $\frac{1}{8}$ inch thick.

The smoke-box tube plate and the back end plate are stayed to the shell by long gusset stays, and they are also stiffened by T-irons. The sides of the inner fire-box are stayed to the outer fire-box by screwed stays. The crown of the inner fire-box is stayed by girder stays, which are placed transversely as shown.

The boiler is supported at the fire-box end on a cast-iron frame, not shown in the illustrations, which serves as an ash-pit. The smoke-box end is supported on a hollow cast-iron pillar, into which the exhaust steam from the engine is led on its way to the chimney.

There is a total heating surface of 290·5 square feet, and a grate area of 11·2 square feet.

The working steam-pressure is 140 lbs. per square inch.

This boiler, which is spoken of by the makers as a 20 nominal horse-power boiler, is capable of supplying steam to a compound engine having high and low pressure cylinders of 9 inches and 14 inches diameter respectively, with a piston stroke of 16 inches, the crank shaft making 135 revolutions per minute.

314. Strength of a Cylindrical Boiler Shell.—A boiler shell subjected

Fig. 665.

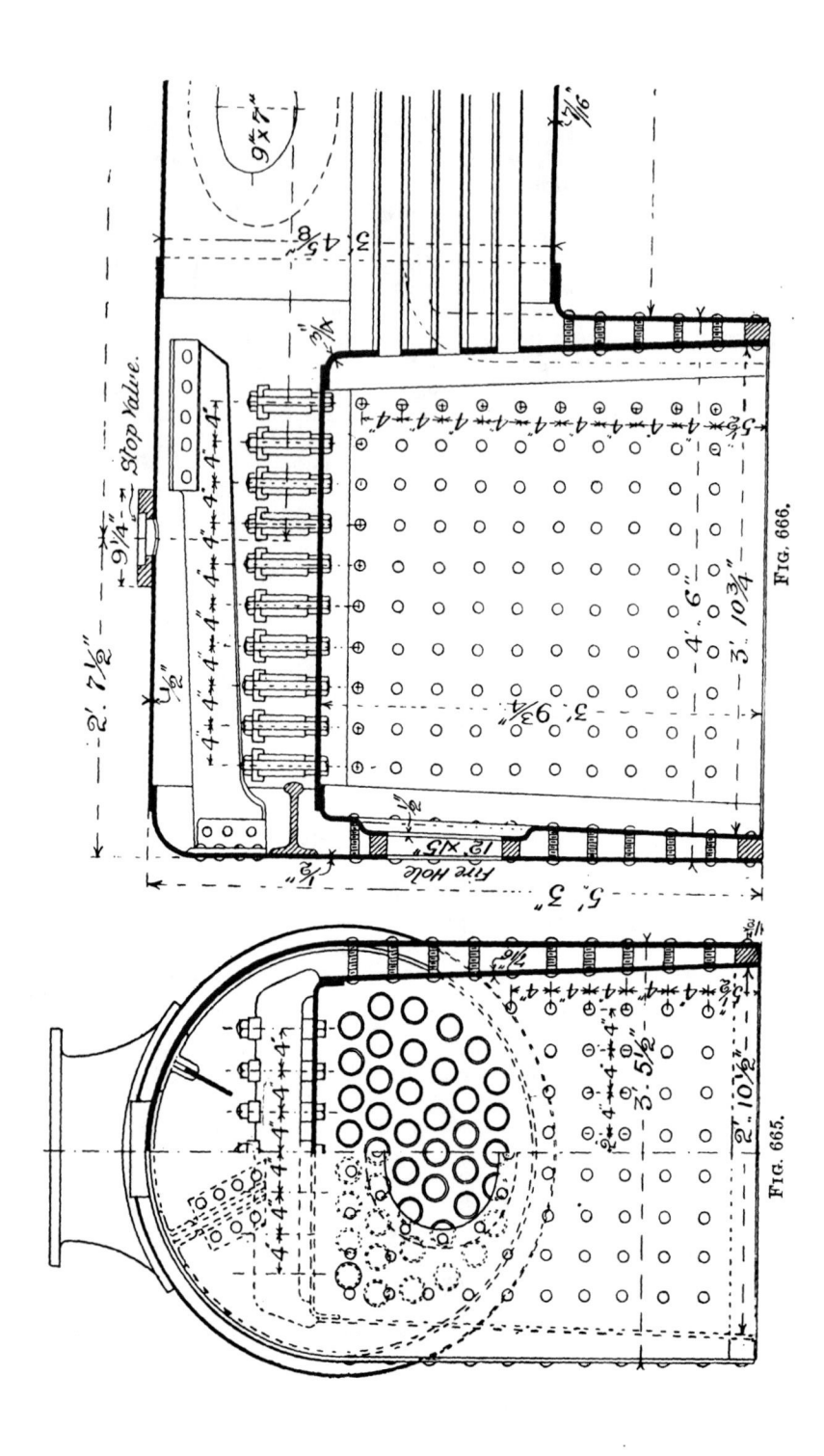

Fig. 666.

to internal pressure may give way either at a longitudinal section or at a transverse section. We will first consider the strength at a longitudinal

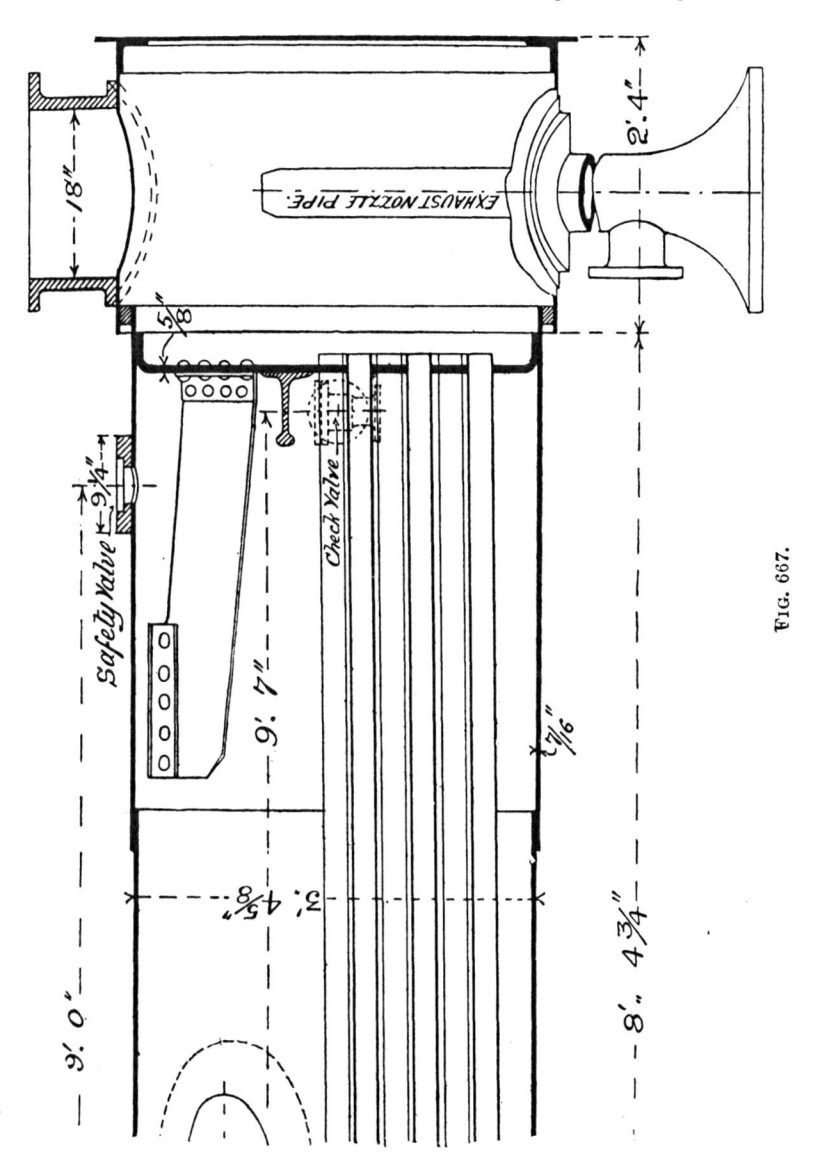

FIG. 667.

section. Let D = internal diameter of shell in inches, l = length of shell in inches, t = thickness of shell in inches, f = stress at a longitudinal section in pounds per square inch, P = pressure of steam in pounds per square inch. The resultant R (Fig. 668) of the whole pressure on one half of the shell is equal to PDl, and this is the magnitude of

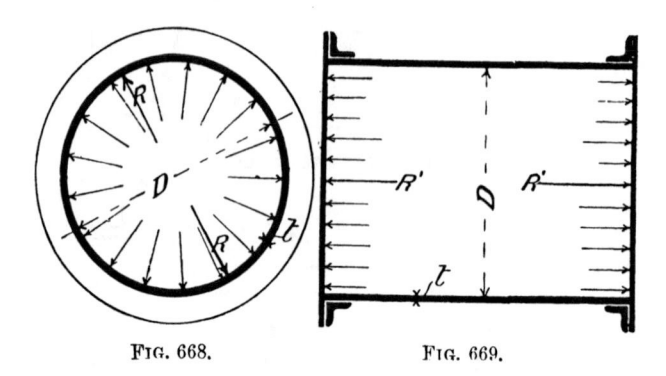

FIG. 668.   FIG. 669.

the force tending to rupture the shell at a longitudinal section. The area of the longitudinal section of the shell is 2tl, and the resistance to rupture is 2tlf. Therefore PDl = 2tlf; or PD = 2tf. The foregoing formula supposes that the shell is solid or without joint. In practice, however, there are longitudinal riveted joints, and the minimum area of a longitudinal section of the shell is only $2tl\left(\dfrac{p-d}{p}\right)$ where p is the pitch and d the diameter of the rivets. Hence for a riveted shell PD = $2tf\left(\dfrac{p-d}{p}\right)$.

Next consider the resistance at a transverse section. The resultant R′ (Fig. 669) of the pressure on one end is equal to ·7854D^{2}P, and this is the magnitude of the force tending to rupture the shell at a transverse section. The area of a transverse section of the shell is 3·1416Dt, and the resistance to rupture is 3·1416Dtf. Therefore ·7854D^{2}P = 3·1416Dtf; or PD = 4tf. This shows that a cylindrical shell is twice as strong at a transverse section as at a longitudinal section.

The tenacity of boiler plates in pounds per square inch may be taken at 65,000 for steel, and at 47,000 in the direction of the fibre, and 40,000 in a direction at right angles to the fibre for wrought-iron. For boiler shells the safe stress should not exceed one-fifth of the tenacity.

315. Strength of a Spherical Shell.—Let D = internal diameter of a spherical shell in inches, t = its thickness in inches, P = internal pressure in pounds per square inch, and f = the stress. The resultant R (Fig. 670) of the pressure on one half of the shell is equal to ·7854D^{2}P, and this is the magnitude of the force tending to rupture the shell at a section by a plane passing through the centre of the shell. The area of the section of the shell by a plane passing through its centre is 3·1416Dt, and the resistance to rupture is 3·1416Dtf. Therefore ·7854D^{2}P = 3·1416Dtf;

or PD = $4tf$. As in the case of cylindrical shells, the right-hand side of the foregoing formula must be multiplied by the efficiency of the riveted joint when the shell is riveted.

The strength of dished ends for boiler shells is determined by the formula PD = $4tf$, where D is the diameter of the spherical shell, of which the dished end may be considered to be a portion. From the formulæ for the strength of cylindrical and spherical shells it is obvious that a dished end will have the same strength as the cylindrical shell if

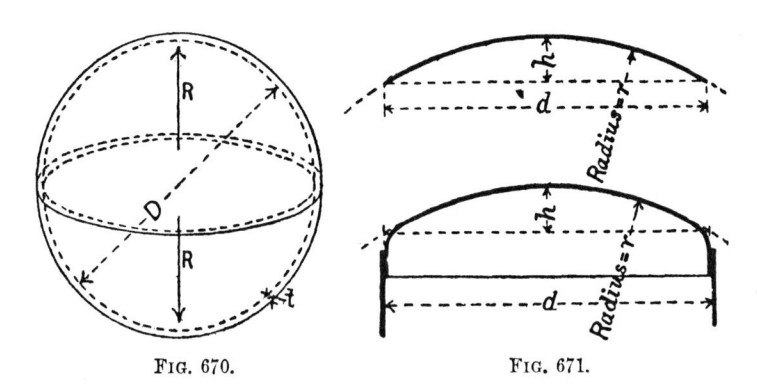

FIG. 670.   FIG. 671.

it is curved with a radius equal to the diameter of the cylindrical shell. If h (Fig. 671) is the amount of camber in a dished end, d the diameter of the cylindrical shell to which it is fitted, and r the radius of curvature, then $r = \dfrac{d^2 + 4h^2}{8h}$, and $h = r - \dfrac{\sqrt{4r^2 - d^2}}{2}$. If $r = d$, then,

$$h = d\left(1 - \dfrac{\sqrt{3}}{2}\right) = \cdot 134 d.$$

316. Furnace Tubes.—So long as a furnace tube which is subjected to external pressure remains truly circular in cross section, the material of the tube is subjected to compression, and the compressive stress is given by the formula $f = \dfrac{PD}{2t}$, where f = compressive stress in pounds per square inch, P = steam pressure in pounds per square inch, D = external diameter of the tube in inches, and t = thickness of tube in inches. The foregoing formula cannot, however, be applied in proportioning ordinary furnaces, owing to the difficulty of guaranteeing that they shall remain truly circular.

The following are the best known rules for the strength of plain furnace tubes:—

Fairbairn's Rule:—P = $\dfrac{806300 t^{2 \cdot 19}}{LD}$, where P = the collapsing pressure in pounds per square inch, t = thickness in inches, D = diameter in inches, and L = length in feet. This formula is based on the assumption that the strength of the tube is inversely proportional to its length,

U

which is not strictly true. To correct this, the variable factor of safety $F = \sqrt{\dfrac{300}{L}}$ is used in finding the working pressure. Hence

$$\text{Working pressure} = \dfrac{806300 t^{2 \cdot 19}}{LD\sqrt{\dfrac{300}{L}}} = \dfrac{46552 t^{2 \cdot 19}}{D\sqrt{L}}$$

$t =$	$\frac{1}{4}$	$\frac{9}{32}$	$\frac{5}{16}$	$\frac{11}{32}$	$\frac{3}{8}$	$\frac{13}{32}$	$\frac{7}{16}$	$\frac{15}{32}$	$\frac{1}{2}$	$\frac{17}{32}$	$\frac{9}{16}$
$t^{2 \cdot 19} =$	·048	·062	·078	·096	·117	·139	·164	·190	·219	·250	·284

Board of Trade Rule:—For circular furnaces with the longitudinal joints welded or made with a butt strap double riveted, or double butt straps single riveted:—

$$\dfrac{90\,000 \times \text{the square of the thickness of the plate in inches}}{(\text{length in feet} + 1) \times \text{diameter in inches}}$$

= working pressure per square inch, provided it does not exceed that found by the following formula:—

$$\dfrac{9000 \times \text{thickness in inches}}{\text{diameter in inches}} = \text{working pressure per square inch.}$$

The second formula limits the crushing stress on the material to 4500 lbs. per square inch.

The length is to be measured between the rings if the furnace is made with rings.

If the longitudinal joints, instead of being butted, are lap-jointed in the ordinary way and double riveted, then 75,000 should be used instead of 90,000; but where the lap is bevelled, and so made as to give the flues the form of a *true* circle, then 80,000 may be used.

For plain steel furnaces the above constants may be increased 10 per cent.

Lloyd's Rule:—

$$\dfrac{89600 T^2}{LD} = \text{working pressure in pounds per square inch,}$$

where T = thickness of plates in inches.
D = outside diameter of furnace in inches.
L = length of furnace in feet. If rings are fitted, the length between rings to be taken.

The pressure in no case to exceed $\dfrac{8000 T}{D}$.

Strengthening Rings.—A long furnace tube is very much strengthened by making it in short pieces, usually about 3 feet long, and connecting these by Adamson's flanged seam, shown in Fig. 672, or by the Bowling hoop, shown in Fig. 673. In the Adamson flanged seam the ends of the different lengths of the furnace tube are flanged outwards and riveted together with a flat ring between them. The ring adds strength to the joint, but it is introduced to allow of the joint being caulked on the inside as well as on the outside, and is frequently called a caulking

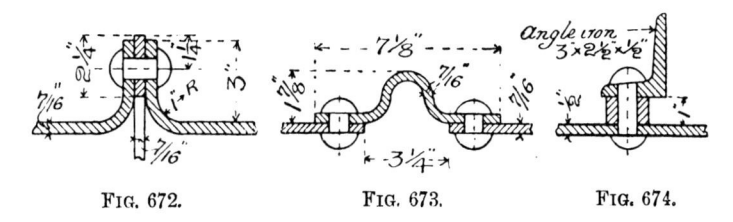

FIG. 672. FIG. 673. FIG. 674.

ring. The Adamson flanged seam has the merit of permitting of a slight longitudinal expansion or contraction of the furnace tube should this be necessary on account of any inequality in the temperatures of the boiler shell and the tube. To secure this longitudinal spring in the furnace tube the inside or smaller curve of the flange should have a radius not less than twice the thickness of the plates.

The Bowling hoop, which is rolled to the desired shape and size without welding, not only gives great strength to the tube, but its form allows of considerable longitudinal spring. It is, however, open to the objections that it presents two rows of rivet heads to the fire and double thicknesses of plates, so that overheating of the rivet heads and plates is more likely to occur. The Bowling hoop, like the Adamson flanged seam, is used to connect the different lengths of which the furnace tube is made up.

If it is desired to strengthen a furnace tube at points other than the joints, this is best done by an angle iron ring, Fig. 674, made in halves, which are riveted with cover-straps after the ring is in position. This ring does not lie close to the tube, but is separated from it by ferrules at intervals of about 6 inches, so that there is a water space about 1 inch wide between the ring and the tube, which prevents overheating. The ring is secured to the tube by rivets, studs, or bolts, which pass through the ferrules already mentioned.

317. Corrugated Furnace Tubes.—A very large number of furnace tubes are now made of a corrugated form, the corrugations running in a circumferential direction round the tube. These tubes have no longitudinal joint, and they are rolled by special machinery to truly circular form. In Fox's corrugated furnace tubes the corrugations, Fig. 675, have a pitch of 6 inches, and a total depth of 2 inches for all sizes of tubes over 24 inches in diameter inside the corrugations. These furnace tubes are made of Leeds Forge Siemens mild steel by the Leeds Forge Company.

FIG. 675.

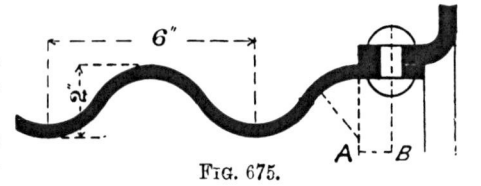

They possess great strength to resist collapse, combined with longitudinal elasticity. Compared with plain tubes they also present a greater heating surface for contact with the products of combustion, but the virtual heating surface for the reception of *radiant* heat is the same in both.

Board of Trade Formula for Fox's Corrugated Furnaces:—

$$\frac{14000T}{D} = \text{working pressure in pounds per square inch.}$$

Provided that the plain parts at the ends do not exceed 6 inches in length (measured from A to B, Fig. 675), and the plates are not less than 5-16ths inch thick.

T = thickness in inches.

D = least outside diameter in inches.

Lloyd's Formula for Fox's Corrugated Furnaces:—

$$\frac{1259(T-2)}{D} = \text{working pressure in pounds per square inch.}$$

T = thickness in 16ths of an inch.

D = *greatest* diameter in inches.

In *Morison's Suspension Furnaces*, also made by the Leeds Forge Company, the corrugations are of the form shown in Fig. 676. Here

FIG. 676.

the corrugations consist of a series of comparatively sharp ridges, which give strength and stiffness to the furnace, and between these ridges the corrugations take the form of catenary curves. A catenary curve is the curve in which a chain of uniform weight hangs when it is suspended from two points, and if a thin rectangular plate be supported at two opposite edges and loaded uniformly, it will bend and form a curve which will, approximately, be a catenary. If this plate is curved before being loaded, the load will tend to bend it into a catenary form. Hence a plate which is bent to a catenary form, before being subjected to a uniform pressure, will be of more uniform strength than one which is bent to any other curve. It is therefore claimed for the Morison suspension furnace that it is of more uniform strength than any other form. The working drawing from which the illustration, Fig. 676, was prepared showed the corrugations with a pitch of 8 inches and a total depth of 2 inches, the internal diameter of the furnace being 3 feet 11½ inches.

The formula allowed by "Lloyds" for the Morison suspension furnace is the same as for Fox's corrugated furnace. The Board of Trade has allowed the provisional formula $\frac{13500T}{D}$ = working pressure in pounds per square inch, where T = thickness in inches, and D = least outside diameter in inches.[1]

---

[1] For much valuable information on furnace tubes, see paper by Mr. D. B. Morison on "Marine Boiler Furnaces," read before the North-East Coast Institution of Engineers and Shipbuilders, December 1892.

Note.—The constant 13500 in the Board of Trade formula for Morison's furnace is now increased to 14000.

318. Boiler Stays.—In steam boilers the flat surfaces which are subjected to pressure require to be supported by *stays*. The principal kinds of stays in use are : (1) direct stays, which are usually round bars placed at right angles to the flat surfaces supported by them ; (2) diagonal and gusset stays, used for supporting a flat surface by tying it to another surface inclined to the first ; (3) girder stays, which are placed edgewise on the flat surface to be supported, and bolted to it at intervals.

Direct Stays.—At (*a*) Fig. 677 is shown one end of a form of stay used for tying together flat surfaces, such as the flat ends of a boiler, which are at a considerable distance apart. This form of stay is made of wrought iron or steel, and each end is screwed to receive two nuts. The outside nut has a thickness equal to the diameter of the screwed part of the stay, and the inside or lock-nut has a thickness of five-eighths to three-fourths of that of the outside nut. This form of stay is not usually screwed into the plates. The plates are stiffened by large washers under

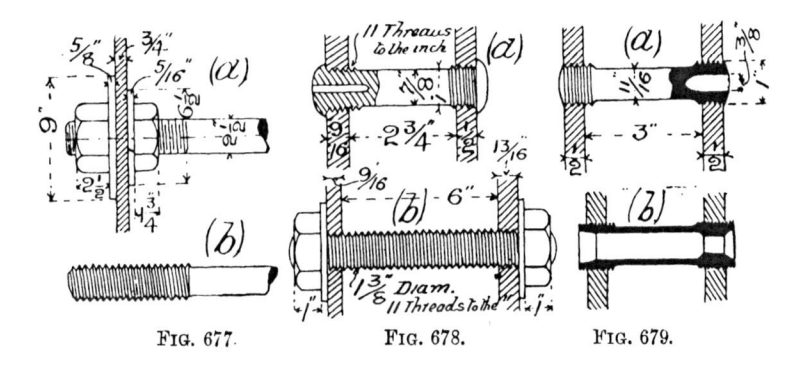

FIG. 677.   FIG. 678.   FIG. 679.

the outside nuts, these washers being nearly always riveted to the plates. When made of wrought-iron, the ends of these stays are generally enlarged as shown at (*b*) Fig. 677, so that the diameter at the bottom of the screw thread is not less than that of the rest of the bar. As the enlarged ends are welded on, and there is always an amount of uncertainty about the strength of a welded joint, this design does not always find favour. With steel, welding is still more unreliable ; hence steel stays are not enlarged at the ends.

In practice these stays do not exceed 3 inches in diameter, and they are spaced from 13 inches to 19 inches apart in horizontal and vertical rows.

The flat parallel surfaces surrounding the fire-boxes of locomotive and marine boilers are stayed by means of *screwed stays*, forms of which are shown in Figs. 678 and 679. These stays are called screwed stays because they are screwed into the plates. In English locomotive practice the screwed stays are generally made of copper, and after being screwed into the plates their ends are riveted over as shown at (*a*) Fig. 678. These copper stays are usually screwed at the ends only, the middle part being turned smaller as shown. The fracture of copper stays is frequently detected by the escape of steam through the small holes which are sometimes drilled through the screwed parts as shown at (*a*) Fig. 678.

The screwed stays for locomotive boilers are usually placed about 4 inches apart, centre to centre, and they vary in diameter from ¾-inch to 1 inch.

In marine boilers the screwed stays are made of steel, and they vary in diameter from 1¼ inches to 1⅝ inches. They are provided with washers and nuts at each end as shown at (*b*) Fig. 678. The nuts have a thickness from five-eighths to three-fourths of the diameter of the stay.

Steel screwed stays of the form shown at (*a*) Fig. 679 were introduced into locomotive practice by Mr. J. C. Park, locomotive superintendent of the North London Railway. The ends of these stays are drilled as shown, and after they are screwed into place a steel drift is driven into the holes; two slight blows on this drift are said to be sufficient to expand the ends so tightly into the plates that a perfectly steam-tight joint is obtained, and no riveting over of the ends is necessary. The steel used for these stays is soft and ductile, and has a tenacity of 26 tons per square inch. The factor of safety used in calculating their diameter is 5 *

Messrs. Yarrow & Co. of London have introduced the modification of Park's screwed stay shown at (*b*) Fig. 679. In this form the two ends of the stay may be expanded by drifts driven from one end only. This is of great advantage where a number of boilers of the locomotive type are placed close together, as is sometimes the case in warships. In such a case the stays may be expanded from the inside of the fire-box.

The diameter of a direct stay may be determined by the formula—

$$\cdot 7854 D^2 f = AP \; ; \; \text{or} \; D = \sqrt{\frac{AP}{\cdot 7854 f}} = C_1 \sqrt{AP}$$

where D = smallest diameter of stay in inches ;
    A = area of plate supported by one stay in square inches ;
    P = working pressure of steam in pounds per square inch ;
    f = safe stress allowed on stay in pounds per square inch.

If the diameter of the stay and working pressure are given, then $A = \dfrac{\cdot 7854 D^2 f}{P} = C_2 \dfrac{D^2}{P}$, and the pitch of the stays, horizontally and vertically, will be equal to $\sqrt{A}$.

If the diameter of the stay and the area of the plate supported by it are given, then $P = \dfrac{\cdot 7854 D^2 f}{A} = C_2 \dfrac{D^2}{A}$.

The safe stress f may be from 4000 to 5000 lbs. per square inch for copper stays.

The Board of Trade allows the following values of f:—For wrought-iron stays which have been welded or worked after heating, $f = 5000$; for

---

* *Note to Fourth Edition.*—The North London Railway Company have discontinued the use of steel for screwed firebox stays. The same design is still used, but the stays are made of copper, and they have a smaller hole than was made in the steel stays.

wrought-iron stays from solid bars, $f = 7000$; for steel stays which have not been welded or otherwise worked after heating, $f = 9000$. The Board of Trade also specifies that the tensile strength of steel stay bars should be from 27 to 32 tons per square inch, with an elongation of about 25 per cent., and not less than 20 per cent. in a length of 10 inches.

"Lloyd's" rules allow the following values of f:—For wrought-iron stays not exceeding $1\frac{1}{2}$ inches effective diameter, and for all stays which are welded, $f = 6000$; for unwelded wrought-iron stays above $1\frac{1}{2}$ inches effective diameter, $f = 7500$; for steel stays not exceeding $1\frac{1}{2}$ inches effective diameter, $f = 8000$; for steel stays above $1\frac{1}{2}$ inches effective diameter, $f = 9000$. No steel stays to be welded.

The following table gives values of C_1 and C_2 for various values of f:—

$f =$	4000	4500	5000	5500	6000	6500	7000	7500	8000	8500	9000
$C_1 =$	·0178	·0168	·0160	·0152	·0146	·0140	·0135	·0130	·0126	·0122	·0119
$C_2 =$	3142	3534	3927	4320	4712	5105	5498	5890	6283	6676	7069

Diagonal Stays.—When it is not possible or desirable to carry a bar stay direct from one flat surface to another parallel to it, the stay may be taken in a diagonal direction, and be secured to a surface at right angles to the first as shown in Fig. 680. The stay shown in Fig. 680 has one end of the form shown in Fig. 677 for a direct stay, while the other has a palm forged on it which is riveted to the plate. This is the

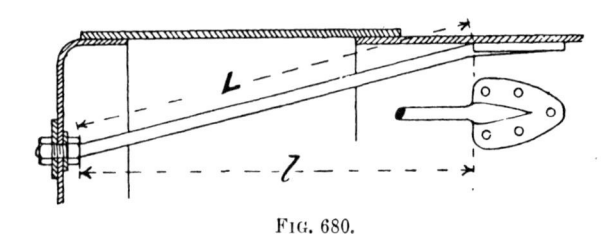

FIG. 680.

most common arrangement, but another design is sometimes adopted, namely, one in which eyes are forged on the ends of the stay to receive pins which are supported by angle irons riveted to the plates. In the latter arrangement the stay is subjected to a simple tension, whereas in the arrangement shown in Fig. 680 there is a bending action at the ends which cannot be determined.

The area of the cross section of a diagonal stay is found as follows:— Find the area of a direct stay to support the same area of plate, and multiply it by the length of the diagonal stay, and divide by the length of its projection on a plane at right angles to the surface supported. Or, using symbols, let—

S = area of cross section of diagonal stay.
s = area of cross section of direct stay to support the same area of surface.
L = length of diagonal stay.
l = length of projection of diagonal stay on a plane perpendicular to the surface to be supported.

Then $S = \dfrac{s \times L}{l}$.

Gusset Stays.—A gusset stay is a diagonal stay in which a flat plate is used instead of a bar or rod.

Perhaps the most important application of the gusset stay is to be found in Cornish and Lancashire boilers for staying the flat ends to the cylindrical shell.

The following description of the arrangement of the gusset stays in a Lancashire boiler, 7 ft. 6 in. in diameter, for a working pressure of 100 lbs. per square inch, is that given by Mr. W. H. Fowler in a specification published by the Technical Publishing Co., Manchester :—

"To arrange the staying so that it shall give adequate support to the flat ends without rendering them unnecessarily rigid, is a point of considerable importance in boilers of the single or double flued internally fired class. If the ends be made too stiff, there is a serious risk of grooving being set up, either in the end plate at the edge of the upper part of the furnace tube angle iron, or at the root of the angle iron itself. This is more particularly the case at the front end, and arises from the varying thrust of the upper portion of the furnace tube upon the end plate. To avoid the risk of grooving, therefore, a certain breathing space should be allowed between the rivets of the furnace tube angle iron and the lowest rivets of the gussets. This is a point in which a mistake is often made, and, with a laudable desire to render the end plate sufficiently strong, the gussets are extended until they almost touch the furnace tube angle iron, the result being that a grooving action is set up when the boiler is put to work, and, although the defect under such circumstances may not be a source of danger, yet it gives rise to considerable annoyance and inconvenience, and is sure sooner or later to necessitate extensive repairs.

"The arrangement of staying shown in Figs. 681 and 682 is the outcome of extensive experience, and represents, in fact, the most modern practice of the best makers of this class of boiler.

"The arrangement of the gussets, it will be seen, follows a simple geometrical construction. The gussets are pitched uniformly at a distance of 21 inches apart, measured along the circumference of the shell, except at the bottom of the front end, where the pitch is reduced to 19 inches, with a view to bringing them as close to the mudhole mouthpiece as possible, and thus assist in supporting the flat space below the tubes. At the back end, as there is no mudhole to interfere with the arrangement, a single central gusset stay is adopted. At the front end, all the gussets, except the two adjacent to the central gusset, are set so as to radiate from the centre, while those adjacent to the

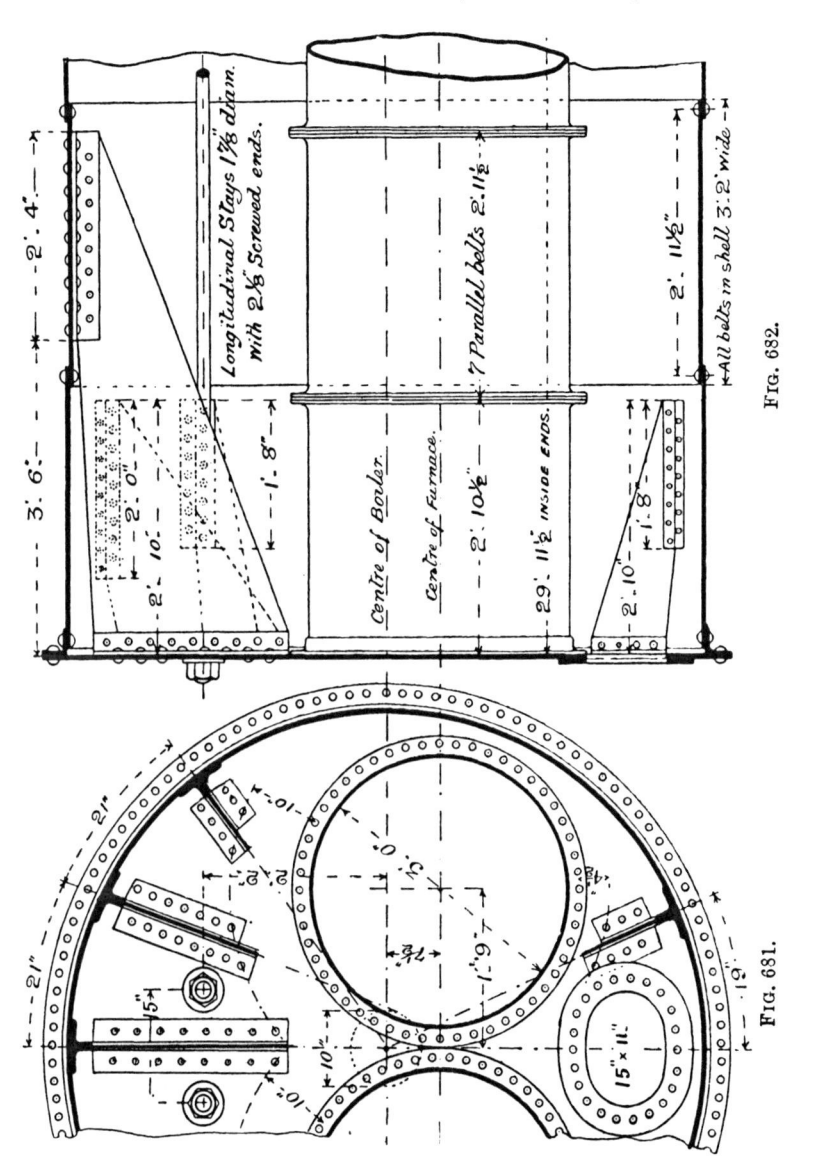

FIG. 682.

FIG. 681.

central gusset are set tangentially to a circle 10 inches in diameter. By this arrangement, it will be noticed, the vertical centre line of the furnace tube bisects, as nearly as may be, the space between the two outer gussets, and thus affords additional breathing space for the furnace tube where it is most required. At the back end the diameter of the furnace tube is reduced from 3 ft. to 2 ft 6 in., and all except the lower and upper central gussets are arranged as tangents to the 10-inch circle.

"It will be observed that the toe rivets of all the gussets at the upper part fall on a circle concentric with and pitched at a distance of 10 inches from the rivets in the furnace tube. Experience over a number of years has shown that, while this amount of breathing space may be allowed without weakening the end plate, it is desirable that it should not be reduced, if grooving is to be avoided. As grooving hardly ever occurs beneath the furnace tubes, the question of breathing space is not of so much importance, and the gussets, in order to afford the requisite support, may be brought closer to the flues, as shown in the drawing.

"A word or two of caution may be given with regard to the two longitudinal bolt stays. These, it may be explained, are not absolutely essential on the score of strength, but rather play the part of sentinels, ready to come into action in the event of anything going wrong. They should, therefore, be left a little slack, so as to afford an inch or two of sag at the middle. If the stays are tightened up taut like fiddle-strings, elasticity of the end plates is unduly interfered with, and we have known of more than one case where serious trouble from straining and breakage has arisen from this cause."

Girder Stays.—The flat crowns of the combustion-chambers or fire-boxes in marine and locomotive boilers may be supported by direct screwed stays from the shell of the boiler; but as the shell of the boiler is usually at a considerable distance from the crown of the fire-box, girder stays are generally preferred. A girder stay consists of a flat

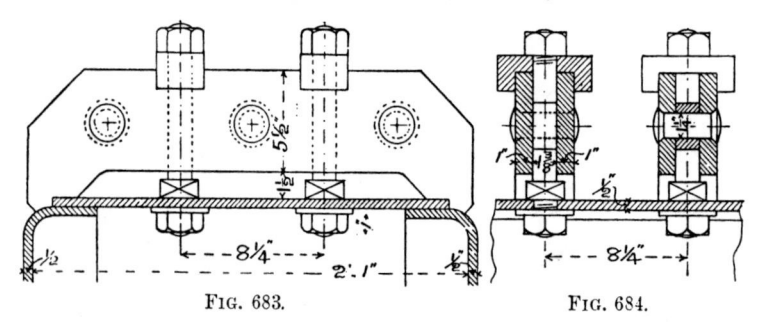

FIG. 683.   FIG. 684.

solid bar or of two plates riveted together with a space between them. This girder is supported at its ends on the vertical plates, forming two opposite sides of the fire-box, and the flat crown is suspended from the girder at intervals by bolts.

Figs. 683 and 684 show a common form of girder stay used in marine boilers. In this example each stay bolt is screwed tightly into the crown-plate and secured with a washer and thin nut on the under side. Each bolt has a solid square collar to receive a spanner, by means of which the bolt is screwed into the crown-plate, the collar pressing hard on that plate.

The following is the rule given by the Board of Trade for proportioning solid rectangular girder stays for marine boilers:—

$$\frac{C \times d^2 \times T}{(W - P)D \times L} = \text{working pressure in pounds per square inch.}$$

W = width of combustion-box in inches.
P = pitch of supporting bolts in inches.
D = distance between the girders from centre to centre in inches.
L = length of girder in feet.
d = depth of girder in inches.
T = thickness of girder in inches.
n = number of supporting bolts.

$$C = \frac{1000\, n}{n+1} \text{ when } n \text{ is odd.} \quad C = \frac{1000\,(n+1)}{n+2} \text{ when } n \text{ is even.}$$

The above values of C apply to wrought-iron girders. If the girders are made of steel, the values of C given above should be increased 10 per cent.

The diameter of the stay bolts is determined by the rule for ordinary stays.

The following is Lloyd's rule for girder stays:—

$$\frac{C \times d^2 \times T}{(L - P)D \times L} = \text{working pressure in pounds per square inch.}$$

L = length of girder in inches.
P = pitch of stays in inches.
D = distance apart of girders in inches.
d = depth of girder at centre in inches.
T = thickness of girder at centre in inches.
C = 6000 if there is one stay to each girder.
C = 9000 if there are two or three stays to each girder.
C = 10,200 if there are four stays to each girder.

When a girder stay is of considerable length, it is very often suspended from the shell of the boiler by sling stays. An example of the application of sling stays is shown in Figs. 685 and 686, which show the design of cast-steel girder stay introduced by Mr. Worsdell, the locomotive superintendent of the North-Eastern Railway. These illustrations have been prepared from the working drawing of the boiler for a compound locomotive, the working pressure being 175 lbs. per square inch.

319. Strength of Flat Plates Supported by Stays.—The following is the Board of Trade rule for the working pressure on the stayed wrought-iron flat plates of marine boilers:—

$$\frac{C(T + 1)^2}{S - 6} = \text{working pressure in pounds per square inch.}$$

T = thickness of the plate in sixteenths of an inch.
S = surface supported in square inches.

FIG. 686.

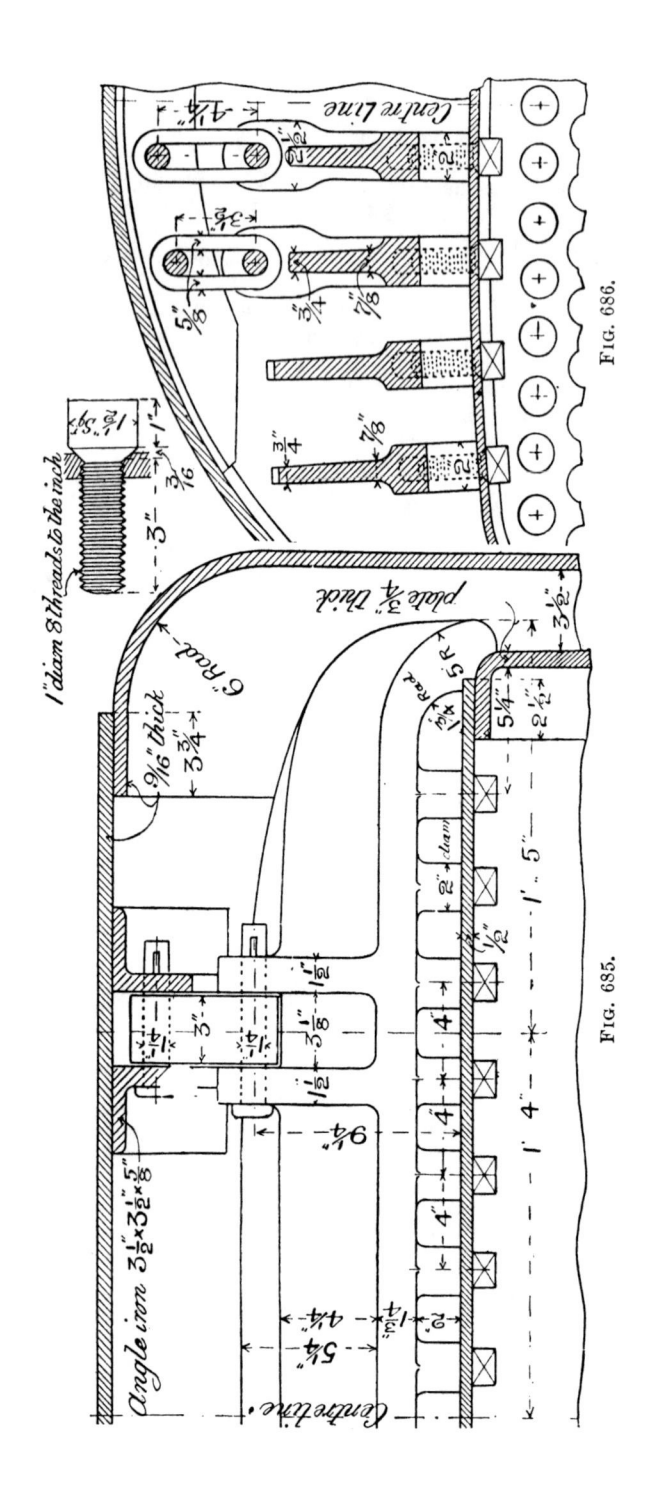

FIG. 685.

STEAM BOILERS.

$C = 100$ when the plates are not exposed to the impact of heat or flame, and the stays are fitted with nuts and washers, the latter being at least three times the diameter of the stay and two-thirds the thickness of the plates they cover.

If the diameter of riveted washers be at least two-thirds the pitch of the stays, and the thickness not less than the plates they cover, the constant may be increased to 150.

When doubling plates are fitted of the same thickness as the plates they cover, and not less in width than two-thirds the pitch of the stays, the constant may be increased to 160.

- $C = 90$ when the plates are not exposed to the impact of heat or flame, and the stays are fitted with nuts only.
- $C = 67\frac{1}{2}$ when the plates are not exposed to the impact of heat or flame, and the stays are screwed into the plates and riveted over. nuts and washers, the latter being at least three times the diameter of the stay and two-thirds the thickness of the plates they cover.
- $C = 54$ when the plates are exposed to the impact of heat or flame, and steam in contact with the plates, and the stays fitted with nuts only.
- $C = 80$ when the plates are exposed to the impact of heat or flame, with water in contact with the plates, and the stays screwed into the plates and fitted with nuts.
- $C = 60$ when the plates are exposed to the impact of heat or flame, with water in contact with the plates, and the stays screwed into the plates, having the ends riveted over to form substantial heads.
- $C = 36$ when the plates are exposed to the impact of heat or flame, and steam in contact with the plates, with the stays screwed into the plates, and having the ends riveted over to form substantial heads.

For steel plates the constant C in the above formula may be increased 10 per cent. when they are supported by stays screwed into the plates and riveted, and when they are supported by stays screwed into the plate and nutted, or when the stays are nutted in the steam space, 25 per cent. This is also applicable to the constants for flat surfaces stiffened by riveted washers or doubling strips, and supported by nutted stays.

The following is Lloyd's rule for the working pressure on the stayed flat plates of marine boilers :—

$$\frac{CT^2}{P^2} = \text{working pressure in pounds per square inch.}$$

T = thickness of plates in sixteenths of an inch.
P = greatest pitch of stays in inches.

$C = 90$ for plates 7-16ths inch thick and under, fitted with screw stays with riveted heads.

$C = 100$ for plates above 7-16ths inch thick, fitted with screw stays with riveted heads.

$C = 110$ for plates 7-16ths inch thick and under, fitted with screw stays and nuts.

$C = 120$ for plates above 7-16ths inch thick, fitted with screw stays and nuts.

$C = 140$ for plates fitted with stays with double nuts.

$C = 160$ for plates fitted with stays with double nuts, and washers at least half the thickness of the plates, and of a diameter 2-5ths of the pitch, riveted to the plates.

In the case of front plates of boilers in the steam space, these numbers should be reduced 20 per cent., unless the plates are guarded from the direct action of the heat.

CHAPTER XXIII.

STEAM ENGINES.

GENERAL DIMENSIONS.

320. Indicated Horse-Power.—The term "indicated horse-power" denotes the power developed in the cylinder of an engine.

Let D = diameter of cylinder in inches.
  L = length of stroke of piston in feet.
  N = number of strokes per minute.
  S = speed of piston in feet per minute.
  p = mean effective pressure of steam on piston in pounds per square inch, as determined from the indicator diagram.
  IHP = indicated horse-power.

$$\text{IHP} = \frac{\cdot 7854 D^2 p L N}{33000} = \frac{\cdot 7854 D^2 p S}{33000}.$$

If the engine has more than one cylinder, the IHP of each is first determined by the foregoing formula. The IHP of the whole engine is then obtained by adding together the IHP's of the separate cylinders.

321. Diameter of Cylinder.—If the indicated horse-power, IHP, the speed of the piston S, and the mean effective pressure, p, be given, the diameter of the cylinder is found from the formula—

$$D = \sqrt{\frac{33000\ \text{IHP}}{\cdot 7854 p S}}.$$

322. Length of Stroke of Piston.—If the speed of the piston, S, and the number of revolutions, R, of the crank shaft per minute are given, then the length of the stroke, L, in feet is given by the formula $L = \dfrac{S}{2R}$.

No definite rule can be given for the length of the stroke. In single cylinder engines the stroke of the piston is generally made to depend on the diameter of the cylinder, but the relation is very variable; some engineers make the length of the stroke equal to twice the diameter of the cylinder. In locomotive engines the stroke of the pistons varies from 1·2 to 1·55 times the diameter of the cylinders.

323. Speed of Piston.—The speed of a piston is usually given in feet per minute, and is the distance in feet which it goes in one minute.

This is the mean speed. The speed varies from nothing at the beginning and end of each stroke to a maximum near the middle of the stroke. The mean speed of the piston is determined by multiplying the length of the stroke in feet by the number of strokes per minute.

In the case of a locomotive engine, if L = length of stroke of pistons in feet, D = diameter of driving wheels in feet, M = speed of train in miles per hour, and S = speed of pistons in feet per minute, then—

$$S = \frac{M \times 5280 \times 2 \times L}{D \times 3\cdot1416 \times 60} = \frac{56\cdot02 ML}{D}.$$

For engines of the same class the speed of piston is generally greater, the greater the length of the stroke, and may be taken as proportional to the cube root of the stroke.

The following table gives speeds of pistons to be met with in ordinary practice:—

Class of Engine.	Speed of Piston in Feet per Minute.
Ordinary direct-acting pumping engines (non-rotative)	90 to 130
Ordinary horizontal engines	200 to 400
Horizontal compound and triple-expansion mill engines	400 to 800
Ordinary marine engines	400 to 650
Engines for large high-speed steam-ships	700 to 900
Locomotive engines (express)	800 to 1000
Engines for torpedo-boats	1000 to 1200

324. Clearance and Clearance Volume.—The term "clearance" as applied to steam cylinders may mean the distance between the cylinder cover and the piston when the latter is nearest to the cover, or it may mean the volume of the space between the piston and the steam valve when that space is least—that is, when the piston is at the beginning of its stroke. When the first meaning above stated is intended, we shall use the term *clearance*, and when the second meaning is intended, we shall speak of *clearance volume*.

Clearance in a steam cylinder is necessary (1) to provide for any slight inaccuracy in the setting of the cylinder in relation to the crank shaft; (2) to provide for inequalities on the surfaces of the piston and the cylinder cover; (3) to provide for the slight errors which may occur in the lengths of the piston-rod, connecting-rod, and crank arms; and (4) to provide for the wear which takes place at the cross-head, crank-pin, and crank-shaft bearings.

The amount of the clearance varies with the size of the engine, being about three-eighths of an inch in small engines, and seven-eighths of an inch in large engines. In horizontal engines the clearance is generally the same at both ends of the cylinder, but in inverted cylinder engines, such as are now used in steamships, the clearance at the lower end is usually about one and a half times the clearance at the upper end.

The clearance volume is usually expressed as a percentage of the volume swept through by the piston in one stroke, and it varies from 2 to 15 per cent. in different cases. Examples of the amount of clearance volume, taken from actual practice, are given in the following table.

STEAM ENGINES.

Examples of Clearance Volume in Steam-Engine Cylinders.

Diameter of Cylinder.	Length of Stroke.	Description of Engine and Cylinder.	Type of Valve.	Clearance Volume.
inches.	inches.			Per cent.
18	48	Single cylinder condensing engine	Corliss valves	2·65
18	48	,, ,, ,, ,,	,, ,,	1·93
7	10	Compound portable engine, H.P. cylinder	Slide valve	13·8
10½	10	,, ,, ,, ,, L.P. ,,	,, ,,	8·8
20	60	Compound condensing engine, H.P. cylinder	,, ,,	8·3
34	60	,, ,, ,, L.P. ,,	,, ,,	6·8
27	72	,, ,, ,, H.P. ,,	Corliss valves	2·2
45	72	,, ,, ,, L.P. ,,	{ Slide valve at each end }	3·2
27·35	33	Compound marine engine, H.P. cylinder	Slide valve	8·5
50·3	33	,, ,, ,, L.P. ,,	,, ,,	5
30	36	,, ,, ,, H.P. ,,	Piston valve	9·39
57	36	,, ,, ,, L.P. ,,	Slide valve	6·23
21·88	39	Triple expansion marine engine, H.P. cylinder	,, ,,	12·41
31·02	39	,, ,, ,, ,, I.P. ,,	,, ,,	10·11
56·95	39	,, ,, ,, ,, L.P. ,,	,, ,,	7·64
26·03	42	,, ,, ,, ,, H.P. ,,	Piston valve	14·51
42·03	42	,, ,, ,, ,, I.P. ,,	,, ,,	9·25
68·95	42	,, ,, ,, ,, L.P. ,,	Slide valve	5·1
29·37	47·94	,, ,, ,, ,, H.P. ,,	Piston valve	12·4
44·03	47·94	,, ,, ,, ,, I.P. ,,	,, valves (2)	9·3
70·12	47·94	,, ,, ,, ,, L.P. ,,	,, ,, (2)	8·02

325. Theoretical Diagram of Work done in a Steam Cylinder.—A line AB (Fig. 687), which represents the length of the stroke of a piston, will also represent the volume swept through by the piston in one stroke. Let AO represent, on the same scale, the clearance volume. Draw AC at right angles to AB to represent the *absolute* steam pressure per square inch on the piston at the beginning of its stroke. Suppose that the steam valve remains open till the piston has reached the point N. Draw ND at right angles to AB, and CD parallel to AB. The horizontal line CD will show that the steam pressure is constant while the piston moves from A to N. There is now enclosed in the cylinder behind the piston a volume of steam represented by ON, and as the piston moves forward this steam will expand and consequently fall in pressure. Assuming that the steam expands according to Boyle's law, its pressure when the piston reaches any point, Q, is determined by the following simple construction. Draw QR perpendicular to AB and DR parallel to AB. Draw OR meeting ND at S. Draw SE parallel to AB to meet QR at E. QE will represent the absolute pressure of the steam in the cylinder when the piston has reached the point Q. A number of points such as E having been determined in the same way, a fair curve, DEF, drawn through them will show the variation of the pressure of the steam from the point of cut off N to the end of the stroke B.

When the piston reaches the end of its forward stroke, the opening

X

of the exhaust valve will cause the pressure to fall from BF to some other pressure, BH. The piston now commences its return stroke, and so long as the exhaust valve is open the pressure of the steam will remain constant. Let the exhaust remain open until the piston reaches the point T, at which point it closes. From B to T the pressure is uniform, as shown by the horizontal line HK. There is now in the cylinder a quantity of steam of volume OT and pressure TK. As the piston moves onwards to the left this steam will be compressed, and consequently rise in pressure. Assuming that the pressure of the steam, as it is compressed, follows Boyle's law, its pressure, when the piston reaches any point, U, is determined as follows :—Draw UV at right angles to AB, and KV parallel to AB. Join OV and produce it to meet

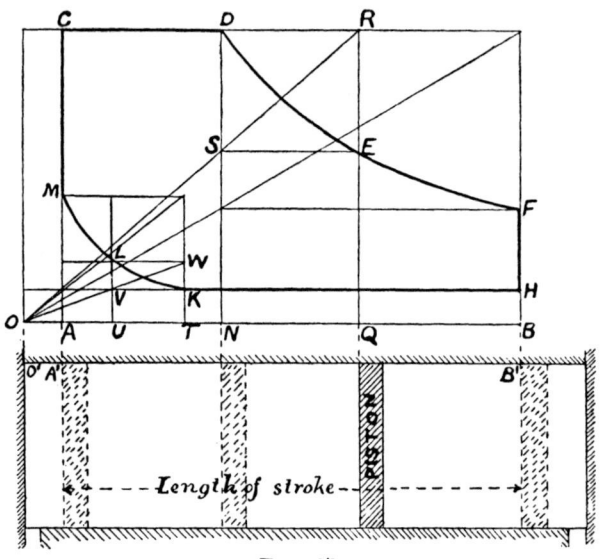

FIG. 687.

TK, produced at W. Draw WL parallel to AB to meet UV produced, at L. UL will represent the absolute pressure of the steam in the cylinder when the piston has reached the point U. A number of points such as L having been determined in the same way, a fair curve, KLM, drawn through them will show the variation of the pressure of the steam from the point of compression T to the end of the stroke A, where the steam valve is again opened, and the pressure rises at once from AM to AC.

The complete diagram CDEFHKLM represents the variation in the pressure of the steam on one side of the piston for two strokes or one revolution. A similar diagram may be drawn for the steam on the other side of the piston.

Confining our attention to the steam on one side of the piston, the average absolute pressure during the forward stroke will be the mean height of the line CDEF above the line AB, measured with the pressure scale, and the work done will be this mean height multiplied by the length of the stroke AB. But the area of the figure ACDEFB is equal to its mean height multiplied by its length, AB. The area of the figure ACDEFB therefore represents the work done by the steam on the left-hand side of the piston during the forward stroke. During the return stroke the pressure of the steam on the left-hand side of the piston will act as a resistance to its motion, and work will be done upon it. This amount of work will be represented by the area AMLKHB. This may be called negative work. Hence the total effective work done during two strokes, or one revolution, by the steam on the left-hand side of the piston will be represented by the area of the figure CDEFHKLM, which is the difference between the areas of the figures ACDEFB and AMLKHB.

The work done on the other side of the piston is determined in the same way, and the two results added together gives the total work done during two strokes or one revolution.

In practice the diagrams which represent the work done on the two sides of the piston do not differ much from one another, and for the purpose of the designer of an engine it will be sufficient if we assume that they are exactly alike, and that the work done in one revolution is twice that represented by the area of the figure CDEFHKLM.

326. Theoretical Mean Effective Steam Pressure on a Piston.—The mean height of the diagram CDEFHKLM (Fig. 687) will represent the mean effective pressure which is used in determining the horse-power of an engine, and it is for the purpose of determining this mean pressure that the designer constructs the diagram explained in the preceding article.

Having shown how to draw the diagram of work done in a steam-engine cylinder, we will now work out an example showing how to determine the mean effective pressure. The particulars for the example are, initial absolute steam pressure, 50 lbs. per square inch, cut off at $\frac{3}{10}$ths of stroke, exhausting into the atmosphere at a pressure of $15\frac{1}{4}$ lbs. per square inch, point of compression at $\frac{8}{10}$ths of stroke. Clearance volume, $\frac{1}{10}$th.

The diagram of work, CDEFHKLM (Fig. 688), is first carefully drawn by the construction explained in the preceding article. The length AB is next divided into ten equal parts, and vertical lines or ordinates drawn across the diagram through the middle points of these parts as shown. The lengths of these lines intercepted by the diagram are then measured by the pressure scale and noted down as shown. The sum of these, 178·3, divided by 10, gives 17·83, which is the mean effective pressure in pounds per square inch.

327. Indicator Diagram and Actual Mean Pressure.—The diagram of work which we have considered in the two preceding articles only represents approximately what takes place in the steam cylinder of an engine. But, by means of the steam-engine "indicator," a diagram of work is drawn which shows exactly the pressure of the steam at every point of the stroke, and from this diagram the actual mean effective steam pressure on the piston may be obtained in the same way as explained for the theoretical diagram.

It does not fall within the scope of this work to discuss the various forms of the indicator diagram and the causes which produce the differences between it and the theoretical diagram. It is of importance, however, to know approximately the relation between the mean pressure as obtained from the indicator diagram, and that obtained from the theoretical diagram. If P_m is the theoretical mean pressure, and p_m the probable actual mean pressure, then $\dfrac{p_m}{P_m} = e$ is called the diagram factor.

For simple engines working expansively e varies from ·75 to ·95. The higher value being applicable to large cylinders provided with steam jackets, and wide ports and passages, and fitted with valves

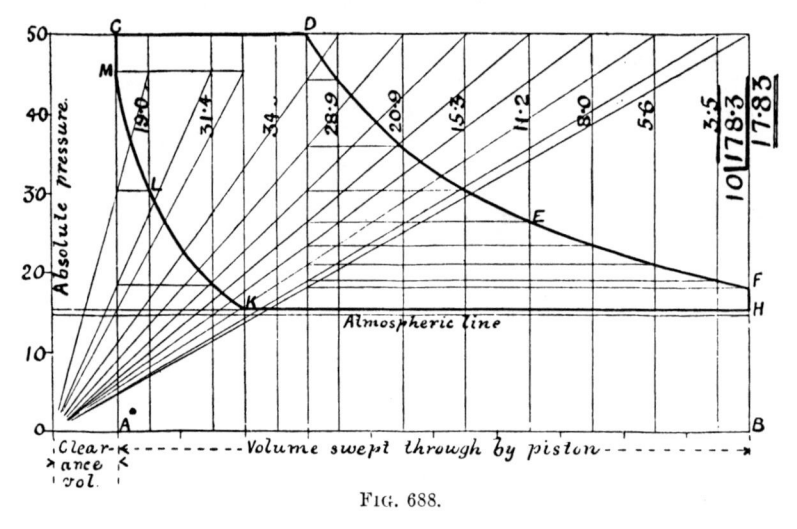

FIG. 688.

which open and close promptly, so that there is little wire-drawing of the steam. The lower value of e would be applicable to small cylinders, unjacketed, and fitted with ordinary slide valves worked by eccentrics in the ordinary way.

328. Compound and Multiple Stage Expansion Engines.—The greater the amount of the expansion of the steam in a steam cylinder, the greater, theoretically, is the amount of work obtained from a given weight of steam. But when steam expands its temperature falls, consequently the steam cylinder is heated by the high pressure steam at the beginning of the stroke, a result which is followed by a partial condensation of that steam. Then, as the steam expands, its temperature falls, and so does that of the cylinder, part of the heat of the latter going to warm the steam and re-evaporate the water deposited at the beginning of the stroke. It is, therefore, evident that with a great range of temperature in the cylinder, consequent on a high rate of expansion, a portion of steam may be condensed at the beginning of the stroke, and reconverted into steam towards the end of the stroke, and thus pass through the cylinder without doing work.

To obtain the advantage following a high rate of expansion, and to obviate to some extent the loss due to the fall in temperature of the expanded steam, the compound engine was introduced. In the compound engine the steam is allowed to expand to a certain extent in one cylinder, and is then passed into a larger cylinder and allowed to expand still further. In this way the range of temperature in each cylinder is much less than it would be if the whole expansion was carried out in one cylinder, and any condensed steam, re-evaporated at the end of the stroke in the first cylinder, does work in the second. But, neglecting the losses due to condensation and wire-drawing of the steam, the total work done in the two cylinders is the same as if the steam was used in the larger or low-pressure cylinder only, the total amount of expansion being the same in both cases.

The range of temperature may be still further reduced by expanding the steam successively in three cylinders as in triple expansion engines, and still further by expanding successively in four cylinders as in quadruple expansion engines.

329. Mean Pressure Referred to Low-Pressure Cylinder.—Let P_1 denote the absolute initial steam pressure in the high-pressure cylinder, and P_2 the absolute terminal pressure in the low-pressure cylinder. Then the total ratio of expansion is $P_1 \div P_2$. Assuming that the whole of the expansion is carried out in the low-pressure cylinder, the theoretical diagram for the work done may be drawn as explained in article 325, p. 321. Referring to Fig. 687, an arbitrary length AB is taken to represent the volume of the low-pressure cylinder, and a portion AO is added on to represent the clearance volume. BF is then made equal to P_2, $AC = P_1$ and $ON = \dfrac{P_2}{P_1} \times OB$. The back pressure BH may be taken at from 2 to 5 lbs. for condensing engines, and from 15 to 16 lbs. for non-condensing engines. The terminal pressure P_2 may be taken at from 7 lbs. to 10 lbs. for condensing engines. From this diagram the theoretical mean effective pressure P_m may be obtained as explained in article 326, p. 323. The probable actual mean effective pressure is then obtained by multiplying P_m by the diagram factor e.

For compound or two stage expansion engines e varies from ·7 to ·9, and for triple expansion engines, neglecting the effects of clearance and cushioning on the theoretical diagram, e varies from ·6 to ·7.

330. Diameter of Low-Pressure Cylinder.—Having obtained the probable mean effective pressure reduced to the low-pressure cylinder, the size of that cylinder may be obtained in the same way as for a single cylinder engine.

EXAMPLE.—*To find the diameter of the low-pressure cylinder for a set of triple expansion engines to indicate 800 horse-power with a piston speed of 450 feet per minute. Initial pressure 165 lbs., terminal pressure 11 lbs., and back pressure 3 lbs., all per square inch absolute.*

From the diagram, Fig. 689, which is drawn without taking account of clearance or cushioning, the theoretical mean effective pressure is

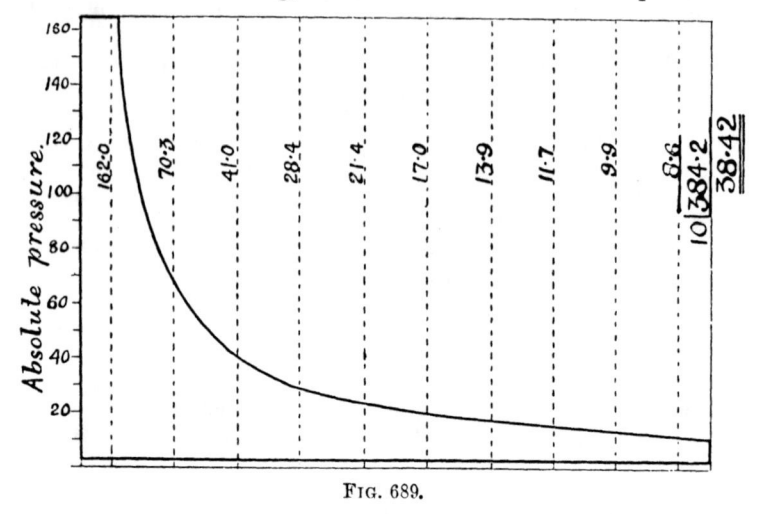

FIG. 689.

found to be 38·42. Taking the diagram factor as ·65, we get the probable actual mean effective pressure equal to 38·42 × ·65 = 24·973, say 25 lbs. per square inch. Then by the formula, $D = \sqrt{\dfrac{33000 \, \text{IHP}}{·7854 p S}}$ given in article 321, p. 319, we get—

$$D = \sqrt{\dfrac{33000 \times 800}{·7854 \times 25 \times 450}} = 54·66 \text{ inches, say 55 inches.}$$

For compound locomotives, in which there is one high-pressure cylinder and one low-pressure cylinder, Mr. Von Borries[1] gives the following rule for finding the diameter of the low-pressure cylinder:—

$$d = \sqrt{\dfrac{2ZD}{ph}},$$

where d = diameter of low-pressure cylinder in inches, D = diameter of driving-wheel in inches, p = mean effective steam pressure per square inch (after deducting internal machine friction), h = stroke of piston in inches, Z = tractive force required, usually ·14 to ·16 of the adhesion. The value of p depends upon the relative volumes of the two cylinders and from experience may be taken as follows :—

Class of Engine.	Ratio of Cylinder Volumes.	p in Percentage of Boiler Pressure.	p for Boiler Pressure of 176 lbs.
Large tender engine	1 : 2 or 1 : 2·05	42	74
Tank engines	1 : 2·15 or 1 : 2·2	40	71

[1] *Engineering*, vol. 50, p. 39.

331. Ratios of Cylinder Volumes in Compound and Multiple Stage Expansion Engines.

—When steam is expanded in two or more cylinders successively, the considerations which determine the relative volumes of the cylinders are: (1) The distribution of the power between the cylinders; (2) the distribution of the initial loads on the pistons; and (3) the range of temperature in each cylinder. When there is a separate crank for each cylinder it is very desirable that the total power should, as nearly as possible, be equally divided between the cylinders, and also that the initial or maximum loads on the pistons should be nearly equal. In all cases also the range of temperature should be as nearly as possible the same in each cylinder. To show how nearly these conditions are sometimes complied with in practice, we give the following particulars of the triple expansion engines of the ss. *Iona*, with a few of the results of the trials conducted by the Research Committee of the Institution of Mechanical Engineers :—

		Cylinders.		
		H.P.	I.P.	L.P.
Diameters of cylinders	inches	21·88	34·02	56·95
Length of stroke of pistons	inches	39	39	39
Speed of pistons	ft. per min.	397·15	397·15	397·15
Mean effective pressures	lbs. per sq. in.	46·65	20·44	7·16
Indicated horse-power		205·6	221·2	218·6
Highest absolute pressure of steam	lbs. per sq. in.	163	59	14·7
Lowest absolute pressure of steam	lbs. per sq. in.	48	13	2
Highest temperature	degrees Fah.	365	292	212
Lowest temperature	degrees Fah.	278	206	126
Range of temperature	degrees Fah.	87	86	86
Maximum loads on pistons	lbs.	43,600	41,800	31,900

No hard and fast rules can be given for the relative volumes of the cylinders of compound and multiple stage expansion engines. In actual practice the ratios of the volumes of the cylinders vary considerably. The following rules and tables are based on the averages of a large number of examples from recent practice, and may therefore be of service in determining, provisionally, the relative sizes of the cylinders.

P = Pressure of steam in boiler by gauge in pounds per square inch.

Compound or two stage expansion condensing engines.

$$\frac{\text{Vol. of L.P. cylinder}}{\text{Vol. of H.P. cylinder}} = \frac{4P + 40}{100}.$$

The following examples have been worked out by the above rule, the volume of the high-pressure cylinder being taken as 1.

Boiler Pressure by Gauge.	60	70	80	90	100	110	120
Volume of H.P. cylinder	1	1	1	1	1	1	1
Volume of L.P. cylinder	2·8	3·2	3·6	4·0	4·4	4·8	5·2

Triple or three stage expansion condensing engines.

$$\frac{\text{Vol. of I.P. cylinder}}{\text{Vol. of H.P. cylinder}} = \frac{P + 100}{100}.$$

$$\frac{\text{Vol. of L.P. cylinder}}{\text{Vol. of H.P. cylinder}} = \frac{4P + 50}{100}.$$

The following examples have been worked out by the preceding rules, the volume of the high-pressure cylinder being taken as 1.

Boiler Pressure by Gauge.	120	130	140	150	160	170	180
Volume of H.P. cylinder	1	1	1	1	1	1	1
Volume of I.P. cylinder	2·2	2·3	2·4	2·5	2·6	2·7	2·8
Volume of L.P. cylinder	5·3	5·7	6·1	6·5	6·9	7·3	7·7

332. Thickness of Cylinder Barrel—Cylinder Liner.—Steam cylinders are made of cast-iron. The thickness of the barrel should not be less than $\frac{pD}{2500}$, and is generally equal to $\frac{pD}{3500} + \frac{1}{2}$, where p is the steam pressure in pounds per square inch and D the diameter of the cylinder in inches. If the cylinder is fitted with a liner the thickness of the latter may be $\frac{pD}{3500} + \frac{1}{2}$ for cast-iron, and $\frac{pD}{3000}$ for steel. Fig. 690 shows one method of fitting a liner to a cylinder. The joint at the top is, in this design, made steam tight by caulking a copper ring into a dovetailed recess. The proportions of the various parts are marked on the figure, the unit being the thickness of the liner. Frequently the liner is carried right out to meet the cover as shown in Fig. 694. The space between the liner and the cylinder barrel may be used as a steam-jacket.

FIG. 690.

333. Cylinder Cover.—In order to diminish the clearance space the surface of the cover inside the cylinder is made to follow the shape of the outer surface of the piston.

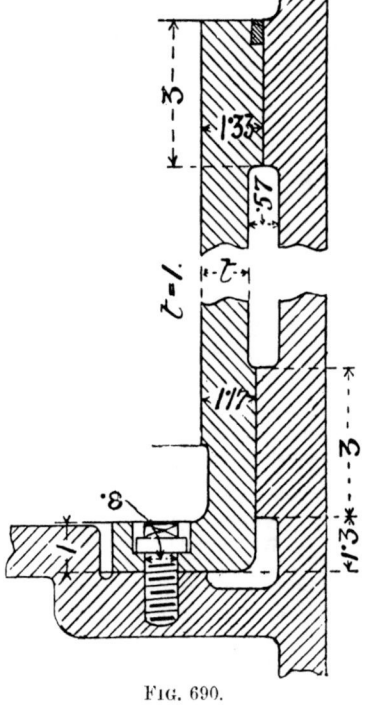

If the cover consists of a flat plate firmly secured to the cylinder at its edge, as shown in Fig. 691, the thickness of the cover is given by the formula $t = \dfrac{D\sqrt{p}}{\sqrt{6f}}$, where t is the thickness of the cover in inches, D the diameter of the cylinder in inches, p the pressure of the steam in pounds per square inch, and f the stress in pounds per square inch. Taking $f = 3000$ for cast-iron, $t = \cdot0075 D \sqrt{p}$, or taking the unit $= \dfrac{D\sqrt{p}}{100}$ then $t = \cdot 75$. Generally the cover is let into the cylinder as shown in Fig. 692. This is equivalent to diminishing the diameter of the cover,

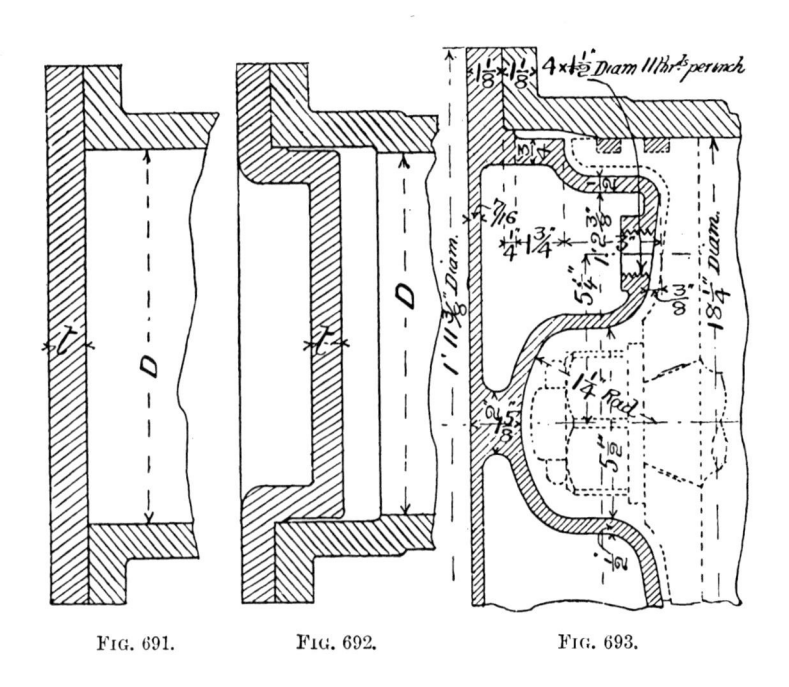

FIG. 691.   FIG. 692.   FIG. 693.

and the thickness may therefore be diminished. From examples in practice we find that for the dished cover (Fig. 692) the thickness t varies from ·4 to ·9, the average of a number of examples giving $t = \cdot 6$, the unit being $\dfrac{D\sqrt{p}}{100}$.

For small cylinders the cover is made with a single thickness of metal. Covers for medium-sized cylinders may have a single thickness strengthened by radial ribs on the outside, or they may be cast hollow. Covers for large cylinders are always made hollow with internal radial stiffening ribs.

Fig. 693 shows a hollow cylinder cover for a locomotive cylinder.

Figs. 694 and 695 show examples of hollow covers suitable for large cylinders. The proportions are marked on the figures, the unit being as

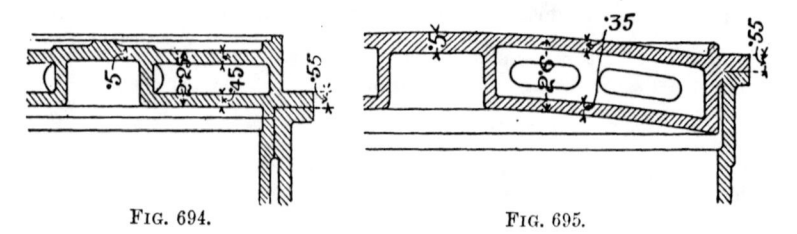

FIG. 694.

FIG. 695.

before $\dfrac{D\sqrt{p}}{100}$. In compound and triple expansion engines the value of the unit, $\dfrac{D\sqrt{p}}{100}$, may be taken as the same for each cylinder.

334. Cylinder Flange.—The thickness of the cylinder flange varies in practice from 1·2 to 1·4 times the thickness of the cylinder barrel. The width of the flange is determined by the size of the bolts for securing the cover to the cylinder. The distance from the centre of the bolts to the outside of the flange should not be less than $d + \tfrac{1}{4}$ inch, and need not be more than $1\tfrac{1}{2}d$, where d is the diameter of the bolt over the threads.

335. Bolts for Cylinder Cover.—The bolts used for securing the cylinder cover to the cylinder are nearly always stud bolts. Stud bolts are preferred to bolts with heads for two reasons: first, the cylinder flange may be narrower when stud bolts are used; second, bolts with heads cannot be removed without interfering with the lagging round the cylinder.

Let d = nominal diameter of bolts in inches.
d_1 = diameter of bolts at bottom of screw thread in inches.
n = number of bolts.
f = stress on bolts in pounds per square inch of net section.
D = diameter of cylinder.
p = pressure of steam in pounds per square inch.

Then $\cdot 7854 d_1^2 n f = \cdot 7854 D^2 p$

and $d_1 = D\sqrt{\dfrac{p}{nf}}$.

Bolts of less than $\tfrac{5}{8}$-inch nominal diameter should not be used for cylinder covers, and the stress f should be less for small bolts than for large ones. f may be taken at $4000d$, but should not exceed 6000.

The number of bolts depends on their distance apart, which, again, depends on the thickness of the flange of the cover and the pressure of the steam. One rule gives the maximum pitch of the bolts equal to $40\sqrt{\dfrac{t}{p}}$, where t is the thickness of the flange in inches.

336. Areas of Steam Pipes, Ports, and Passages.—Let V be the mean velocity of the steam in feet per minute, A the area of the pipe, port, or passage at right angles to the direction of flow, in square inches, D the diameter of the piston in inches, and S its mean speed in feet per minute. Then, $A = \dfrac{\cdot 7854 D^2 S}{V}$. If the pipe, port, or passage be circular and of a diameter d inches, then $\cdot 7854 d^2 = \dfrac{\cdot 7854 D^2 S}{V}$, and, $d = D\sqrt{\dfrac{S}{V}}$.

The following table gives the value of V usually taken in different cases:—

Main steam-pipes	5,000 to 8,000
Exhaust pipes, ports, and passages	4,000 ,, 6,000
Stop and throttle valves	4,000 ,, 6,000
Steam ports and passages	4,000 ,, 7,000
Steam port opening. Ordinary slide valve	6,000 ,, 9,000
,,　　,,　　,,　　Quick cut-off valve	9,000 ,, 12,000

337. Jet Condensers—Quantity of Injection Water.—The terminal pressure in the low-pressure cylinder of a steam-engine may be taken at from 5 to 10 lbs. per square inch absolute, so that every pound weight of steam as it passes into the condenser will contain from 1131 to 1140 British thermal units of heat above that contained by 1 lb. of water at 32° F. This steam is condensed to water having a temperature of 100° to 120°, by mixing it with water of the ordinary temperature. If the temperature of the injection water is 50° and that of the hot well 110°, then every pound of injection water will take up 110 − 50 = 60 units of heat. If the total heat in each pound of steam as it leaves the cylinder be taken at 1140 above that contained by water at 32°, then each pound of steam in condensing must give up 1140 + 32 − 110 = 1062 units. Hence weight of water to condense 1 lb. of steam = $\dfrac{1062}{60}$ = 17·7 lbs.

Let H = number of units of heat in 1 lb. weight of steam above that contained by 1 lb. of water at 32°.
T = temperature of hot well.
t = temperature of injection water.
W = weight of injection water (in pounds) required for each pound of steam.

Then $W = \dfrac{H + 32 - T}{T - t}$

A common rule is, weight of injection water = 25 to 30 times the weight of the steam to be condensed.

The volume of a jet condenser should be proportioned according to the volume of the low-pressure cylinder. The ratio of the volume of the condenser to that of the low-pressure cylinder varies greatly in practice. It should not be less than $\frac{1}{4}$, but we know of examples from recent practice where it is as high as $1\frac{1}{4}$. A common rule is to make the volume of the condenser half that of the low-pressure cylinder exhausting into it.

The area of the injection orifice is given by the formula $A = \dfrac{W}{26v}$, where—

    A = area of valve in square inches.
    W = weight of injection water required per minute in pounds.
    v = velocity of water through the orifice in feet per second.

In many cases in practice v may be taken at 20 feet per second, then, $A = \dfrac{W}{520}$.

338. Surface Condensers.—Since the feed water for the boiler is taken from the hot well of the engine, it is evident that if the injection water is of a character which would be injurious to the boiler if pumped into it, there is then a serious objection to a jet condenser. The difficulty is got over by the using a surface condenser. In marine engines where the injection water is taken from the sea, surface condensers are always adopted. Surface condensers are either cylindrical or rectangular shells, and are made of brass, cast-iron, wrought-iron, or steel. The end or tube plates are generally made of rolled brass. This form of condenser contains a large number of tubes, which pass from one end of the condenser to the other. The tubes are made of brass, and are solid drawn, and they are sometimes tinned outside and inside. They vary in diameter from $\frac{1}{2}$ inch to 1 inch, but generally they are $\frac{3}{4}$ inch in diameter outside. The thickness of the tubes is from 16 B.W.G to 19 B.W.G. For $\frac{3}{4}$-inch tubes, the pitch, or distance from centre to centre, varies from $1\frac{1}{16}$ inch to $1\frac{3}{4}$ inches. The length of the tubes varies greatly, depending on the size and design of the condenser. In some cases the length is over 18 feet.

The most common methods of securing the tubes to the tube plates are shown in Figs. 696, 697, and 698. The arrangement shown in Fig. 696 is a very simple one, and consists of a soft wood ferrule, which, after having been compressed, is driven in tight over the tube into the

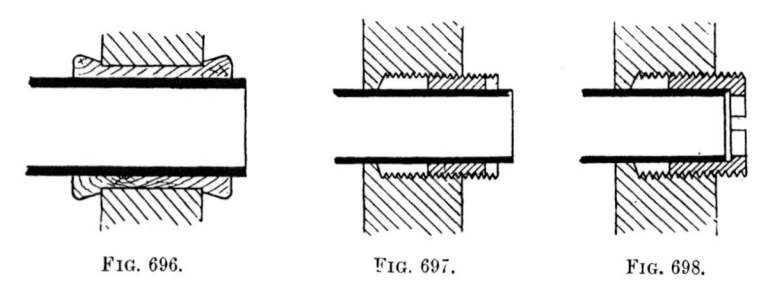

    Fig. 696.   Fig. 697.   Fig. 698.

hole in the tube plate. The moisture in the condenser caused the ferrule to swell at its ends as shown, and there is then no danger of it dropping out. The other designs shown are ordinary stuffing-boxes with screwed glands, having slotted ends to receive a suitable "screwdriver." The gland shown in Fig. 698 has a projection on the inside at its outer end to prevent any longitudinal displacement of the tube. The packing consists either of soft cotton cord, or of a piece of tape made into a ring with a sewed joint.

In modern triple expansion marine engines working with steam having a boiler pressure from 150 to 180 lbs. per square inch, the amount of cooling surface in the surface condenser averages 2 square feet per indicated horse-power; it varies, however, from 1·1 to 3·5 square feet per indicated horse-power.

At a meeting of the American Society of Mechanical Engineers in 1888, Professor Whitham gave the following formula for the area of cooling surface, $S = \dfrac{17W}{180}$, where S = area of surface in square feet, and W = weight of steam condensed per hour in pounds.

The circulating or cooling water in most cases passes through the tubes, and the steam surrounds the tubes. In the navy, and in a few cases in the mercantile marine, the water is outside the tubes and the steam inside.

The amount of condensing water required is determined in the same way as for the jet condenser, except that it must be noted that the temperature of the water obtained by condensing the steam is not necessarily the same as that of the condensing water as it leaves the condenser.

Let H = number of units of heat in 1 lb. of steam above that contained by 1 lb. of water at 32°.
T = temperature of condensed steam or feed water.
t = temperature of circulating water as it enters the condenser.
t_1 = temperature of circulating water as it leaves the condenser.
W = weight of circulating water (in pounds) required for each pound of steam condensed.

Then, $W = \dfrac{H + 32 - T}{t_1 - t}$.

EXAMPLE.—*In a marine engine the circulating water enters the condenser at a temperature of* 60° *and leaves it at* 85°. *The temperature of the feed water (condensed steam) is* 120°, *and the steam enters the condenser at an absolute pressure of* 8½ *lbs. per square inch* ($H = 1138$). *To find the weight of circulating water per pound weight of steam condensed.*

$$W = \dfrac{H + 32 - T}{t_1 - t} = \dfrac{1138 + 32 - 120}{85 - 60} = \dfrac{1050}{25} = 42 \text{ lbs.}$$

339. Circulating Pumps.—The cooling water for surface condensers is circulated by means of pumps. Circulating pumps for marine engines are generally either double-acting piston pumps or centrifugal pumps. The piston pump may form part of the main engine, or it may have an engine of its own. Centrifugal pumps are always worked by separate special engines.

Let D = diameter of piston of circulating pump in inches.
L = length of stroke in inches.
n = number of strokes per minute.
w = total weight of circulating water required per minute in pounds.
64 = weight of 1 cubic foot of sea-water.

Then $\dfrac{\cdot 7854 D^2 L n}{1728} = \dfrac{w}{64}$,

$$D^2 L = \frac{34\cdot 38 w}{n} \text{ and } D = 5\cdot 86 \sqrt{\frac{w}{nL}}.$$

The discharging capacity of the circulating pump per minute is generally from $\frac{1}{40}$th to $\frac{1}{30}$th of the volume swept through by the low-pressure piston per minute.

340. Air-Pumps.—In an engine fitted with a jet condenser the function of the air-pump is to remove from the condenser the condensed steam, the water used for condensing the steam, and the air which enters with the water. In a surface-condensing engine the air-pump has to remove the condensed steam and any air that may leak into the engine, but it has nothing to do with the condensing water.

The most efficient and most common form of air-pump is the vertical single-acting bucket pump. When it is desirable that the air-pump should be horizontal it is usually a double-acting piston pump.

For a jet-condensing engine the capacity of the vertical single-acting pump, that is, the area of the bucket multiplied by the length of its stroke, varies from $\frac{1}{5}$th to $\frac{1}{10}$th of the capacity of the low-pressure cylinder. A horizontal double-acting pump for the same type of engine would have a capacity of $\frac{1}{8}$th to $\frac{1}{16}$th of that of the low-pressure cylinder.

For a surface-condensing engine the single-acting air-pump would have a capacity of $\frac{1}{10}$th to $\frac{1}{18}$th of that of the low-pressure cylinder, and the capacity of the double-acting pump would be $\frac{1}{15}$th to $\frac{1}{25}$th of that of the low-pressure cylinder.

The above proportions are for pumps whose buckets or pistons make the same number of strokes per minute as the piston of the low-pressure cylinder.

341. Consumption of Steam per Indicated Horse-Power.—The weight of steam per indicated horse-power varies greatly in different types of engines, and even in engines of the same type there are often great differences in the quantity of steam consumed per indicated horse-power.

The following table shows approximately the weight of steam, or weight of feed water, consumed per indicated horse-power per hour in various types of engines:—

Type of Engine	Weight of Steam per 1 H.P. per Hour in Lbs.
Simple non-condensing engines	22 to 40
Simple condensing engines, with steam at 60 lbs. pressure, and fitted with expansion gear	19 ,, 22
Compound condensing engines, with steam at 60 lbs. pressure	18 ,, 20
Compound condensing engines, with steam at 100 lbs. pressure	$16\frac{1}{2}$,, $18\frac{1}{2}$
Triple expansion condensing engines, with steam at 160 lbs. pressure	14 ,, 16

All the above, except the first, are given on the authority of Mr. Michael Longridge.

342. Example of a Compound Marine Engine.—Fig. 699 is a plan, and Figs. 700 and 701 are elevations of a two-cylinder compound marine

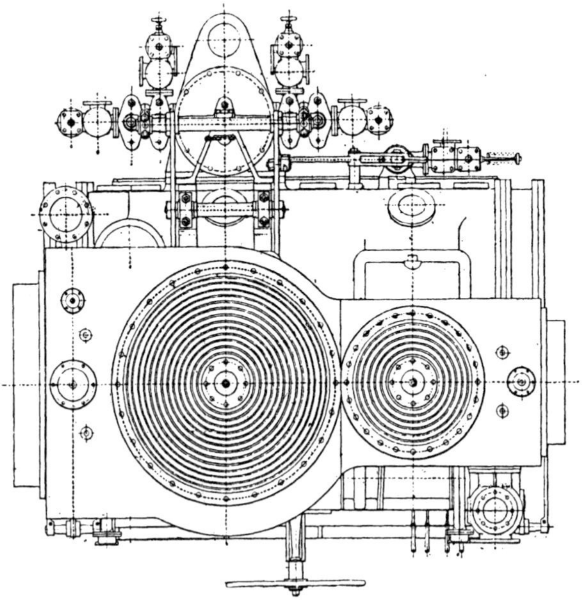

FIG. 699.

engine made by Messrs. T. A. Young & Sons, London. The diameters of the cylinders are—high pressure, 23 inches; low pressure, 42 inches, and the stroke of the pistons is 27 inches. The table which follows gives the principal dimensions of this engine—

PRINCIPAL DIMENSIONS OF A COMPOUND MARINE ENGINE.

High-Pressure Cylinder.

	Ft.	In.
Diameter	1	11
Stroke	2	3
Distance between centre lines of cylinder and valve spindle	1	$11\frac{1}{4}$
Thickness of liner . . .	0	$1\frac{1}{16}$
Thickness of cylinder . .	0	$1\frac{1}{4}$
Thickness of valve casing .	0	$1\frac{1}{8}$
Number of studs in cylinder cover		24
Diameter of studs in cylinder cover	0	$\frac{7}{8}$
Width of steam ports . .	0	$1\frac{1}{2}$
Width of exhaust port . .	0	$1\frac{7}{8}$
Length of ports	1	9
Width of bars between ports .	0	$1\frac{9}{16}$
Diameter of steam-pipe . .	0	$5\frac{1}{4}$
Diameter of exhaust-pipe (copper)	0	$6\frac{1}{2}$

Low-Pressure Cylinder.

	Ft.	In.
Diameter	3	6
Stroke	2	3
Distance between centre lines of cylinder and valve spindle	2	$9\frac{1}{4}$
Thickness of liner . . .	0	$1\frac{1}{16}$
Thickness of cylinder . .	0	$1\frac{1}{4}$
Thickness of valve casing .	0	$1\frac{1}{8}$

336   MACHINE DRAWING AND DESIGN.

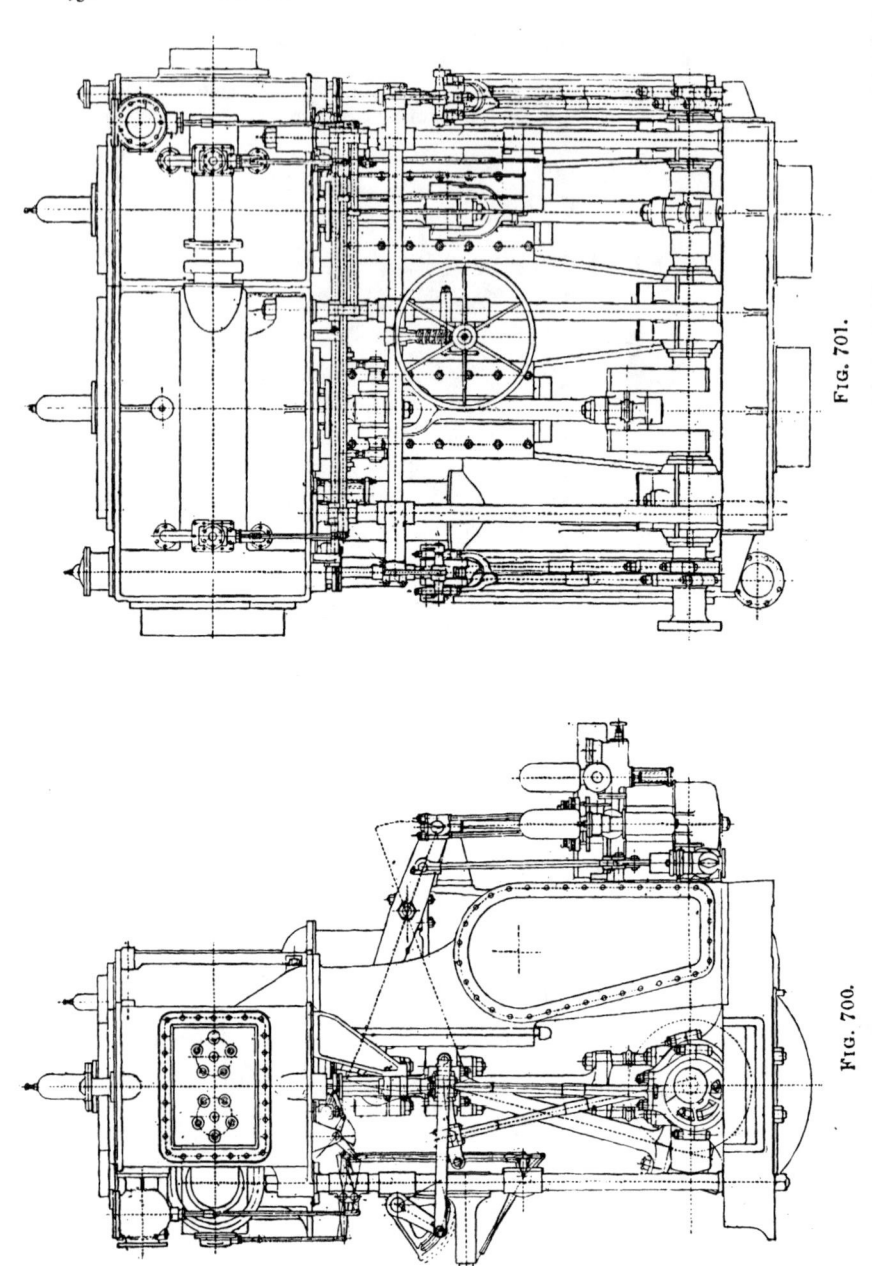

FIG. 701.

FIG. 700.

STEAM ENGINES.

	Ft.	In.
Number of studs in cylinder cover	24	
Diameter of studs in cylinder cover	0	$0\frac{7}{8}$
Width of steam ports (double-ported)	0	$1\frac{1}{2}$
Width of exhaust port	0	$3\frac{3}{4}$
Length of ports	3	1
Width of bars between steam ports	0	$4\frac{7}{8}$
Width of bars between steam and exhaust ports	0	$1\frac{3}{8}$
Distance between centre lines of cylinders	3	4
Depth of pistons	0	$6\frac{1}{2}$
Diameter of piston-rods	0	4
Diameter of connecting-rods at crosshead end	0	$3\frac{7}{8}$
Diameter of connecting-rods at middle and at crank end	0	$4\frac{3}{8}$

Surface Condenser.

	Ft.	In.
Number of tubes		557
Diameter of tubes (outside)	0	$0\frac{3}{4}$
Distance between tubes, centre to centre	0	$1\frac{1}{8}$
Distance between tube plates	7	5
Thickness of tube plates	0	$1\frac{1}{8}$
Diameter of steam inlet	0	$11\frac{1}{2}$
Diameter of condensing water inlet	0	7
Diameter of condensing water outlet	0	7
Thickness of metal in body of condenser	0	$1\frac{1}{8}$
Two columns on top of condenser to support cylinders and carry guide plates 20 inches × $12\frac{1}{4}$ inches × $1\frac{1}{8}$ inches thick at smallest part.		

Air Pump.

	Ft.	In.
Diameter	1	4
Stroke	1	$1\frac{1}{2}$
Depth of bucket (brass)	0	3
Diameter of rod (including sheathing of brass $\frac{1}{4}$ inch thick)	0	$2\frac{1}{4}$
Thickness of barrel (brass)	0	$0\frac{1}{2}$
Width of suction opening	1	4
Depth ,, ,, ,,	0	$3\frac{3}{4}$
Diameter of discharge pipe	0	7
Number of delivery valves		4
Number of bucket valves		4
Number of foot valves		4
Diameter of valves (india-rubber)	0	$5\frac{1}{2}$

Feed Pumps, two.

	Ft.	In.
Diameter	0	3
Stroke	1	$1\frac{1}{2}$
Thickness of metal (brass) in barrel	0	$0\frac{3}{8}$
Thickness of metal (brass) in plunger	0	$0\frac{3}{8}$

Bilge Pumps, two, same as feed pumps.

Crank Shaft.

	Ft.	In.
Diameter of shaft	0	$7\frac{1}{4}$
Diameter of crank pin	0	$7\frac{1}{4}$
Length of crank pin	0	8
Thickness of crank arms	0	$4\frac{1}{2}$
Width of crank arms	0	$8\frac{1}{2}$

Thrust Shaft.

	Ft.	In.
Diameter of shaft	0	$6\frac{3}{4}$
Number of thrust collars		5
Diameter of thrust collars	0	11
Thickness of thrust collars	0	$1\frac{1}{2}$
Distance between thrust collars	0	$2\frac{1}{2}$

CHAPTER XXIV.

EXAMPLES OF TRIPLE EXPANSION MARINE ENGINES.

343. General Description of Engines.—We give in this chapter a very complete series of illustrations of a set of triple expansion marine engines made by Messrs. Ross & Duncan, Whitefield Works, Govan, Glasgow. Figs. 703 and 704 show elevations of the engines complete, and these are followed by fully dimensioned illustrations of all the details, which have been specially prepared from the working drawings kindly lent by the makers. The student of machine drawing will find excellent practice in first drawing all the details separately to a large scale, and then from his own detail drawings to make a plan, and front, back, and end elevations of the engines complete.

The cylinders are placed in the order, high, intermediate, low, and their diameters are 8 inches, 13 inches, and 21 inches respectively. All the pistons have a stroke of 16 inches. The cylinders are in one casting, and are supported at the back of the engine on three hollow cast-iron columns cast on the condenser. At the front the cylinders are supported by two turned wrought-iron columns, which are flanged at their upper ends, and firmly secured at their lower ends to the bed of the engine.

The guides for the crossheads are formed on the columns on the condenser.

The condenser is of cast-iron in one casting, and contains 228 tubes, giving 220 square feet of cooling surface. The tube plates are made of brass, and are 4 feet 11 inches apart. The circulating or cooling water enters the condenser at the bottom, and passes through the lower nest of 114 tubes, and back through the upper nest of 114 tubes. The cover on one end of the condenser has cast on it, on the inside, an air vessel, which serves to reduce the shock due to any irregularity in the flow of the water which enters at that end. The chamber at the same end of the condenser is divided horizontally by a partition, which is cast on the cover and round the air vessel just mentioned. The steam from the low-pressure cylinder passes through the hollow column under it to the condenser, where it passes round the tubes and is condensed and drawn off at the bottom by the air-pump.

The air-pump is a vertical single-acting bucket pump 9 inches in diameter, and the circulating pump is a double-acting piston pump 5 inches diameter, also vertical.

There is one bilge pump and one feed pump, each $1\frac{3}{4}$ inches in diameter.

All the pumps have a stroke of 8 inches.

The crank shaft is forged and in one piece. The three cranks are equally inclined to one another. The main and crank pin bearings are each 4 inches in diameter. The thrust is taken by a single collar 8 inches diameter forged on the shaft. The area of the surface which receives the thrust is $·7854\,(8^2 - 4^2) = 37·7$ square inches.

The eccentric sheaves or pulleys are forged on the crank shaft; they are each $7\frac{1}{2}$ inches in diameter, and have an eccentricity of $1\frac{3}{4}$ inches.

The valve gear is that invented by Mr. G. A. C. Bremme of Liverpool, and is known as the Bremme valve gear. Only one eccentric is used for each engine, and this eccentric serves to give motion to the slide valve whether the engine is going "ahead" or "astern." A diagram of the gear is shown in Fig. 702. AB represents the crank, and AC the eccentric. (It will be observed that the direction of the eccentric radius coincides with that of the crank radius.) CFD is the eccentric rod. DE is the valve rod link jointed to the valve rod at E. The swinging link FK is jointed to the eccentric rod at F, and to the upper end K of the radius arm HK. H is a fixed centre, and the radius arm HK is at rest, except when the gear is being shifted to reverse or to cut off earlier. As shown in Fig. 702 the engine is going full speed ahead. By moving the radius arm nearer the vertical the steam is cut off earlier, and by moving it to the other side of the vertical the engine is reversed. The angle which the radius arm makes with the vertical should never exceed 25°; in the example before us this angle does not exceed 20°. The radius arm is shifted by means of the reversing handle MN, the wyper LM, and the drag-link KL.

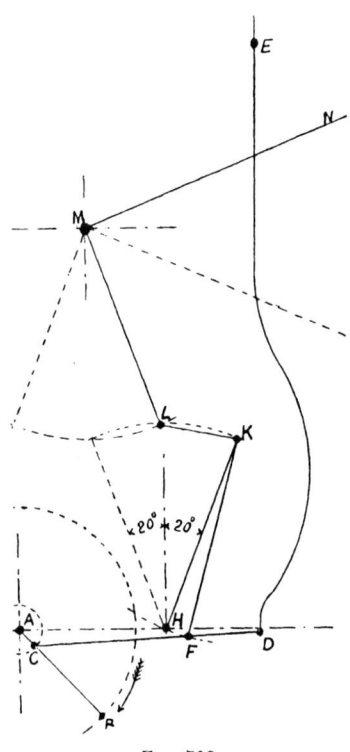

FIG. 702.

The length of the swinging link is equal to that of the radius arm, and it follows that the point F will move in an arc of a circle which passes through H.

The student will understand the action of this valve gear much better if he draws it in a number of different positions corresponding to different positions of the eccentric. He should also draw the curve traced by the point D.

The following table gives particulars of the working of the slide-valves when the radius arms are inclined at 20° to the vertical.

		H.P.	M.P.	L.P.
Top	Lead	inches $\frac{3}{32}$	inches $\frac{3}{16}$	inches $\frac{1}{4}$
	Port opening	$\frac{7}{16}$	$\frac{7}{8}$	$1\frac{1}{16}$
	Cut off	$10\frac{3}{4}$	11	$11\frac{1}{2}$
	Compression	$2\frac{1}{2}$	$2\frac{1}{2}$	$2\frac{1}{2}$
Bottom	Lead	$\frac{3}{16}$	$\frac{5}{16}$	$\frac{3}{8}$
	Port opening	$\frac{5}{8}$	$\frac{7}{8}$	$1\frac{5}{16}$
	Cut off	$11\frac{1}{4}$	$11\frac{1}{2}$	12
	Compression	$2\frac{1}{2}$	$2\frac{1}{2}$	$2\frac{1}{2}$

As to the power of the engines we have been describing we are unfortunately not able to give exact particulars, as the makers have no data of actual performance, but they state that these engines would probably indicate about 147 horse-power at 170 revolutions per minute. The mean effective pressures in pounds per square inch in the cylinders would probably be, 66 in the H.P., 28 in the M.P., and $11\frac{1}{2}$ in the L.P. The horse-power would be distributed approximately as follows, 45 in the H.P., 50 in the M.P., and 52 in the L.P. cylinder. The vacuum would be about 27 inches of mercury.

344. Table of Reference for Illustrations. — The names of the different parts are in most cases given on the illustrations, but for convenience of reference they are here tabulated:—

    Fig. 703. Front elevation of engines complete.
      „   704. End elevation of engines complete.
      „   705. Crank shaft, engine bed, and main bearings.
      „   706 and 707. Front and end elevations of condenser with dotted sections.
      „   708 and 709. Plan, and horizontal and vertical sections of condenser.
      „   710. Sectional plan of cylinders.
      „   711. Vertical section of high-pressure cylinder.
      „   712. Vertical section of medium- or intermediate-pressure cylinder.
      „   713. Vertical section of low-pressure cylinder.
      „   714 and 715. Escape or relief valves for cylinders.
      „   716. Piston for high-pressure cylinder.
      „   717. Pistons for intermediate- and low-pressure cylinders.
      „   718. Piston-rod glands and stuffing-box bushes.
      „   719. Valve-rod glands and stuffing-box bushes.
      „   720 and 721. Piston-rods and crossheads.
      „   722. Connecting-rods.
      „   723, 724, and 725. Slide valves and valve-rods.
      „   726. Wyper shaft.
      „   727. Reversing shaft.
      „   728. Wrought-iron columns.
      „   729. Supports for wyper shaft.

Fig 730. Wypers.
„ 731. Drag links.
„ 732. Radius arms.
„ 733. Swinging links.
„ 734. Eccentric straps and eccentric rods.
„ 735. Return lever for high-pressure slide valve, also reversing handle.
„ 736. Valve-rod links.
„ 737. Locking gear.
„ 738. Details for starting gear.
„ 739. Starting cock.
„ 740. Throttle valve.
„ 741. Quadrant and handle for throttle valve.
„ 742. Air and circulating pump casings.
„ 743. Air and circulating pump casings; also valves, piston, and piston-rod for circulating pump.
„ 744. Air-pump bucket and rod.
„ 745. Air-pump head or delivery valve.
„ 746. Air-pump foot or suction valve.
„ 747 and 748. Bilge and feed pumps.
„ 749. Engine links
„ 750. Pump links.
„ 751. Crosshead for pumps.
„ 752. Pump levers and pedestal.
„ 753. Turning gear.

345. Engines of the R.M.S. "City of Paris."—By way of contrast to the engines described in the two preceding articles, we here give the principal dimensions and a brief description of the engines of the Inman and International Steamship Co.'s Royal Mail steamship *City of Paris*. This famous ship has twin screws, driven by two separate and distinct sets of triple expansion engines, together capable of developing over 20,000 indicated horse-power.

The diameters of the cylinders are: H.P., 45 inches; I.P., 71 inches; L.P., 113 inches. Each piston has a stroke of 5 feet. All the valves are piston valves, there being one to the H.P. cylinder, two to the I.P. cylinder, and four to the L.P. cylinder.

Steel is largely used in the construction of these engines. Many parts, which in older engines were made of cast-iron, are made of cast-steel, and the piston-rods, connecting-rods, and principal moving parts are made of ingot steel.

The crank shaft has a diameter of $20\frac{1}{4}$ in. at the journals, and the crank pins are 21 in. in diameter. Through each crank pin there is a hole $3\frac{1}{2}$ in. in diameter. The thrust shaft has a diameter of $19\frac{1}{4}$ in., enlarged to $20\frac{1}{2}$ in. between the collars. The tunnel shaft is $19\frac{1}{4}$ in. in diameter, and the tube or propeller shaft is $20\frac{1}{4}$ in. in diameter.

Steam is supplied to the engines by nine steel boilers 15 feet 6 inches in diameter and 19 feet long. The boilers are double ended, and each contains six furnaces. The working pressure is 150 lbs. per square inch. The boilers are worked on Howden's forced draught system. The total grate area is 1026 square feet, and the total heating surface is 50,265 square feet.

Fig. 703.

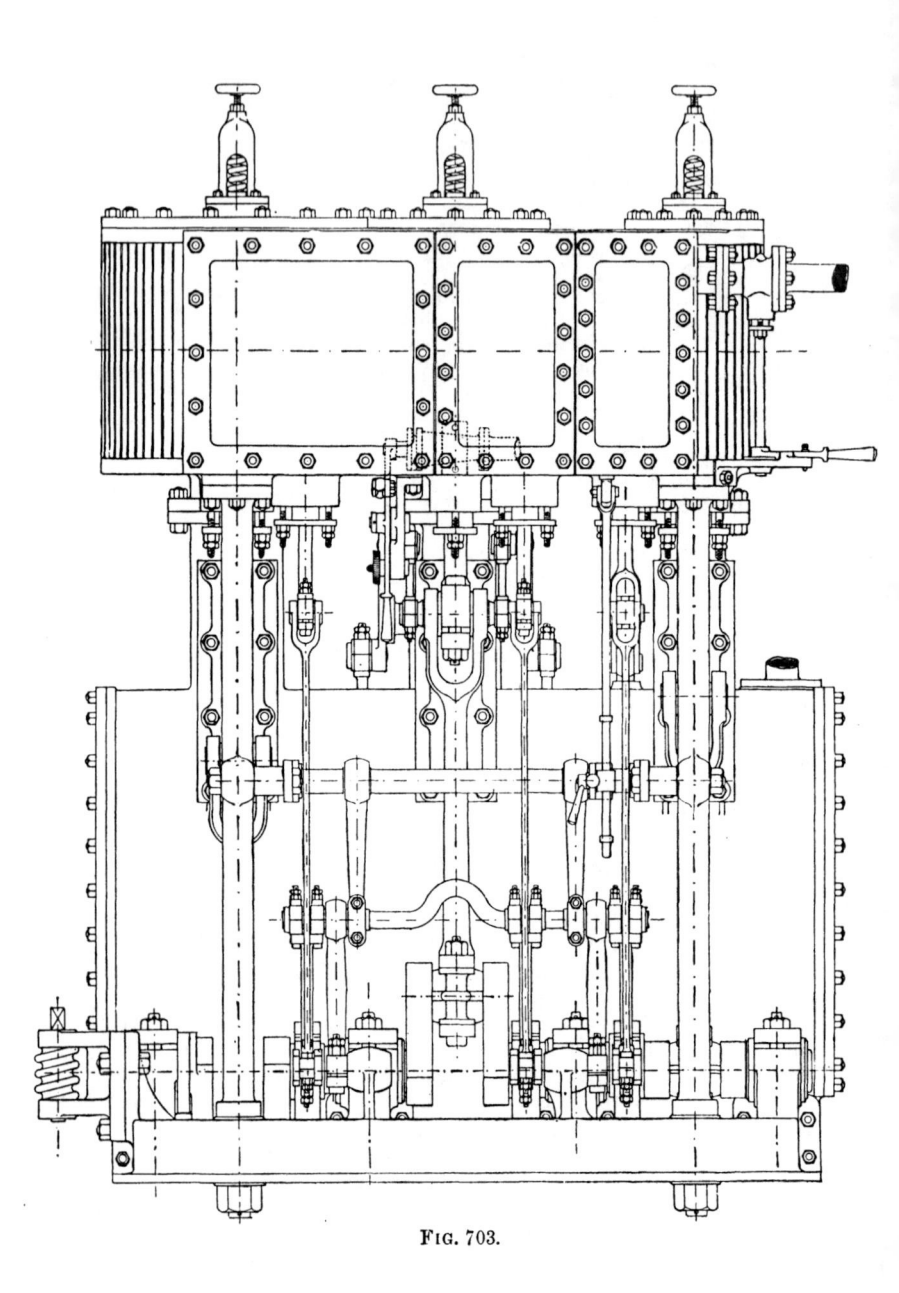

FIG. 704.

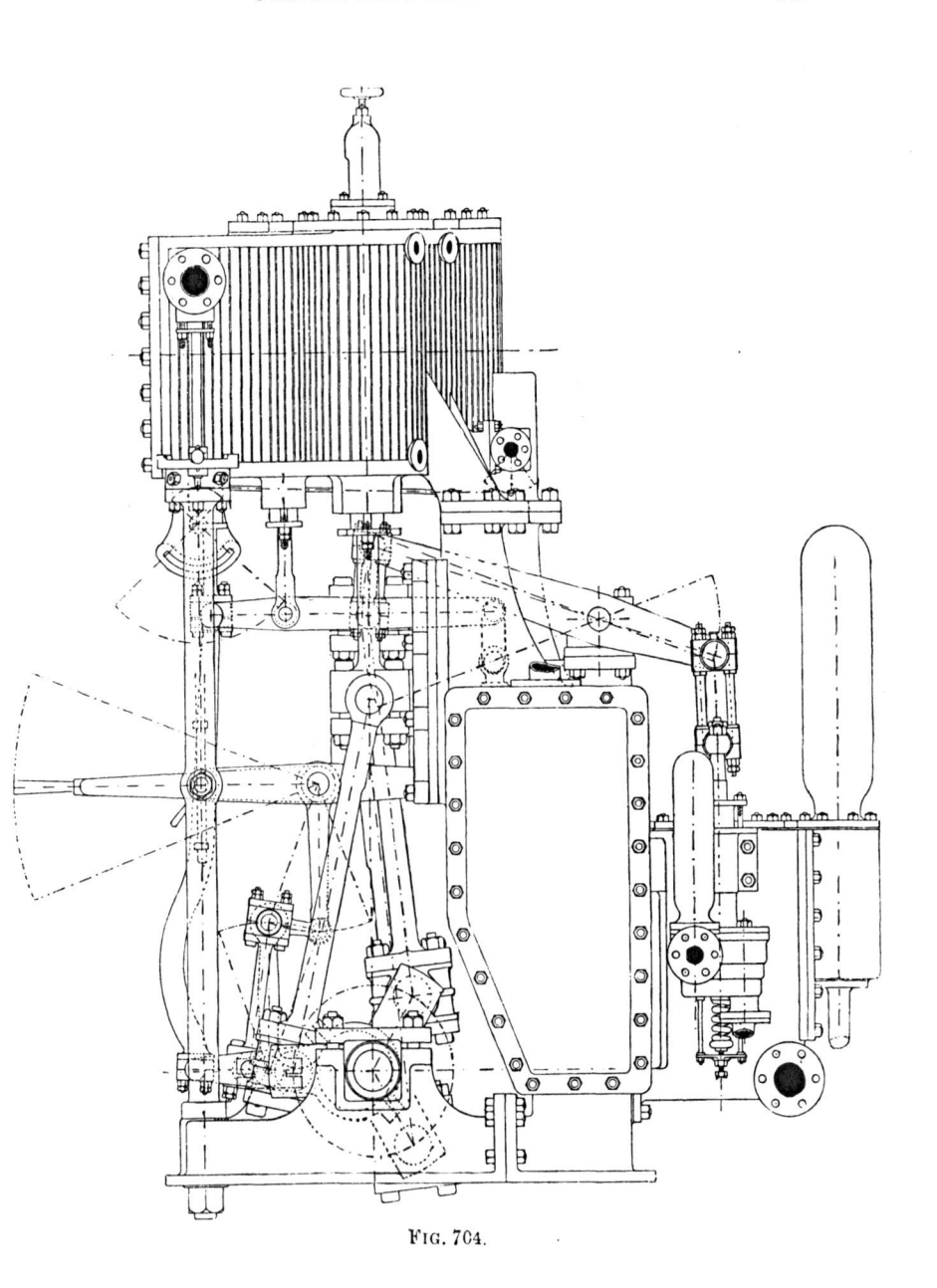

344 MACHINE DRAWING AND DESIGN.

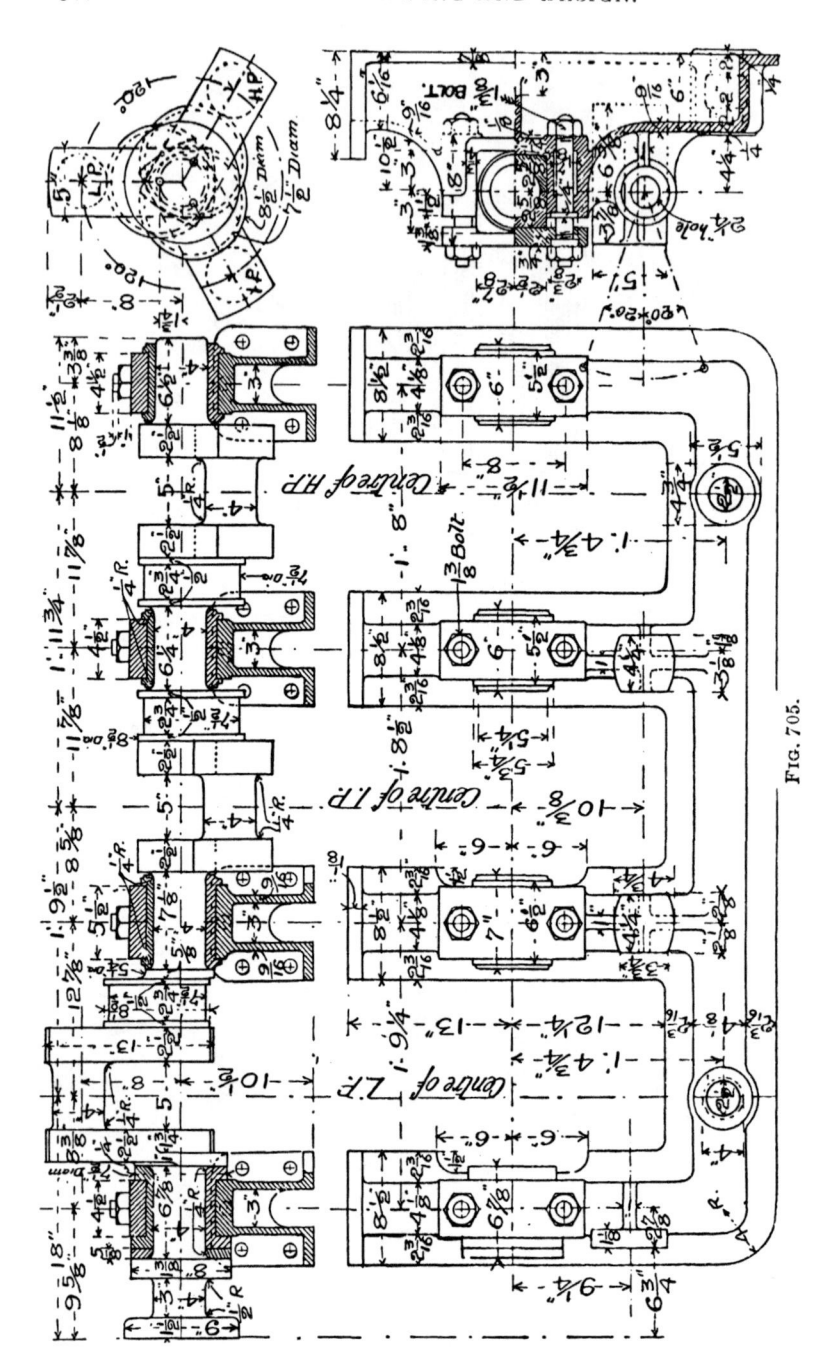

FIG. 705.

TRIPLE EXPANSION ENGINES. 345

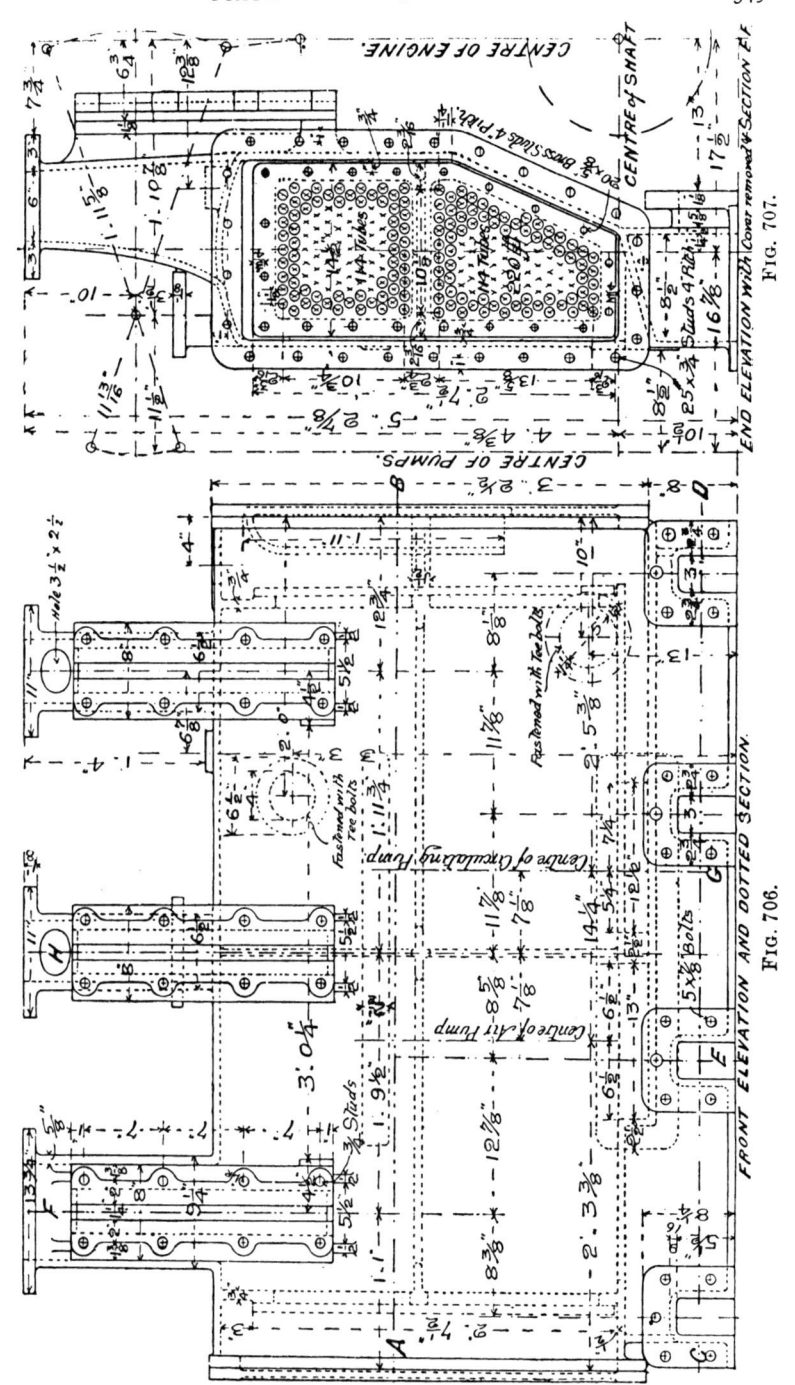

FIG. 706.—FRONT ELEVATION AND DOTTED SECTION.

FIG. 707.—END ELEVATION with Cover removed & SECTION E F.

346   MACHINE DRAWING AND DESIGN.

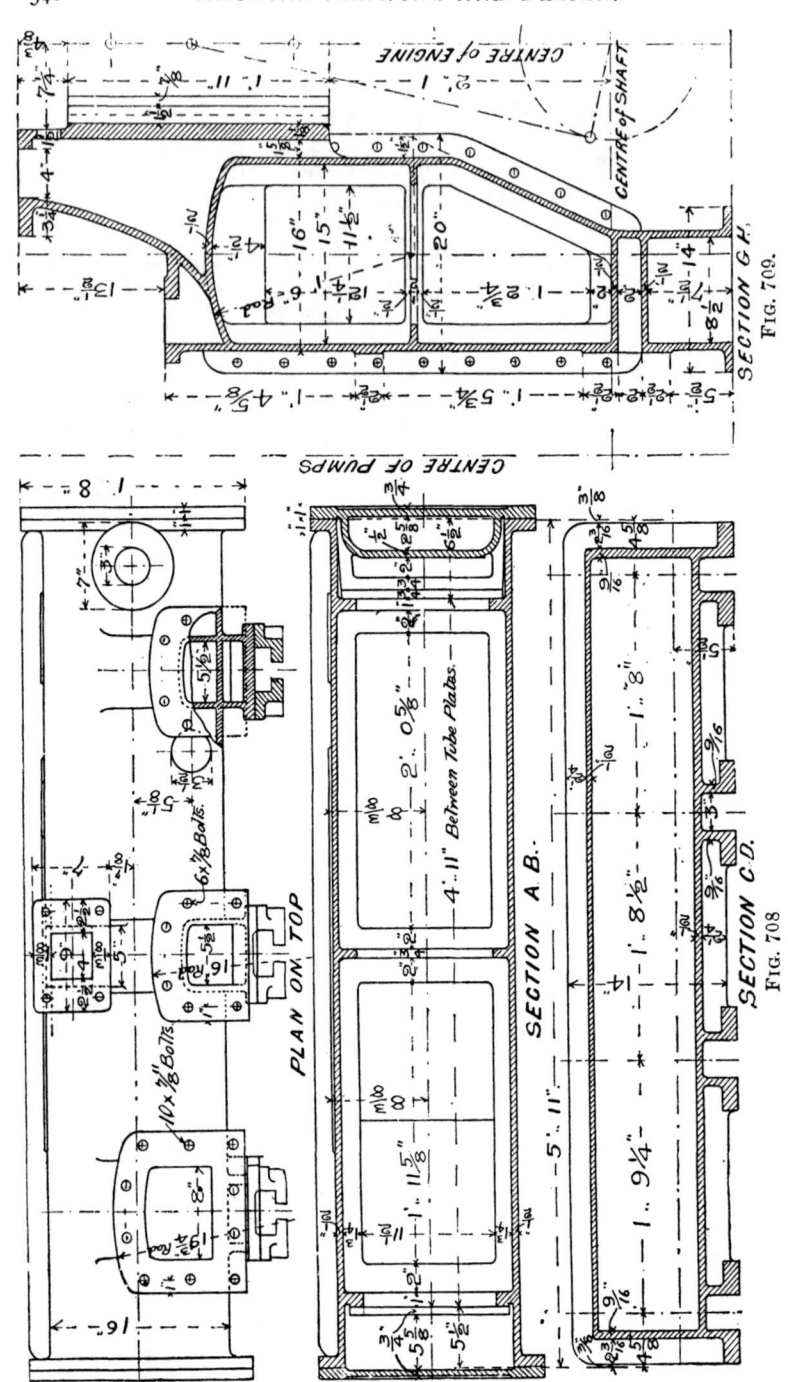

FIG. 710.

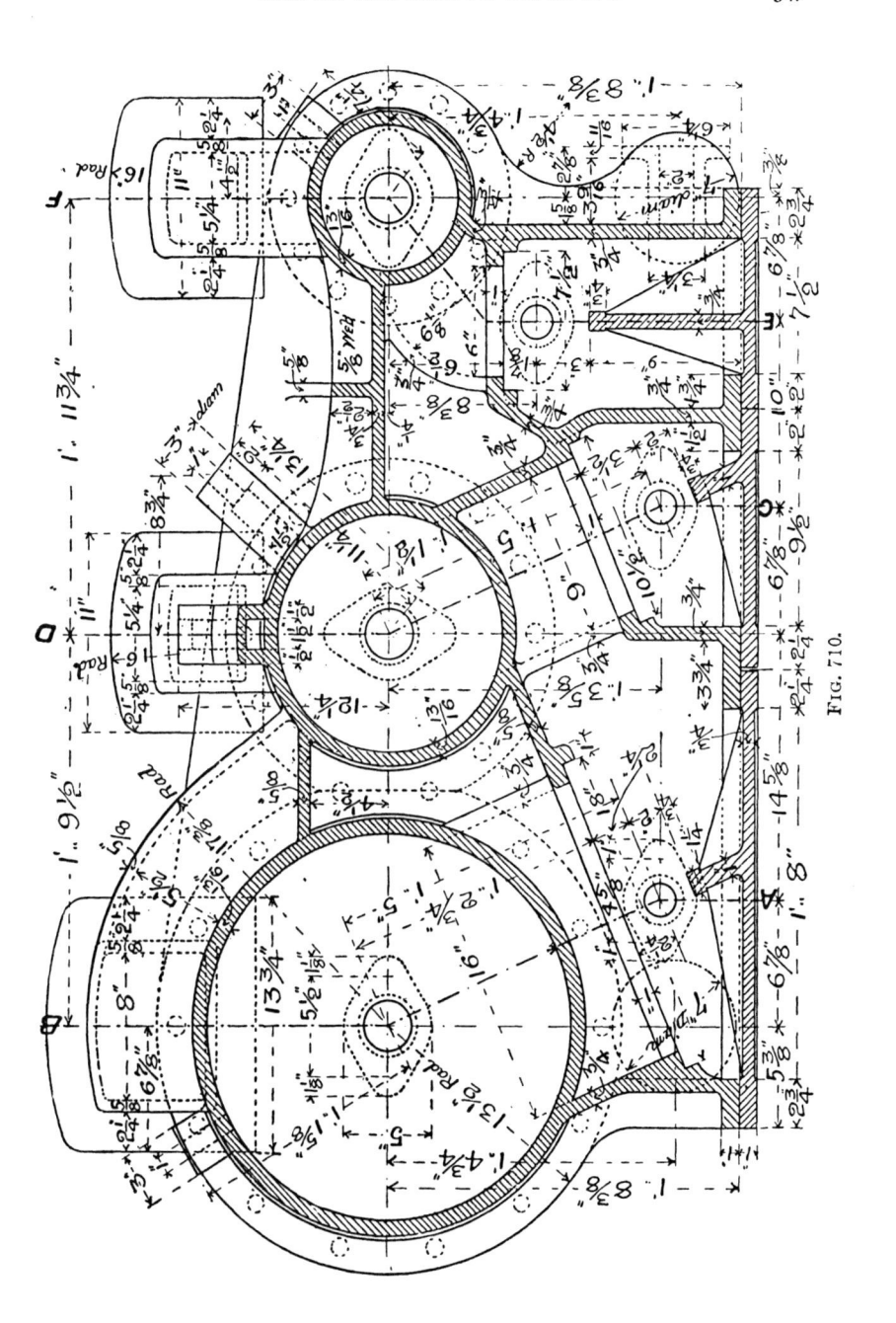

348   MACHINE DRAWING AND DESIGN.

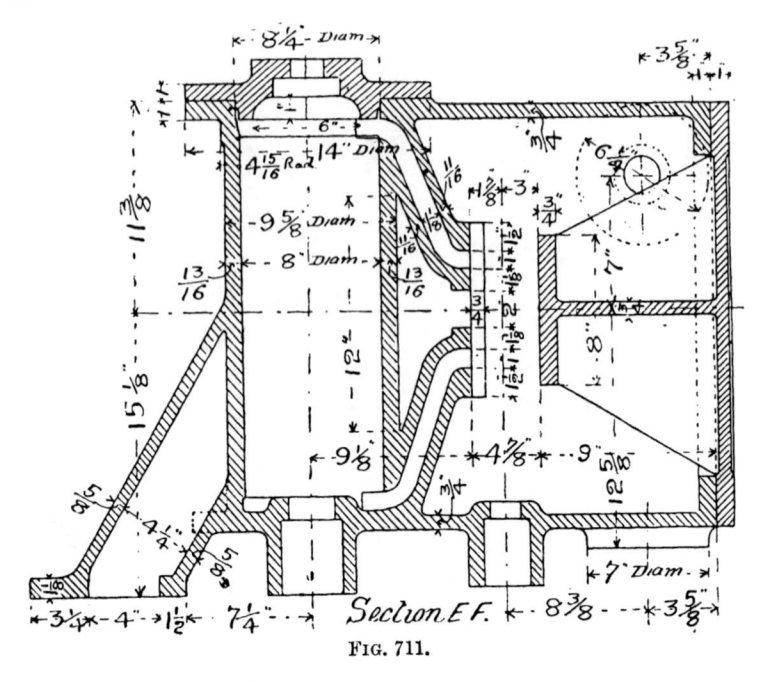

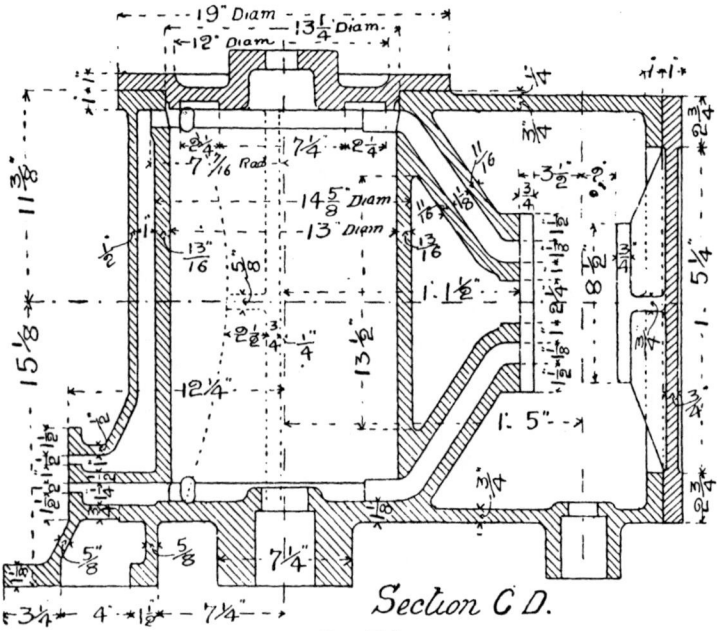

FIG. 711.

FIG. 712.

TRIPLE EXPANSION ENGINES. 349

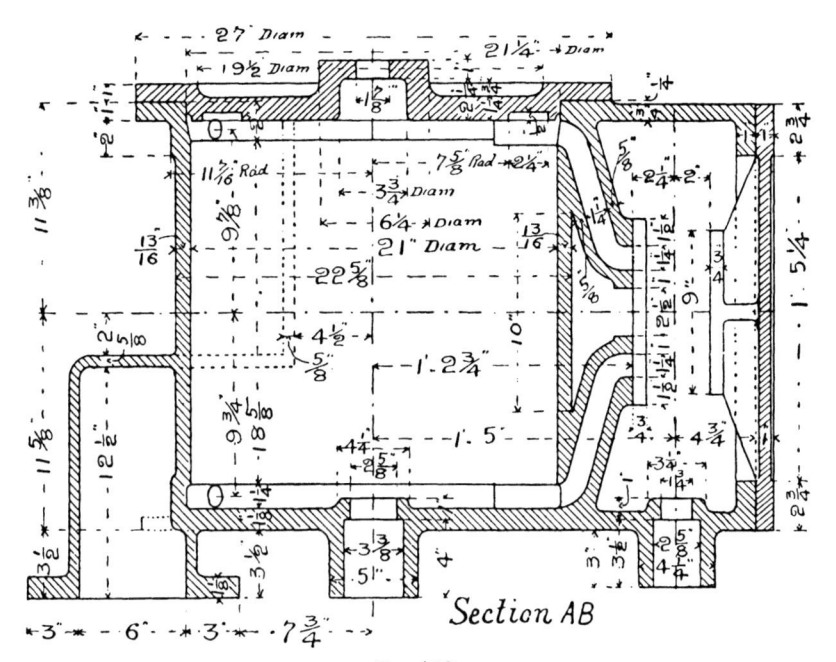

Fig. 713.

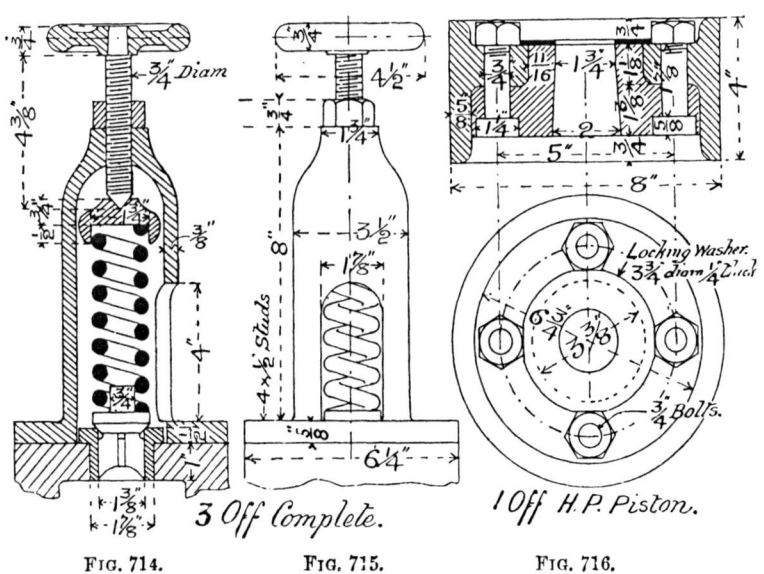

Fig. 714. Fig. 715. Fig. 716.

Fig. 717.

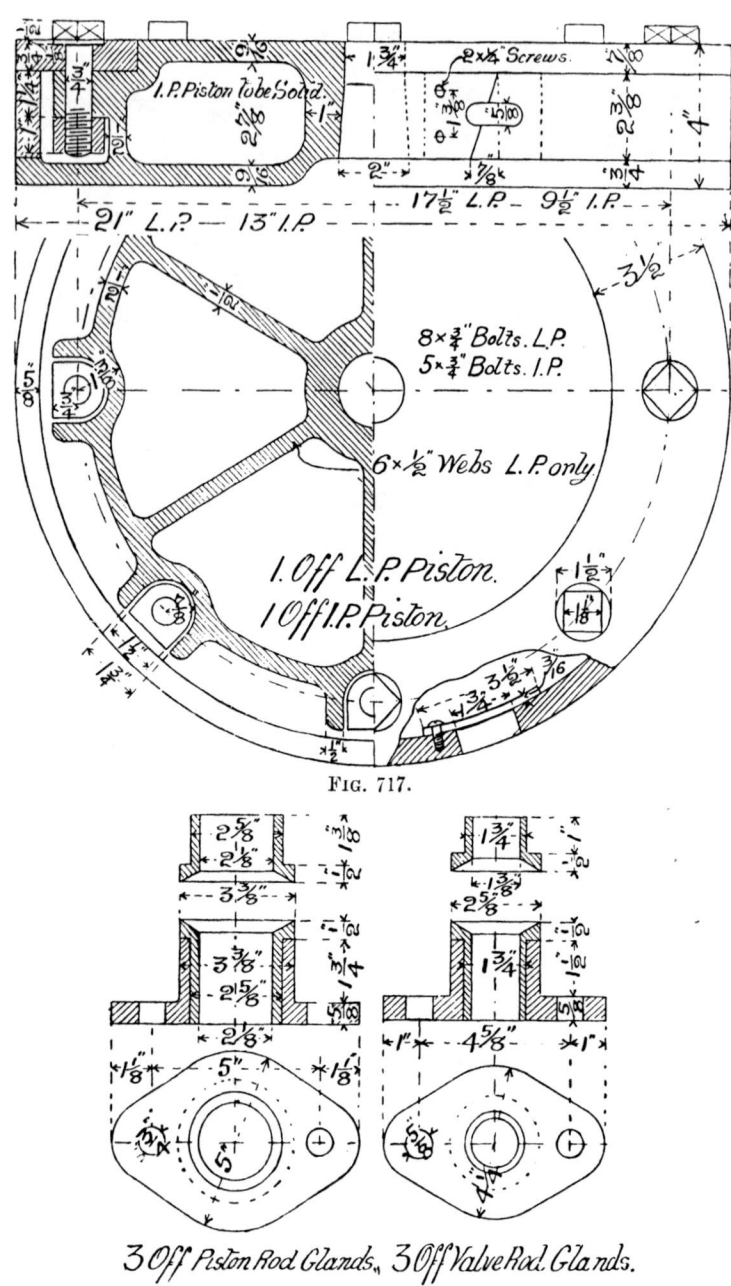

Fig. 718. 3 Off Piston Rod Glands.

Fig. 719. 3 Off Valve Rod Glands.

TRIPLE EXPANSION ENGINES. 351

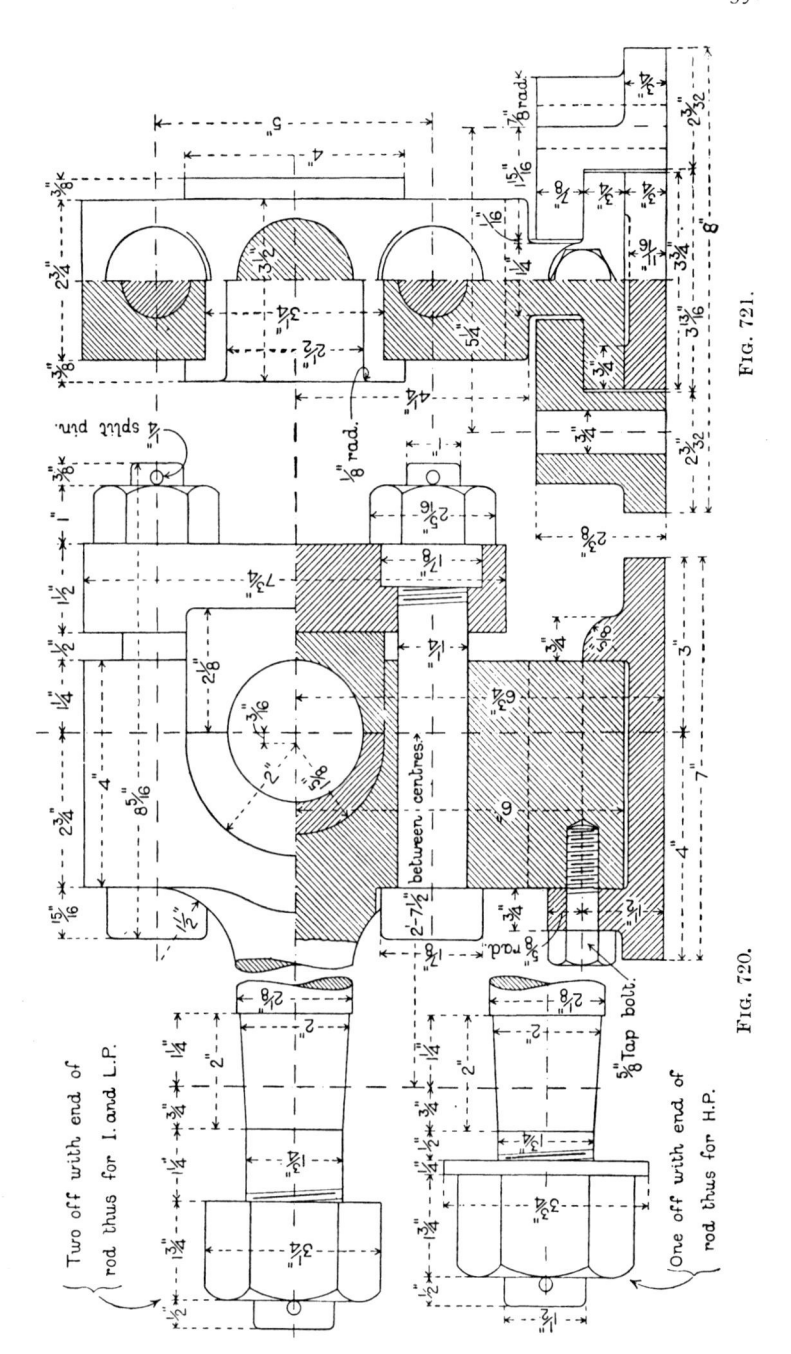

FIG. 721.

FIG. 720.

352   MACHINE DRAWING AND DESIGN.

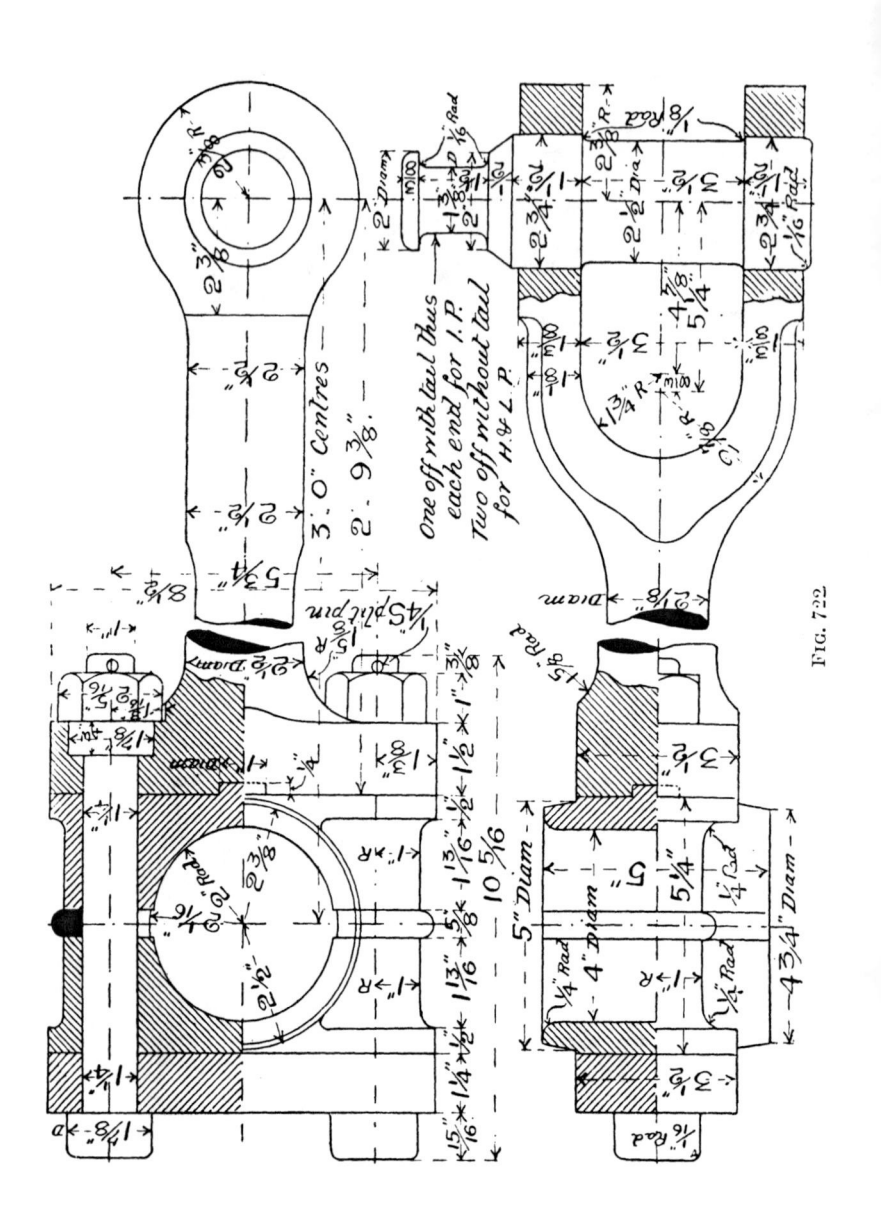

FIG. 722.

TRIPLE EXPANSION ENGINES.

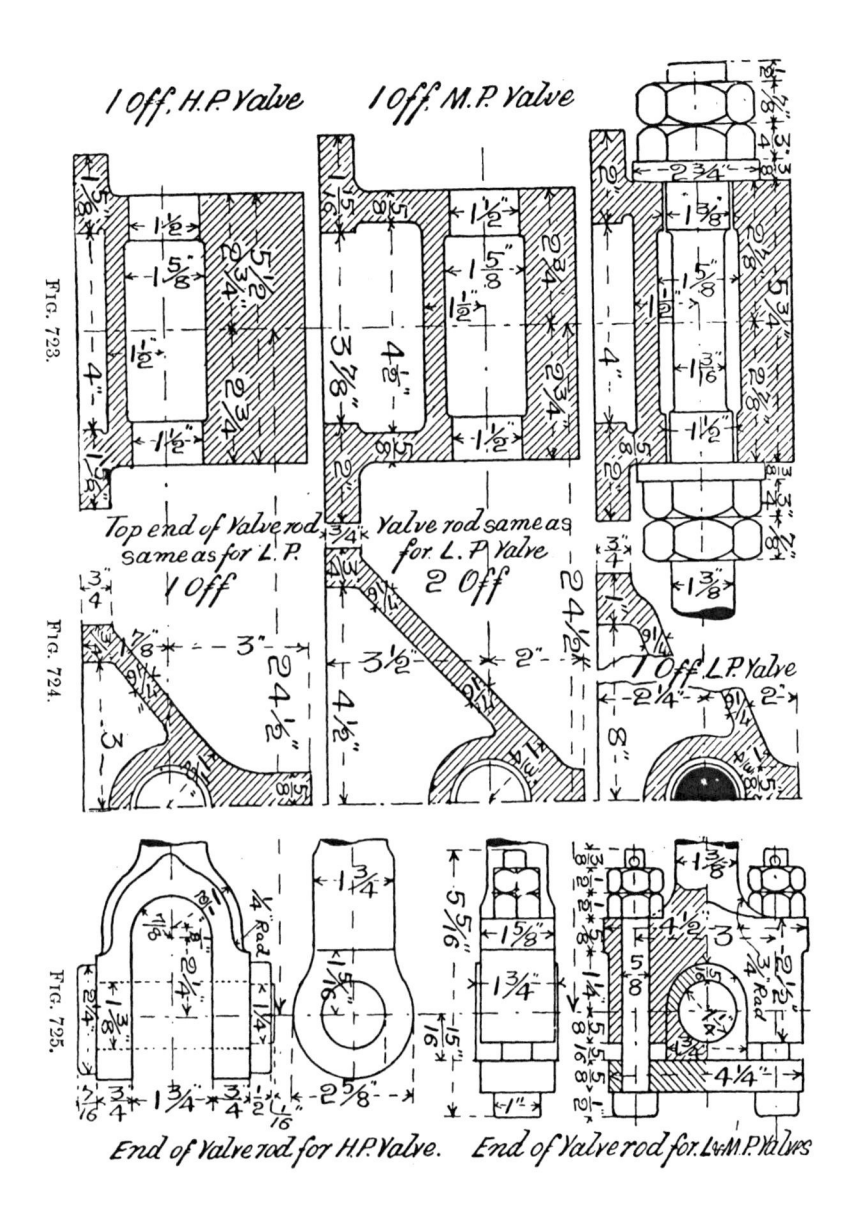

FIG. 723.

FIG. 724.

FIG. 725.

End of Valve rod for H.P. Valve.   End of Valve rod for L.&M.P. Valves.

FIG. 726.   FIG. 727.   FIG. 728.

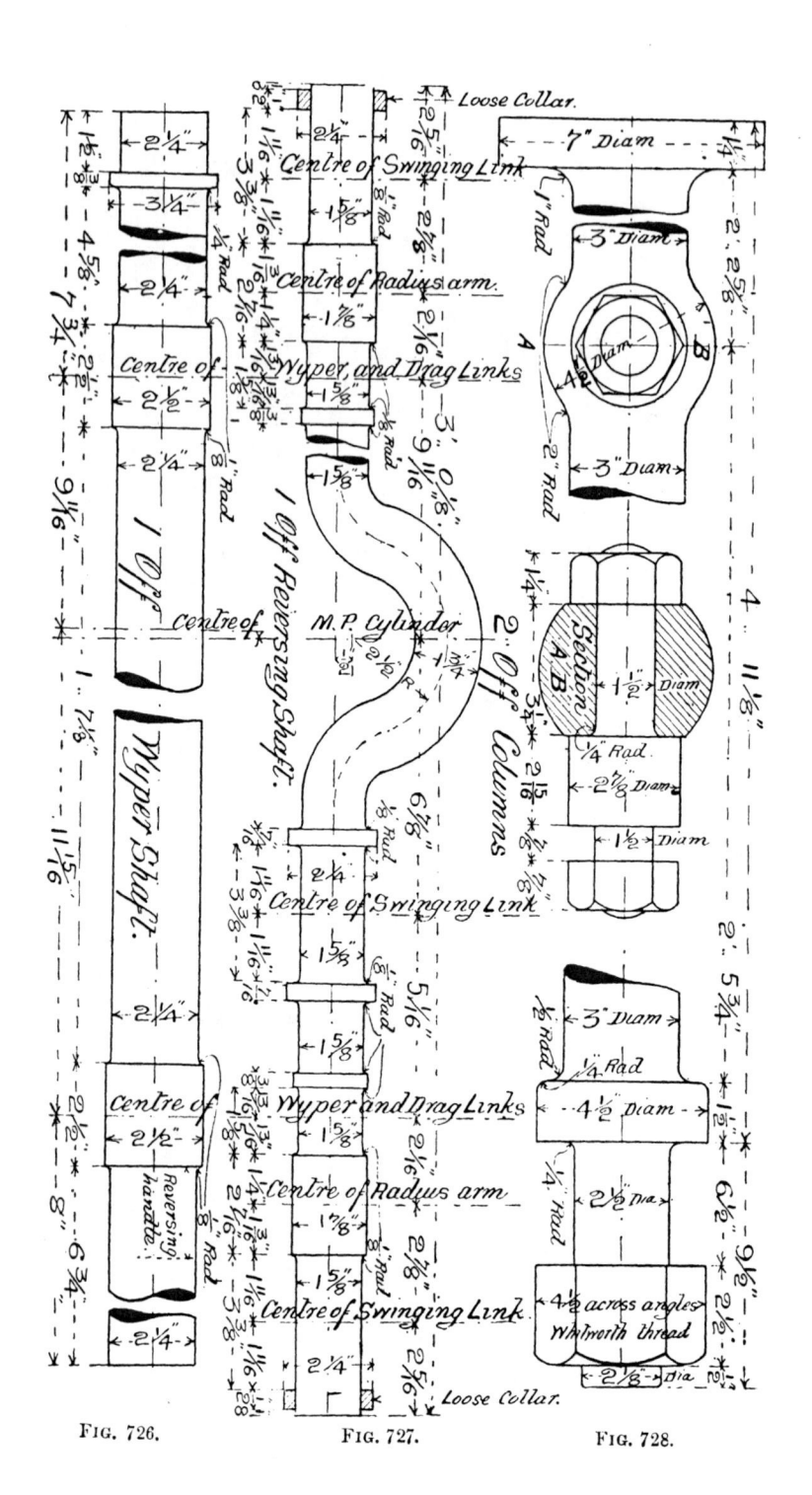

FIG. 729.

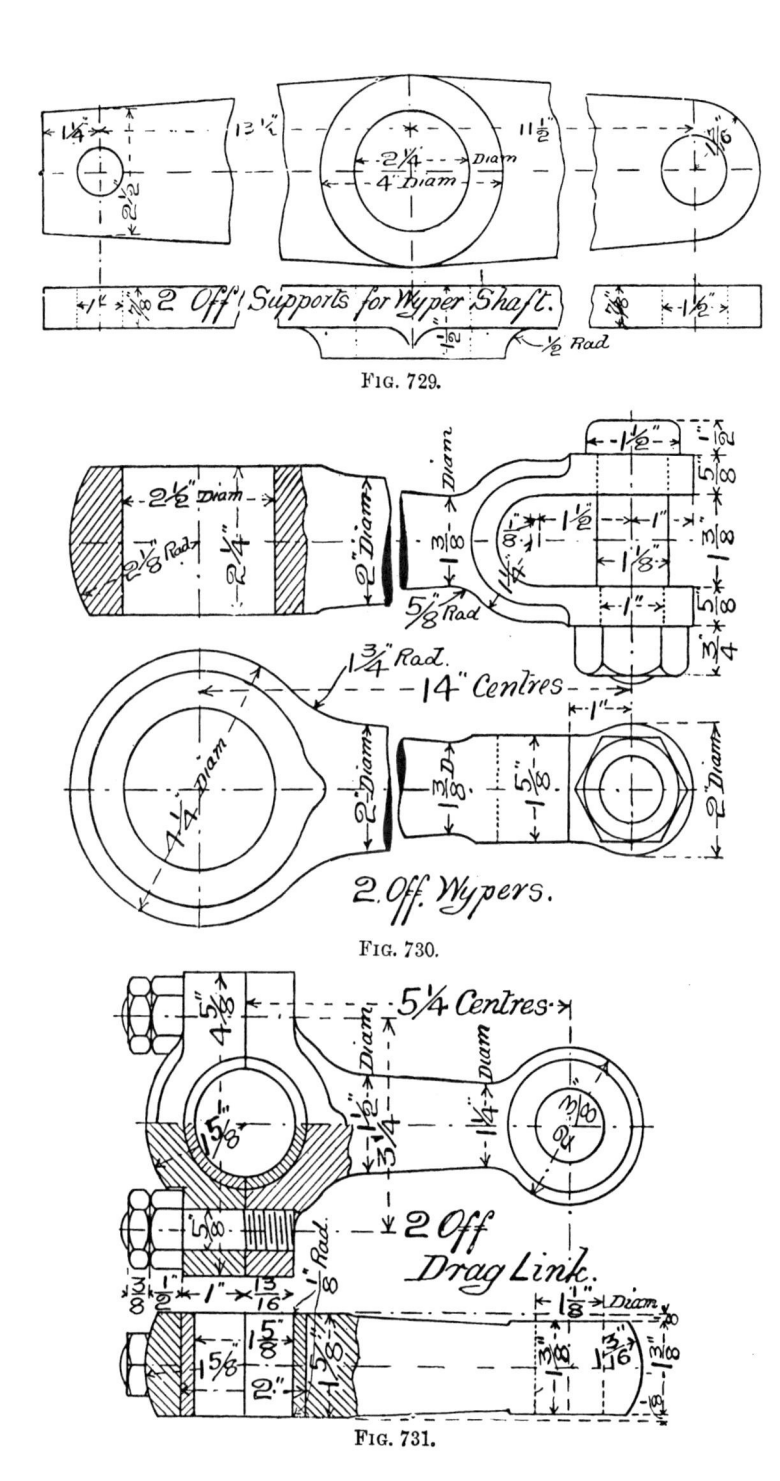

FIG. 730.

FIG. 731.

356   MACHINE DRAWING AND DESIGN.

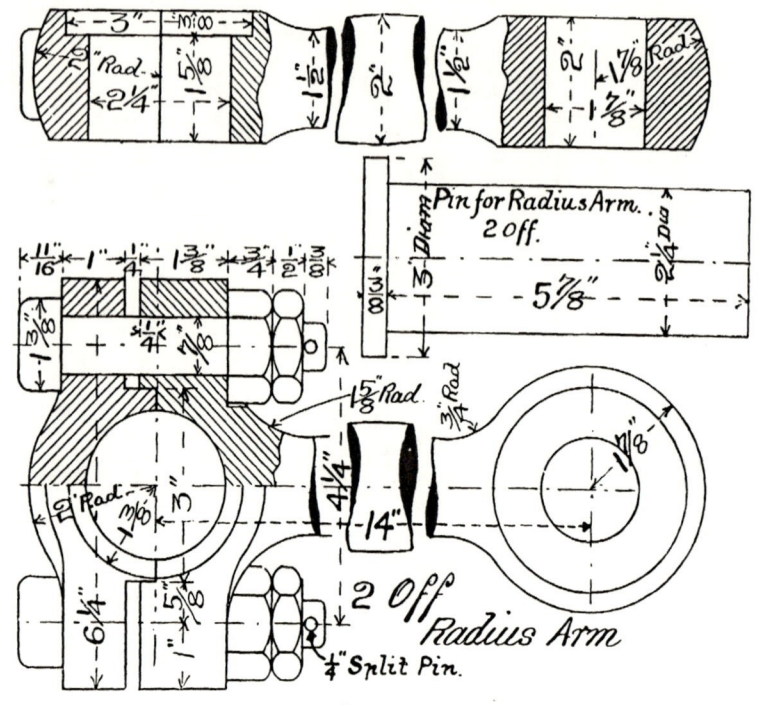

Fig. 732.

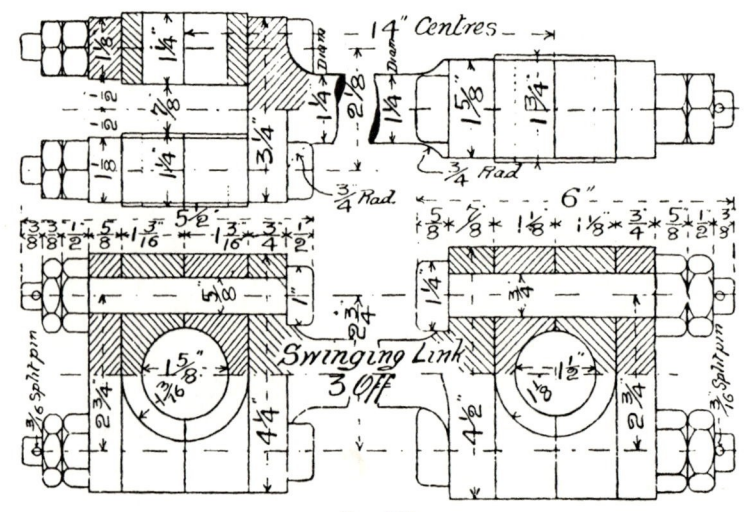

Fig. 733.

TRIPLE EXPANSION ENGINES.

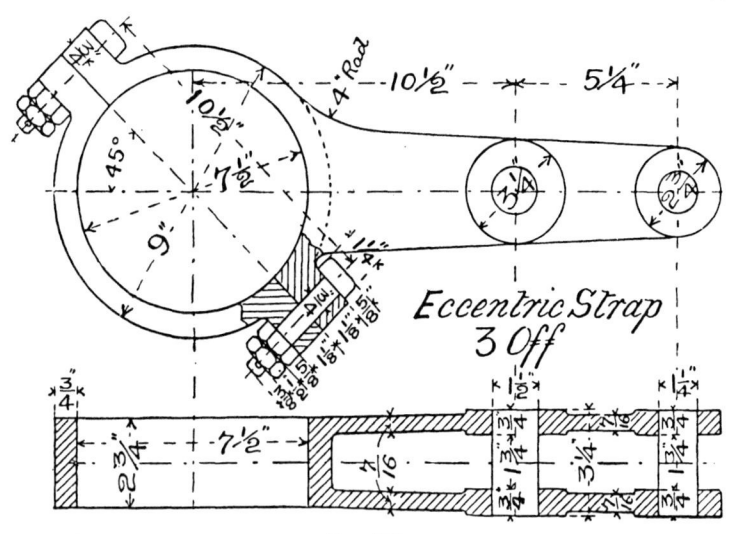

Fig. 734.

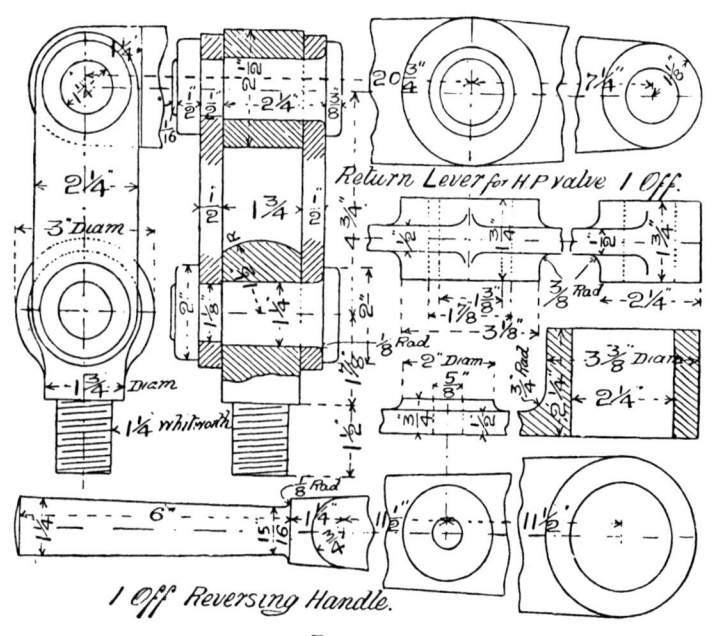

Fig. 735.

358   MACHINE DRAWING AND DESIGN.

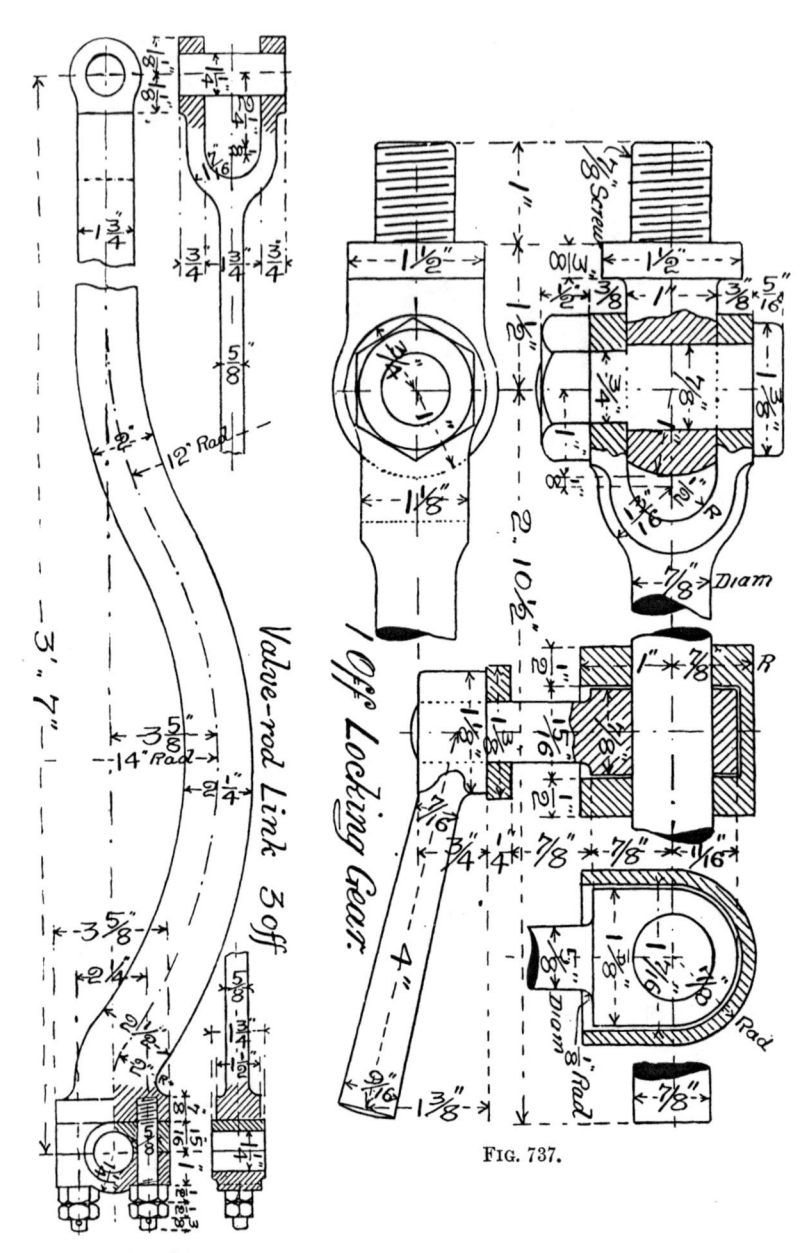

Fig. 736.

Fig. 737.

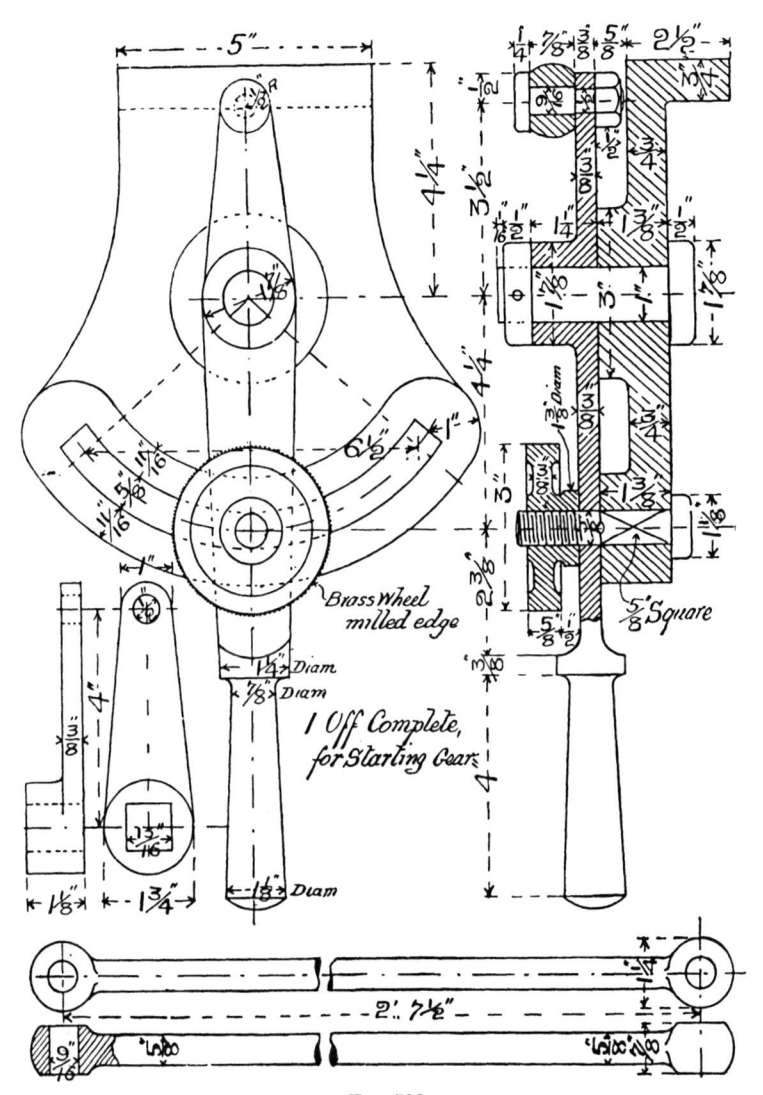

FIG. 738.

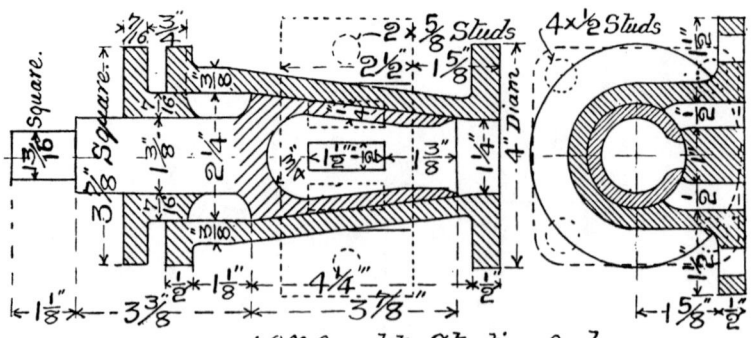

1 Off Complete. Starting Cock.
Fig. 739.

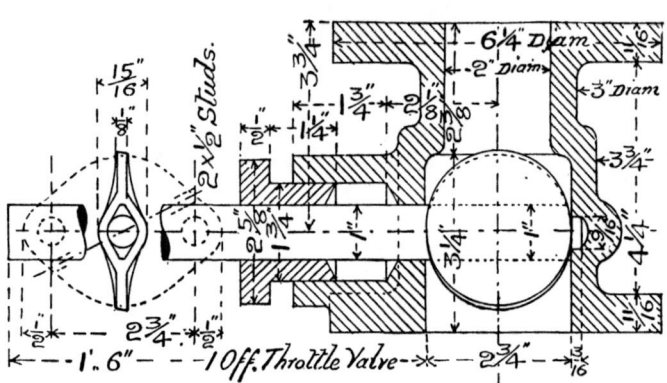

1 Off Throttle Valve
Fig. 740.

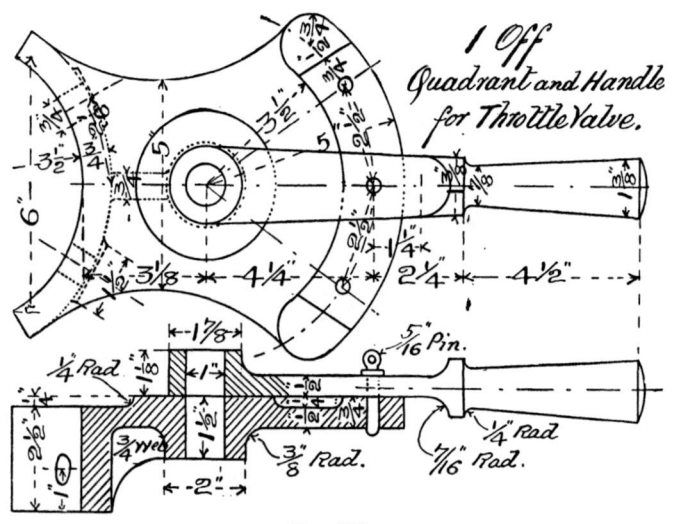

1 Off Quadrant and Handle for Throttle Valve.
Fig. 741.

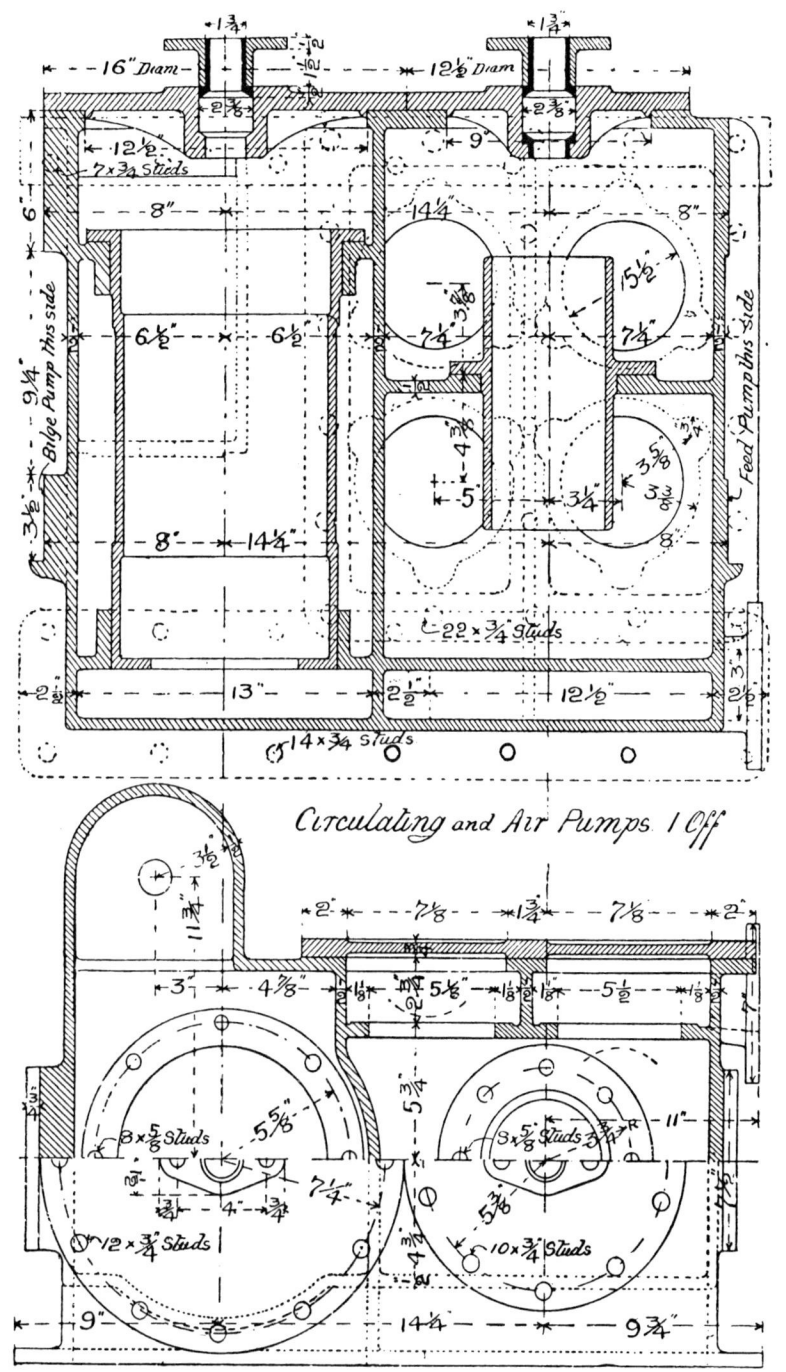

FIG. 742.

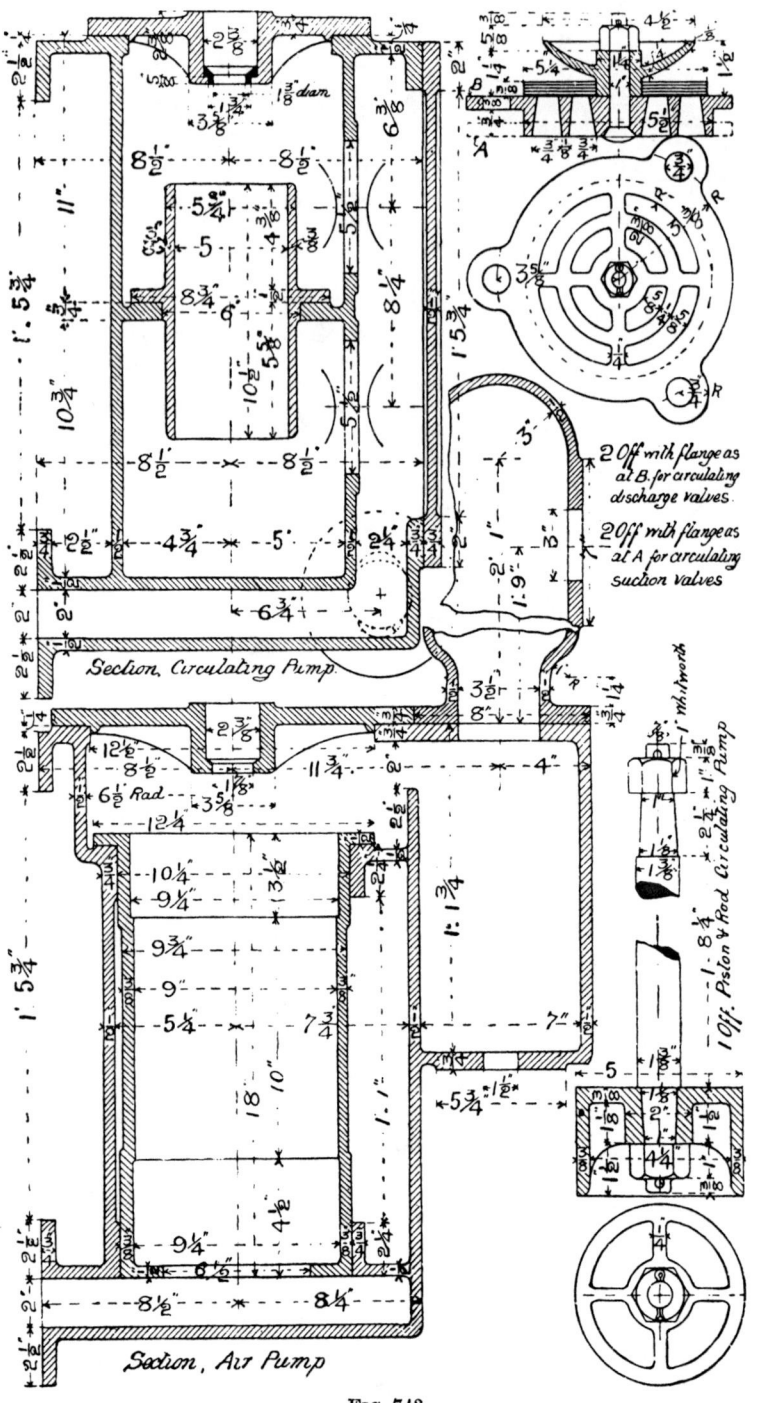

FIG. 743.

TRIPLE EXPANSION ENGINES.

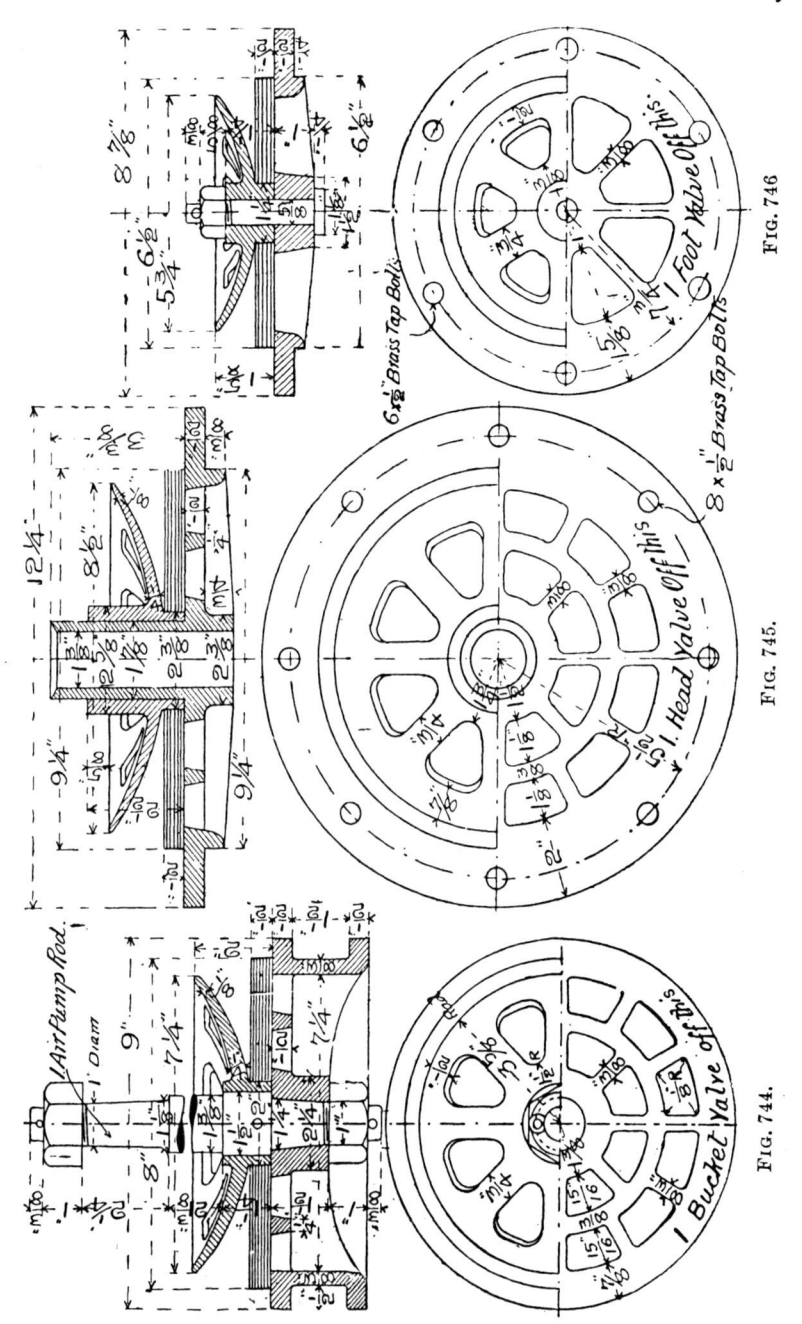

FIG. 746.

FIG. 745.

FIG. 744.

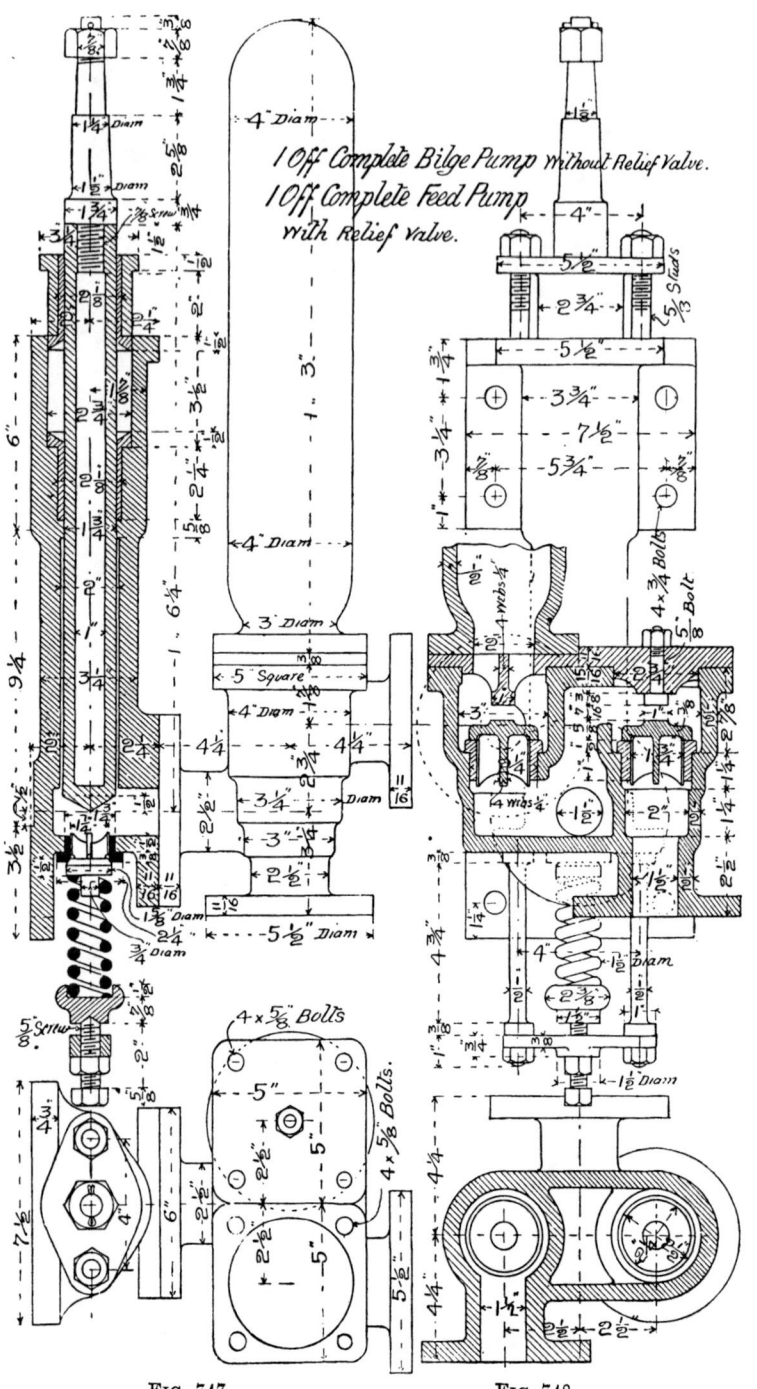

FIG. 747.   FIG. 748.

FIG. 749.

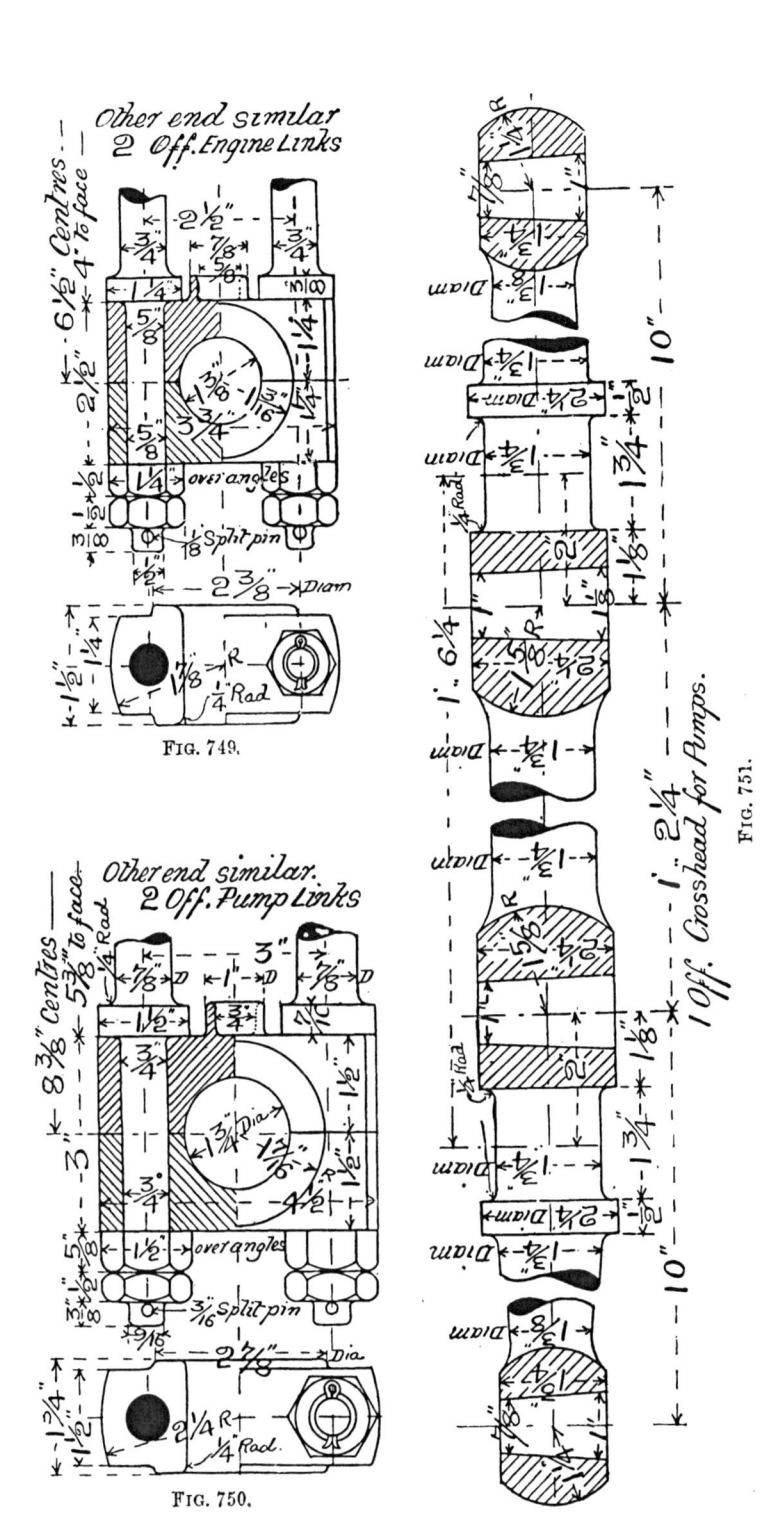

FIG. 750.

FIG. 751. — 1 Off. Crosshead for Pumps.

Fig. 752

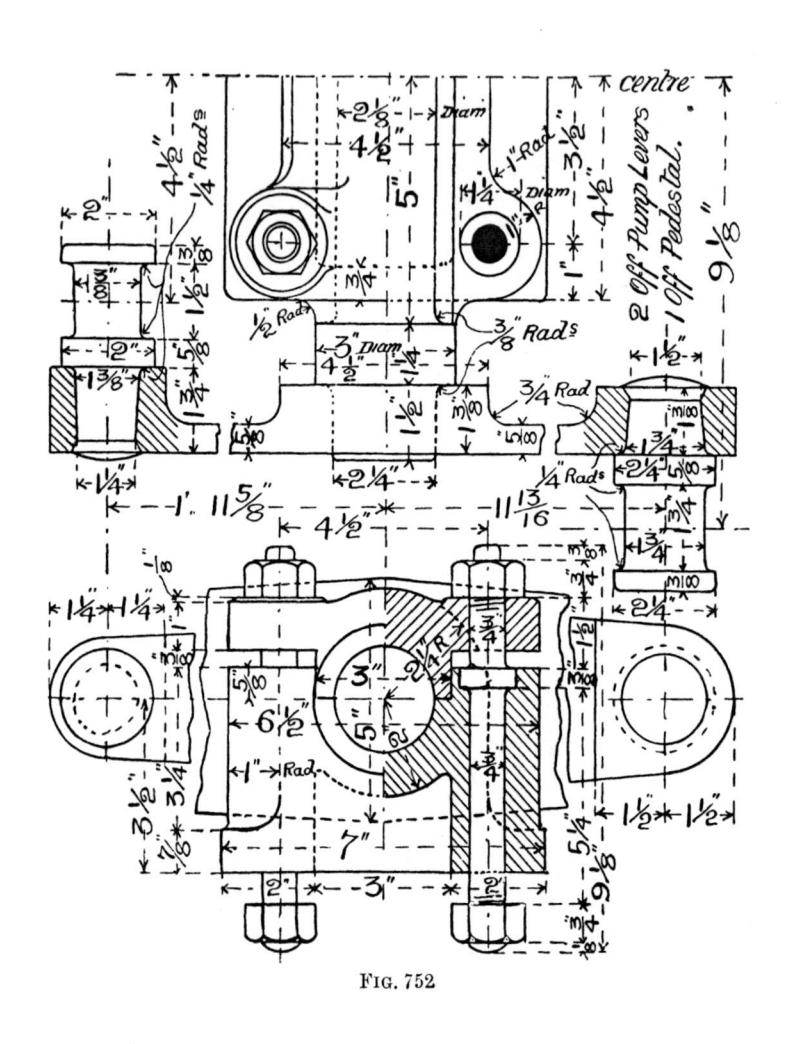

TRIPLE EXPANSION ENGINES. 367

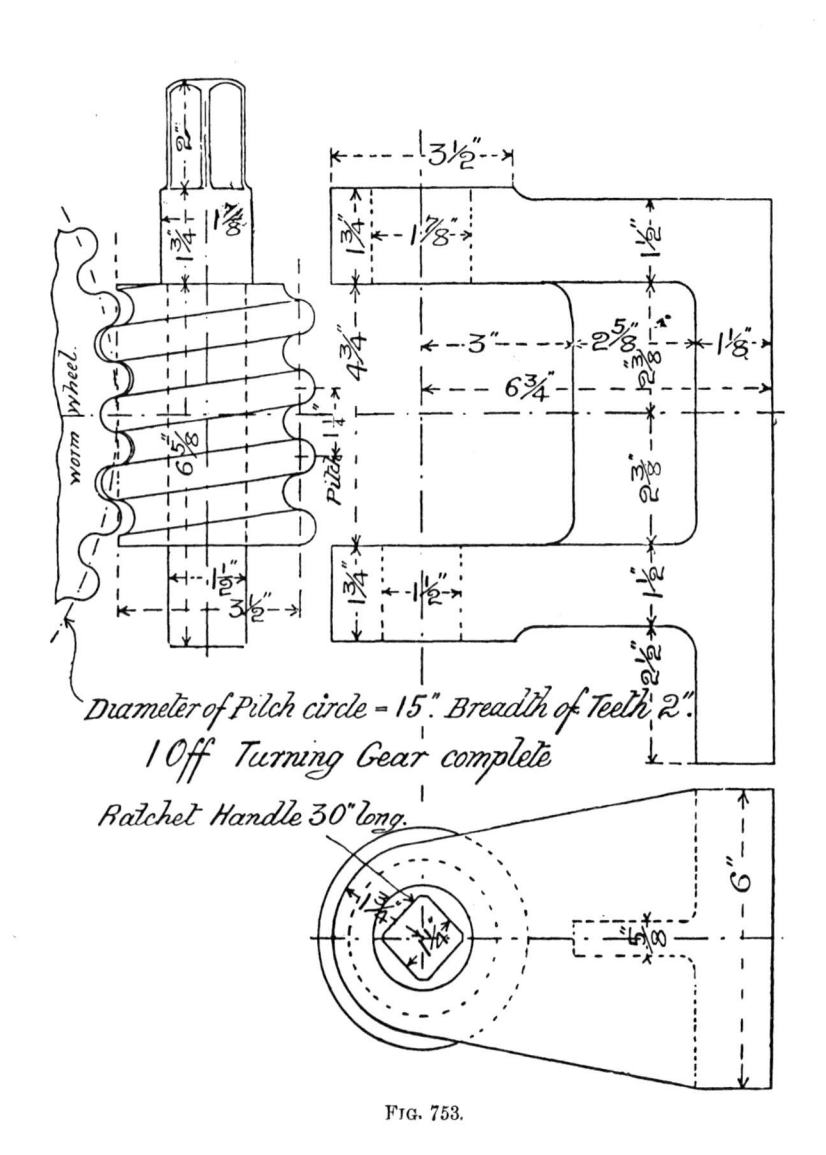

Diameter of Pitch circle = 15". Breadth of Teeth 2".
1 Off Turning Gear complete
Ratchet Handle 30" long.

FIG. 753.

CHAPTER XXV.

EXAMPLE OF LOCOMOTIVE ENGINE.

346. General Description of Engine.—Through the courtesy of Mr. James Holden, M.I.C.E., Locomotive Superintendent of the Great Eastern Railway, we are able to give in this chapter a large number of illustrations showing most of the details of an express locomotive of recent design.

The engine illustrated is one of the type known as the "Claud Hamilton" class. The first of this class, No. 1900, was exhibited at the Paris Exhibition in 1900, and gained the distinction of a Grand Prize. These engines are four-coupled bogie engines for hauling heavy express trains. They are fired with liquid fuel on Mr. Holden's well-known system; but they may be converted into coal-burning engines by lifting out the liquid fuel tank from the tender, and filling the space thus gained with coal.

The oil fuel, as it passes from the tank in the tender, is heated in the tubular chamber shown in Fig. 804, p. 399, by the exhaust steam from the air-brake pump which circulates through this chamber. The heated liquid fuel then passes to the burners in the fire-box, which are in the form of injectors worked by steam.

The smoke-box is fitted with two nests of tubes for heating the air which atomises or sprays the oil in the burners. The air enters from the atmosphere just under the smoke-box door, passes up through one of the nests of tubes, crosses over and down through the other nest, and is then led back to the burners in the fire-box. The air is made to travel through the heater in the smoke-box by the exhausting action of the central jets of the liquid fuel burners, and also by the forward motion of the engine. The heated air reaches the burners at a temperature of about 400° F., and, as the steam is only at about 350° F., the resultant spray is dry and hot. Previous to the application of the air-heating apparatus it was found that there was a tendency to condensation of a portion of the steam used for injection, due to its coming in contact with the cold liquid fuel at the burners, which caused a lowering of the temperature in the fire-box. By heating the air as described, it has been found that the combustion is improved, and the consumption of fuel reduced by about $1\frac{1}{2}$ lbs. per mile.

These engines are fitted with double sight feed lubricators, exhaust steam injector, and steam sanding apparatus. The reversing gear is worked by means of the compressed air from the Westinghouse brake system.

The tender has six wheels, 4 ft. 1 in. diameter, on the tread, and has a wheel base of 12 ft. The tender is fitted with a scoop for refilling the tank with water from the track while running; this, like the reversing gear, is worked by compressed air from the Westinghouse brake.

347. Dimensions of the Engine.—The illustrations which follow are very fully dimensioned, but for convenience of reference the principal dimensions are here given in tabular form:—

Table of Principal Dimensions.

Cylinders.	Ft.	In.
Diameter	1	7
Stroke of pistons	2	2
Length of ports	1	4¾
Width of steam ports	0	1⅝
Width of exhaust ports	0	3½
Distance apart of cylinders, centre to centre	2	2
Wheels.		
Diameter on tread, coupled	7	0
Diameter on tread, bogie	3	9
Thickness of tyres on tread	0	3
Width of tyres on tread	0	5⅜
Crank Axle.		
Diameter of journals	0	7⅝
Length of journals	0	9
Diameter of crank pins	0	8⅛
Length of crank pins	0	4½
Trailing Axle.		
Diameter of journals	0	7½
Length of journals	0	9
Bogie Axles.		
Diameter of journals	0	5¾
Length of journals	0	9
Frames.		
Total length	30	7
Distance apart, inside	4	1½
Thickness	0	1
Wheel Base.		
Distance between centres of bogie axles	6	6
Centre of driving axle to centre of bogie	11	3
Centre driving axle to centre of trailing axle	9	0
Total wheel base	23	6
Boiler.		
Height of centre of barrel above rails	8	3
Height of chimney above rails	12	11
Largest diameter of barrel (inside)	4	8
Length of barrel	11	9
Thickness of barrel plates	0	0½
Thickness of smoke-box tube plate	0	0⅝
Diameter of rivets	0	0⅞
Working pressure, 180 lbs. per sq. in.		
Fire-box Shell.		
Length (outside)	7	0
Breadth at bottom (outside)	4	0½
Depth below centre of barrel (front)	4	10½
Depth below centre of barrel (back)	4	1
Thickness of plates, throat and back	0	0 9/16
Thickness of plates, sides and top	0	0½
Inside Fire-box.		
Width of water space round fire-box at bottom	0	3
Thickness of tube plate	0	1
Thickness of other plates	0	0½
Tubes.		
Number of tubes, 274.		
Diameter of tubes (outside)	0	1¾
Length between tube plates	12	1

Heating Surface.	Sq. Ft.
Tubes	1516·5
Fire-box	114·0
Total	1630·5
Grate area	21·3

Weight of Engine in Working Order.

	Tons.	Cwt.	Qrs.
On bogie	17	2	2
On driving axle	16	12	1
On trailing axle	16	11	2
Total	50	6	1

	Tons.	Cwt.	Qrs.
Weight of tender in working order	35	1	0
Weight of tender, empty	19	10	0
Weight of engine, empty	46	10	0

Capacity of Tender.	
	Gals.
Water	3420
Oil	750
	Cwt.
Coal	30

348. The materials employed in the construction of these engines are indicated on the detail drawings by letters according to the following table:—

Letters.	Signification.	Letters.	Signification.
S.	Steel.	P.B.	Phosphor Bronze.
S.C.	Steel Casting.	S.B.	Stone's Bronze.
W.I.	Wrought Iron.	W.M.	White Metal.
C.I.	Cast Iron.	I.R.	India-rubber.
M.C.I.	Malleable Cast Iron.	L.	Leather.
G.M.	Gun Metal.	C.	Copper.
B.	Brass.	B.M.	Brazing Metal.

LOCOMOTIVE ENGINE.

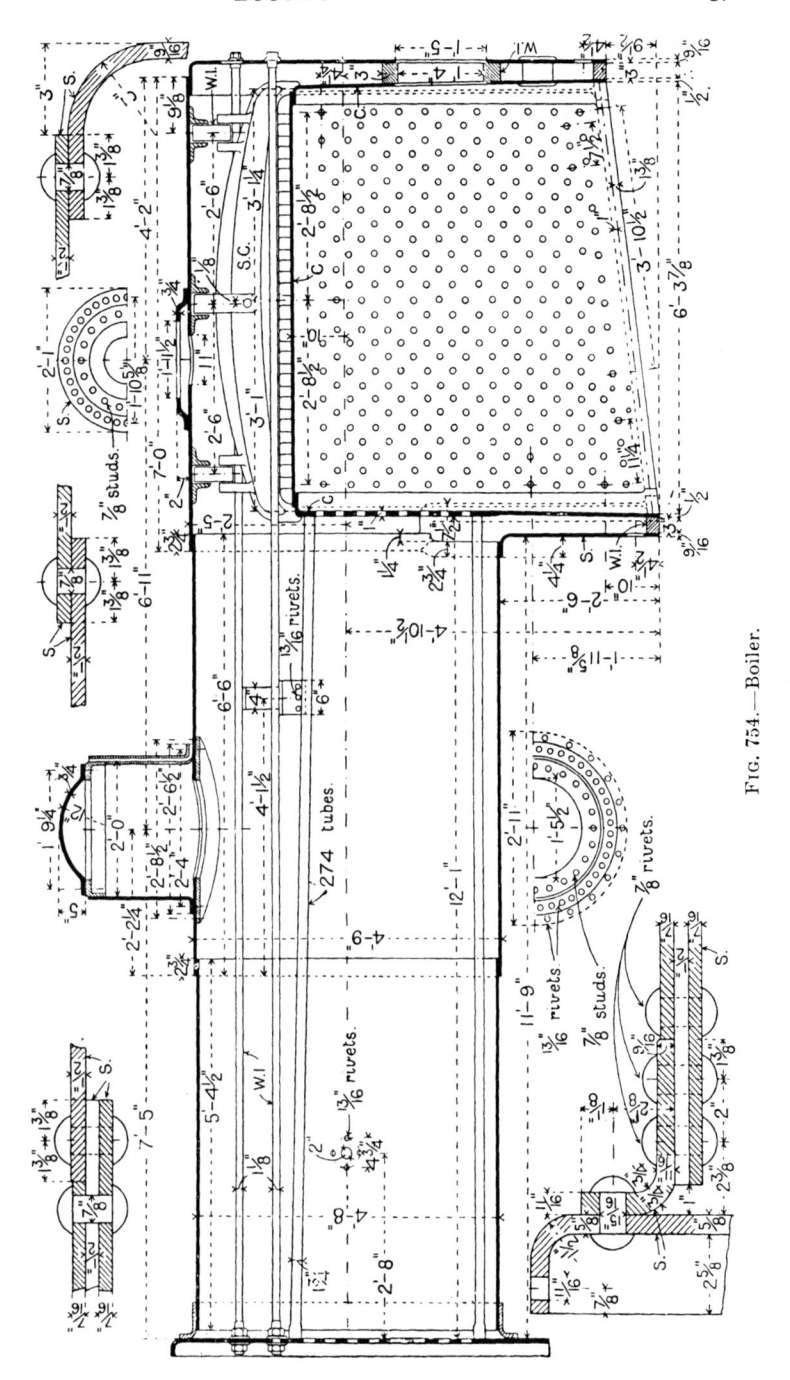

FIG. 754.—Boiler.

372 MACHINE DRAWING AND DESIGN

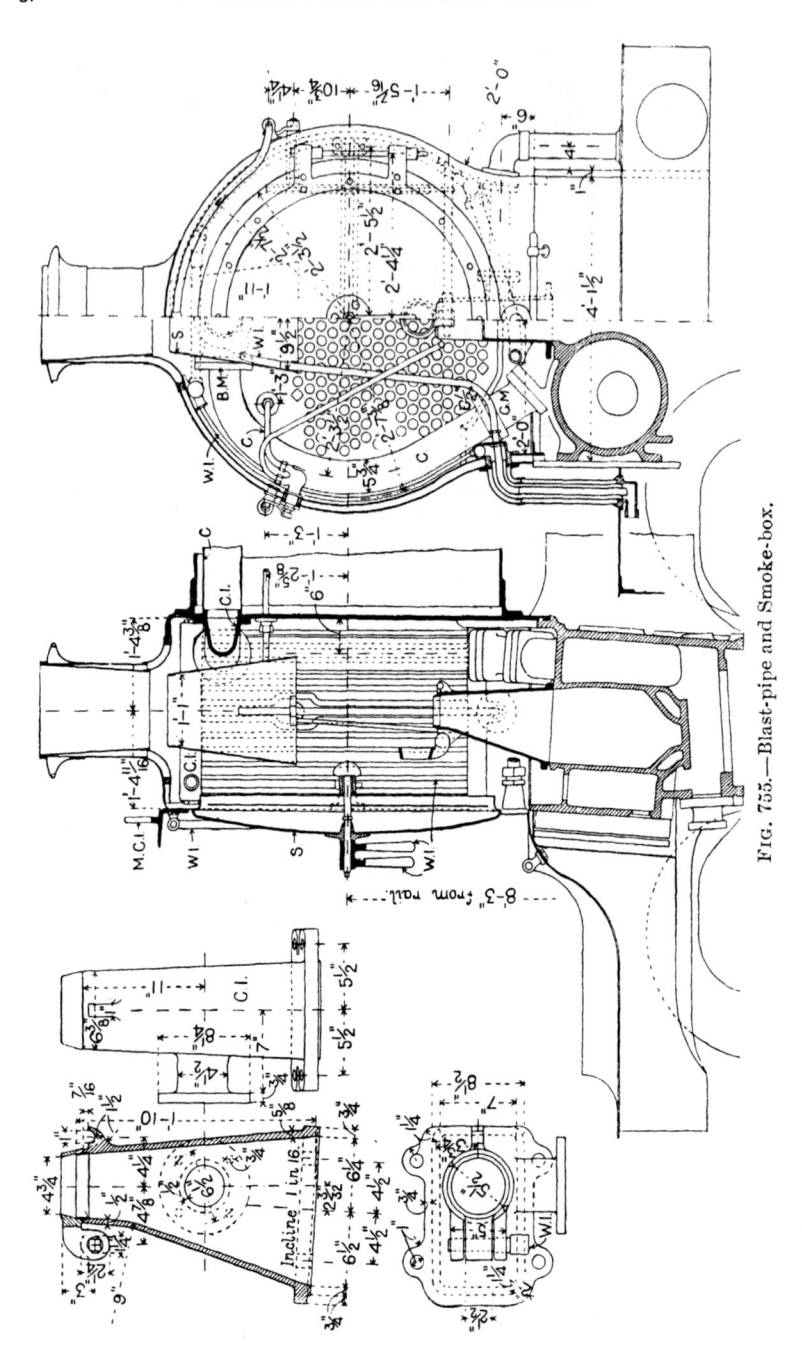

FIG. 755.—Blast-pipe and Smoke-box.

LOCOMOTIVE ENGINE. 373

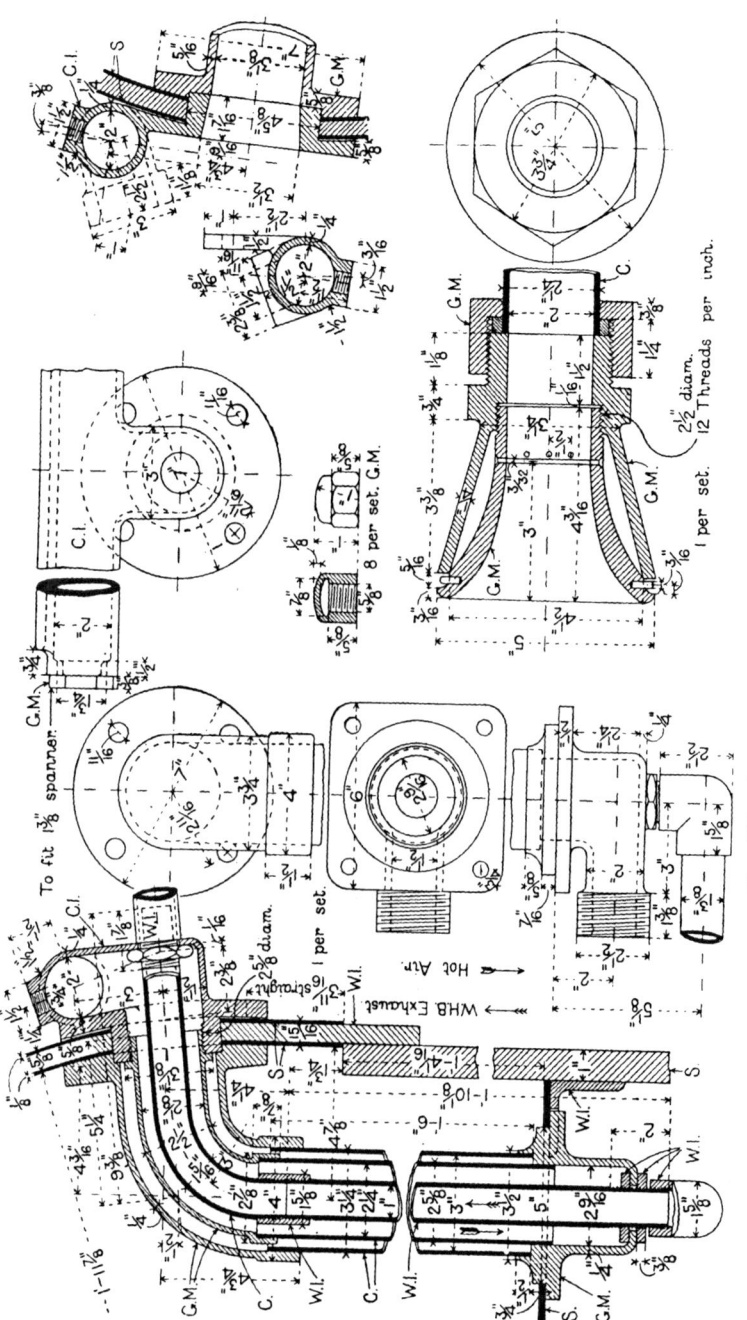

Fig. 756.—Details of Air-heating Apparatus.

374 MACHINE DRAWING AND DESIGN.

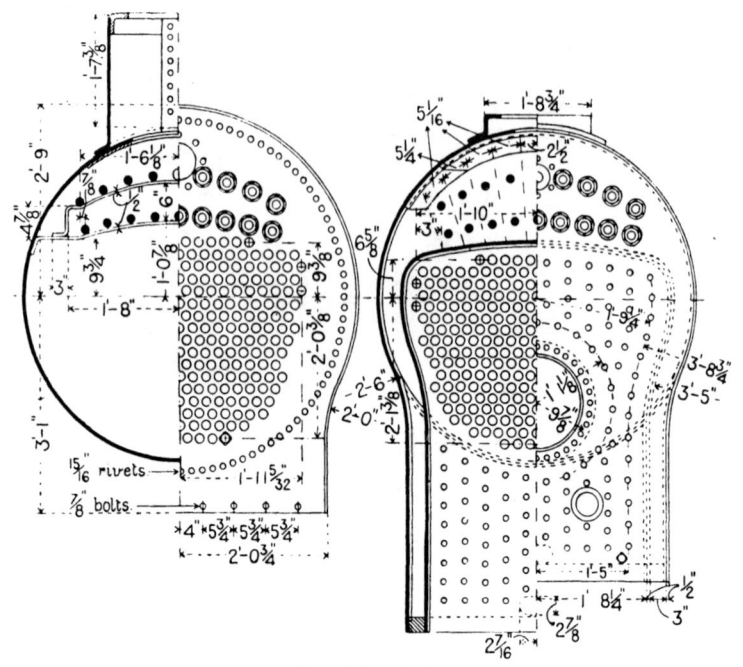

FIG. 757.—Boiler.

In the engine, No. 1900, sent to the Paris Exhibition, the tubes (274 in number) were of copper, but in all the other engines of this class they are of steel.

The fire-box screwed stays are of bronze, made flexible by means of longitudinal saw-cuts. These stays are $1\frac{5}{8}$ inch diameter, except the two top rows in the back plate, which are 1 inch diameter. Pitch of stays $3\frac{1}{2}$ inches.

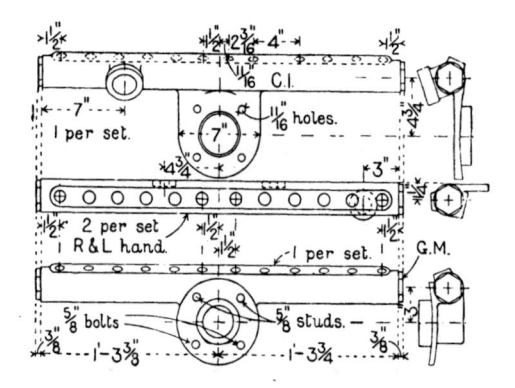

FIG. 758.—Details of Air-heating Apparatus in Smoke-box.

LOCOMOTIVE ENGINE. 375

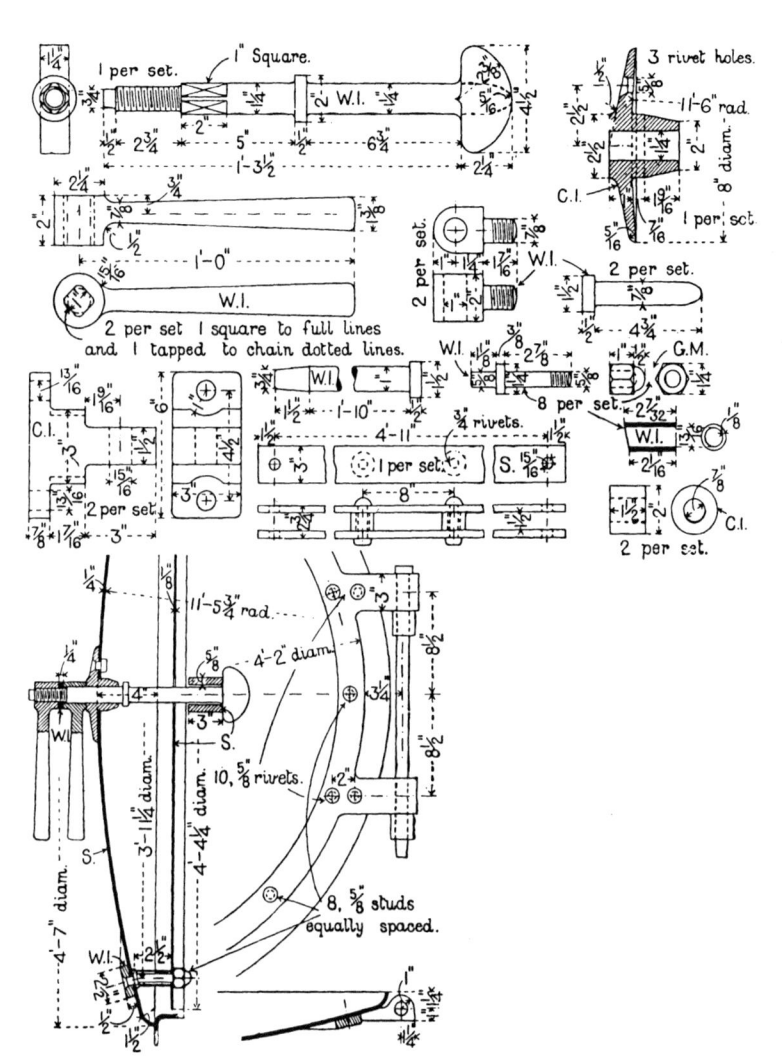

FIG. 759.—Smoke-box Door and Details.

376 MACHINE DRAWING AND DESIGN.

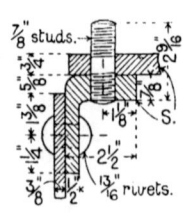

Fig. 760.—Dome Cover Joint

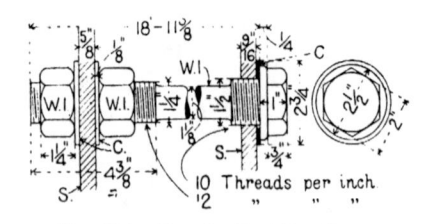

Fig. 761.—Longitudinal Bar Stay.

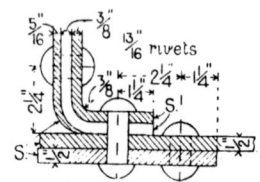

Fig. 762.—Joint between Dome and Barrel.

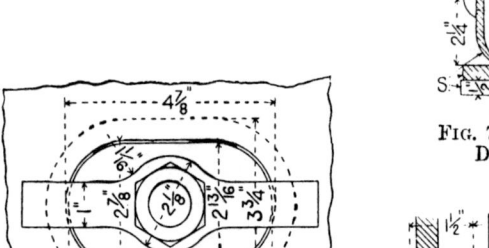

Fig. 763.—Mud Door.

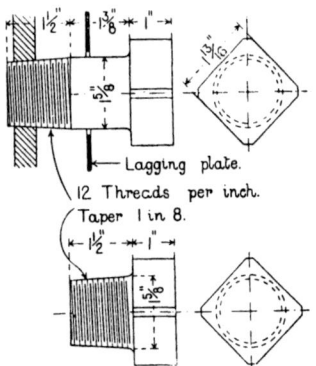

Fig. 764.—Plugs.

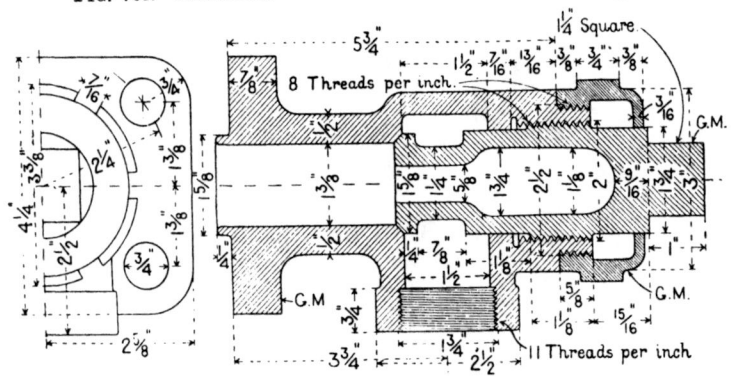

Fig. 765.—Blow-off Cock.

LOCOMOTIVE ENGINE.

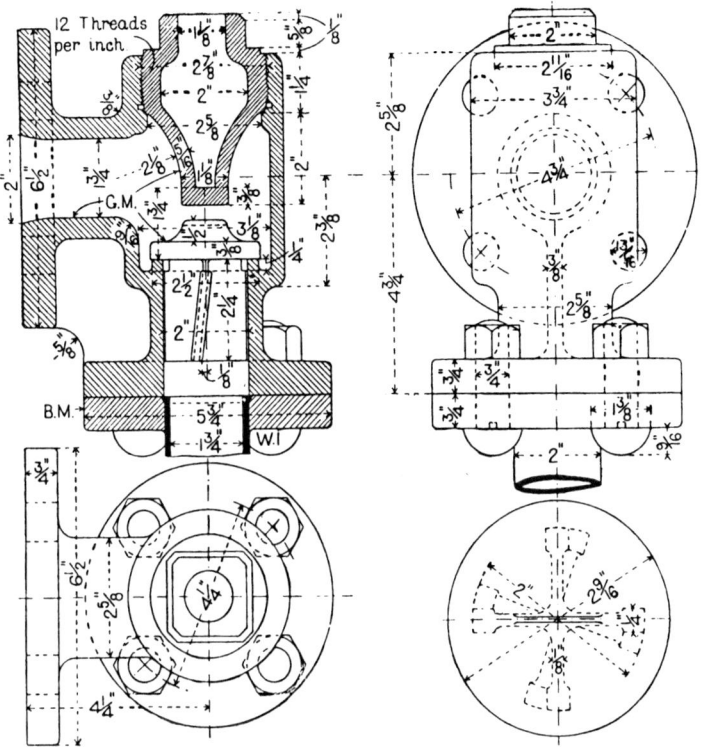

FIG. 766.—Clack-box.

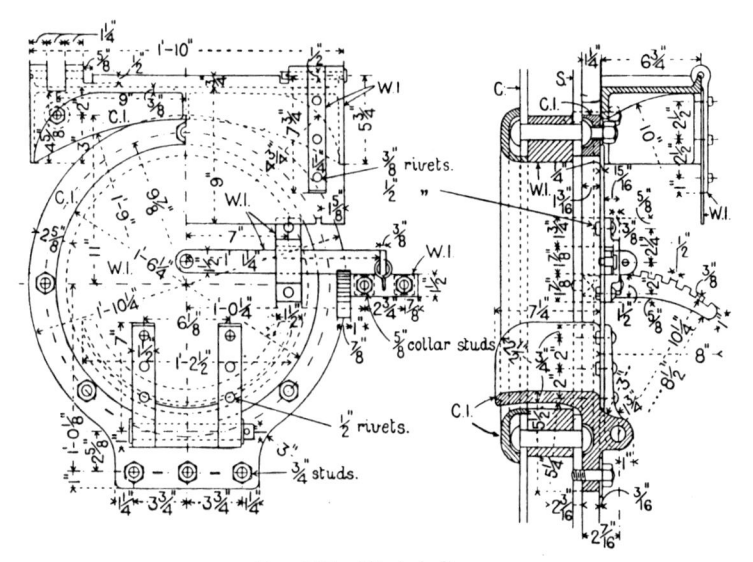

FIG. 767.—Firehole Door.

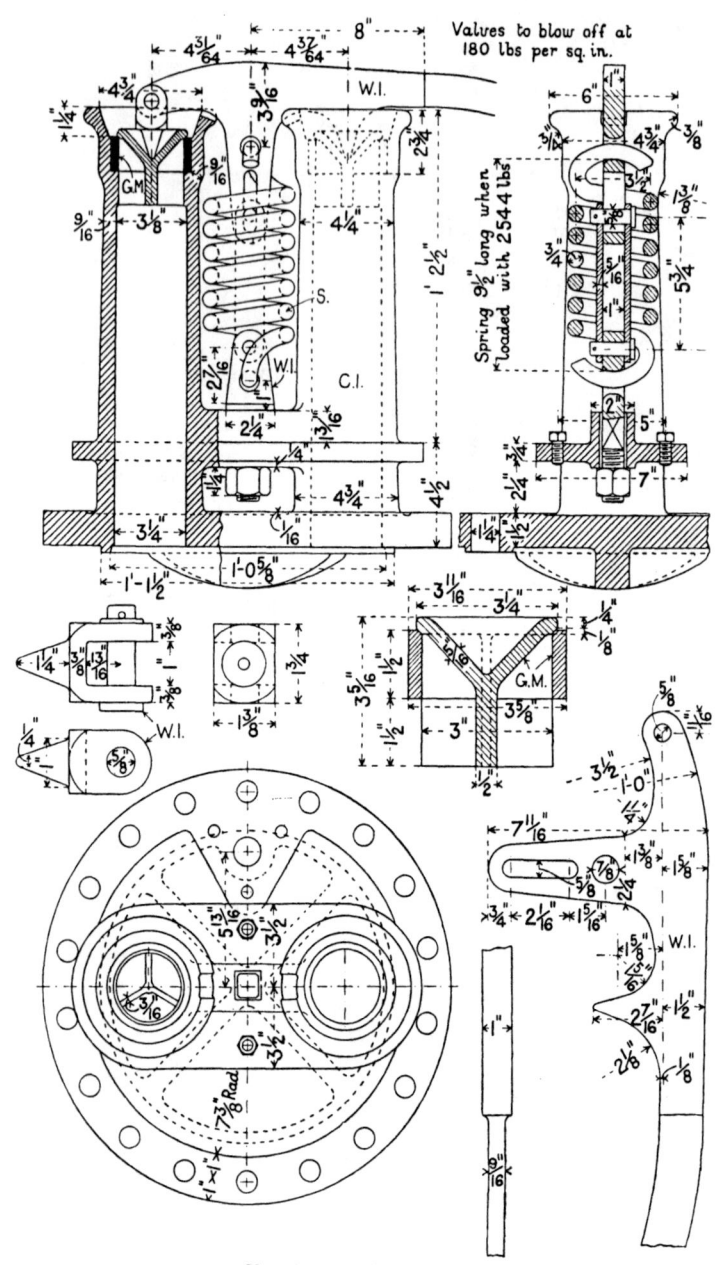

FIG. 768.—Safety-valves.

FIG. 769.—Cylinders.

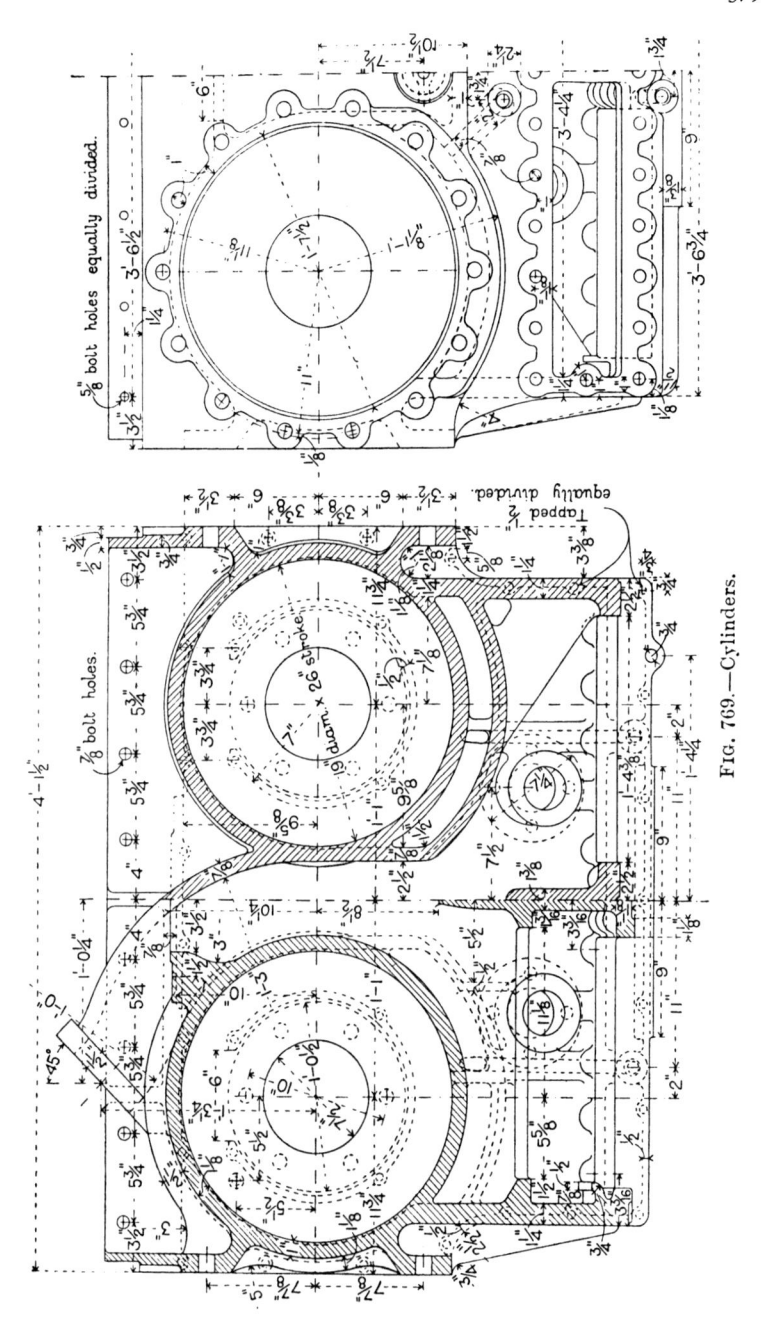

380   MACHINE DRAWING AND DESIGN.

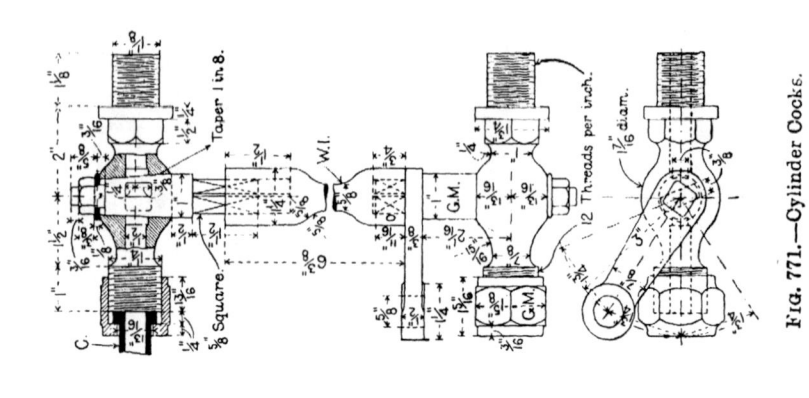

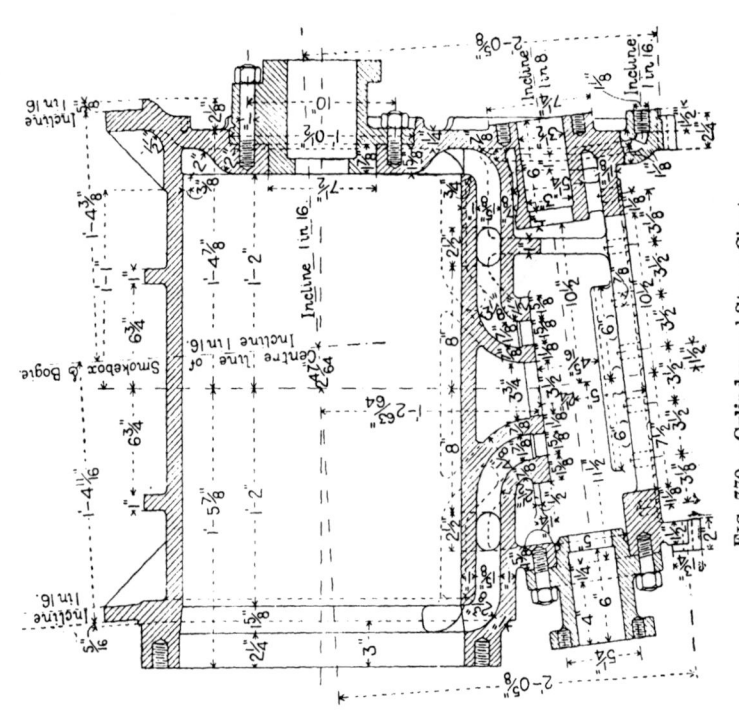

FIG. 771.—Cylinder Cocks.

FIG. 770.—Cylinder and Steam Chest.

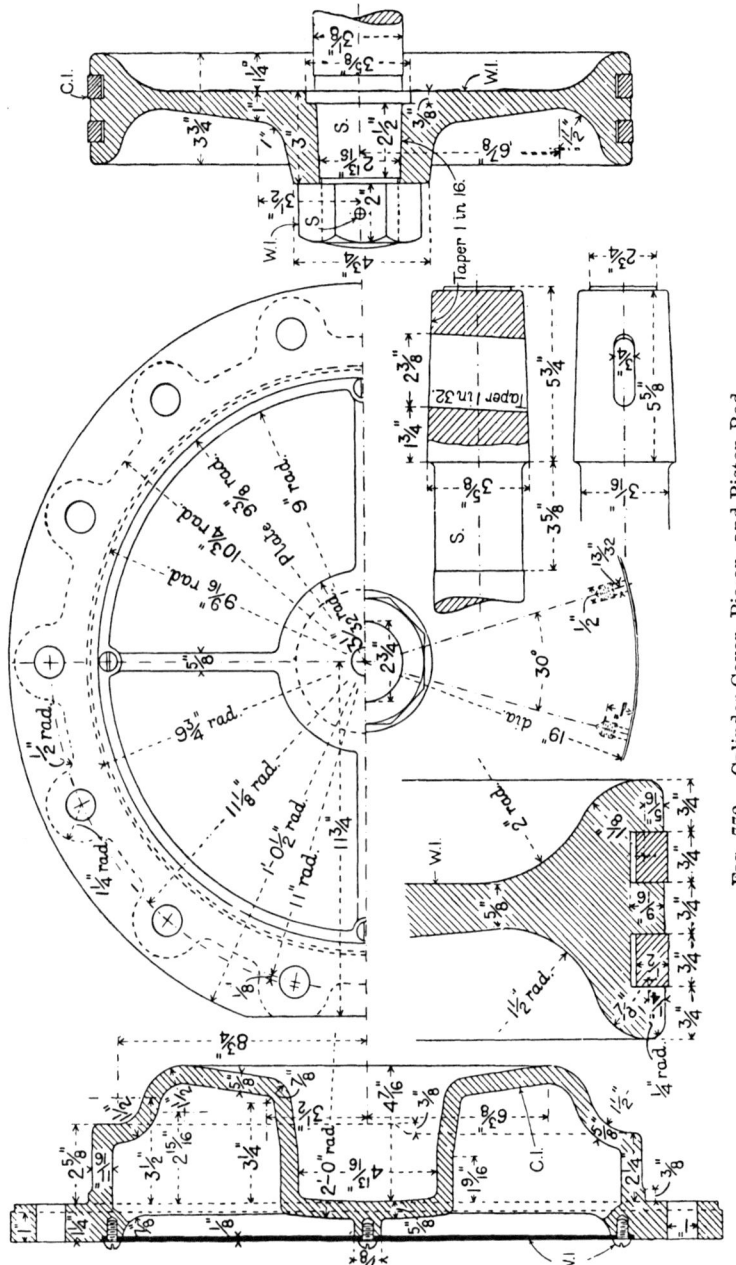

FIG. 772.—Cylinder Cover, Piston, and Piston Rod.

382 MACHINE DRAWING AND DESIGN.

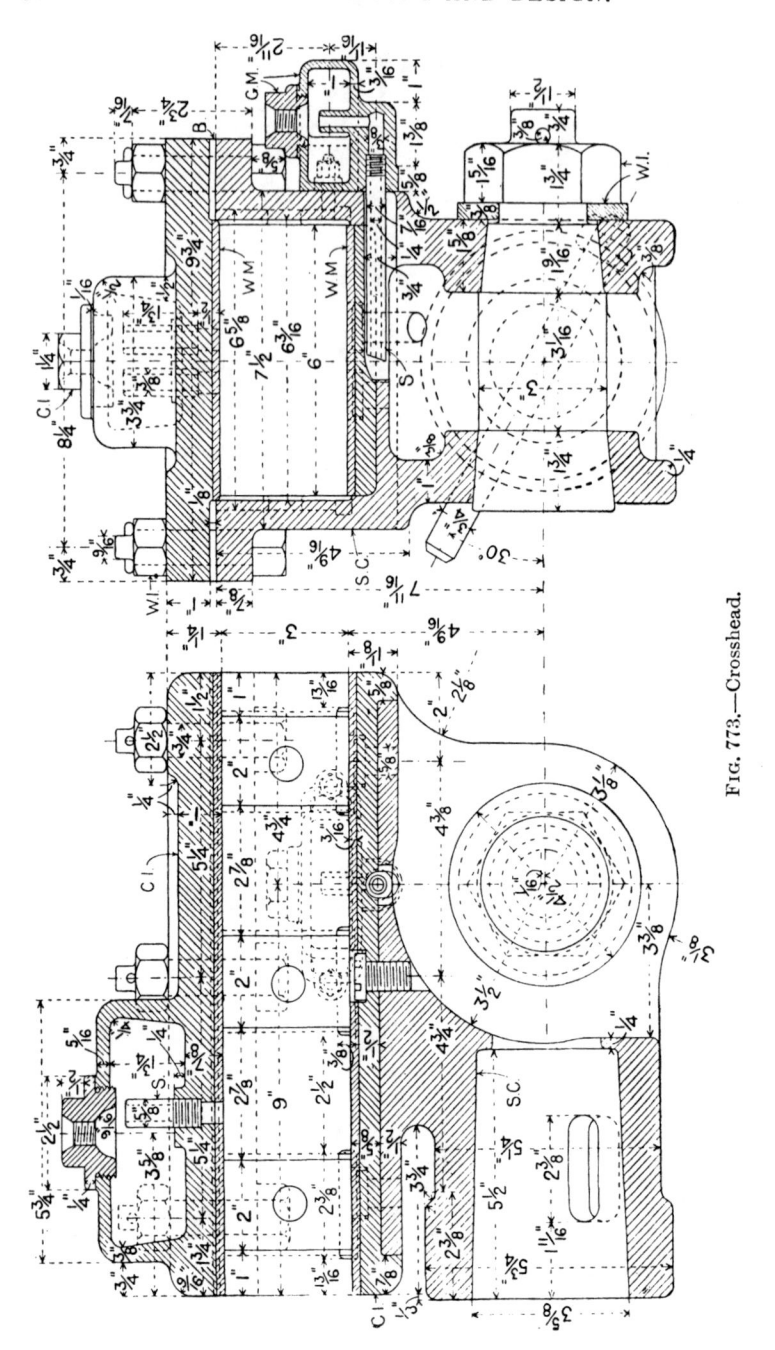

Fig. 773.—Crosshead.

LOCOMOTIVE ENGINE. 383

FIG. 776.—Details of Valve Spindle Packing.

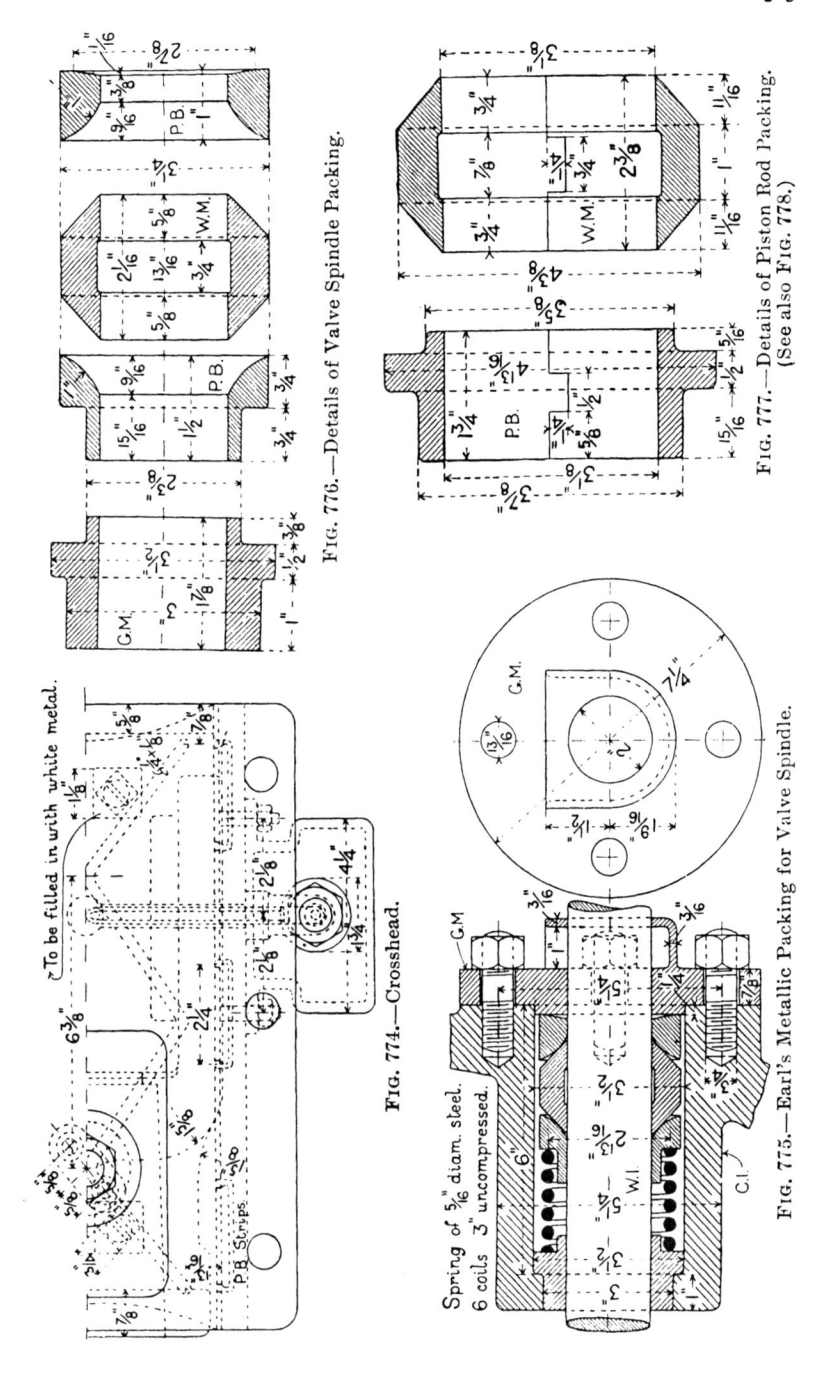

FIG. 777.—Details of Piston Rod Packing. (See also FIG. 778.)

FIG. 774.—Crosshead.

FIG. 775.—Earl's Metallic Packing for Valve Spindle.

384   MACHINE DRAWING AND DESIGN.

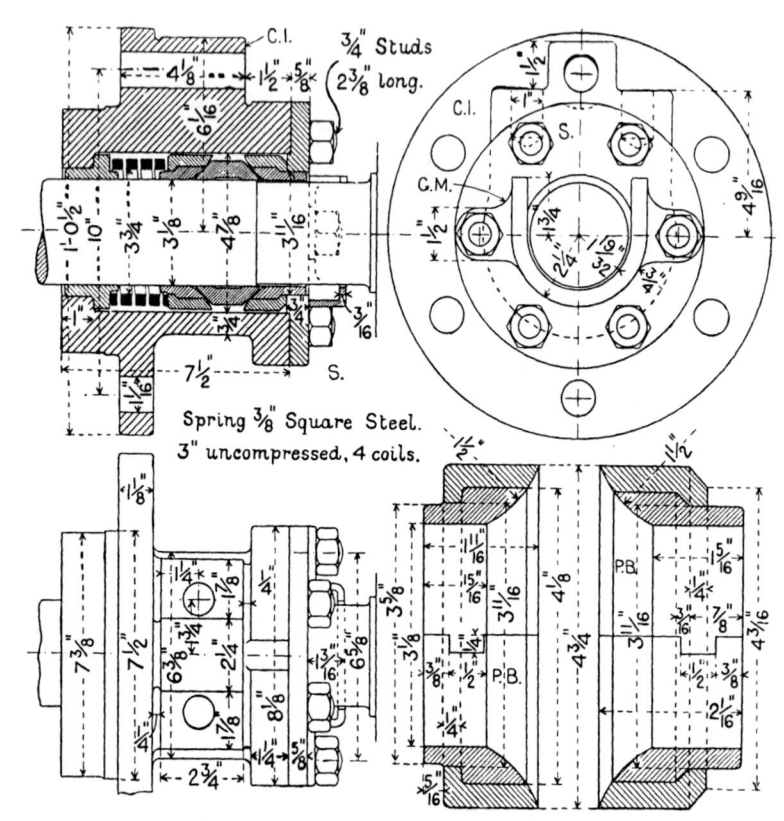

FIG. 778.—Earl's Metallic Packing for Piston Rod.   (See also FIG. 777.)

LOCOMOTIVE ENGINE. 385

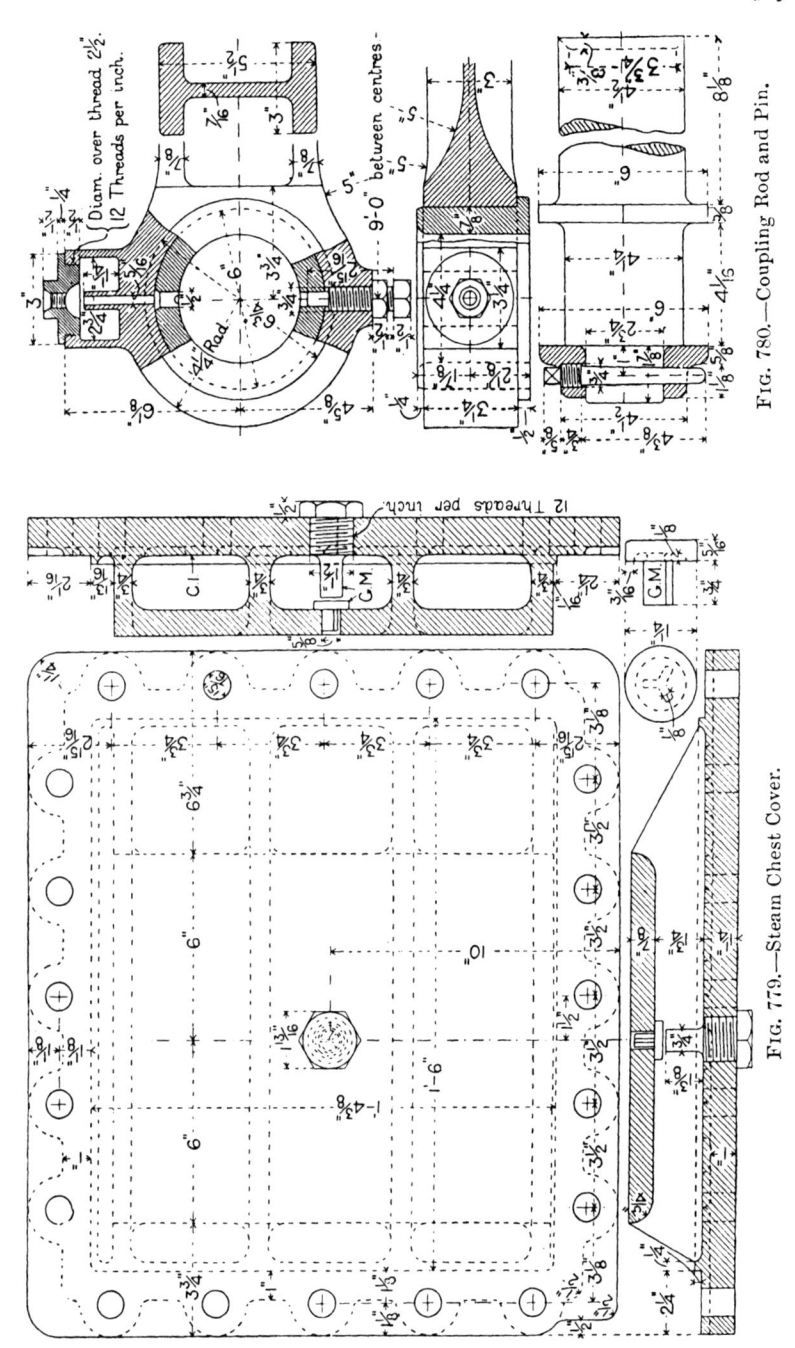

FIG. 780.—Coupling Rod and Pin.

FIG. 779.—Steam Chest Cover.

2 B

386 MACHINE DRAWING AND DESIGN.

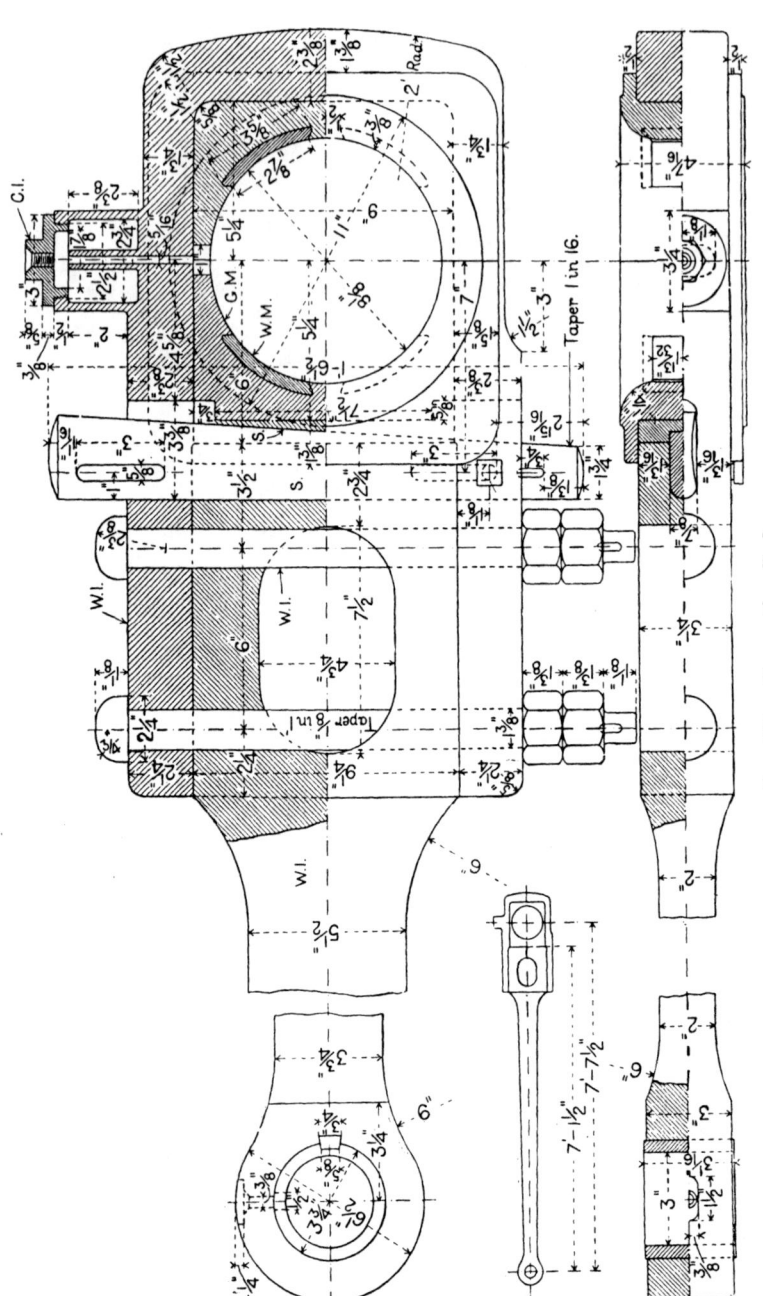

FIG. 781.—Connecting Rod.

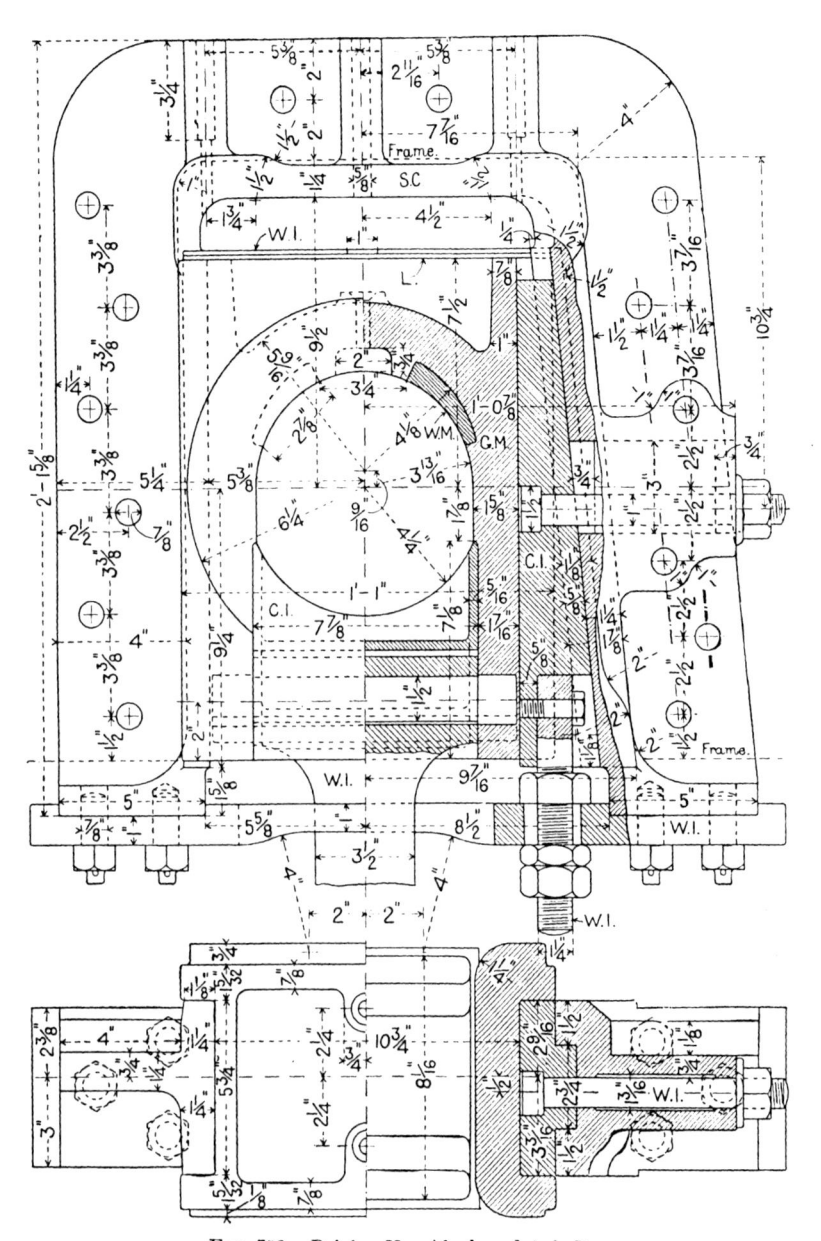

FIG. 782.—Driving Hornblock and Axle Box.

388   MACHINE DRAWING AND DESIGN.

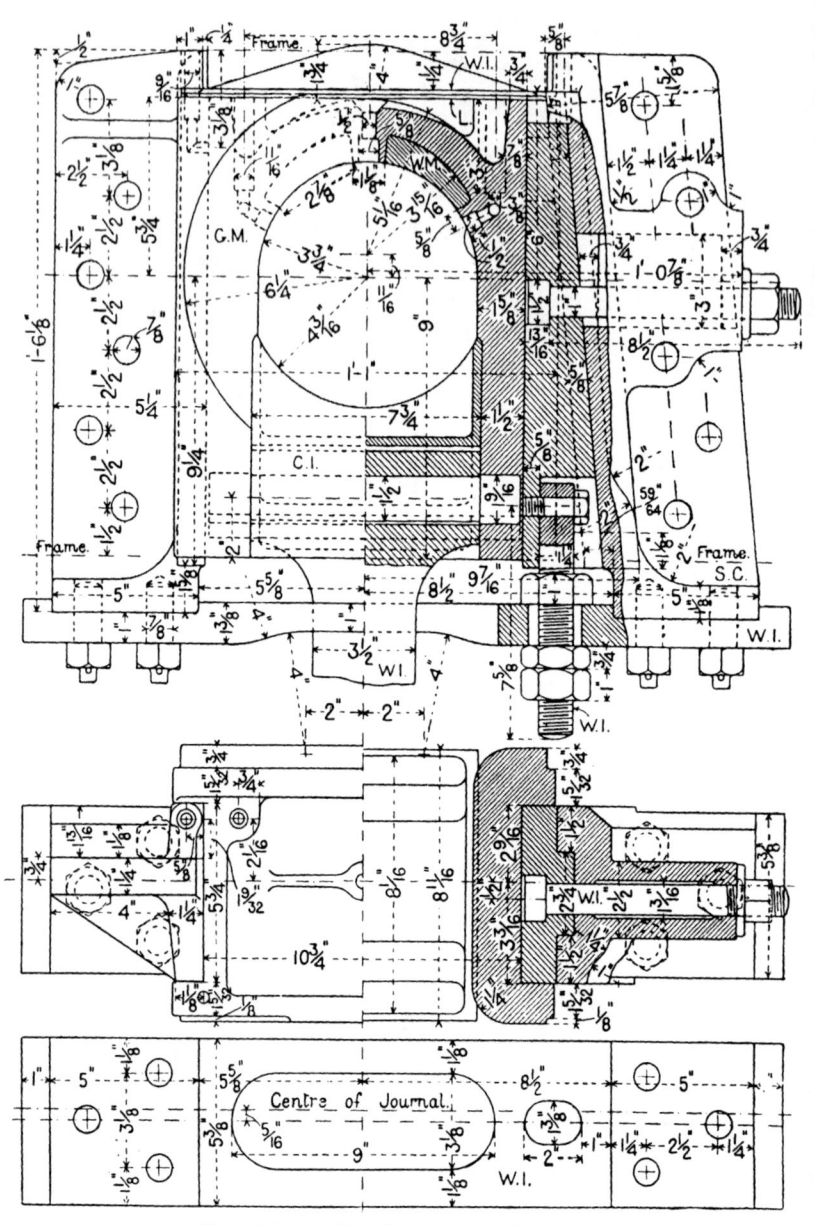

FIG. 783.—Trailing Hornblock and Axle Box.

LOCOMOTIVE ENGINE. 389

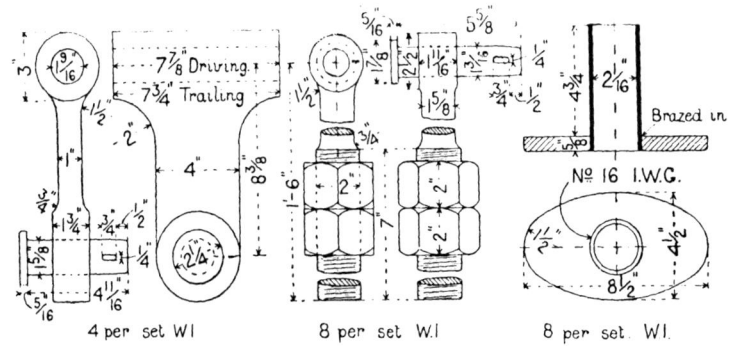

FIG. 784.—Driving and Trailing Spring Gear.

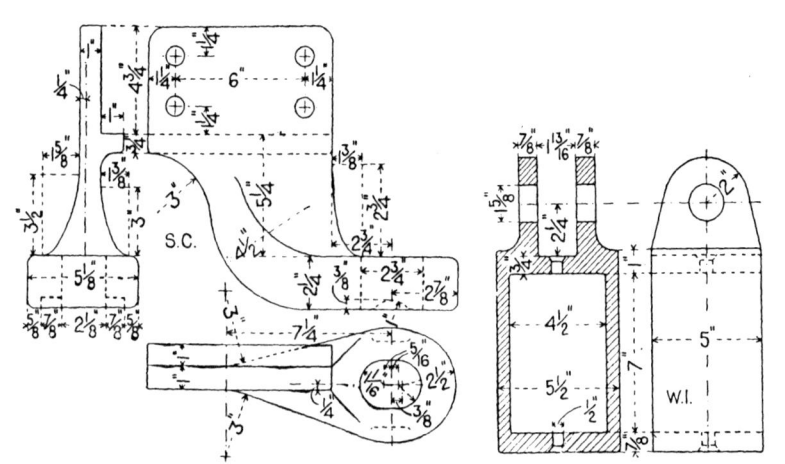

FIG. 785.—Driving and Trailing Spring Bracket.

FIG. 786.—Driving and Trailing Spring Buckle.

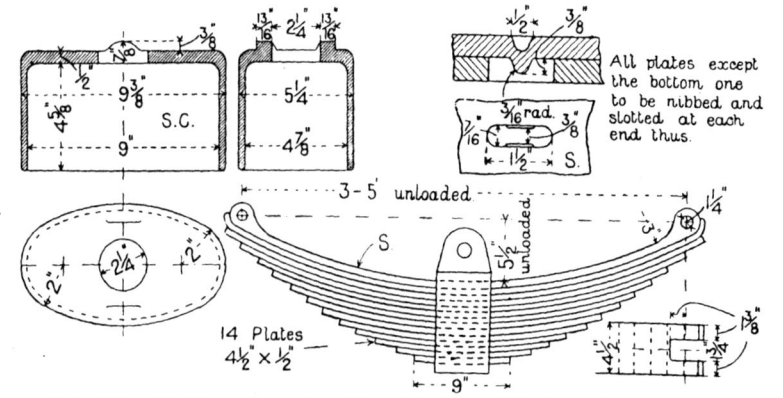

FIG. 787.—Shield.

FIG. 788.—Driving and Trailing Spring.

390 MACHINE DRAWING AND DESIGN

FIG. 789.—Trailing Axle.

FIG. 790.—Driving and Trailing Wheels.

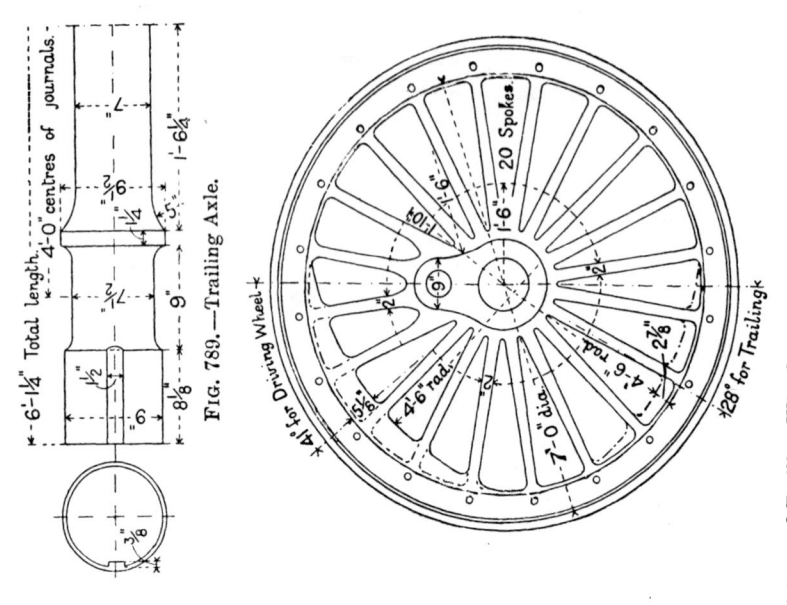

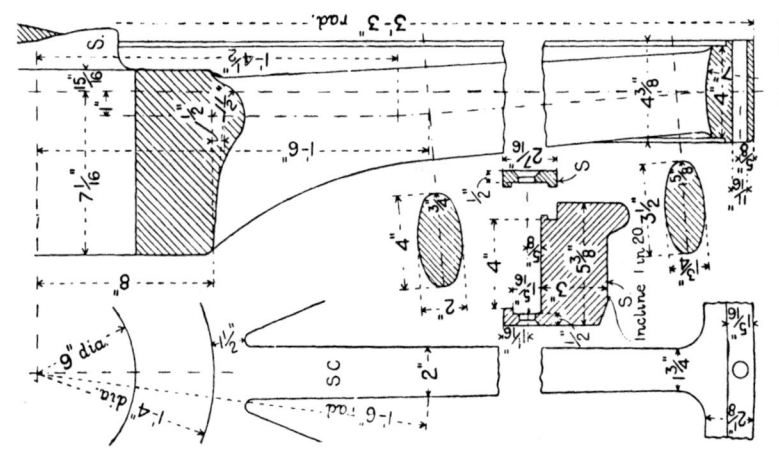

LOCOMOTIVE ENGINE. 391

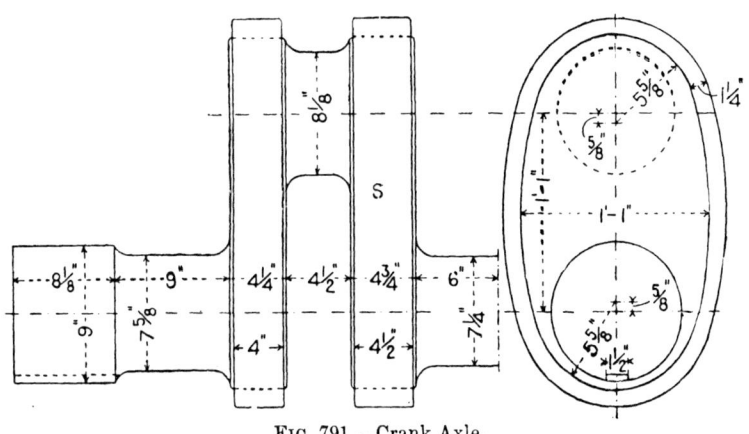

FIG. 791.—Crank Axle.

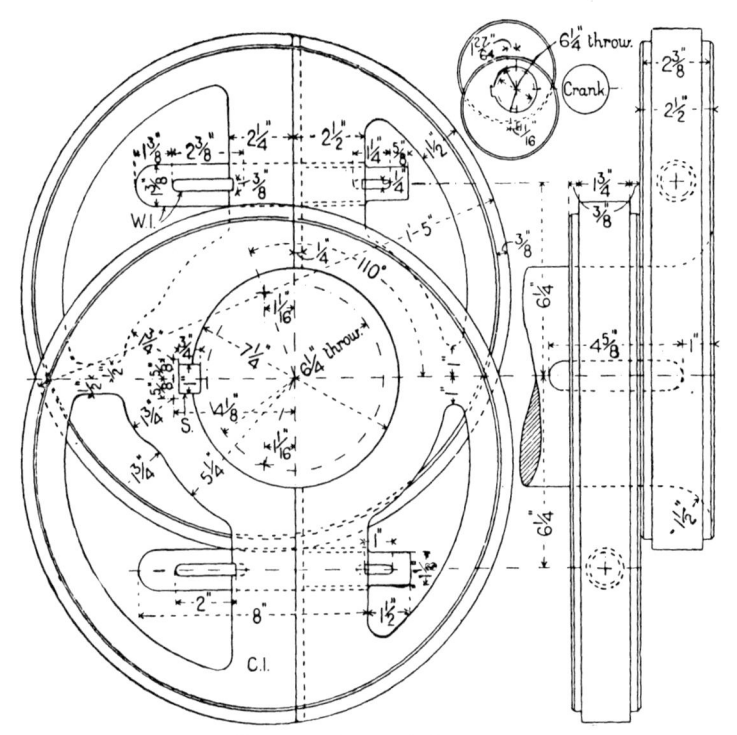

FIG. 792.—Eccentric Sheaves

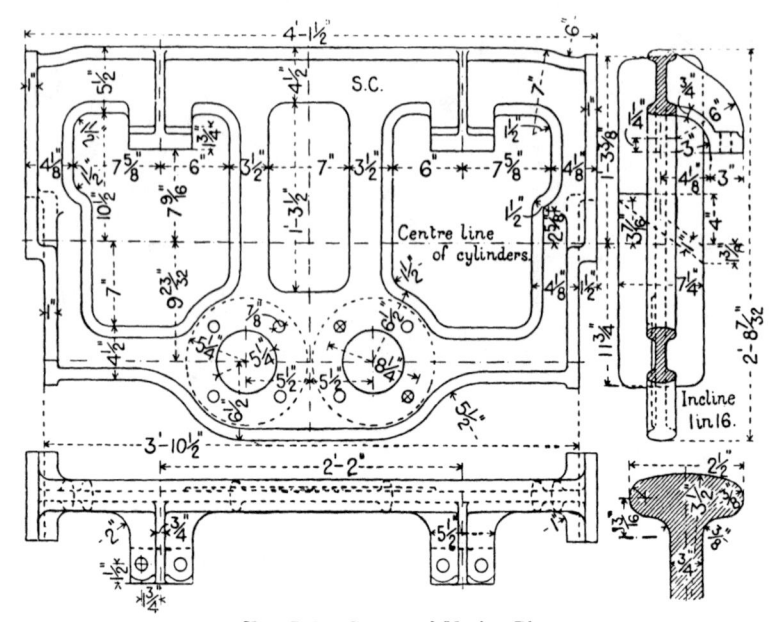

FIG. 793.—Cast-steel Motion Plate.

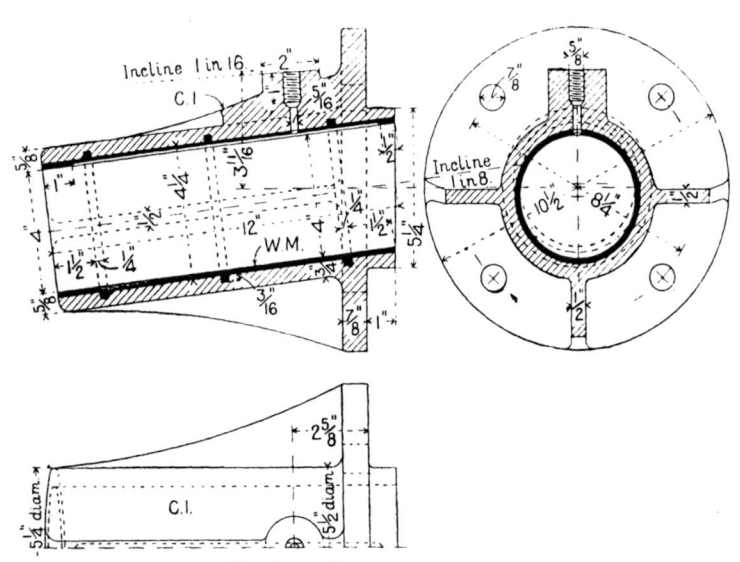

FIG. 794.—Valve Spindle Guide.

LOCOMOTIVE ENGINE. 393

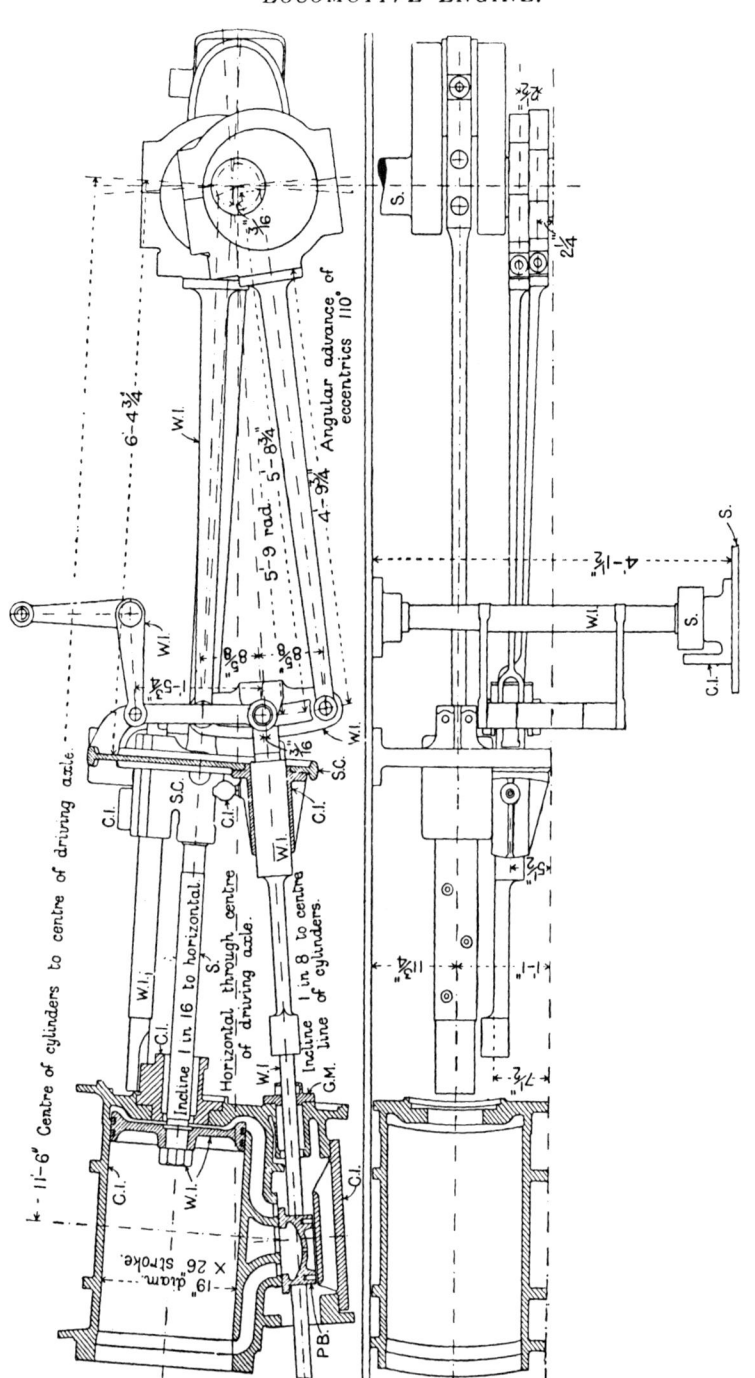

FIG. 795.—Valve Motion Arrangement.

394 MACHINE DRAWING AND DESIGN.

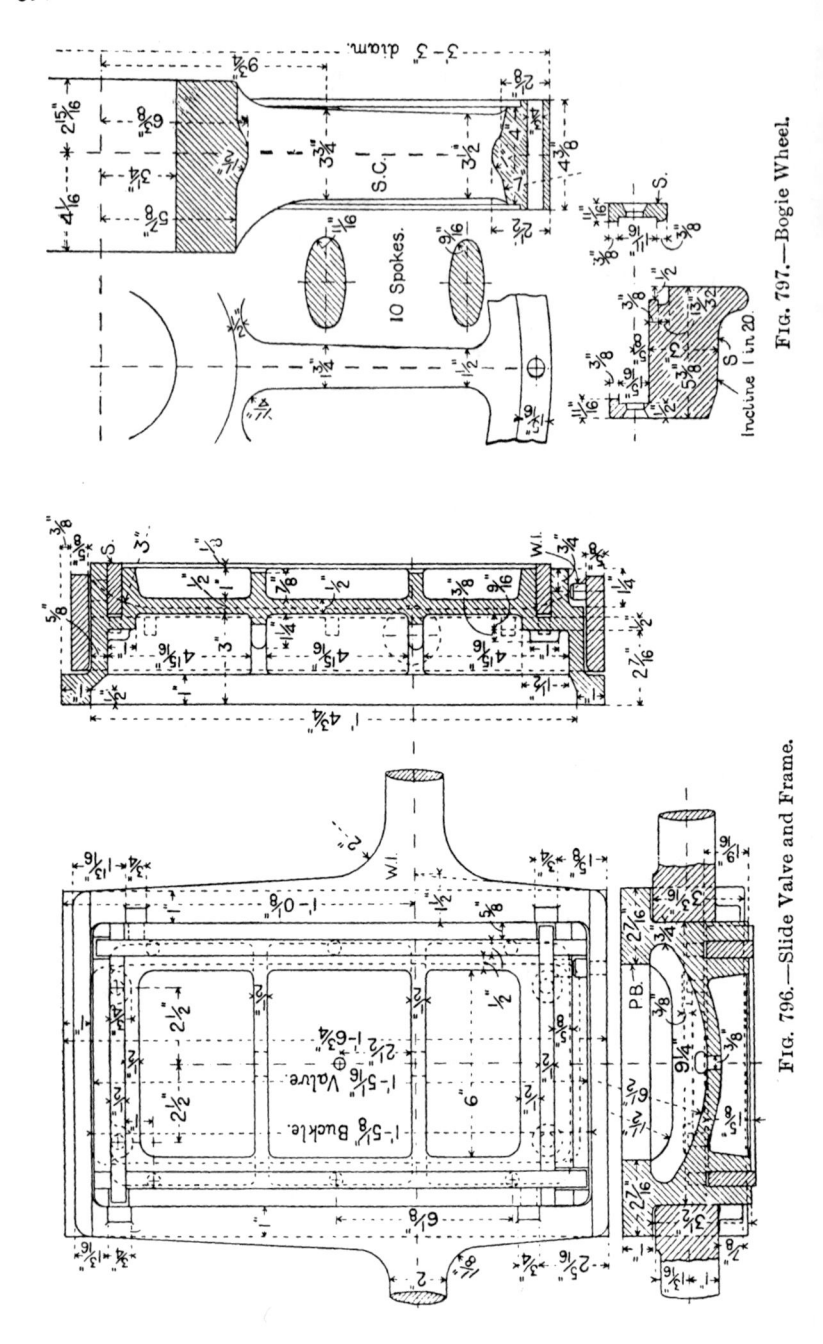

FIG. 797.—Bogie Wheel.

FIG. 796.—Slide Valve and Frame.

LOCOMOTIVE ENGINE.

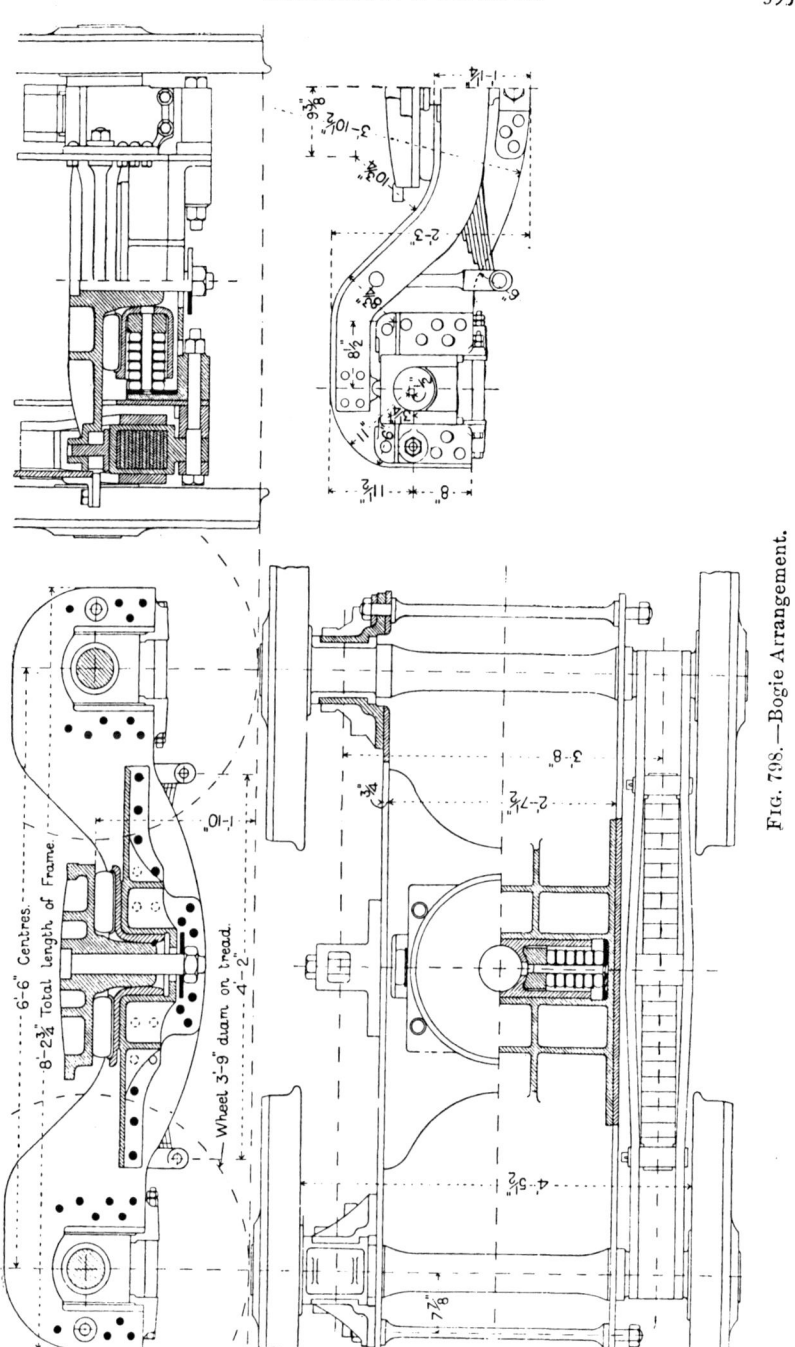

FIG. 798.—Bogie Arrangement.

396 MACHINE DRAWING AND DESIGN.

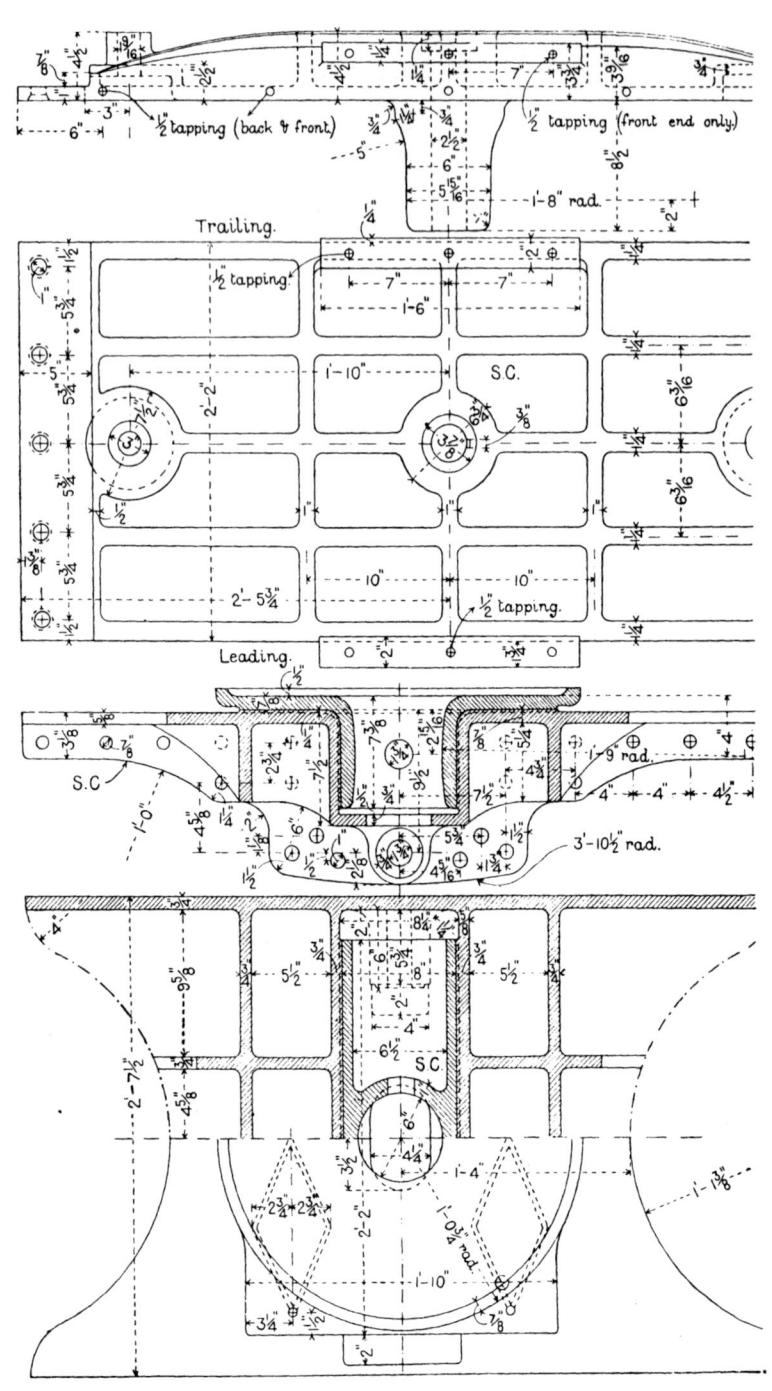

FIG. 799.—Bogie Centre and Slide.

LOCOMOTIVE ENGINE. 397

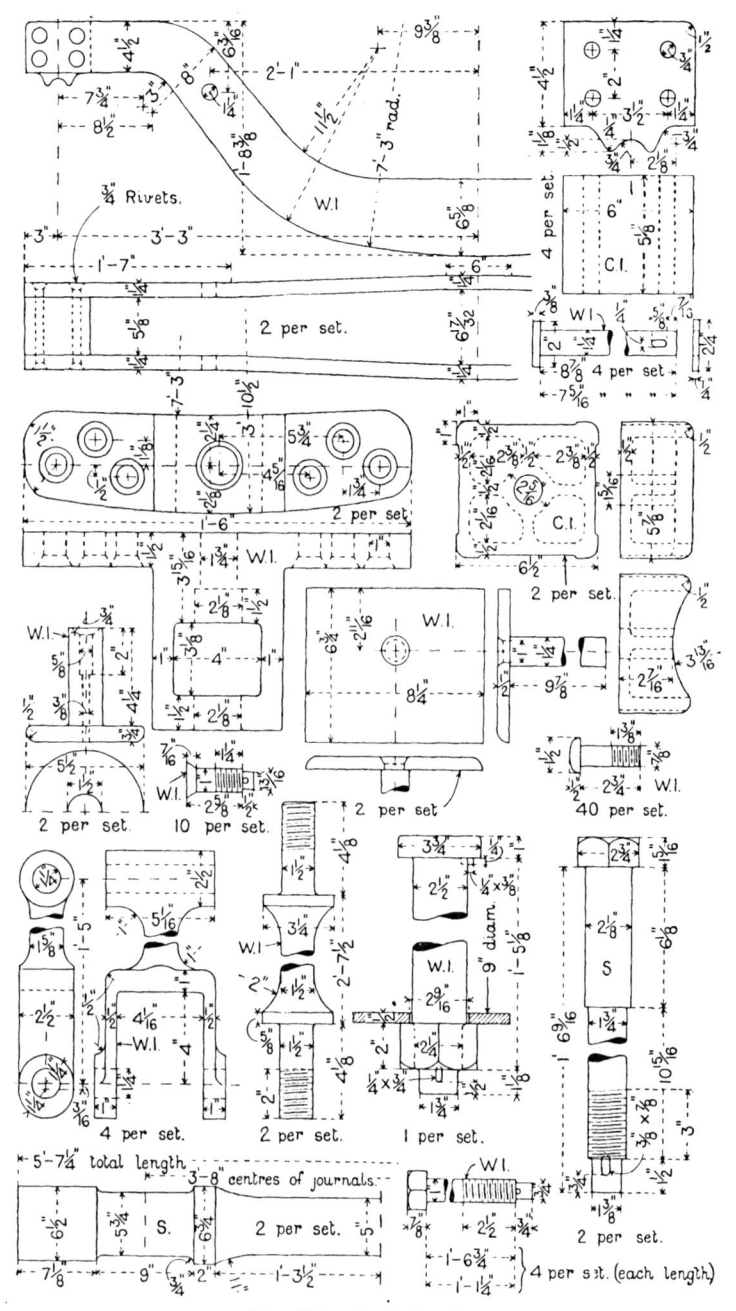

FIG. 800.—Bogie Details.

Fig. 801.—Bogie Axle Box.

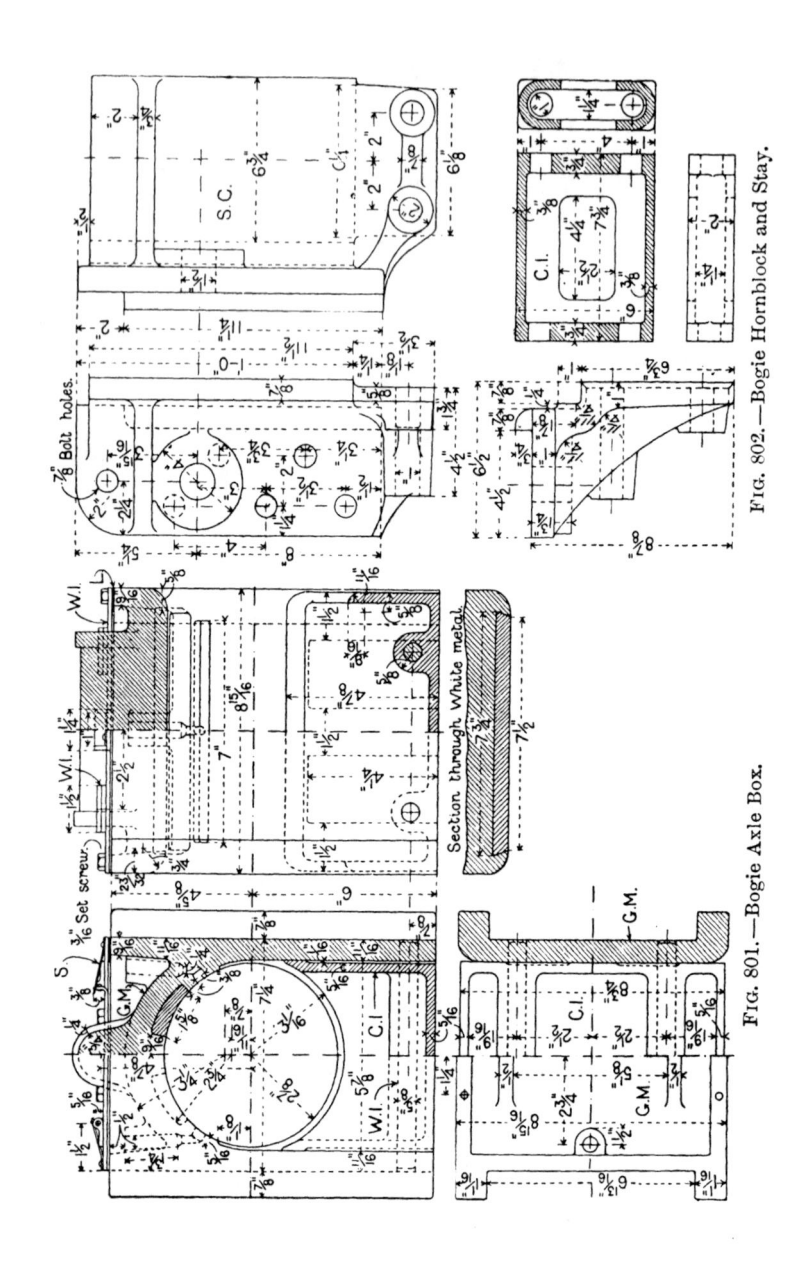

Fig. 802.—Bogie Hornblock and Stay.

LOCOMOTIVE ENGINE.

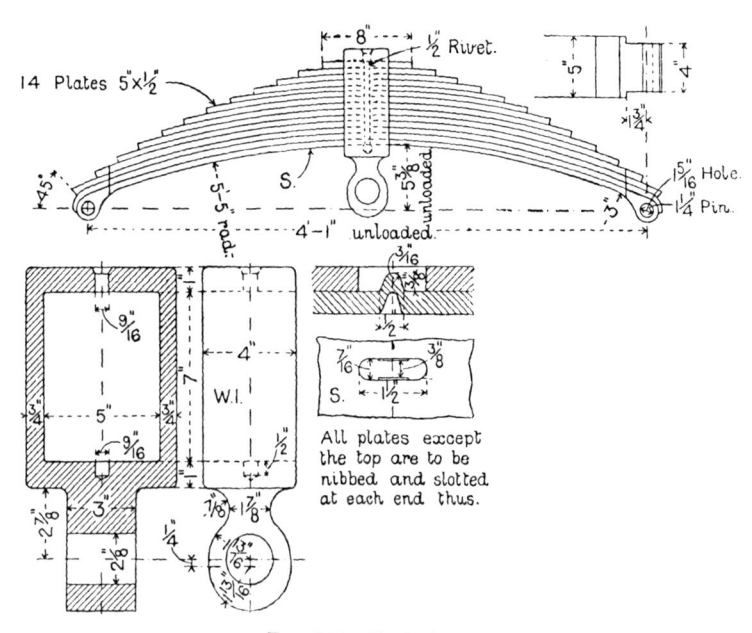

FIG. 803.—Bogie Spring.

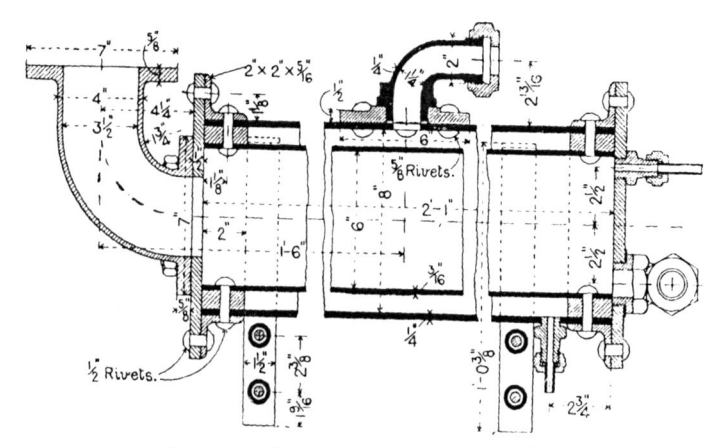

FIG. 804.—Liquid Fuel Heater for Tender.

INDEX.

Numbers refer to Pages.

ACCUMULATED work, 21.
Adamson flanged seam, 307.
Air-pumps, 334.
Alley's flexible coupling, 113.
Alloys, 42.
Angle of advance of eccentric, 254.
Angle of torsion, 98.
Angle, to bisect an, 7.
Arc of circle through three points, 7.
Area of a bearing, 115.
Areas of steam pipes, ports, and passages, 331.
Arms of pulleys, 147.
Arms of wheels, 185.
Armstrong's pipe joint, 83.
Ash, 44.

BABBITT'S metal, 44.
Backlash in screws, 64.
Ball valve, 249.
Bands, transmission of motion by, 139.
Bands, transmission of power by, 142.
Beams, continuous, 35.
Bearing, area of, 115.
Bearing surface of thrust collars, 133.
Bearings, collar, 115.
Bearings, crank shaft, 122.
Bearings for turbine shaft, 129.
Bearings, journal, 115, 117.
Bearings of shafts, span between, 99.
Bearings, pivot, 115, 120.
Bearings, pressure on, 116.
Beech, 44.
Belt fasteners, 135.
Belt joints, 135.
Belting, cotton, 134.
Belting, india-rubber, 134.
Belting, leather, 134.
Belting, link, 135.

Belting, paper, 135.
Belting, strength of, 141.
Belts, creep in, 145.
Belts, speed of, 144.
Bending moment, 31.
Bending moments on beams, 34, 35.
Bending, stresses induced by, 31.
Bends for tubes, 86.
Bessemer steel, 39.
Bevel wheels, teeth of, 178.
Blake's belt fastener, 137.
Boards, drawing, 1.
Boiler, Cochran's, 288.
Boiler plates, strength of, 271.
Boiler shell, strength of, 301.
Boiler stays, 309, 374, 376.
Boilers, locomotive, 298, 371–377.
Boilers, marine, 292.
Boilers, semi-portable, 301.
Boilers, vertical cross-tube, 287.
Bolt heads, forms and proportions of, 51.
Bolts for cylinder cover, 330.
Bolts, forms of, 62.
Bolts of uniform strength, 53.
Bournemouth pipe joint, 89.
Bowling hoop, 307.
Boxwood, 44.
Brackets for shaft bearings, 123.
Brass, 43.
Brasses, 117.
Brasses for connecting-rod ends, 214.
Bremme valve gear, 339.
Bronze, 43.
Brown's socket and spigot joint, 88.
Buckets, pistons, and plungers, 229.
Buckley's piston packing, 231.
Built-up cranked shafts, 198.
Built-up wheels, 188.
Butler's frictional coupling, 106.

INDEX.

Butt joints, double riveted, 277.
Butt joints, single riveted, 277.
Butt joints, treble riveted, 279.
Butt joints with single cover straps, 277.
Buttress screw-thread, 49.

CASE-HARDENING, 39.
Cast-iron, 36.
Cast-iron, malleable, 37.
Cast-iron, strength of, 36.
Caulking of riveted joints, 270.
Centrifugal force, 20.
Centrifugal force, tension in pulley rim due to, 149.
Chilled castings, 37.
Circle, involute of, 12.
Circle, tangent to, 8.
Circle through three given points, 7.
Circle to touch two given circles, 8.
Circles, inscribed and escribed, 9.
Circles, tangents to two, 8.
Circlip, Carver's, 78.
Circulating pumps, 333.
Clack valves, 247.
Claw coupling, 108.
Clearance and clearance volume, 320.
Clutch couplings, shifting gear for, 112.
Clutch, Edmeston's friction, 110.
Clutch, Mather and Platt's friction, 110.
Coating for pipes, 92.
Cochran's vertical multitubular boiler, 288.
Cocks, 259, 376, 380.
Collar bearings, 115.
Colouring, 3.
Combined twisting and bending, 96.
Combustion, rate of, 287.
Compound and multiple stage expansion engines, 324.
Compound marine engine, example of, 335.
Compression, resistance to, 30.
Condensers, jet, 331.
Condensers, surface, 332.
Cone, development of, 16.
Cone keys, 69.
Conical friction coupling, 112.
Connecting-rod brasses, 214.
Connecting-rod end, marine, 212.
Connecting-rod, length of, 206.
Connecting-rod, solid end for, 209, 386.

Connecting-rod, strap end for, 208, 209.
Connecting-rod, strength of, 206.
Connecting-rod, thrust on, 205.
Connection of parallel plates, 284.
Connection of plates at right angles, 284.
Consumption of steam per I.H.P., 334.
Continuous beams, 35.
Converse pipe joint, 89.
Copper, 40.
Copper pipes, flanges for, 82.
Corrugated furnace tubes, 307.
Cotters, 73.
Cotters, strength and proportions of, 74.
Cotters, tightening and locking arrangements for, 76.
Cotton belting, 134.
Coupling, Alley's flexible, 113.
Coupling, Butler's frictional, 106.
Coupling, cast-iron flange, 102.
Coupling, claw, 108.
Coupling, conical friction, 112.
Coupling, Oldham's, 107.
Coupling rods, 216, 385.
Coupling, Sellers' cone, 105.
Coupling, solid flange, 103.
Coupling, split, 101.
Coupling, universal, 107.
Couplings, box or muff, 100.
Cover, cylinder, 328, 381.
Crank axles, locomotive, 196, 391.
Crank discs, 200.
Crank effort, diagram of, 22.
Crank pin, connection of, to crank arm, 195.
Crank pins, strength and proportions of, 193.
Crank shaft bearings, 122.
Cranked axle, Worsdell's, 198.
Cranked shaft, Dickinson's, 199.
Cranked shaft, Foster's, 199.
Cranked shafts, built up, 198.
Crank shafts, forged, 196.
Creep in belts, 145.
Cross-head pins, diameter of, 219.
Cross-heads, forms of, 220, 382, 383.
Cross-heads for four guide bars, 220.
Cross-heads for two guide bars, 222.
Cross-heads, guides for, 227.
Cross-heads with slipper slides, 225.
Curves of teeth of wheels, 174.
Cycloid, 11.

INDEX. 403

Cycloidal curves, 11.
Cycloidal teeth, 176.
Cylinder cover, 328, 381
Cylinder cover, bolts for, 330.
Cylinder, development of, 13.
Cylinder, diameter of, 319.
Cylinder, flange, 330.
Cylinder, thickness of, 328.
Cylinders, intersection of, 16.

DEAD load, 26.
Design for spur wheel, 187.
Development of cone, 16.
Development of cylinder, 13.
Diagonal stays, 311.
Diagrams of crank effort and twisting moment, 22.
Diagram of work in steam cylinder, 321.
Diagrams, stress, 19.
Dickinson's cranked shaft, 199.
Dimensions on drawings, 4.
Disc valve, 249.
Discs, crank, 200.
Double-beat valves, 250.
Double-ported slide-valve, 254.
Double riveted butt joints, 277.
Double riveted lap joints, 273.
Drawing boards, 1.
Drawing instruments, 2.
Drawing paper, 1.
Drawing pins, 3.
Drawings, working, 5.

ECCENTRIC, angle of advance of, 254.
Eccentric sheaves, proportions of, 203.
Eccentric straps, 214, 215.
Eccentrics, 202, 391.
Edmeston's friction clutch, 110.
Effect of live loads, 28.
Efficiency of a riveted joint, 272.
Elasticity, 26.
Elastic strength, 28.
Elbows for tubes, 86.
Electro-deposited copper, 41.
Ellipse, 9.
Elm, 44.
Elmore process, 41.
Energy of rotating body, 21.
Engines, compound and multiple stage expansion, 324.
Epicycloid, 11.

Equivalent evaporation from and at 212°, 286.
Escribed circles of a triangle, 9.
Evaporative performance of steam boilers, 286.
Expansion in pipes, provision for, 92.

FACTOR of safety, 29.
Feather keys, 68.
Fir, 44.
Flange couplings, 102, 103.
Flange, cylinder, 330.
Flanges for cast-iron pipes, 79.
Flanges for copper pipes, 82.
Flanges for pipe joints, Pope's, 83.
Flap valves, 247.
Flat key, 65.
Flat plates, strength of, 315.
Flexible coupling, Alley's, 113.
Fly-wheel, 21.
Fly-wheel, weight of, 25.
Footstep bearings, 126.
Force to keep friction wheels in gear, 170.
Forked ends for connecting-rods, 216.
Foster's cranked shaft, 199.
Foundation bolt heads, 52.
Fox's corrugated furnace tubes, 307.
Frictional coupling, Butler's, 106.
Friction clutch, Edmeston's, 110.
Friction clutch, Mather and Platt's, 110.
Friction coupling, conical, 112.
Friction gearing, ordinary, 169.
Friction gearing, Robertson's, 169.
Friction gearing, transmission of power by, 170.
Friction of a band on a pulley, 143.
Friction of a rope on a grooved pulley, 159.
Friction of slide valves, 258.
Friction wheels, Shaw's experiments with, 171.
Fullering of riveted joints, 270.
Furnace tubes, 305.

GIB-HEADS on keys, 68.
Gibs and cotters, 73.
Girder stays, 314.
Greene's belt fastener, 137.
Grooved pulley, friction of rope on, 159.
Gudgeon, cross-head, 219.
Guides for cross-heads, 227.

Gun-metal, 43.
Gun-steel, 40.
Gusset stays, 312.
Gyration, radius of, 21.

HALF-LAP coupling, Fairbairn's, 100.
Hall's corrugated disc pulley, 153.
Hangers, 126.
Harris's belt fastener, 138.
Hat-leather packing, 246.
Helical teeth, 191.
Helix, 14.
Hollow key, 65.
Hollow shafts, 97.
Hooke's joint, 106.
Hornbeam, 44.
Horse-power, 21.
Horse-power, indicated, 319.
Hydraulic leather packing, 245.
Hydraulic pipes, swivel joints for, 85.
Hydraulic pistons, 236.
Hyperbola, 10.
Hypocycloid, 11.

INCLINED plane, 21.
Indian-ink, 3.
India-rubber belting, 134.
India-rubber disc valves, 248.
Indicated horse-power, 319.
Injection water, 331.
Inking-in, 3.
Inscribed circle of a triangle, 9.
Instruments, drawing, 2.
Intersecting riveted joints, 281.
Intersection of cylinders, 16.
Involute of a circle, 12.
Involute teeth, 177.

JET condensers, 331.
Joint, Hooke's, 106.
Joint rings for pipes, 92.
Joints for wrought-iron and steel pipes, 86.
Journal bearings, 115, 117.

KEYS, 65.
Keys, proportions of, 71.
Keys, strength of, 70.
Kimberley pipe joint, 86.
Knuckle screw thread, 48.

LANCASHIRE boiler, staying of, 312.

Lancaster's piston packing, 231.
Lap joints, double riveted, 273.
Lap joints, quadruple riveted, 277.
Lap joints, single riveted, 273.
Lap joints, treble riveted, 275.
Lap of slide valve, 253.
Leather belting, 134.
Leather packings, hydraulic, 245.
Lettering, 6.
Lignum-vitæ, 44.
Limit of elasticity, 26.
Liner, steam cylinder, 328.
Linings for steps, 118.
Link belting, 135.
Live load, 26.
Live loads, effect of, 28.
Load, 26.
Locking arrangements for cotters, 76.
Locking arrangements for nuts, 57.
Lock-washer, 60.
Locomotive boilers, 298, 371-377.
Locomotive crank axles, 196, 391.
Locomotive, regulator valve for, 254.
Low-pressure cylinder, diameter of, 325.

MACBETH'S pulley, 152.
Mackie's pulley, 152.
Mahogany, 44.
Malleable cast-iron, 37.
Manchester waterworks, pipes for, 81.
Manganese-bronze, 43.
Marine boiler, example of a, 295.
Marine boilers, 292.
Marine connecting-rod end, 212.
Marine coupling, 103.
Mather and Platt's friction clutch, 110.
Mean pressure on piston, 323.
Mean pressure referred to low-pressure cylinder, 325.
Medart's pulley, 152.
Metallic packing for stuffing-boxes, 243.
Modulus of elasticity, 27.
Modulus of section, 32.
Moment, bending, 31.
Moment of a force, 19.
Moment of resistance, 32.
Moment, twisting, 94.
Moments, principle of, 19.
Morison's corrugated furnace tube, 308.
Motion in a circle, 20.
Moxon's belt fastener, 138.

INDEX.   405

Mudd's piston packing, 231.
Muff couplings, 100.
Multiple threaded screws, 50.
Muntz metal, 44.

NAVES of pulleys, 148.
Naves of wheels, 186.
Normal pitch of involute teeth, 178.
Nut, common lock, 57.
Nuts, divided, 64.
Nuts, forms and proportions of, 54.
Nuts, locking arrangements for, 57.
Nuts, necessity for locking, 56.

OAK, 44.
Oldham's coupling, 107.
Oscillating cylinder, stuffing-box for, 241.
Overhung crank pins, strength and proportions of, 193.
Overhung cranks, proportions of, 194.

PACKINGS for pistons, 230.
Paper belting, 135.
Paper, drawing, 1.
Parabola, 10.
Parallelogram of forces, 18.
Park's screwed stay, 310.
Path of contact, 176.
Pedestal, shaft, 119.
Pencils, 2.
Pendulum, 20.
Phosphor-bronze, 43.
Pig-iron, 36.
Pillow-block, 119.
Pillow-block, Sellers', 121.
Pin keys, 67.
Pine, 44.
Pins, drawing, 3.
Pins, split, 78.
Pipes for Manchester waterworks, 81.
Pipes, provision for expansion in, 92.
Pipes, thickness of, 79.
Piston packings, 230, 381.
Piston, speed of, 319.
Piston, stroke of, 319.
Piston valves, 258.
Piston-rod, connection of, to piston, 237.
Piston-rod, diameter of, 236.
Pistons, buckets and plungers, 229.
Pistons, construction and proportions of, 232.

Pistons, hydraulic, 236.
Pistons, steel, 235.
Pistons without packing, 229.
Pistons, wrought-iron, 233.
Pitch lines of toothed wheels, 173.
Pitch of helix, 14.
Pitch of teeth, 173.
Pitch surfaces of toothed wheels, 173.
Pivot bearings, 115, 126.
Plane sections of solids, 12.
Plates with scolloped edges, 284.
Plummer block, 119.
Plungers, pistons, and buckets, 229.
Polygon, construction of, 9.
Polygon of forces, 18.
Pope's flanges for pipe joints, 83.
Pressure on bearings, 116.
Principle of moments, 19.
Prior's piston packing, 231.
Proof strength, 28.
Propeller shafts, thrust bearings for, 131.
Properties of saturated steam, 285.
Pulleys, arms of, 147.
Pulleys for rope gearing, 156.
Pulleys for wire ropes, 162, 167.
Pulleys for wire ropes, weight of, 167.
Pulleys, naves of, 148.
Pulleys, rims of, 145.
Pulleys, split, 151.
Pulleys, wrought-iron, 151.
Pumps, air, 334.
Pumps, circulating, 333.

QUADRUPLE riveted lap joints, 277.

RADIUS of gyration, 21.
Rate of combustion, 287.
Ratios of cylinder volumes in multiple stage expansion engines, 327.
Reducing socket for tubes, 86.
Regulator valve for locomotive, 254.
Resistance to combined twisting and bending, 96.
Resistance to compression, 30.
Resistance to shearing, 31.
Resistance to tension, 29.
Resistance to twisting, 31, 95.
Riley's steel socket for pipes, 88.
Rims of pulleys for flat bands, 145.
Rims of toothed wheels, 183.
Rivet heads, forms and proportions of, 268.

INDEX.

Rivet holes, 269.
Riveted joint, efficiency of, 272.
Riveted joint, strength of, 270.
Riveted joints for pipes, 89.
Riveted joints, intersecting, 281.
Riveting, 270.
Rivets, strength of, 272.
Robertson's friction gearing, 169.
Rope gearing, pulleys for, 156.
Rope gearing, ropes for, 155.
Rope gearing, system of, 154.
Ropes, power transmitted by, 161.
Ropes, speed of, 155.
Ropes, weight of, 155.
Ropes, wire, 162.
Round keys, 67.

Saddle key, 65.
Safety, factor of, 29.
Saturated steam, properties of, 285.
Scales, drawing, 3.
Screw threads, 14, 46.
Sections of solids, 12.
Sellers' pillow block, 121.
Sellers' screw thread, 47.
Sellers' shaft coupling, 105.
Set-screws, 63.
Set-squares, 2, 6.
Shade lines, 5.
Shafts, Board of Trade rules for, 201.
Shafts, hollow, 97.
Shafts, strength of, 96.
Shaw's experiments with friction wheels, 171.
Shaw's transmitter, 109.
Shearing forces on beams, 34.
Shearing load, 32.
Shearing, resistance to, 31.
Shepherd's perforated pulley, 153.
Sherwin's disc pulleys, 153.
Shrouding of wheel teeth, 182.
Siemens-Martin steel, 39.
Slide block, pressure on, 218.
Slide valve diagram, Zeuner's, 264.
Slide valve, double ported, 254.
Slide valve, friction of, 258.
Slide valve, simple, 252, 394.
Sliding keys, 68.
Smith's screwed bush, 70.
Span between bearings of shafts, 99.
Specification for steel pipes, 91.

Speed of belts, 144.
Speed of piston, 319.
Speed of ropes in rope gearing, 155.
Speed of toothed gearing, maximum, 183.
Spherical shell, strength of, 304.
Spigot and socket joints for pipes, 81.
Split pins, 78.
Split pulleys, 151.
Spur wheel, design for, 187.
Square screw thread, 48.
Stays, boiler, 309, 374, 376.
Steam pipes, ports and passages areas of, 331.
Steel, 39.
Stepped teeth, 191.
Steps, 117.
Steps for connecting-rod ends, 214.
Steps, linings for, 118.
Stop plates for nuts, 59.
Strain, 26.
Strap end for connecting-rod, 208, 209.
Strength, 28.
Stress, 26.
Stress diagrams, 19.
Stud-bolt, 62.
Stuffing-box for oscillating cylinder, 241.
Stuffing-box for pipes, 93.
Stuffing-box, screwed, 243.
Stuffing-box, use of, 239.
Stuffing-boxes and glands, proportions of, 240.
Sunk key, 65.
Surface condensers, 332.
Swivel joints for pipes, 85.

Tangent to a circle, 8.
Tangent to two circles, 8.
Tap-bolt, 62.
Taper of cotters, 76.
Taper of keys, 72.
Teak, 45.
Tee-squares, 2.
Teeth, cycloidal, 176.
Teeth, helical, 191.
Teeth, involute, 177.
Teeth of bevel wheels, 178.
Teeth of wheels, curves of, 174.
Teeth, pitch of, 173.
Teeth, proportions of, 173.
Teeth, stepped, 191.
Teeth, strength of, 180.

INDEX.

Telodynamic transmission, 162.
Tensile strength of materials, 30.
Tension, resistance to, 29.
Tensions in suspended rope, 163.
Thrust bearings for shafts, 131.
Thrust collars, bearing surface of, 133.
Thrust on connecting-rod, 205.
Toothed gearing, maximum speed of, 183.
Toothed wheels, pitch surfaces and pitch lines of, 173.
Toothed wheels, rims of, 183.
Toothed wheels, transmission of power by, 182.
Torsion, angle of, 98
Transmission of motion by bands, 139.
Transmission of power by bands, 142.
Transmission of power by friction gearing, 170.
Transmission of power by toothed wheels, 182.
Transmitter, Shaw's, 109.
Treble riveted butt joints, 279.
Treble riveted lap joints, 275.
Triangle of forces, 18.
Triple-expansion engines, example of, 338.
True length of a line, 12.
Tubes, details of wrought-iron and steel, 85.
Turbine shaft, bearings for, 129.
Twisting moment, 94.
Twisting moment diagram, 22.
Twisting, resistance to, 31, 95.

U-leather packing, 245.
Ultimate strength, 28.
United States metallic packing, 243.
Universal coupling, 106.

Valve, double-ported slide, 254.
Valve, simple side, 252, 394.
Valves, classification of, 247.
Valves, double beat, 250.
Valves, flap or clack, 247.
Valves, india-rubber disc, 248.
Valves, piston, 258.
Valves, single beat direct lift, 249, 377
Velocity, ratio of pulleys, effect of thickness of band on, 141.
Vertical cross tube boilers, 287.
Vertical multitubular boiler, Cochran's, 288.
Volume, clearance, 320.

Weight of fly-wheel, 25.
Weight of large rope pulleys, 159.
Weight of materials, table of, 45.
Weight of ropes, 155.
Weight of steam required by an engine, 285.
Weight of wire-ropes, 162.
Weight of wire-rope pulleys, 167.
Whitworth screw threads, 46.
Whitworth steel, 40.
Wire ropes, 162.
Wood, 44.
Work, 20.
Work, accumulated, 21.
Working drawings, 5.
Worm gearing, 192.
Worsdell's cranked axle, 198.
Worsdell's steel girder stay, 315.
Wrought-iron, 38.

Yarrow's screwed stay, 310.

Zeuner's slide valve diagram, 264

Printed by Ballantyne, Hanson & Co.
Edinburgh & London

WORKS BY D. A. LOW.

AN INTRODUCTION TO MACHINE DRAWING AND DESIGN.

With numerous Illustrations. Crown 8vo, 2s. 6d.

"One of the best little books on the subject we have met with; and students and apprentices will find it an excellent guide in their efforts to master machine drawing. Mr. Low is a successful and experienced teacher, and this little work gives evidence of his skill and of his intimate acquaintance with modern engineering practice."—*Mechanical World.*

TEXT-BOOK ON PRACTICAL, SOLID, OR DESCRIPTIVE GEOMETRY.

In Two Parts. Crown 8vo. Part I., 2s.; Part II., 3s.

"A handbook which should be on the shelf of every engineering or architectural draughtsman. Not a carpenter and joiner, not a workman in any branch of mechanism, but would profit by a perusal of this work."—*The Schoolmaster.*

"The definitions and explanatory matter throughout merit some notice for their clearness and simplicity."—*Journal of Education.*

"Mr. Low seems to us to have done his work with great clearness, fulness, and method. He has, it is evident, sound ideas as to the methods to be pursued by the student who would obtain a practical knowledge of his subject."—*The Saturday Review.*

IMPROVED DRAWING SCALES.

Printed, from carefully engraved plates, on strips of thin cardboard, and varnished.

Sixpence net per Set, in Case.

IMPROVED DRAWING APPLIANCES:

Set-Squares, Adjustable Protractor Set-Squares, Tee-Squares, Protractors, Scales, &c.

** A detailed and Illustrated Prospectus will be sent on application.

LONGMANS, GREEN, AND CO.
39 PATERNOSTER ROW, LONDON,
NEW YORK AND BOMBAY.

AN ELEMENTARY TEXT-BOOK OF APPLIED MECHANICS.

With numerous Illustrations. Fcp. 8vo, 2s.

"Mr. Low has produced a work that, as an elementary treatise on this subject, is unequalled."—*Mechanical World.*

"We commend it highly for its clear and concise arrangement and method of exposition, and for the large number of well-chosen examples which are to be found in it."—*Industries.*

BLACKIE AND SON, LIMITED,
LONDON AND GLASGOW.

Printed in the United States
97720LV00005B/71/A